TURNBERRY
AIRFIELD

Margaret Morrell

carn

© Margaret Morrell, 2020.
First Published in Great Britain, 2020.

ISBN – 978 1 911043 09 6

Published by Carn Publishing Ltd.,
Lochnoran House,
Auchinleck,
Ayrshire, KA18 3JW.

www.carnpublishing.com

Printed by Bell & Bain Ltd.,
Glasgow, G46 7UQ.

Contents

Foreword

I have known Margaret Morrell for many years and can bear witness to the admirable commitment and dedication which she has shown writing and researching this book, charting the development of the flying school at Turnberry, that meticulously documents the planes which flew, and movingly recounts the stories of the men and women who served while training at the airfield, many of whom sadly also lost their lives.

The Kennedy family has been indelibly linked for over 700 years through our estates at Cassillis and Culzean to this quiet corner of Ayrshire and south west Scotland and, in particular, to this story of our land holding which was to become Turnberry Golf Course and Turnberry Hotel, the Turnberry School of Aerial Gunnery and Fighting and later RAF Turnberry, principally while under the ownership and management of my great grandfather, Archibald Kennedy, 3rd Marquess of Ailsa.

Never before has the history of Turnberry been so extensively and comprehensively chronicled. Margaret Morrell explains in fascinating narrative the role Turnberry and the surrounding area played in both World War One and World War Two. Margaret brings the runways, hangars, and classrooms so vividly to life that it is almost possible to smell the aviation fuel and hear the roar of those historic aircraft as they soar in the skies above the villages of Maidens, Turnberry and Kirkoswald.

Most poignant are the accounts of the young men and women, both air crews and support teams, from all over the Commonwealth, who trained at Turnberry Airfield, requisitioned by the British Government to play its role in the war efforts. I was humbled by their stories of bravery and companionship at a time of international conflict.

The memory of those who served lives on, with the recent addition to the Turnberry War Memorial of a further 89 names, which now commemorates all 172 souls who are thus far known to have perished at Turnberry. This memorial, erected by the local people and first unveiled by my great grandfather, in 1923, stands prominently and serenely overlooking the twelfth green on the famous Ailsa Course.

Equally this wonderful book serves as a lasting testament of courage, innovation, daring and triumph. We all owe a debt of gratitude to Margaret Morrell.

9th Marquess of Ailsa

Introduction

What exactly happened at Turnberry Aerodrome? This was the question I asked some years ago and found no answer, which prompted me to begin the research which led to this book. It was a question that had been asked many times by locals and visitors alike, but no one seemed to know very much: apart from the odd snippet of information or occasional rumour. It appears all memories of the airfield had been lost. Physically, all that is left today are a few derelict buildings and lonely strips of concrete nestling amongst the fairways of the famous Turnberry golf course, but Turnberry's legacy, I have found, is much more important than this scant evidence would suggest.

Most books on aviation history fail to recognise the importance of Turnberry and the major role it played during both world wars. This is probably because it was never an operational airfield, but rather a training ground, making it a less glamorous subject for historical research. Turnberry's units played a vital role in training thousands of desperately needed aircrew to fly and fight during both world wars. Loyal to Britain, they came from the many countries of the Commonwealth and other allied nations, experiencing enormous hardship and in many cases losing their lives before ever seeing active service. The memorial to some of those who died at Turnberry is situated on the Ailsa Golf Course and bears a long list of casualties, most of whom were little more than boys when they died. Of the many thousands of trainees who came to Turnberry, very few returned. Some fortunate men came back to claim their brides, or sadly arrived on a pilgrimage to visit the graves of their pals who were left behind. But many airmen who found love and happiness at Turnberry perished in the sea or came to rest in an unknown grave. Others lie in the cemeteries at Dunure and Girvan.

I have been deeply honoured by the gift of friendship of many ex-aircrew, and these friendships have yielded a host of stories of the aerodrome and its personnel – some amusing, some heroic, some tragic – and some, it must be said, I dare not publish! The personal memories entrusted to me have, where possible, been written as told, and if some inaccuracies occur allowance must be made for the passage of time. I hope that the recollections gathered here will provide a glimpse into Turnberry's years of operation, particularly during the Second World War.

In addition to the personal narratives, I have searched through official documents and archives, dusted off many old photographs and logbooks, and been able to put names to some long-missed faces. In a couple of cases emotional reunions have ensued. I write this book as a tribute to those who died whilst at Turnberry, during both world wars. This information has been gathered from many sources including Operations Record Books, Official Crash Cards, Commonwealth War Graves Commission, and the memories of survivors. In many cases records are incomplete, or facts from official sources contradict each other, and whilst I have attempted to document all those who lost their lives at Turnberry, I must apologise for any omissions or errors.

It must be said that this is not a technical volume – details of the organization of the Royal Air Force, aircraft, and the progress of the wars, can be found elsewhere.

This is a book about the people and the place, the men, and women of Turnberry Airfield. This is their story.

Margaret Morrell
Kirkoswald, 2020.

1
Lands of Turnberry

Situated in the historic landscape of Carrick, built on a rocky outcrop, surrounded on three sides by the sea, Turnberry Castle once held an imposing position overlooking the Firth of Clyde. Its origins are unknown, although the name is probably from Norman French, *taurnei, tornei, tounrey* or tournament and from the Old English, *byrig, burg,* or *burg* – fort, or 'castle of the tournaments'.

It was here that one of Scotland's most famous kings, Robert the Bruce, was born on 11 July 1274. Turnberry was the home of his mother, Marjory, Countess of Carrick.

Scotland at this time was unified by King Alexander III - it was under his reign that Scotland emerged as a separate kingdom. On 19 March 1286, Alexander III was killed in a fall from his horse, and with no male heir, this ended the royal dynasty and threatened Scotland's survival as an independent nation. Thus began

the struggle for the throne of Scotland. The prime claimants were two noblemen, John Balliol and Robert Bruce. Both being descendants of the daughters of Alexander III's great-uncle, the rival claims were exceedingly complex.

It was at Turnberry, in September 1286, that an assembly of Scottish nobles gathered to proclaim Bruce their king, and by making such a proclamation, announced the intent of resisting English dominance.

Asked to arbitrate in the question of succession, King Edward I of England grasped the chance to seize power and established garrisons across Scotland, coming down in favour of Balliol, who was King in name only, being controlled by the English monarch. The new King of Scotland faced increasing unrest from the people, who in 1295 begged the support of the Pope, and of the French, who were then enemies of England, in an endeavour to escape

Ruins of Turnberry Castle

English control. The growing opposition and discontent led to local revolts and rebellions against the English governors and garrisons, an uprising which the English army defeated in just nine weeks, removing Balliol as king, demanding the surrender of the royal regalia, and the Stone of Destiny, the iconic symbol of Scottish kingmaking was taken to England to signify the end of Scotland's independent monarchy. Edward thinking his conquest complete returned to England, leaving behind his administrators and garrisons. In 1297 the Scottish people again revolted against this foreign rule, led by Ayrshire man William Wallace. A long and bloody guerrilla war followed, Bruce who had taken over from Wallace, was forced to make peace after the siege of Stirling Castle in 1304, which act led to the submission of the Scots.

With the betrayal, capture and horrific execution of William Wallace in August 1305 an ordinance for Scottish administration was drafted the next month, appointing John of Brittany to serve as Guardian, aided by a council of 22 Scottish nobles, giving the Scots some degree of self-government. It left the Kingdom of Scotland a conquered nation.

In 1306, after killing his main rival for the throne, Robert Bruce was crowned King of Scotland, causing a split in the Scottish aristocracy, many taking sides with the English against the 'Rebel King'. Many skirmishes followed and Bruce was forced to flee into exile after being defeated at the battle of Methven. He spent this time gathering more men to his cause and planning further rebellion.

In January 1307, he landed at Maidens with his small army of 300 men and began a new offensive when he tried to recapture his own castle at Turnberry which was held by a strong English force under the command of Henry Percy, Lord of Northumberland. Although his assault failed, the occupying force at Turnberry took fright and retreated to the garrison at Ayr for safety. In 1310 Bruce himself ordered the demolition of Turnberry castle to prevent the English again taking command, the destruction of this once imposing fortress has been complimented throughout the centuries by the effects of severe coastal weather and erosion. There can still be seen traces of masonry showing what would appear to be a drawbridge and portcullis, and on the shoreline suggestion of ruinous vaults and cellars and it would seem that the castle had its own harbour.

These historic events are remembered to this day in the local place names of Jameston and Douglaston (named after two of his commanders), while Bruce himself is commemorated by the King's Field, where he reputedly raised his standard.

Bruce continued his guerrilla warfare against the English forces, winning victories at Glen Trool and Loudoun Hill, and with his small band of men managed to evade the army of King Edward that had been sent north to crush the rebellion. The war of independence continued for many years. Edward I died, and his heir Edward II carried on the punitive occupation of Scotland, a campaign which culminated in the Battle of Bannockburn in 1314 when a great English army marched north in an attempt to relieve the besieged Stirling Castle. Bruce and his smaller force won the day and victory for Scotland, forcing Edward II to flee for his life.

Several centuries later, in the heart of the ruined castle site, the lighthouse was built by the company of John Barr & Co. of Ardrossan. The surveyors were David and Thomas Stevenson, from the famous family of Scottish engineers (Thomas was the father of author Robert Louis Stevenson). Just off this part of the Ayrshire coast is the treacherous Bristo Rock which was responsible for many shipwrecks throughout the centuries. In 1869 it was proposed by the Receiver of Wrecks at Ayr to the Board of Trade, that a light should be erected on this rock. The surveyors reported to the Commissioners of Lighthouses that it was inadvisable and dangerous to place a lighthouse on the rock itself and recommended that the best placement would be on Turnberry Point. Sanctioned by the Board of Trade that year, building did not actually begin until 1871, the delay being caused by disputes over how big the lantern should be and what

Turnberry Lighthouse

type of oil was to be used to light it. Costing £6,576 the light was first lit up on 30 August 1873, showing one flash every twelve seconds, which has a nominal range of twenty-four miles. The lighthouse is an iconic part of the Turnberry landscape today, with its imposing white tower standing watch over the Clyde, although there are no longer lighthouse keepers, it being automated in 1986 and remotely monitored from the offices of the Lighthouse Board in Edinburgh.

Along the coast to the north lies the small village of Maidens, a scattering of houses along the shoreline around Maidenhead Bay. There is evidence of a settlement here in Prehistoric times with the existence of a standing stone, probably erected during the Bronze Age, on Bain's Hill, overlooking Maidens. The name Maidens (as applied to the village) first appears in records dated 1847; prior to this the name on the map was Douglaston. The first road map of Ayrshire, printed in 1828, uses the name Maidenhead to describe the rocks, which project out beyond the village's harbour. Timothy Pont's map of 1590-1614 names these same rocks as Maidens of Turnberry and shows Port Murray or Morrow next to Maidenhead Bay, which would suggest that this was the main port then in use. To this

day there exists a milestone at Turnberry pointing the way to Dunure by Port Murray, with no mention of the village of Maidens.

In the late nineteenth century, the village was made up of typical Scottish, thatched, small two room dwellings situated around a sandy port, and like most other rural and coastal communities, life in Maidens revolved around agriculture and fishing. Prior to the harbour being built, the small boats were launched from the beach, fishing for whiting and plaice in the winter and cod, skate and turbot in the summer months.

Mention must be made of the great influence the Marquis of Ailsa had on the development and care of the area. Archibald Kennedy, 14th Earl of Cassillis, 3rd Marquis of Ailsa, was born at Culzean Castle in 1847. He was 22 when he inherited the title and estate after his father (also Archibald Kennedy) was killed in a hunting accident in 1870. Educated at Eton, he joined the Coldstream Guards, attaining the rank of Captain, which he resigned on the death of his father. The Marquis was an actively considerate landlord and improver of the Culzean Estate, and aided and abetted by his factor, Tom Smith, took a keen interest in the welfare of his tenant farmers and fishermen. He set about upgrading the old

WWI motorboat shed at Maidens harbour

housing in Maidens and like his father before him the 3rd Marquis of Ailsa was also an enthusiastic yachtsman, having learned to sail with the fishermen at Maidens, and had a keen interest in boatbuilding. After the success of his small boatbuilding yard at Culzean harbour where he designed and supervised the construction of small racing yachts and steam launches, leading to larger orders, he had built a new boatyard at Port Murray (to the south of Maidens) in 1883. This new venture went under the name of the Culzean Shipbuilding and Engineering Company. He built a tenement block called Ailsa Buildings, known locally as 'The Block', to accommodate the workers, and although this was completed, the Culzean Shipbuilding and Engineering Company proved unprofitable due in part to the high cost of transportation of materials. The project was moved to Troon in 1886 becoming the Ailsa Shipbuilding Company, with the Marquis as chairman. The Marquis was also responsible for building the first pier and breakwater and laying down chain and anchor moorings at Maidens. Fishing methods had also changed, ring nets had

been introduced and herring became the main catch during the early twentieth century. The fishing fleets of Maidens, Dunure and Ballantrae were considered the best operators of ring net fishing in Scotland. The harbour and the fishing fleet were the centre of the community.

The Marquis was also a member of the Ayrshire Agricultural Association and was deeply involved in all aspects of farming in Carrick and he played a major part in the growth and development of the early season potato industry along the coast. In 1892 he became involved with the G & SWR (Glasgow and South Western Railway Company) when he was elected a director of the company. Keen to promote his beloved Carrick, he was quick to see the advantages of having a railway through his estate and the potential benefits this would bring to the agriculture and fisheries in the area, facilitating the transport of produce to market faster and fresher, and encouraging the growth of tourism, bringing in extra revenue and thereby increasing the prosperity of his tenants. In his role as director, he therefore proposed in 1896 that the

Maidens Railway Station

railway company should finance a new railway line to run along the Carrick coast between Ayr and Girvan. The line was to be called the Maidens & Dunure Light Railway and work commenced in 1902.

Ayr Advertiser 20 November 1896
Will Alloway remain a terminus for the new branch railway? Lord Ailsa, one of the most enterprising peers in the country, is a railway Director. He is known also to be in favour of railway communication along the Carrick shore, of which he is the largest proprietor. The Maidens and Turnberry present a locality which is unequalled anywhere for summer residents, golfers, etc, etc, and the whole district is productive and could be made remunerative for a railway. An extension of the new line from Alloway to the shore would take up much ground left out by the present Maybole and Girvan railway; and Ayr at least would have no reason to regret the improvement it would make in its position as a railway centre.

Although coming to fruition at Turnberry, it could be said that the idea of a golf course was conceived at Pau, a spa town in the French Pyrenees. As was usual for the landed gentry at the turn of the century, winters were spent abroad. Quickly bored at the holiday resort, with nothing to do but cycling, the Marquis wrote home 'that he had been obliged to take up golf, and was receiving lessons from a French professional. Becoming enamoured of the game, he arranged for Dominique, the professional from Pau, to come and work on the laying out of the Turnberry links as a private course. Changing his mind, the Marquis decided to lay out a course in the grounds of Culzean Castle, which would be far more convenient for him than Turnberry. Although the proposal to lay out a course at Turnberry had been first suggested in 1892, this was postponed by the Marquis, who considered it would 'spoil the only good bare ground' locally. As with other great estates in Scotland at the time, finances, or the lack of, were always a concern. Seeing the potential of his lands as premier golfing country, the Marquis now conceded that he could raise capital and increase estate revenues with the bonus of having his golf course. Dominique was once again summoned from France to advise on the laying out of a full-size course at Turnberry. In his role as director of G & SWR, the Marquis then persuaded them

Turnberry Hotel main entrance

to commission Will Fernie, the professional at Troon Golf Club, to design and lay out the course. This was completed in 1901 and instantly became recognised as one of the best in Scotland. In 1902 Turnberry Golf Club was instituted and within three years had over 260 members. A second course was added in 1909.

The peace of the small seaside village was to be rudely disturbed by the prospect of tourism. Squads of Irish navvies descended on the area to build the luxurious 100-bedroom Turnberry Hotel and railway. Kirkoswald Parish Church provided an Outreach Mission for the labourers. Taking four years to complete, the hotel, opening on 17 May 1906, attracted a flock of well-to-do visitors, wishing to partake in the pleasures of the magnificent new golf links while the railway brought throngs of campers to the beach during the summer months.

The new G & SWR Company, Turnberry Hotel, was designed by Scottish architect James Miller. It was an imposing structure which dominated the rural landscape. Equipped with the most modern facilities, the entire building was lit by electricity, with lifts connecting each of the floors, the luxurious rooms were opulently but tastefully furnished, with suites of bathrooms incorporating plunge baths, sprays and showers, with a selection of hot, cold, fresh or saltwater. Service was of the highest standard. Recreational facilities included croquet, tennis, bowling and billiards and, of course, golf.

The railway station at Turnberry was connected to the hotel by a covered walkway which led guests via a large conservatory into the hotel's entrance lounge. Although not facing the sea, the railway side of the hotel was officially the front, while the 'back' provided spectacular views beyond the golf course, over the Firth of Clyde to Ailsa Craig, Arran, with the Mull of Kintyre in the distance.

The behaviour and manners of many of the visitors proved too much for the locals of the parish, with the result that a petition was handed in to the management of Turnberry Hotel in 1908. The main objection was the flagrant desecration of the Sabbath by golfers. The management responded by putting up a notice requesting that golfers refrain from teeing-off on a Sunday. It wasn't very effective, and local people had to get used to tourism, with all its sacrilegious practices. Little did they know that the summer holidaymakers were soon to be replaced by a completely different type of visitor!

During the desperate days of the First World War, in response to the severe need for trained airmen, Turnberry was requisitioned by the government for use as an airfield.

2
WWI Airfield

There is some confusion as to when Turnberry Aerodrome became active during WWI - evidence points to its use as a landing ground on an 'ad hoc' basis as early as 1915. The man chosen to establish the School of Aerial Gunnery here was Lt.-Col. Louis Arbon Strange. Lt.-Col. Strange had already been tasked with setting up the School of Aerial Gunnery at Hythe in 1916, where he complained of the difficulty of obtaining equipment of any sort from the Air Ministry or assistance in securing suitable accommodation. In his memoirs he comments, 'Those were the days when I learnt the full meaning of the verb "to scrounge". At Turnberry we were short of everything and met with the usual trouble when we tried to get anything through proper channels out of the Air Ministry'. It would seem, that the RFC in their urgent need to establish these schools were annexing civilian property without any official authority. Louis Strange strongly held the belief that better trained pilots and accurate gunnery would win the battle for the air and was therefore determined to get these schools working as soon as possible. Before he could be held to account for his unofficial commandeering, Lt.-Col. Strange had been posted to Loch Doon, and shortly thereafter was Assistant Commandant of the Central Flying School at Upavon.

In a minute of the Air Council dated 27 December 1916 it is recorded that:

> Owing to the increase in the number of pupils that are now required to be trained in aerial gunnery, it has been found necessary to form two additional schools of Aerial Gunnery at Hythe and Turnberry, general sanction for which has been requested.
> The establishment proposed for these

schools is the same as for the School of Aerial Gunnery, Loch Doon.
> It is essential that these schools should start working almost immediately, it is requested that early sanction may be given.

SCHOOL OF AERIAL GUNNERY, LOCH DOON

No history of wartime aviation in Ayrshire can fail to allude to the Loch Doon Scheme and its connection to Turnberry. As previously mentioned, in response to the urgent requirement for training establishments during WWI, the War Office authorised the development of specialist Schools of Aerial Gunnery, one of these was at Loch Doon. The necessity for providing our pilots with the highest possible form of gunnery instruction was vital, both for their own safety and for their efficiency.

In June 1917 a letter from Headquarters, Training Brigade, RFC, to the Air Board Office states:

> School of Aerial Gunnery, Loch Doon, which is in the process of construction, will be in the nature of higher training in Aerial Gunnery for Pilots who will pass on to that School from No.2 Aux. School of Aerial Gunnery, Turnberry. Special types of targets are being constructed at Loch Doon to approximate as near as possible to the targets which will be met by a Pilot on actual service. To attain this end, Motorboats, Seaplanes and Railways are required, which necessitate special types of personnel by trades, not required at the two Auxiliary Schools, and for this reason it is impossible to co-ordinate this

establishment with either or both of the Auxiliary Schools.

The man behind this ambitious project was Col. W. S. Brancker, who while serving as Commander of a Wing in France had witnessed the effectiveness of the French Aerial Gunnery School at Lake Cazaux near Bordeaux. Brancker is quoted in *Sir Sefton Brancker* by Capt. Norman Macmillan:

I had come back from France obsessed with the importance of giving pilots a proper training in aerial fighting. Here again we studied French methods and I received full reports about the most efficient aerial fighting school at Cazaux on one of the big lakes near Arcachon. A suitable place for an aerial gunnery school is difficult to find. A land-locked lake presenting a large surface of smooth water in an uninhabited country offers the best possible site. We tried something of the sort in the British Isles. I had officers reconnoitring all over Scotland, Ireland and the Norfolk Broads to find a suitable place in which to develop our first school of aerial gunnery. The most promising site we could find anywhere was at Loch Doon, and there we set to work to develop a school of aerial gunnery.

And so, in early 1916, Loch Doon, part of the Craigengillan Estate was selected, situated some three miles from the small Ayrshire mining village of Dalmellington. The lands of Craigengillan had been the property of the McAdam family for over 400 years.

The Chief Engineer of the Scottish Command stated he was in favour of the scheme and Lt.-Gen. David Henderson wrote. 'Loch Doon fulfils all our requirements and the engineering difficulties are not insuperable, I would like it proceeded with at once', and Brancker added, 'Loch Doon could not be bettered'. It seemed the perfect place to set up the training establishment. Loch Doon is an isolated, land-locked body of water some 700ft above sea-level with the

Loch Doon

prerequisite steep hills of Craigencolon and Cullendoch Hill which were required for the principle feature of rail mounted targets. The proposed airfield was to be established on the shore opposite Cullendoch Hill south of the Garpol Burn, and that floatplanes would land on the loch. The site chosen for the airfield was a moss, (in Scotland a moss is equivalent to a morass or peat bog), but the RFC considered that this could be easily drained.

A local man, Mr J. Smith, was hired to transport the officers to and from their inspection of the proposed airfield site. When he overheard the plans for the scheme, he mentioned that as a local he could offer his knowledge of the loch, such as that in dry weather, there were several rocks just below the surface of the water that would be hazardous to floatplanes taking off and landing. They were informed that in winter, the level of Loch Doon sometimes rose to an extraordinary height, so enormous was the volume of water descending from the hills and that a total of thirteen lochs also emptied their excess water into Loch Doon. In times of excessive rain or melting snow, some sections of the approach road were liable to be flooded to a depth of five or six feet. These floods on several occasions were known to wash out the schoolhouse garden at the head of the loch. On one instance in recent years the water had risen over the parapets of the bridge carrying the road across the outlet, and this section of the road was submerged to a depth of ten feet - this flooding sometimes existed for days. It was further pointed out by a local tenant who had seen every abnormal flood there for the past forty years, what effect those floods would have on the approaches from the lower end of the loch up to the camp. It was also common knowledge in the area that during severe winter months the loch was completely frozen over and had been in the recent past, cut off completely by 10 feet snowdrifts.

The RFC Officers scoffed at these opinions and told them they didn't know what they were talking about.

In his book *Ayrshire 1745-1950*, James

E. Shaw mentioned his involvement with the planning of the airfield at Loch Doon. Shaw was at the time the County Clerk of Ayrshire. He writes:

> I was recalled from a holiday at Windermere to a conference at the Station Hotel at Ayr. I found a large number of uniformed officials from the War Office, whose names are quite forgettable and whose manners are better so, interviewing several local people of experience called to advise, and anxious to help. It was clearly explained by engineers and others with knowledge of the ground that it would be impossible to land planes near the Loch, because of the deep moss. The officials were disappointed and said that it could be drained. It was suggested that they ask an old farmer who was present and who had been a tenant of the land for many years. He was very deaf and apparently nervous about being brought to the head of the table, not having heard what the meeting was about. After they all shouted both in concert and individually, someone he knew told him that they proposed to drain the moss. He looked up with a sweet smile that one might give to children, shook his head and said, 'Drain the moss. Ye couldna drain the moss'. The officials from London went up to Loch Doon, viewed the land through field glasses and returned to London.

Another visit to the Loch was made on 5 September 1916, by officials from the Department of Fortifications and Works, Lands Branch, RFC and RNAS. Also, in attendance were representatives of the proposed civilian contractors, Robert McAlpine & Sons, and the county surveyor.

Col. W. McAdam of the Directorate of Fortifications and Works wrote in his subsequent report to Lt.-Gen. David Henderson, who was then the Director-General of the Military Aeronautics Directorate: '... Although

Camlarg

flying conditions have nothing to do with this directorate, the conditions would seem questionable, and it might be well to make certain that this has been fully considered.'

This view was backed up by McAdam's superior, Maj.-Gen. Sir George Scott-Moncrieff who added an addendum to the report, 'This is a most serious matter, and I think that the scheme ought to be deferred until it is definitely ascertained whether no alternative is possible.'

Brancker, who was by now Director of Air Organisation under Lt.-Gen. David Henderson, sent a scathing reply, 'I am not apprehensive regarding mists, bumps, snow or wind, which are equally more prevalent in any uninhabited part of the country', which showed a complete ignorance of the severe weather that can be experienced at Loch Doon.

Lt.-Gen. David Henderson, with the recommendation of Brancker, decided to go forward with the scheme, the Chief Engineer of the Scottish Command advised him that it was quite possible to drain the moss for the proposed airfield within three months.

The firm of Robert McAlpine & Sons were given the order to begin the works at the end of September 1916. This civilian labour force was supplemented by a detachment of the Royal Engineers and Royal Defence Corps with staff of the RFC. Land was summarily requisitioned from Mrs McAdam of Craigengillan Estate (4,483 acres), The Marquis of Ailsa (8,320 acres), Major F. Cathcart (2,784 acres) and Mr W. C. MacMillan (1,000 acres). Half a dozen sheep farms were commandeered, and the sheep removed from the hills. The tenants of these farms, Lambdoughty, Craigmalloch, Back Starr, Low Starr, and Beoch in the heart of the proposed camp, had to leave, as did the occupants of seven cottages, mostly shepherds' shielings, overlooking the loch.

In command of the School of Aerial Gunnery was Lt.-Col. Louis Arbon Strange. A pre-war pilot, having served with distinction in France, Lt.-Col. Strange had been tasked with setting up the schools of aerial gunnery at Hythe and Turnberry, after which Brancker sent him to visit the Gunnery School at Cazaux for ten days, returning in November 1916 to take command at Loch Doon. Lt.-Col. Strange took off from

Turnberry and flew over Loch Doon on a reconnaissance flight that same month.

Local lad David Smith was employed to check the stores arriving at Dalmellington station whilst working for the RFC in a civilian capacity. He writes in his diary entry for 19 January 1917: 'The first aeroplane (Short 827 8560) arrived at Dalmellington station in two large L&NW vans. It was unloaded onto a long lorry and taken up to Loch Doon.'

Establishing the RFC Headquarters at Camlarg House, on the eastern side of Dalmellington, Lt.-Col. Strange took command of the proposed school. Work began in earnest; the height of the loch was raised by six feet with the installation of a dam at the northern end, this was to provide extra water for a hydro-electric scheme needed to provide power to the camp (This was jointly financed by the War Office and the Dalmellington Iron Company.) At the same time some of the problematic rocky islets were blasted away. The effect of these enormous explosions set in force a mini tsunami which swept away the new dam and caused a torrent to pour down Ness Glen almost stripping it bare of vegetation.

One of the chief features of the establishment was the widening, remaking, and improving of the old road from Mossdale, on the Dalmellington-Carsphairn road, up to and along the loch nearly as far as Lambdoughty, a distance of not less than five miles. The old road, good enough for the occasional motor car, horse vehicles, and pedestrians, was found totally inadequate for the military camp. It was wide enough for only one vehicle, it broke down frequently, and was unable to bear the extra weight and volume of the heavy traffic used to transport men and materials to the Loch Doon site. One of the first construction tasks to be undertaken was to make it fit for the new burdens it had to bear. This work was done in an exceptionally thorough manner. The road was widened to twice its former breadth on a solid foundation, and the result was the road, as far as it goes that is, to within three miles of the head of the loch, joining the numerous short

lengths of road communicating with the various departments, leading from the main road up to the loch. These roads, leading down the extensive moss to the loch side beyond the Garpel Burn, were 'made in a manner and quality that, at the time, were the best in Ayrshire and beyond.'

On the east shore a long row of moving targets was installed in the form of heavy iron structures resting on massive concrete foundations forming a monorail. Large hangars, a seaplane shed, a motorboat dock, and a slipway were built as well as a fully fitted hospital and cinema. Other buildings such as stables, storehouses and accommodation blocks to house 1,300 men were erected, consisting of prefabricated wooden huts built by John Laing & Son of Carlisle, along with a sewage plant capable of coping with a population of 1,500 people. A new water supply was carried in iron pipes from a reservoir among the hills and distributed via pipelines all over the camp. A great amount of underground electric cable was laid down and telephone wires were all connected. Land reclamation was undertaken, and great masses of concrete were laid down for the foundations and footings of the various large buildings.

One of the most serious problems that the engineers had to face was that some of the lower-lying camp buildings had to be placed so close to the loch as to bring them well within the high-water mark and liable to be flooded. Accordingly, an elaborate system of supplementary overflows was designed and completed at a considerably lower level than the already existing overflow, but a good deal above the old tunnel outlet of the water into the Ness Glen. The new overflow was partly regulated by half-a-dozen iron sluices and provided so large a channel of outlet to the water that it was unlikely the loch would ever again reach its previous high levels.

The water that was discharged by way of the new overflows found its way into the Ness Glen along a tailway, under a higher new concrete bridge built to carry the road over it. The old discharge continued to flow through the old bridge, the roadway across which was raised to the level of that across the new bridge.

These works considerably changed the aspect of the loch at this point. What they did to counteract the floods was to open up a much larger waterway, 25 feet wide, alongside the old, by blasting out the whinstone rock. Across this they carried the road by a square-shaped concrete bridge, the same length by about 14 feet high.

But, although they undoubtedly prevented the loch from rising to its former flood levels, no thought seemed to be given to the effect of the excess flood water pouring through the Ness Glen and into the River Doon. The new works would aggravate the floods in the river by more quickly draining off the surplus water from Loch Doon, which has a wide catchment area and where the annual rainfall is over 60 inches, twice that of the average for Scotland.

With a view to minimising this increased risk of flooding on the River Doon, north of the village of Dalmellington the engineers attempted to straighten out a bend in the river, but it was never completed.

A large hill, including iron and lead mines, was bought from the landowners, A drainage ditch was cut around the proposed airfield site and 56 miles of pipe were laid in an attempt to drain the moss; thousands of tons of top soil was spread over the peat and grass seed was sown. When the authorities assumed control they placed an embargo on the use by the public of the road up the west side of the loch, along both sides of which the camp was laid out, and non-military visitors required a pass to enable them to go to the head of the loch by that road or over a wide extent of the moors on both sides, the whole of Loch Doon became a 'prohibited area'. As far as the War Office was concerned, this was to have been a permanent undertaking.

Lieut. Gwilym Hugh Lewis of the Northamptonshire Regiment, attached to the RFC, wrote in a letter home dated 13 February 1917:

Monday, I spent looking around the place. The loch itself is perfectly glorious and entirely taken over by the RFC. The

Loch Doon School of Aerial Gunnery Map

whole place is in a state of construction and is going to be on a very large scale. Contractors, engineers and Flying Corps are all mixed up and twelve hundred Huns will be here as soon as accommodation can be built for them. At present the work consists of loading trucks at Dalmellington and sending them up to the loch on perfectly appalling roads and transforming the loads into wooden huts or anything that is wanted. The aerodrome is rotten and will take a great deal of knocking about; at present it is being drained. Colonel L. A. Strange is in charge of the school; he is a pre-war pilot, only twenty-six but everybody swears by him down here, especially as regards efficiency. At present there is one Short seaplane; they will probably have a number of seaplanes down here later on. At present the loch is covered with ice, but there is little snow about; it is quite cold enough, but I can tell you the blue skies are often conspicuous. The RFC fellows down here are mostly crashed Huns (RFC slang usually referring to pupil pilots). Today I have been made assistant in charge of loading the lorries; sounds a simple job but I had no idea there was so much to be done. The least is to keep about one hundred men working hard, every plank has to be checked and every truck kept separate. Every dump has to have its own special stuff and lorry and, when you have a whole train of stuff to clear, I find it keeps one going with practically no NCOs. Today I have been working under a fellow; tomorrow I have to take over sole command as his results are not good enough. I have an awful wind-up; there is heaps of stuff to clear, especially as regards checking and unloading. I shall have to learn by experience – always very costly.

Promoted to Officer in Charge of Convoy, Lieut. Lewis improved on his forerunner's total of

deliveries from the railhead to the loch from the average of 90 to 95 tons to 150 tons per day.

Loch Doon Hangar

On 1 March 1917 David Smith again writes that, 'Col. Strange flew an aircraft very low over the area'. This may have been a reconnaissance flight with the consideration of building the new airfield on the fringes of Dalmellington, three miles over the hills from Loch Doon. Not yet having a landing ground ready at the Loch Doon establishment, it was decided to build a satellite airfield at Bogton, which was the nearest alternative site close to the loch. This was another poorly drained parcel of land, as the very name would imply, right next to Bogton Loch and the flood plain of the River Doon. The civilian contractors, McAlpine & Sons, started to lay a railway spur to the Bogton airfield site on 4 March, and in just under two weeks, using their own No.9 locomotive (formerly Hudswell Clarke's No.492, built in 1898), pulling a six-wheeled van, they delivered the first aircraft.

As with Loch Doon, the Bogton airfield construction spared no effort or expense, having five large hangars heated by radiators, workshops, stores, vehicle sheds and accommodation consisting of eighteen brick-built huts measuring sixty feet by twenty feet, each capable of housing thirty men and built using the best type of bricks, cemented and pointed with meticulous care. These structures were built to last half a century or more.

Due to the remoteness of the site, the transportation of goods and materials was a major problem. It was calculated that the cost of a single trip for a lorry from the railhead at Dalfarson to the Loch was £2 17s, which probably took into account the expense of maintaining the narrow approach road. The logistics and expenditure of transporting his men from Ayr at a cost of £37 per day gave Lt.-Col. Strange another challenge. Many of the civilian McAlpine contractors were billeted in Ayr, travelling each day to Dalmellington by train and then transported to Loch Doon by lorry. The Royal Engineers were encamped in large huts near Beoch farm. Lt.-Col. Strange requested the requisition of Craigengillan House itself, Mrs McAdam, the owner was unsuccessful in her objections.

An advert appearing in the *Dumfries and Galloway Standard* on Saturday 24 March 1917 reading:

> LABOURERS, Stone-Breakers, Drainers Wanted. 9d Per Hour, Free Hut. Apply D. Kirkland, Loch Doon Road, Dalmellington.

On 1 April a new Commanding Officer of the School of Aerial Gunnery, Loch Doon, arrived - Lt.-Col. Evelyn Boscawen Gordon DSO of the 5th Northumberland Fusiliers. He was seconded to the RFC and was known to his junior officers as 'Uncle Boss'.

Lieut. Lewis again writes home on 8 April:

> As expected, Col. Strange has developed measles the day before his departure, with the result that we are all isolated until the 19th, though we are allowed to carry-on out of doors. The school is getting more into shape, we have several hundred Hun prisoners down here and eventually we shall get twelve hundred. It has been a big task building huts and accommodation for them. They are excellently looked after and are quite a branch of their own. They have a special camp commandant and a fossilised guard consisting of the Royal Defence Corps; these old soldiers are the most unholy rabble ever seen. One of the

POWs was heard to say when looking at his guard, 'Is this what England has come to then?'

> There is a very efficient Hun Sgt.-Maj. who has been made camp Captain; he has his own orderly room and all orders are given through him. There is also a large number of Royal Engineers down here, together with an equal quantity of RFCs. This is the army! Then comes the civilian contractor with his Sinn Feiners from Ireland. They are the only people worth a fig. The contractor started with a very small job down here and now has gradually absorbed the whole thing under his wing, thanks to a very able civil engineer in charge. If you want anything done it is quite hopeless to go to the army position for it, you simply go to the civilian portion and things start to whistle. The contractor is simply fed up with our motor transport system and said, 'I'm going to lay on a railway' and the next day the railway was started. For some reason the military authorities will not have the railway going more than half the distance. We are making quite a nice temporary aerodrome just outside Dalmellington; it is a bit small and rather marshy but will improve in both respects in time. There are two hangars up already and we have the best part of two flying animals here. One is a two-seater very much like a B.E.2c made by Armstrong Whitworth and the other a D.H. Scout. I hope we shall get them into the air in a week or two; it is a bit difficult starting without any spares as they will take some time to get. Newling and I are working on the aerodrome, the D.H. is not very nice for this part of the world unless you have very skilled mechanics in case of a forced landing. A new Colonel has taken over and is much better than expected and is trying to learn his job before dictating. If I can succeed in gaining his respect, I feel I could make a few useful suggestions.

By 20 April 1917 McAlpine's men had also completed a standard gauge railway from Dalmellington railhead through Craigengillan Estate to Dalfarson, directly opposite Craigengillan House. Mrs McAdam, as with her house, was unsuccessful in her opposition to the railway, complaining that she already had 4,483 acres under requisition plus the further 88 acres for Bogton Airfield. This railway line was forced to stop short of the Loch, the hilly terrain meaning the railway could go no further without blasting a tunnel through nearly 1,200 feet of solid rock. This new line was served by a larger Hudswell Clarke (formerly No. 1011, built in 1912) locomotive which McAlpine's had purchased and which was delivered to Dalmellington in a well-wagon on 27 April.

Some of the Royal Flying Corps officers who had been posted to Loch Doon by the end of May 1917 included: Lieut. A. M. Scott, 2nd Lieut. D. C. Ellis, and Maj. E. L. Conran MC, who with Captain H. C. Jackson, had the distinction of carrying out the first bomb attack by a British aircraft at Lessines on 24 August 1914. Further officers posted in later included: Lieut. Edward Addington Hargreaves Pellew, Viscount Exmouth, Lieut. F. L. Herbert, Lieut. R. J. Cowan, 2nd Lieut. R. H. Tweedy, 2nd Lieut. A. H. Waterman, 2nd. Lieut. George Eric Brookes and Capt. Edwin Louis Benbow MC.

The Establishment list dated 31 May 1917 shows a total of almost 1,500 personnel required for the School of Aerial Gunnery at Loch Doon, later lists give an amended total of 2,007, not including the attached two Field Companies of the Royal Engineers and three Works Companies.

The RFC Establishment was broken down into sections as follows: -

Headquarters

Commandant	1
2nd in Command	1
Adjutant	1
Equipment Officer	1
Warrant Officers	2
Clerks General	27
Clerks Pay	7

Clerks Ledger	4
Clerks Stores	6
Clerks Tally Card	1
Grooms	2
Shoemakers	3
Tailors	3
Storemen	1
Telephonists	8
Officer i/c Messing	1
Batmen	97
R.A.M.C.	9

Instructional Section	135
Range and Workshop Section	133
Motorboat Section	29

Maintenance Section

Equipment Officer	1
Sergeants	2
Electricians	5
Carpenters	4
Painters	1
Telephonists	4
Labourers	13
Road Workers	13
Glazier	1
Plumber	1
Masons	2
Sanitary Workers	5
Builder	1
Railwaymen	22
Level Crossing Attendant	1
Sluices Attendant	1

Mechanical Transport Section

Equipment Officer	1
Flight Sergeants	1
Sergeants	2
Clerks	1
Carpenters	1
Drivers MT	51
Fitters	7
Motorcyclists	15
Storemen	1
Turners	1

Aeroplane Repair Section

Equipment Officer	2
Warrant Officer	1
Flight Sergeants	2
Sergeants	4
Acetylene Welders	1
Blacksmiths	2
Carpenters	36
Clerks General	2
Clerks Pay	1
Clerks Ledger	1
Clerks Stores	1
Clerks Tally Card	1
Two Squadron Headquarters	136
Four Flights	248
Seaplane Flight	71
Quartermasters	27

By May 1917, the firm of Robert McAlpine & Sons had processed some 50,000 tons of materials and used 25,000 tons of stones for road making.

By mid-1917 there were upwards of 3,000 men working on the Loch Doon project, including the addition of German prisoners of war as extra labour, who were kept in a large temporary camp north of Lambdoughty farm. It being against the Hague Convention II (1899 Article 6) to use POWs on military works, these prisoners were supposedly only used to build and maintain the access roads and were guarded by No. 213 Protection Company.

In October 1917 The Duke of Connaught paid a visit to the Loch Doon School of Aerial Gunnery. He was received by Lt.-Col. Gordon as officer commanding, and a guard of honour of the RFC, under Captain Benbow, MC. After his inspection of the guard of honour, His Royal Highness inspected the scheme and took a keen interest in the targets and ancillary buildings in the course of erection.

Three large crates were offloaded at Dalmellington railway station on 1 November, forwarded to the School of Aerial Gunnery, Loch Doon, from Paris by the British Aviation Commission. Each crate contained a high-speed hydrofoil type craft, powered by a 220hp Salmson (Canton-Unne) engine, driving a pusher air propeller. These de Lambert Hydro Glisseur were numbered 39, 40 and 42 (engine numbers 5045, 5032 and 5029 respectively), and were possibly ordered by Lt.-Col. Gordon following his visit to France in May 1916. This visit was part of an English fact-finding mission, along with pilot Major Prettyman, Captain Chaney commanding the Royal Flying Corps Machine Gun School, and Sergeant Gray NCO, Training Instructor at the RFC Gunnery School. Exchanges of information were beneficial to all parties, and it was agreed that close cooperation would be established between the Cazaux Ecole de Tir Aérien à la Base Aérienne and the British Schools of Aerial Gunnery.

At Loch Doon itself, more aircraft were delivered including B.E.2c 4721 which was transferred from 'Y' Squadron to the seaplane flight on 8 November 1917. Taken to Bogton airfield where it was fitted with a central float and smaller side stabilising floats, the B.E.2c was then physically carried to Bogton Loch from where it was flown on to Loch Doon. On 7 November Lt.-Col. Gordon was further appointed Commandant of Bogton Aerodrome. It was also declared a 'prohibited area'.

In the same month Lt.-Col. Gordon wrote to the factor of Cassillis Estate (owned by the Marquis of Ailsa) enquiring about the meteorological conditions at Loch Doon. The factor replied that although he personally did not have much experience of the year-round weather conditions, he had enquired of the gamekeepers, 'herds and others. It was reported that 1917 was abnormal, a gamekeeper employed on the shootings at Loch Doon for the past fourteen years stated that the weather conditions this year were the worst he had ever seen in all his time there. He further stated that the severe gales and rain would likely make flying impossible in winter, but the usual conditions prevailing in summer should not be unfavourable.

By the end of 1917 the estimated cost of the project had risen considerably over £350,000, far more than the initially sanctioned £150,000, and the proposed airfield site was still not drained. A request for more funding to blast a tunnel through the solid granite hillside from Dalfarson and extend the railway to the head of the loch now attracted the scrutiny of the Director of Fortifications and Works, who instigated an independent survey to be undertaken by the institute of Civil Engineers. This survey formed a damning report which found its way to the newly formed Air Ministry and the desk of Major John Lawrence Baird, the Parliamentary Under-Secretary of State.

The famous fighter Ace, Capt. William Avery Bishop (later Air Marshal) VC, DSO, and MC, was posted to Loch Doon as an instructor. This came about as the result of unfortunate remarks he had made during a tour of Canada following his investiture with the VC, in which he was very critical of the United States aircraft industry. This caused an outcry in the press and to get him out of the public eye and diffuse the situation Maj.-Gen. Trenchard informed Bishop that he would be posted as Chief Instructor to the Aerial Gunnery School, Loch Doon, graded as a Squadron Commander until the fuss had died down. But before he could take up his new post, the Loch Doon School of Aerial Gunnery was closed down.

In January 1918 Major Baird along with Sir John Hunter, the Administrator of Works and Buildings arrived at the Loch Doon site to inspect the project. On receiving their subsequent report, the Government immediately called a halt to all further work, closed the site and convened a Select Committee to investigate what was to become known as the 'Loch Doon Scandal'.

The findings of the Committee noted that several considerations combined had led to the decision to abandon the site:

1. The Commandant who had overseen the proposed school for nearly year, Lieut.-Col. Gordon, reported that flying would be possible on only half the days of the year on account of climatic conditions. On many of the other days the conditions would not be good for instruction on account of the low clouds.

2. It was stated that turbulence was very bad over the targets, and that this would often interfere with the value of the practice.

3. With the much faster machines which had come into use during the year and half since Loch Doon was first planned, it was doubtful whether it would be safe to fly at hill targets at all.

4. For the same reason the speed of the movable targets, which varied from 25 to 60 miles hour, had become too slow to represent fighting conditions.

5. There was no place in the neighbourhood of the loch where forced landing could be made without crashing except the aerodrome itself, and flying would frequently be at a height insufficient to allow the aerodrome to be reached in the event of engine stoppage.

6. It was doubtful whether the drainage of the aerodrome would ever succeed in providing a place on which aeroplanes could land without danger and 'take off' without difficulty. After a spell of wet weather, it might still be found little better than a bog.

7. Great sums had already been spent, and further large sums for the extension of Bogton Aerodrome and for the railway were being asked for to complete the scheme. Materials and labour were urgently needed elsewhere. The date of completion seemed still remote. It was thought better to decide at once to sacrifice what had been spent than to throw good money after bad.

The Air Board wrote a letter dated 7 January 1918 to Major Baird:

I understand that you have come to the conclusion that it will be better to stop the whole of the works at Loch Doon and not to proceed with any of the schemes, and that you wish for my views respecting the retention of the lands. I have considered the matter and briefly my opinion is as follows: -

On the whole it would be better to face the loss, reinstate where necessary, and pay the compensation to which Mrs McAdam would be entitled for direct loss incurred which would be assessed by the Defence of the Realm Losses Commission and which, apart from reinstatement, ought not to exceed £10,000.

It is obvious that the retention of any portion of the property entirely depends upon the price which we would have to pay, and we should only be justified in buying assuming we could resell at a profit, thus reaping some benefit for the expenditure which has been incurred. I do not think there is much hope that we shall be able to buy at a price which would justify my recommending the Ministry to adopt any other course than that suggested above.

It does not appear to me that the compensation which would be awarded to Mrs McAdam can be large and, in any case, she is entitled to be paid for the loss which she has incurred for the period during which her property has been occupied under the Defence of the Realm.

Such payments as she can claim beyond this will be for reinstatement or for any permanent loss which she can prove she has suffered owing to the works carried out. It is difficult to say what her claim might amount to in this respect as I am unaware of the position of the different works at the present moment and in order to give you an estimate it would be necessary to go to Loch Doon and consider the position on the spot. I am assuming however that the railway would be removed from the drive and reinstatement effected, and that everywhere else where the residential and sporting amenities have been damaged

the damage will be made good so far as is possible. For example, trees have been felled.

We shall also have to settle the claims of the agricultural tenants and arrange for the farmers to come back.

Whilst there is no doubt that the Craigengillan Estate has many attractions in view of its historical associations and residential and sporting amenities, and its accessibility to Glasgow, I do not feel that the Government should take upon themselves any risk at a time like the present, and as Mrs McAdam's advisors, in deciding the price they would accept would naturally take into account the very considerations which I have mentioned above, there appears to me to be small probability of acquiring the estate upon favourable terms, and the work carried out is such that would only slightly increase the value in view of the character of the property.

Bearing in mind therefore that we have no compulsory powers of purchase of the policies or home farm and that Mrs McAdam is aided by a solicitor who has always held out for double what the property is worth, I think it would be better at once to decide to close down altogether, have the loss assessed by the Commission and come to the best arrangement we can in regard to reinstatement, endeavouring to reap some benefit from such of our expenditure as we can induce Mrs McAdam to admit has improved the value of her property.

Lord Balfour, speaking in the House of Lords, stated:

Hitherto the Defence of the Realm Act has stopped any public criticism of this scheme, but, now that it is abandoned, the circumstances are changed, and I think we are free to ask questions. At any rate, I cannot see that there is any public interest that would be contravened by a full statement of the facts and discussion of them. I know, from my own personal knowledge and the testimony of friends, that if the advice of those with real local knowledge had been taken this scheme would never have been begun at all, and I do not think it is unfair to ask who was responsible for the selection of the site, what inquiries were made before the money was expended, and what engineering advice was taken.

The real point of the whole matter is this. Was the place really suitable? Was proper local information and advice appealed to before the expenditure was undertaken? I have read the discussion which took place in the other House, and if I understand that discussion rightly the Royal Engineers were the advisers.

And so, the military and their contractors departed almost as suddenly as they had arrived. For a while the whole site was left deserted and unguarded. All of the fine buildings with their fixtures and fittings were left open and abandoned. It was soon brought to the attention of the authorities that under the requisition agreement with the land owners, it was required for the land to be restored to its prior state. Teams of men were brought in to demolish the buildings with little concern for salvage. Masses of expensive fittings and equipment were literally just thrown into the loch. Although the buildings were removed, the concrete bases were left in situ, as were many other concrete structures, including the substantial pier and boat-landing stage, these latter on the west side of the loch, midway between Beoch farm and the Garpel Burn.

The road, a curious anomaly in such surroundings, was also left intact for generations to come, as a memorial of misplaced zeal. Some of the most important buildings of the encampment were at the shore end of this road, and they, with the exception of their floors and foundations, were completely cleared off the

site. The whole of the encampment was confined to the west side of the loch, but the important constructions on the east side in the form of the mono-rail targets, the solid concrete under-structures were left. The iron rail was removed to be utilised elsewhere. For the purposes of the particular training to be undertaken, considerable operations were carried out on the declivity between Cullendoch Hill and Black Craig, overlooking the loch from its east side. What constructive work was carried out there was removed, but left long scars up and down, and across the base of the hills. Cullendoch Hill is about half-way up the loch, and beyond this point to the head no permanent operations were carried out.

One of the principal parts of the work in connection with the scheme was the making of the railway connecting with the G&SW line at Dalmellington, leading to past the avenue of Craigengillan House, to the cottage at Dalfarson. The railway was, of course, also removed.

The electric power building and plant at the Dunaskin Ironworks, about three miles on the Ayr side of Dalmellington, which was intended to supply power and light to the camp, was in the course of completion, and had been to some extent utilised for these purposes at Loch Doon. This was left intact and was taken over by the Dalmellington Iron Company.

Up to thirty small boats which were moored on the loch were scuttled and remain on the bottom, a small motor boat was also sunk on the loch. Instantly regretting the folly of their action, divers and salvage equipment were ordered from Glasgow, but the expense involved in the recovery of the vessel was much greater than the original cost of the boat and the attempt to bring it to the surface was abandoned.

There was severe criticism in parliamentary debates of the Department of Military Aeronautics, especially directed at Lt.-Gen. Sir David Henderson and his deputy, Maj.-Gen. Brancker. They were condemned, not for initiating the scheme, but for allowing for it to carry on after the summer of 1917 when it was apparent that the problems encountered really were unsurmountable. (Henderson had resigned as its Director-General in October 1917 and Brancker was posted to an overseas command, he was later killed in the R101 Airship disaster)

In the House of Commons, Mr Bonar Law replying to Mr Joynson-Hicks, said the government had given most careful consideration to the report of the Select Committee on the Loch Doon 'Scandal'. It was their duty to weigh against any errors of judgment in this case, the great services - and they could not be rated too highly - which had been rendered by the officers in question, to the rapid development our air forces. They did not consider that any action against these officers was called for but both in parliament and the press frequent demands were made for the 'culprits' of the scheme to be brought to justice. Lord Balfour commented in a House of Lords debate on 21 March 1918, 'If I can trust to the testimony of a private friend, a man whom I have known for fifty years, the person who was really responsible was an English engineer of a town in the south of England, who had no practical knowledge of the particular conditions he would have to meet. My friend tells me that he interviewed this gentleman—I do not mention his name because I do not want to be unjust—told him the scheme was impossible and gave him the reasons, but that all objections were repelled, and no advantage of local knowledge was taken at all. Beyond all question a very large amount of local damage has been done, and the forecast of those who knew the circumstances has, if my information is right, been absolutely fulfilled. My friend goes on to say— In my opinion there never has been a madder scheme or a more shameful waste of public money. I am told that about £3,000,000 has passed through the bank at Dalmellington. I know all traces of the muddle are being destroyed as fast as possible, and one correspondent says he thinks there is enough cement buried at Loch Doon to build a wall round Jerusalem, and all this is being covered up so as to hide it as a monument of the stupidity of those concerned'.

'Loch Doon and the country around it,' the Committee commented, 'will soon return to the solitude and silence from which they were roused by the introduction of thousands of men. employed, over period fifteen months, at a cost of hundreds of thousands of pounds of public money, on an enterprise which was misconceived from the beginning, and which, even once begun, ought never have been continued. This name will be remembered as the scene of one of the most striking instances of wasted expenditure that our records show'.

National and International Newspapers were quick to pick up on the story such as the *Dumfries and Galloway Standard*, Saturday 23 February 1918:

AVIATION SCHEME IN AYRSHIRE
IS IT ANOTHER COSTLY FAILURE?

The acerbic debate rumbled on in parliament throughout 1918, in reply to Sir S. Roberts (C.U. Eccleshall). The Under-Secretary for Air (Major General Seely) said the total expenditure in the Loch Doon Scheme up to February 19 last was £484,433. In that figure the cost of land and compensation to tenants was not included. The saleable plant and other assets had already been largely removed, being either transferred to Government purposes elsewhere or sold to other purchasers. The value of the material disposed of in this way would amount, when removal was completed to approximately £210,000.

Similarly, other newspapers reported on the 'Scandal', such as the *Pall Mall Gazette* of 22 March 1918:

A TYPICAL INSTANCE OF
OFFICIAL WASTE

The debate last night in the Lords on the Loch Doon aviation base, which, after half a million of the public's money has been wasted upon it, is found to be quite unsuitable for the purpose it was intended to fulfil, exposed a typical case of the way in which the nation's business has been conducted. Can we wonder that the cost of the war is so great? It may, we think, be taken that Loch Doon is only a particularly glaring instance of the incapacity and ineptitude which have for three years obstructed the conduct of the air branch of the Service, and that Lord Rothermere will have an uphill fight to clear the new Department of the old incompetents. It is characteristic of departmental procedure that for the waste of public money on the abortive Loch Doon experiment no one is responsible. The men who make these muddles always seem to cover up their tracks, but it is essential that they should be discovered and brought to book. The House of Commons must probe this matter thoroughly, and the guilty persons, when discovered, must be made to pay the penalty of their incompetence. We would strongly advise Lord Rothermere to rely very little at the moment on his experts, or so-called experts. It is quite possible that there are other Loch Doons up and down the country, and we would recommend Lord Rothermere to make a tour of the great aeroplane works and training centres and see for himself how the business of his Department is being conducted. Let him, moreover, pay no respect to names or reputations, and whenever he finds himself obstructed let him make the facts known, and we will guarantee him a public support that shall sweep from his path all opposition, come whence it may.

From the *Northern Whig*, Friday 03 May 1918, we read:

LOCH DOON AERIAL GUNNERY SCHOOL

Replying to question in the House Lords yesterday regarding the proposed aerial

gunnery school at Loch Doon. Earl Curzon said the site was reported on by committee of experts as an almost ideal spot. When the Air Council was formed it was found that the engineering difficulties had been underestimated. The cost incurred was £433,000, against which there were credits of £205,000 for plant and materials, and the adverse balance being £228,000. Some of the experts were of opinion that the scheme ought to have been proceeded with instead of being abandoned.

The *Irish Independent* of 20 May 1918:

While half-a-million of public money was thrown into Loch Doon, Ireland, which could provide dozens of more suitable sites, is put off with a couple of aerodromes, and is still waiting for the long-promised aircraft factory.

Similarly, in the *Sheffield Daily Telegraph*, Monday 20 May 1918:

AN ORGY OF WASTE.
WHEN WILL IT BE STOPPED?

The third Report of the Select Committee on National Expenditure in no way relieves the apprehension of the taxpayer that his interests are being gravely neglected. The appalling story of colossal waste, extravagance, and incompetence, which the present Report carries a step further, is evil reading for the public which knows only too well that only major scandals can possibly be exposed by such an inquiry, however diligently it pressed. Of the many instances of reckless waste to which the Committee calls attention, the Loch Doon scandal is, of course, of outstanding gravity, the actual loss of public money involved, large though it is, being by no means the worst feature of a particularly bad case. What will strike the nation the gravest charge which the Report virtually formulates, is that of the singular incompetence of the experts who passed the scheme for the great Flying School, and long after it should have been known that a mistake had been made, continued to waste time and money and men upon task that was hopeless from the beginning. In spite of the fact that, as the Report points out, Major-General Scott-Moncrieff wrote a minute describing the projected Loch Doon Flying School, as a 'very risky measure to attempt,' and advised the deferring of operations until an alternative had been considered, before the work was proceeded with. The Chief Engineer of the Scottish Command reported that the drainage of the area was practicable, and Sir David wrote that 'Loch Doon fulfils all our requirements, and if the engineering difficulties are not insuperable, I would like it proceeded with at once.' Governments and Departments must, of course, act upon the advice of their experts, but we should like to be informed what measures were taken in this case to punish those who tendered such bad advice. Has anyone suffered in any way for the loss which the country sustained in this particular case? It seems to us that in regard to this specific instance we are provided with some evidence showing us what during his tenure of the post of Air Minister Lord Rothermere must have had to contend with. The matter clearly cannot be left as it is, and Parliament must insist that the Select Committee's exposure leads to punitive action against the principal culprits. Parliament must at once resume its neglected control of the public purse and put stop to this orgy of waste that is crippling the prosecution of the war.

Again from the *Pall Mall Gazette*, Monday 20 May 1918:

THE LOCH DOON FOLLY

The precise amount of money that was wasted over the projected School of Aerial Gunnery at Loch Doon in Ayrshire has not been stated, but the Select Committee on National Expenditure says that it ran to hundreds of thousands. In all the war there has been no more monumental example of waste of public money; and that, as our readers are aware, is saying a very great deal. What makes the folly stand out conspicuously is its amazing lack of prevision. The people who proposed to spend all this money on aviation seem to have known nothing about aviation or money either. One would have thought that in choosing an aerodrome site some attention would have been paid to climate; apparently climate was left out of account. And will it be believed that the absence of suitable landing places was also left out of account? As for money, the whole of the original estimate of cost - £150,000 - was to be spent on a tunnel through which four trucks were to run daily. Who were the wiseacres, who embarked on this project, and persisted in it, at vast expense, even when its difficulties were becoming more and more obvious? They cannot have known much about flying, or they would have thought about landing places; they cannot have been business men, or they would have thought about the cost of boring tunnels. There is a sort of mystery about the affair; anyhow, we are glad it has been shown up, for this gives some sort of safeguard against similar mysteries in future. We cannot afford any more like it.

In a later edition of the *Pall Mall Gazette*, Wednesday 5 June 1918, we read:

THE LOCH DOON SCANDAL. NO DISCIPLINARY ACTION REQUIRED

Major Baird, when asked whether he could state the names of the persons responsible for the inauguration and completion of the scheme regarding the aviation park Loch Doon, and what disciplinary action had been taken, stated in Parliamentary Papers that the matter is not one in which disciplinary action on the part of the Air Council is considered be required.

A report of the ongoings in parliament appeared in *The Scotsman* on Tuesday 2 July 1918:

LOCH DOON .... THE-SPEAKER
AND MR BILLING

Asked by Mr Joynson-Hicks (C.U., Brentford) whether the Government had considered the report of the Select Committee on the Loch Doon scandal, and, if so, what action they intended to take regarding such of the officers responsible as still remained in His Majesty's service.

The CHANCELLOR of the EXCHEQUER said His Majesty's Government have given most careful consideration to the report referred to. It is the duty of the Government to weigh against any errors of judgment in this case the great services, and they cannot be rated too highly which have been rendered by the officers' in question in the rapid development of the Air Forces. After taking every circumstance into account, the Government do not consider that any action against these officers is called for.

Sir R. ESSEX (L., Stafford) asked if the right hon. gentleman would have placed in the library a large-scale ordnance map in order that members might see how much, if any, landing space was available. The CHANCELLOR of the EXCHEQUER—I do not think any useful purpose would be served by that. The object of the hon. member was not to

punish particular officers but to prevent the same thing happening again.

Mr BILLING (I., Hertford)—Is the right hon. gentleman aware that some of the officers referred to are the same officers who were responsible for the hopeless inefficiency of the Air Service two years ago, and, in these circumstances does he not think that some action is most necessary in the public interest?

The CHANCELLOR of the EXCHEQUER—I entirely disagree with the hon. member. Nothing done in this country has been more remarkable than the success which has characterised the development of our Air Service.

Mr BILLING endeavoured to put a further question but was subjected to frequent interruption.

THE SPEAKER at length intervened and asked the hon. member to put the question on the paper.

Mr BILLING. On a point of order, you were proposing to allow me to put the question. 'You have no idea, what the question was. Therefore, you are not in a position to judge'.

The SPEAKER. That is the very reason why I asked the hon. member to hand it in. (Laughter)

Mr BILLING. 'Am I to understand that it is your ruling that if the conduct of members of this House is not compatible with the 'dignity' of the House (laughter) supplementary questions cannot be asked?'

The SPEAKER. I am passing no reflection on the conduct of any hon. member. All I did was to ask the hon. member to hand his question in, surely a very civil and ordinary request.

Later Mr BILLING asked the Secretary to the Air Ministry whether any special inquiry had been instituted or was contemplated into the matter of the Loch Doon aerodrome.

The CHANCELLOR of the EXCHEQUER said the answer he gave to the hon. member for Brentford also applied to this question.

Mr BILLING Is the right hon. gentleman aware that, had it not been for the complete change in the command, the increased efficiency of the Air Service would not have been made possible?

The CHANCELLOR of the EXCHEQUER. I am not in the least aware of it. (Cheers)

Mr BILLING. He is not aware of what takes place in his own Department.

The SPEAKER. The hon. member is not entitled to interject remarks of that kind and I warn him that if he cannot conform to the ordinary rules, I shall ask him to step outside. (Cheers)

Further adverts appeared in the press, such as this from the *Daily Record* - Saturday 17 August 1918:

LOCH DOON SCHOOL OF AERIAL GUNNERY
CATERING DEPARTMENT
FOR SALE, BY TENDER
The stock of the store and dry canteen consisting of groceries and other dry goods and amounting at cost price to £341 13s 0½d, is offered for sale by Tender in one Lot. For Lists and other particulars and Forms of Tender apply to Sir Robert McAlpine & Sons, Contractors, Dalmellington.

Again, from the *Daily Record*, Monday 12 August 1918:

Scrap Timber, 30/40 ton, suitable for firewood only, for Sale; can be seen on work: offers wanted, free on rail, Dalmellington. Apply Sir Robert McAlpine & Sons, Contractors, Loch Doon, Dalmellington.

In his book *Ayrshire 1745 – 1950*, J. E. Shaw states that in 1918 Lord Curzon had asked him if he could enquire locally as to what sums had been expended on construction, in his capacity as Clerk to Ayrshire County. Shaw was only able to establish that more than £3 million (£198 million in today's money) had passed from the small bank at Dalmellington during the building period.

The numerous fine buildings at the loch stood empty and abandoned until after the war, when they were torn down under the original terms of agreement made with the landlords.

One who had condemned the projected Loch Doon site from the very start had been Lt.-Gen. Sir Spencer Ewart of Craigcleugh, Langholm. Having local knowledge from living in the Dumfriesshire area, he was also General Officer Commanding Scotland from 1914 to 1918. Lt.-Gen. Ewart also had a comprehensive knowledge of aviation, having met many of the pioneer aviators (including the Wright brothers) during 1908-09, when he was Director of Military Operations at the War Office. He writes the following in his unpublished autobiography:

I was personally much opposed to Loch Doon being selected as a site for such a range, because I realised what an enormous sum would have to be expended in the making of roads and approaches to it. Moreover, any local idiot could have informed the Army Council that a narrow and confined loch, situated at such a high elevation, was bound to be frozen and therefore useless for hydroplanes in the winter. Nobody in Whitehall would, however, listen to my remonstrations and something over a million pounds, I believe, was wasted before this senseless project was abandoned. No such scandalous waste of public money was equalled, I imagine, anywhere else in the war and those responsible should, in my opinion, have been tried by Court Martial and removed from the service for incapacity. The saddest thing about this melancholy business was the suicide of Mrs McAdam's niece (Miss Ethel Mary Mills, 1880-1917, author) who lived with her at Craigengillan House as a companion. She had appealed piteously but vainly that officers should not be quartered upon the mansion where her aunt had just undergone an operation, and finally, poor girl, made away with herself. How the War Office hushed up this disgraceful episode I do not know. All I do know is that I felt well satisfied that I had refused to have anything to do with the matter. I even declined to sign the order quartering troops upon Craigengillan. Who signed it I never heard. While at Loch Doon I inspected the huts, which had been built for a detachment of the Royal Flying Corps and also the quarters provided for the Royal Engineers at Patna, Waterside and Dalmellington. I discussed with officers on the spot the question of sending a number of German Prisoners of War to act as a working party and selected a suitable site for an encampment.

Among many contemporary accounts from newspapers across the globe, we can gather more information on what happened at Loch Doon, both during construction and in the aftermath. According to *The Scotsman* of Thursday, 19 September 1918, in the Company Report of the Dalmellington Iron Company, 'The report for the year ending 30 June states that the plant erected at the works for the transmission of electric power for Loch Doon has been acquired from the War Office on equitable terms.'

In the *Drogheda Argus and Leinster Journal* of Saturday, 12 October 1918, we read:

WAR OFFICE SAGACITY
(from *Munsey's Magazine*)

In 1916 the Air Board of the British War Office determined that it was necessary to establish in Scotland a school for training airmen in gunnery, and selected a site for the establishment at Loch Doon, in Ayrshire. The expenditures upon the scheme have exceeded half a million pounds, equivalent to nearly two and one-half millions of dollars; and now the undertaking has had to be abandoned because the weather conditions of the locality are utterly unsuitable for the work. Much severe criticism has naturally been aroused in England and Scotland at the neglect of the responsible authorities to ascertain this beforehand. After the money has been spent, it is somewhat exasperating to the British public to be told that 'no flying would he possible at Loch Doon for half the days of the year on account of climatic conditions,' and that 'on many of the other days the conditions would not be good for instruction, on account of the low clouds.' The aerodrome at Loch Doon is as worthless as a white elephant for war purposes, and only serves as a lesson of precaution for the future.

The apologists for the Air Board say that the engineers considered only the question of draining a bog on the proposed site and constructing the necessary buildings and tracks on the premises; they paid no attention whatever to meteorological questions. This does not render the tremendous mistake any more excusable. The suitableness of the prevailing weather is an essential element of fitness in locating a school for aviators; and nowhere in the world has meteorology been more carefully and successfully studied than in Scotland. For many years the British Government, through the Meteorological Office - like our own Government, at first through the Signal Service, and now through the Department of Agriculture — has been collecting daily, and almost hourly, reports concerning the weather in all parts of the country; and no doubt the Air Board could have found out all they know now about the climate of Loch Doon if they had pursued the obvious course of going to the official weather observers for information. This unhappy British blunder is a lesson for our own Government at Washington, as applicable to the location not only of aerodromes, but of any other military establishments. Before a site is finally chosen, find out what sort of weather may be expected there in summer or winter, seed time or harvest. The information can be supplied or obtained by the Weather Bureau; and it is well worth while to make the necessary inquiry if you can thereby save a couple of million dollars or more of the people's money. The calmness with which the announcement of such a loss is received now-a-days shows how the war has accustomed us to regard large sums of money with comparative indifference. We talk about millions now as we used to talk about thousands. Still, two and one-half millions of dollars is a good deal to put into a bog in Scotland with nothing to show for it, and we must see to it here in America that we do not go and do likewise.

An article appeared in *The Scotsman* on Wednesday, 30 October 1918:

ACCESS TO LOCH DOON

Although no announcement has been made by the authorities, the restrictions imposed upon the public against approaching Loch Doon or passing along the road to the head of the loch are apparently no longer in operation. The restrictions were removed shortly after it was decided to abandon the camp, but they were re-imposed again, and were in force until recently. The neighbourhood of the loch is now deserted by the military, and even the contractors and their men have gone. The fishing club have been reinstalled in their fishing lodge, which had been taken over by the military, and

people passing along the road now do so unchallenged. All the buildings have been taken down and the material removed, but there are many evidences of the important operations that were intended to be permanently established along the greater part of the loch on both sides. These remains are mostly in the shape of side roads to the ranges and living huts, workshops, and aerodromes on the west side of the loch, and great masses of concrete in connection with the railway and moveable targets on the east side. The latter extended along practically the whole length of the loch, and up part of the adjacent hills, and the railway was carried on blocks of concrete, which remain as they were put down. Some of the concrete is of permanent benefit, such, for instance as that which was put down to raise the main road above the level of water at the lower end of the loch in times of heavy flood. The new concrete bridge across the new outlet, supplementary to the old bridge and the old outlet remains, as well as the new overflow. This new overflow, which is at a lower level, and more extensive, than the old, will prevent the loch from ever again rising to its former abnormal level. The loch has otherwise been little interfered with except for the piers and artificial coves on its western side.

Advertisements continued to appear in national newspapers for the sale of materials at Loch Doon, such as that which appeared in the *Daily Record and Mail*, Friday, 20 December 1918:

PUBLIC NOTICES. SALE BY TENDER, AT DALMELLINGTON, AYRSHIRE

A Quantity of Building Material, the remains of works at Loch Doon. Compromising the following: Pitch Pine Logs, Whitewood (of various dimensions), Home Grown Timber (squared and round), Cast Iron Pipes, Corrugated Iron Sheets, Scrap Metal, Second-hand Stoves, Large Quantity Cement Sacks, number of small Sectional Buildings, Latrine and other Stores. Offers to be lodged with the Resident Engineer, Dalmellington, on or before 31 December 1918, from whom copies of Conditions of Sale may be obtained.

A detailed report appeared in *The Scotsman* on Tuesday, 25 March 1919:

EXPENDITURE ON LOCH DOON SCHEME
In reply to Sir S. Roberts (C.U. Eccleshall)

The Under-Secretary for Air (Major General Seely) said the total expenditure in the Loch Doon Scheme up to February 19 last was £484,433. In that figure the cost of land and compensation to tenants was not included. The saleable plant and other assets had already been largely removed, being either transferred to Government purposes elsewhere or sold to other purchasers. The value of the material disposed of in this way would amount, when removal was completed to approximately £210,000.

In December 1919 Mrs Charlotte Tilke McAdam of Craigengillan, raised an action in the Court of Session against the Lord Advocate, as representing the War Department, in which she asked for declaration that the whole operations of the War Department in entering upon, taking possession of, using, and evicting her from the estate and house of Craigengillan, and constructing railways, buildings, and other works on the estate, and cutting down trees, levelling fences, and otherwise interfering with the estate, were unwarranted and illegal. She maintained that the War Department had no statutory or other authority to carry out the undertaking, or to take possession to occupy any part of the property, which was a particularly fine agricultural,

pastoral, residential, and sporting estate, and that the scheme was not designed to meet any emergency, but provide a school mainly intended for use after the war. In September 1916 she was informed that her residence would not be taken by the military authority. During that month a considerable number of officers and men arrived, and a large body of German prisoners were housed on the property and employed on the undertaking, and ultimately 5,700 acres were occupied by the military. No formal notice was ever served upon her under any act or authority, and the area was never proclaimed to be one which it was necessary to safeguard in the interests of the training or concentration of the forces. The railway arrangements at Dalmellington and on the roadways were totally inadequate for dealing with the great traffic in material and men which the scheme required, and no steps were taken to make them adequate. In December a private road running through her policies was requisitioned. Instead of repairing the roads which, had broken down, the War Office commenced the construction of a railway from the Glasgow and South-Western system, and it was laid almost entirely in the policies of Craigengillan. She protested, but a formal requisition was sent to her. The only result had been to largely destroy the value to her property, and that the scheme was impracticable, and she submitted a claim for £150,000 compensation. The actions of the War Office, she said, were a gross abuse of authority, and the only result of those operations has been to expend wastefully large sums of public money. In support of these averments she quoted from the third report of the Select Committee of the House of Commons on National Expenditure. Mrs McAdam further stated that in the autumn of 1916 she underwent a serious operation, and it was essential for her recovery that she should have undisturbed possession of Craigengillan, which was her sole residence. In January 1917 intimation was made that Craigengillan House was to be taken over by the military, but the Scottish Command, who had given the previous undertaking against this, knew nothing about the requisition, for which

Craigengillan

there was no necessity. Mrs McAdam said that the officers had stated that it was impossible to consider individuals where the safety of the nation was concerned, and that public opinion would be greatly against her if it was known that she had refused to give up her house. Her niece, Miss Mills, resided with her during the period of convalescence. She said that the disgraceful conduct of the officers, the presence of large numbers of men, and the operations and critical state of Mrs McAdam so wrought upon Miss Mills that she attempted to take her life on March 10, and died two days later. Her action, Mrs McAdam said, was wholly attributable to the distress and anxiety caused by the actions of the War Department and their officers, and her death had seriously affected her own health. In April she removed to Rozelle House, Ayr, of which she took a lease. She gave details as to the damage caused to the mansion-house at Craigengillan, as well as the estate and deals with the further operations at Bogton and Bogton Loch. The work here, she said, cost from £60,000 to £80,000, and the place was never used as an aerodrome. Subsequently, negotiations had commenced for the purchase of the whole or part of the estate, but these were brought to an end by the abandonment of the scheme. The military

then proceeded with the work of reinstatement, but there are acres of concrete floors and a great quantity of other material lying about, and part of the property is in a disgraceful condition; while many picturesque features of the estate have been destroyed. She returned to Craigengillan in September 1918.

The War Department in defence, replied that in taking possession of Mrs McAdam's property, acted throughout strictly under the Defence of the Realm Acts and Regulations of the Prerogative Rights of the Crown. He stated that the establishment of such a gunnery school was necessary, and its future would have depended on the course of the war and future necessities. It was admitted that owing to certain difficulties the original plans were in part departed from as the work proceeded, and the necessity for taking Craigengillan House arose later, and it was with great reluctance that it was taken. The subsequent abandonment of the scheme was due to physical conditions not anticipated. The War Department admitted that the military occupation deprived Mrs McAdam of the full use and enjoyment of her property and that she was not in good health during her stay at Craigengillan after her operation; but pronounced that statements relative to the act of Miss Mills were denied, and also said that the recent military occupation of the estate was greatly overstated. It was pointed out that Mrs McAdam's remedy is to apply to the Defence of the Realm Losses Commission if not satisfied with what the military have done. No legal liability could be attached to the War Department, and, in any event, the sum sued for is excessive.

A considerable account of the scandal at Loch Doon appeared in the *Sunday Post* on 18 February 1923:

THE WASTED MILLIONS AT LOCH DOON MONUMENTS OF OFFICIAL FOLLY AS THEY ARE SEEN TO-DAY

The recent purchase of the army buildings, aeroplane hangars, and offices at Dalmellington, Ayrshire, by a Dundee firm from the Government for what is referred to locally as a mere song, sheds further light on one of the most crazy and impossible schemes ever conceived and carried out by a presumably responsible Government Department. How many millions of pounds were literally thrown away in the Dalmellington undertaking, and the equally colossal blunder at Loch Doon, will never be known. Money was poured out like water on these hare-brained schemes, which local experts, with a knowledge of the peculiarities of the district, warned the Government were foredoomed to absolute failure. If the Government officials did not know that the site at Dalmellington was subject to severe flooding from the River Doon, local residents did, and gave them timeous and earnest warning. But all such hints failed to convince officialdom, who, with a seeming disregard for the spending of public money, hurried on the undertakings, perhaps hoping that the River Doon might by some miracle change its course and leave the Dalmellington site firm and dry.

The buildings erected at Dalmellington were constructed to stand for half a century. The best type of bricks was used, and these were cemented and pointed with a meticulous care, which, now that the structures are being razed to the ground, emphasises the initial futility and absolute foolishness of such an undertaking.

Occupying a site extending to about a quarter of a mile in length by 200 yards or so in depth, the numerous structures were originally intended to form permanent barracks in this little, out of the way Ayrshire town. The tragedy of it all! The eighteen barrack-rooms, each sixty feet by twenty feet, and capable of housing forty to fifty men, are being stripped of the iron and woodwork, leaving the brick walls standing as a monument of Government folly. It won't pay to remove the walls, and consequently', they are left standing, gaunt and dilapidated-looking on their marshy site. Blunders galore were committed at Dalmellington.

Two of the three huge aeroplane hangars were constructed facing the wrong way and on ground totally unsuited for the landing of aeroplanes! These hangars had to be pulled down and huge concrete beds two feet six inches thick constructed to enable aeroplanes to be pulled into the other mammoth building, which is very often waterlogged.

Here are a few items in the cost of this hangar, which was sold to the Dundee firm for £120 — twelve doors, twenty feet high by thirteen feet wide, each costing £50, thirty window-frames at 15s each, 15,000 square feet of sarking for the roof at 10d square foot, 8,860 feet of purlins supporting the roof at 6d foot, 126 iron tie rods at 6d a foot, 6,100 feet of 5 by 2 timber at 5d a foot, 340 feet of moulded guttering at 1s a foot, 144 feet of pipes at 6d a foot, not to mention the thousands of bricks, tons of cement, and the wages of the men engaged in the erection of this veritable white elephant. That is but one concrete instance of how the public money was senselessly squandered. There was never any hope from the first of these buildings being suitable for the purpose, and only too late was this fact realised by the Government of the day.

A Guardroom which was never used, and which cost £300 to build, is offered for a £10 note, if anyone can be got to pay that price. But nobody appears to want these derelict Government buildings. A sectional wooden hut, for which £120 was offered three years ago, is now awaiting a purchaser at £25.

And what of the unpardonable blunder of trying to form a seaplane and aeroplane station at Loch Doon? This beautiful Loch, which is situated about 100 feet higher than the town of Dalmellington, is surrounded by an amphitheatre of hills, and a more unlikely spot for such purpose could hardly be conceived.

Along with two local professional men, who acted as guides, I motored to the scene of what is piquantly referred to in Ayrshire as 'the graveyard of millions.'

When first the Government officials inspected Loch Doon in order to form a school of aerial gunnery there, a rough cart track was the only road up the marshy and mossy hill to the Loch. If ever a gigantic scheme was conceived in panic and born in a spirit of desperate haste, that undertaking was at Loch Doon.

A magnificent road seven miles in length was formed, and tens of thousands of tons of the best road metal were dumped on the surface of this road. What the actual cost of this vast stretch of beautifully-metalled roadway was, only imagination can fancy. But it must have been a stupendous figure. Seven miles! Today it is the finest motor road in Ayrshire, if not in Scotland, but it leads to nowhere, terminating at the edge of the Loch!

Huge buildings including barracks, lecture halls, offices, seaplane hangars, and such like, were rushed up, a veritable army of skilled and unskilled workmen being employed. The tract of moorland designed for the landing of aeroplanes was found to be boggy and uneven. Something like 1,400 German prisoners were employed to level the ground and drain it. Over fifty miles of pipes were used to drain the marsh, but the bog won in the end, as it was bound to do, and the ground is to-day as damp as ever. A fleet of between twenty and thirty small boats were placed on the loch, and many of these were sunk and allowed to remain at the bottom. An instance of official short-sightedness may be mentioned. A small motor boat was sunk on the loch, and divers and equipment were ordered from Glasgow, but the expense involved in the salvage was much greater than the original cost of the boat.

Everything at Loch Doon was carried out on titanic lines. A solid concrete bed the area of George Square in Glasgow, stands today as a landmark at the edge of the Loch. It was designed for the landing of seaplanes and must have cost a huge sum. But perhaps the crowning folly of the many madcap schemes carried out at Loch Doon was the construction of what was known as a 'moving target'. Gigantic blocks of solid concrete were placed in a line at the edge of the loch, and the idea was to have a running target constructed to run along the top of the pillars and give seaplanes practice in hitting a moving target.

This scheme must have taken months to complete, with hundreds of workmen on the job, but, like McCaig's Monument at Oban, it was never finished, and today it stands by the side of that lonely loch, a reproach to ill-advised zeal. Not a single experiment was carried out with the moving target, as one day, with the suddenness of a clap of thunder, peremptory orders to cease all operations at Loch Doon were received.

The folly of throwing away more money on what ought to have been recognised from the first as a hopeless undertaking had at last penetrated the official mind.

Hastily as the numerous structures had been erected, that was nothing to the zeal displayed in the demolition of the buildings once the scheme had been damned. Today nothing but the ruins of the gigantic undertakings are visible. Piles of concrete blocks bearing the date stamped on them as they left the moulds are lying scattered about by the side of the loch and add to the general impression of abandonment and wasted effort. In fact, there appears to have been a panicky scuttle from the place, no effort whatever having been made to clear away these tell-tale memorials. It is a heart-breaking experience for a Scotsman and a rate payer, to wander about this 'graveyard of millions' and muse on the colossal expenditure of public money on a scheme so crazy and obviously futile. Loch Doon will forever remain one of the crowning follies of the late Government during the war.

The cost of the whole debacle was referred to in the *Dundee Evening Telegraph* of 23 March 1923:

A WAR-TIME WHITE ELEPHANT COST OF AERODROME AT LOCH DOON

Sir S. Hoare in Parliamentary Debates states that the expenditure on the aerodrome at Loch Doon was approximately £435,000. Of this amount £320,600 was paid to Messrs. McAlpine. Competitive tenders were not at that time obtainable for these large construction contracts, and the contract was on prime cost basis. The residual value was approximately £167,000 for plant and other material transferred for use on other Government works and £25,000 for sales.

The whole sorry saga of the Loch Doon Fiasco continued to be commented on in the press for years, and the story was told in newspapers far and wide, such as the *Brooklyn Daily Eagle* and the *Singapore Free Press and Mercantile Advertiser*.

I shall leave the final word on Loch Doon to Lt.-Col. Louis Arbon Strange:

A great deal of money was wasted on this scheme because unforeseen difficulties were not tackled resolutely, and so became too much for those who were in charge at the time, the failure was due to a lack of continuity and steadfastness of purpose caused by continual change in the personnel at the head of affairs … There is no doubt in my mind that it would have been a great success if its originators had been allowed to see it through. Local opposition undoubtedly had a good deal to do with the failure, but a strong hand could have put it down and completed the work.

Meanwhile, back at Turnberry, the first hint of the proposed establishment of 2 (Auxiliary) School of Aerial Gunnery received by the local landowners, prompted Tom Smith, the factor to the Marquess of Ailsa to write to the Board of Agriculture for Scotland on 18 December 1916:

The tenant of Little Turnberry farm in the Parish of Kirkoswald called on me this morning and informed me that he had been called upon on Friday last and again yesterday by two different lots of the Royal Flying Corps who inspected the farm with a view to taking portions for landing places for aeroplanes.

The farm of Little Turnberry is cropped on a highly intensive system and the portions selected are most suitable for the growing of early potatoes. The extent proposed to be taken is about 80 acres and this would mean a loss, in potatoes alone, of over 800 tons early potatoes and the loss of the after crop, with a corresponding decrease of land for fattening sheep in the Autumn; and would certainly necessitate a very considerable reduction of the dairy stock. There are considerable areas of land in the immediate vicinity not under crop which, while perhaps not so immediately suitable for the purposes indicated, could easily be made suitable and which are not any use for food production.

Even allowing for the urgent necessities of the Flying Corps, in the present circumstances when farmers and others are being urged to increase the supplies of food, and looking to the shortage of potatoes especially, it would not seem to be in the National interest that any land on this farm should be used for any purpose other than food production, more especially when there are areas in the immediate vicinity which could be readily made available and which are not at present used for that purpose, and I trust that the Board of Agriculture will use their utmost endeavours to have this land retained in its present use.

But the War Department were set on their course of action and Turnberry Golf Club also received intimation that they planned to requisition the golf courses and club house. Members were requested to vacate and remove their personal belongings forthwith, the lockers were then emptied of golf clubs and pressed into service as 'receptacles for machine guns'. The club house building was now to be used for administration, lecture and instructional classrooms. Turnberry Hotel was also commandeered, to be used as the unit's Headquarters, Officers Mess and accommodation.

At the end of 1916 the Military then turned their attention to Ailsa Craig and the Admiralty took possession of several buildings on the island to establish a Coastal Watching Station with the installation of a wireless telegraphy post and the attendant staff.

The civilian construction workers of McAlpine & Sons (dubbed McAlpine's Army) moved in and began removing the farm fences and erecting the technical buildings to the north of Little Turnberry farm. Additional buildings were erected to the west and south west, extending across the Maidens to Turnberry public highway, with the grass landing strip running east of and parallel to the road slightly north of the main aerodrome site, which was just below the hotel. The hutted flight offices bordered the golf course, and the aerodrome surfaced with coarse sea grass, was flanked by the sea on one side and high, wooded ground to the east. An auxiliary hospital building was erected on the lands of Turnberry Lodge, south of the hotel.

Residents in the Turnberry neighbourhood went out of their way to make the men of the Royal Flying Corps feel at home, despite the disturbance caused by the extensive flying activities as the aerodrome grew. It seems the airmen were preferable to the holidaymakers! The railway brought in men, equipment and materials daily, restrictions were placed upon public travel for pleasure trips, allowing the rail services to concentrate their manpower, engines and rolling stock on wartime operations. A branch line of the Maidens and Dunure Light Railway was built westwards to the aerodrome, crossing the public highway and terminating in sidings just beyond the road. This extension was approved by the G&SWR as requested by the Ministry of Works and completed at a cost of £3,015, paid for by the Government.

Turnberry soon settled down to be a well organised and happy training aerodrome. Staff and pupils were luxuriously accommodated in the Turnberry Hotel, a wit at the time joked, 'The Military have a strong connection with the stars. The Army and Marines sleep under them, the Navy navigate by them … and the Royal Flying

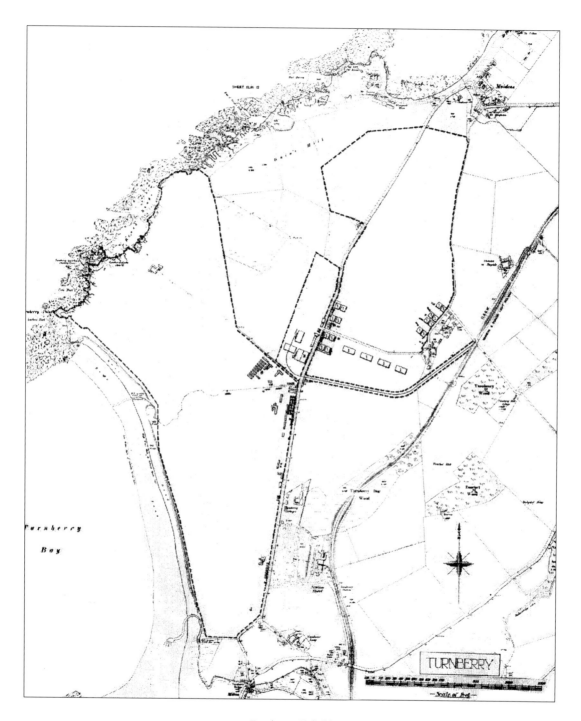

Turnberry Airfield

Corps judge accommodation by them!'

One airman who trained at Turnberry was Frank Best, and the following is his account of his arrival there:

On arriving at Turnberry, one entered the hotel direct from the railway station. We were at once struck by the opulent atmosphere, with its faint aroma of good food, wine and cigars. The place had been taken over by the authorities just as it was - head waiter, chefs, staff, cellars and all. The bedrooms, bathrooms and appointments were what one would expect in a first-class golfing hotel. The bars were attractive, and it was in one of them that we found ourselves on that first night. As we sipped our gin-and-Italians I experienced, once again, that joyous feeling of companionship. After the second we began to get slightly hilarious, more with pleasure and high spirits than with the stimulant. On the way into the dining room we had to control an inclination to laugh uproariously.

Next morning, we lit our pipes as we strolled about on the asphalt terrace in front of the hotel which, we discovered, stood on a low wooded ridge of hills. Below us we saw the Firth of Clyde, with Ailsa Craig on the left, the Isle of Arran to the right and, beyond, the Mull of Kintyre. We learnt that on a clear day the Irish coast was plainly visible. A drive to the hotel wound up the side of the hill from a narrow coast road. Between it and the sea we noticed sand-dunes and two golf courses. A wide flight of steps led direct from the terrace to some tennis courts at the bottom.

A lighthouse near the ruins of Turnberry Castle completed the landscape. We looked at each other and nodded approval of the sporting and recreational prospects. The sun shone brightly. We played a round of golf nearly every day, using our own clubs which had been sent on to us from home.

Immediately to the north, that is to say on our right as we faced the sea, aeroplanes could already be seen taking off and landing. The low wooded ridge on which the hotel stood continued north up the landward side of the aerodrome; it was surrounded by Bessonneaux Hangars. The flying speed of aeroplanes was such, that it was essential to land into the wind. But with a light breeze blowing off the sea the pilot would be forced almost to touch the treetops of the wooded ridge in order to avoid overrunning the narrow aerodrome and crashing into the fences bordering the coast road. Further north, a large potato field provided what golfers call a 'natural hazard' for pilots with engine trouble trying to scrape home. Beyond, and a mile or so away, was a small fishing village called the Maidens.

Another airman to report favourably on being posted to Turnberry was John McGavock Grider, from Mississippi in the United States. 'This is a wonderful place! The flying corps has taken over this wonderful summer hotel and we are billeted in it. My window overlooks the sad sea waves on one side, and on the other, one of the most famous golf courses in the world. Famous I suppose on account of the very high, thick grass that grows all over it. They say it sometimes takes hours to find your ball; it must be a wonder.'

Fellow American Robert Miles Todd was also much impressed with his new training station:

After reporting in we were assigned sleeping quarters in the Turnberry Hotel, a very fashionable vacation resort. The adjoining golf course had been turned into an aerodrome and the hotel sat halfway up the side of a mountain, looking out over the aerodrome and into the bay, a most beautiful spot. The Turnberry Hotel was a very exclusive resort and catered for the aristocratic society of Britain. The room Aldy and I were assigned, was in the front overlooking the aerodrome (golf course) and across the bay to Ailsa Craig, a round mass of rock sticking out of the water several miles away. We had very nice comfortable twin beds, lounge chairs, a cupboard and a dresser. We also had our own bathroom. We were on the second floor which was nice because the elevators were not running. All of our meals were served in the main dining room and we were introduced to several Scottish dishes immediately; namely, kippered herrings with our eggs and treacle for our cereal. We liked both and the service was very good. The English-enlisted men make good servants. They did the maid service around the hotel as well as waiting on tables. The dining room was known as the 'Officers Mess - No.2 Auxiliary School of Aerial Gunnery, RFC' We were billed weekly for our meals and other services that we had.

The aerodrome was the golf course complete with bunkers, sand traps and a dandy water hazard that circled through the whole field. This was a brook; a typical Scots hazard on a golf course. The prevailing wind came off the sea, which was not bad for take-offs; but to land we had to come down the mountain where the hotel was and sideslip for a very short landing strip to avoid the brook and other hazards.

Australian Jack Henry Weingarth, wrote in a letter dated January 1918 from 2 (Auxiliary) Aerial Gunnery School, Turnberry, 'I arrived at the above address, which is in Scotland, on the West Coast, about a week ago. We left St Pancras station in London at about 8:50am and arrived here about 9 o'clock the same night. This place was in peace time, a hotel, one of the most luxurious places I have ever been in. The tariff is 6/- a day, but it will cost us 4/- a day. Of course, it has been commandeered by the RFC.'

In his memoirs, 'Line', Henry Frederick Vulliamy Battle wrote, 'Upon the termination of

or leave, the six of us duly arrived at Turnberry after a dismal rail journey in the pouring rain and were agreeably surprised by the magnificence of our new quarters. As the railway station formed part of the hotel, the whole of which had been requisitioned, as it stood, for an Officers' Mess, we stepped straight out of our train into the mess. It was a delight to live in a house for a change after our tents, and to use real beds and furniture in place of our camp kits.

WWI Airfield Buildings

'The Mess which had been a golfers' paradise, was an enormous building on high ground which swept down to the sea, with the famous golf course spread out over the intervening space. The hotel was entirely staffed by service personnel. I started by sharing a room with one or two other officers, but as time went on, I succeeded in obtaining a room to myself, which was a luxury normally reserved for instructors.'

Captain Sydney Harris writes:

The officers lucky enough to have rooms at the front could see Ailsa Craig, a mass of rock sticking out of the water several miles away. All meals were served in the main dining room with many Scottish dishes on the menu: kippered herring with eggs and they had treacle on their porridge. The dining room was known as the Officers Mess – (2 Auxiliary) School of Aerial Gunnery, RFC. At the hotel they were treated like private visitors and only charged 3s 6d per day, although it was quite difficult to get a bath at night if they wanted one. The hotel had a large palm court full of flowers and was very luxurious and relaxing place to stay. There were no visitors at the hotel, all the guests were officers in the RFC. There were some 200 officers there and the aerodrome was situated on the seafront. They could also go to nearby Girvan, a seaside resort about 6 miles away. There were no trains so they would often walk. Because so many of the pupils were quite young the school was very strict with considerable red tape. Unfortunately, this made it rather boring for the pilots leading to excessive alcohol consumption by some.

Throughout the summer months, RFC Staff who had their families with them applied for housing on the Cassillis and Culzean Estate. But, much as His Lordship was desirous of helping the staff of the Flying Corps as far as was in his power, with regard to cottages, all those belonging to the Marquess of Ailsa in the vicinity were either let or in the occupation of estate workers. Before formally granting permission on any vacancies that may arise, the factor asked the Commandant of Turnberry for references and for his private and confidential opinion as to the character of such prospective tenants and guarantee that they would not abuse any privilege granted to them. The factor ends his letter, 'His Lordship would not trouble you but, as they are quite unknown to anybody in the neighbourhood, will be obliged if you can furnish any information.'

In June 1917, with the aerodrome well established and training underway, a letter was sent from Headquarters, Training Brigade, RFC, to the Air Board Office:

With reference to the establishments for the Schools of Aerial Gunnery, which have been forwarded to you for sanction, I would point out that these have been prepared very carefully and it has been found impossible to produce one establishment to cover all three Schools, or even one establishment to cover No.1 Auxiliary School and No.2 Auxiliary School, as their functions vary very considerably.

The work carried out at the schools is as under:

No.1 Auxiliary School of Aerial Gunnery at Hythe is entirely devoted to the training of Army Observers, whose course of instruction is of one month's duration, 14 days of the course being devoted to groundwork, and 14 days of actual firing from the air. Officers posted to this school for instruction have in most cases little or no knowledge of handling the Lewis Gun and all the instruction at this School is on the Lewis Gun.

The School is divided into two portions, the ground practices being carried out on the local golf links, which adjoin the Imperial Hotel where the pupils are accommodated, and the air practices being carried out from the new Aerodrome near to Dymchurch Redoubt.

Further, these officers have, in hardly any case, ever been into the air at all, consequently the proposed establishment has been prepared to meet the training on these lines.

No.2 Auxiliary School of Aerial Gunnery at Turnberry carries out the actual aerial instruction for all fighting Pilots. This instruction is given on both the Lewis and Vickers Guns, and is given entirely to Flying Officers who have had ample experience in the air before going to the School.

The training is progressive so far as the gunnery is concerned, commencing from the Cadet Battalion and passing through the School of Military Aeronautics at Oxford or Reading, and the elementary and higher training Squadrons, on to No.2 Aux. School of Aerial Gunnery. All the ground firing on both the above guns is carried out before reaching No.2 Aux. School of Aerial Gunnery.

The establishment of this school is calculated on a basis of putting through 150 pilots per fortnight, in air firing with

WWI Airfield showing the canvas hangars

both the Vickers and Lewis Guns.

The situation of the buildings etc. is entirely different to Hythe, the training is on different lines, and the proportion of trades required therefore varies considerably.

School of Aerial Gunnery, Loch Doon, which is in the process of construction, will be in the nature of higher training in Aerial Gunnery for Pilots who will pass on to that school from No.2 Aux. School of Aerial Gunnery, Turnberry. Special types of targets are being constructed at Loch Doon to approximate as near as possible to the targets which will be met by a pilot on actual service. To attain this end, motorboats, seaplanes and railways are required, which necessitate special types of personnel by trades, not required at the two Auxiliary Schools, and for this reason it is impossible to co-ordinate this establishment with either or both of the Auxiliary Schools.

The initial proposal for this establishment was as follows:

TURNBERRY
NO. 1 FIGHTING SCHOOL (NW AREA)

LOCATION - Scotland, Ayrshire, 14 miles southwest of Ayr.

Address - Station Hotel, Turnberry. ½ mile. There is a railway siding to the site (under construction)

Road - A good main road passes the site.

FUNCTION - The School receives as pupils, pilots who have graduated "B" at their Training Station. The instruction given is a finishing course and passes out the pupil as a service pilot entitled to wear Wings and fit to proceed for service work. The course combines Aerial Gunnery and Aerial Fighting instruction.

SUBJECTS TAUGHT -

(i) The advanced use of sights, guns and gears, both on the ground and in the air.

(ii) Advanced formation flying at service heights, and low formation flying in the case of Scout Pilots.

(iii) Fighting in the air, at first dual control with an instructor, then, one pupil against another, and finally pupil v. instructor, and Scouts v. Two-seaters.

CAPACITY - 160 Pupils

LENGTH OF COURSE - 3 weeks with average weather.

MONTHLY OUTPUT - About 200 finished Pilots.

ESTABLISHMENT -

Personnel

Officers	61
Officers under instruction	204
W.O.'s and N.C.O.'s above the rank of Corporal	72
Corporals	68
Rank and File	597
Boys	8
Women	145
Women (Household)	60
TOTAL (exclusive of Hotel Staff)	1,215

Transport

Touring Car	1
Light Tenders	2
Heavy Tenders	3
Motorcycles	3
Sidecars	3
Trailers	5
Workshop Lorries	1
Ambulance	2
Motorboats	3
Motor Boat Dinghies	3
Ford Tenders	5
Charabancs	1
Float Lorry	1
Ford Chassis (Fire Tender)	1
Life Boats	4
TOTAL	38

MACHINES (not yet finally approved) -

Avros	24
Bristol Fighters (R.R.)	8
D.H.4 or 9	16
Camels	20
S.E.5	12
Dolphins	12
TOTAL	92

AERODROME - Maximum dimensions in yards, 1,800 X 1,250. Area 370 acres, of which 35 acres are occupied by the Station Buildings. Height above sea level 30 feet. Soil, sandy loam. Surface, good where it has been prepared for flying, and bad elsewhere. General surroundings: the sea is immediately to the west of the Aerodrome, and the railway close by to the east; low hills (400ft) 1/2 mile to the east.

METEOROLOGICAL - The reports for the winter months, November 1917, to March 1918 inclusive, for 1,564.5 daylight hours observed are as follows:

Low Clouds Hours	Rainfall Hours	Wind Hours	Mist Hours	Fog Hours
213	547.5	734.5	53.5	nil

Possible Flying Hours	Total Hours of Daylight Observed	Ratio of Possible to Daylight Hours	Category Flying
533	1,564.5	34.07%	4

TENURE POLICY - Not at present on the list of permanent stations.

ACCOMMODATION -

Technical Buildings

4	Aeroplane Sheds (each 150' X 80')
16	Canvas Hangars (each 80' X 60')
	A.R.S. Shed
	Salvage Shed (150' X 80')
	M.T. Shed
	M.T. Office
	Workshops:
	Carpenters', 40' x 28'
	Sailmakers', 40' x 28'
	Dope, 40' x 28',
	Smiths', 30' x 10'
2	Technical Stores
	Store
	Petrol Store
	Goods Shed (80' X 150')
	Instructional Offices
4	Instructional Huts
	Power House
	Guard House
	Machine Gun Ranges
	Ground and Sea Ranges

Hangars rebuilt with protective wall after the storm

Regimental Buildings:
Officers' Mess in Station Hotel
Officers' Quarters
2   Regimental Institutes
Dining Rooms
Canteens
2   Regimental Stores
25 Men's Huts
Men's Baths
Men's Latrines and Ablution
Reception Station (In "White House")
Isolation Hut
2   R.E. Offices
R.E. Stores
Y.M.C.A. Hut

With the proposed 92 aircraft and 38 vehicles and boats, to further compliment the equipment of the Aerodrome, a motorboat 'Lynx' was hired from the Marquess of Ailsa, and within a couple of months the Officer in charge reported the loss of the following:

1 Anchor – 15lbs
25 fathom 8" galvanized Anchor chain
2 5'9" boat hooks
2 10'9" oars, and
1 pair galvanised rowlocks

The factor made a claim on behalf of his lordship for the replacement of these articles.

In July 1917 a major fire broke out at Whitestone Sawmill, near Culzean, and in recognition of the services rendered by the members of the Royal Flying Corps at Turnberry in helping to fight the fire, the insurance company sent the sum of £5 as a donation to the Comforts Fund of the Corps. Lord Ailsa also conveyed to the members of the Corps his appreciation of the prompt manner in which the men responded to the call for assistance, and for the services which they rendered in assisting to extinguish the fire, and asked that his thanks be passed on to all ranks concerned.

Capt. Sydney Harris commented on another fire a month later, 'At 3 a.m. on 17 August 1917 there was some excitement for the officers in the hotel, when the fire alarm went off because the cellars were on fire. Fortunately, the floor above was made of concrete and the fire was unable to spread. However, the rest of the hotel rapidly filled with smoke all the way to the top floor, so the men were briefly evacuated while the fire was put out and the smoke cleared.'

The airfield continued to expand, Tom Smith wrote to Lieut.-Col. R. Bell-Irving, Commandant, No. 2 School of Aerial Gunnery:

18 October 1917

Dear Sir,
I have received your letter of 16th inst. as to the extension of petrol store for the School. As I do not know the exact position of the present store perhaps you would be good enough to send me a rough sketch in order that his Lordship might be exactly informed as to the position etc. I will then be able to submit to his first opportunity.

On 5 October 1917 Brig.-Gen. C. Longcroft, Headquarters, Training Division, Royal Flying Corps, wrote altering the number and types of vehicles at Turnberry:

MECHANICAL TRANSPORT. ESTABLISHMENT

It is desired to alter the establishment in Mechanical Transport at the No.2 Auxiliary School of Aerial Gunnery, Turnberry, as follows:

	Present Estab.	Proposed	
Crossley Touring Cars	2	2	
Light Tenders	7	6	-1
Heavy Tenders	6	5	-1
Motor-cycles	10	10	
Sidecars	5	5	
Trailers Mark 1	6	6	
Workshop Lorries	6	6	
Ford Touring Cars	-	3	+1
Char-a-banc	-	1	+1
Ambulance	-	2	+1

It will be noticed this allowed for a reduction of two of the heavier types of vehicles. The Char-a-banc was required for conveying pupils to the various ranges.

The establishment of the aerodrome caused great disruption to the local farmers, who, although realising the desperate need for the school, were also under great pressure themselves to increase food production during wartime. The early potato crop being vital to the local area, that the land be kept working and fertilised was of prime importance. To this end the farmers used the application of 'wrack' (seaweed lifted from the shore) on the soil, but this was prohibited due to the aerial gunnery targets placed along the beach. The factor again wrote to the commandant of the Turnberry Flying School followed by an interview with him, when Lieut.-Col. R. Bell-Irving suggested that wrack lifting might be carried on under the following conditions: namely, 'Dailly before 6.30 a.m. and after 5.30 p.m., the hours on which flying, and target practice begin and end. On occasions when weather conditions prevent flying and there is no firing near the sea, wrack lifting could be conducted from the hotel gate across the golf links, towards Girvan but no further on the coast towards Turnberry Lighthouse. Should the weather on such a day become favourable, warning of the resumption of flying and firing will be given by a long blast on the Hotel Siren, on which your men and horses would have to clear out.

Maidens - The Block, showing wrack lifting

'Wednesday, Thursday and Friday of the current week would be available for lifting on the whole shore from Turnberry Lighthouse towards Girvan, and Saturday, Sunday and Monday of the week following. Thereafter one day per week would be available when the schools are changing.

Information of the days and hours available can always be obtained from the Instruction Officer at the Golf House and you should make enquiry there before arranging to start lifting, practically on all occasions.'

Another issue was the serious curtailment of movement across the requisitioned areas, particularly for the farmers. William Gray in Shanter raised his concerns regarding the access road to Shanter farm, and the possibility of part of that road being used by the public in consequence of the main road being closed for aerial gunnery etc., the farm road being most unsuitable for the extra traffic. The Commander of the Flying School stated that it was not proposed to close the public road more than was absolutely necessary and that, until flying operations were more developed in the Spring and Summer, and approximation of the periods when the road will be closed can be arrived at, and until such is determined, nothing may be done to improve the Shanter road, and therefore it was not likely that the road would be widened in its entire length. The most that would be considered, would be to provide passing places along its length to enable vehicles to pass through. Lieut.-Col. R. Bell-Irving was unable to advise as where the passing places would be constructed but Mr Gray was reassured that he would not suffer very much loss was he to put in his crop as formerly, and in all probability would be able to reap the most of it, if not all, before the work was started on the road.

Overnight between Oct 24/25 1917, a fierce storm blew up which almost swept away the airfield. The hangars were blown down and many aircraft were smashed. It took almost a month to get the aerodrome working again. Lieut.-Col. R. Bell-Irving wrote to Cassillis and Culzean Estate office regarding the reconstruction of the hangars, and their future protection from prevailing winds. The proposal was submitted to the Marquess of Ailsa; his lordship was agreeable that these should be erected, and that the soil for the proposed wall to protect the hangars should be taken from the area on the west side of the public highway. It was agreed that the turf was first to be stripped off and laid aside, and when the soil is removed, the turf to be re-laid in the area. It was also agreed that further Technical buildings be erected on the portion of ground on the west side of the public road and north of the Turnberry Warren – Lighthouse Road.

With the aerodrome rapidly expanding and evolving, land continued to be annexed piecemeal and correspondence between the estate factor and the War Office and Ministry of Munitions at Whitehall became a regular feature of the Factor's working day:

17 November 1917

Dear Sir,
Turnberry Aerial School of Gunnery
I have received your letter with formal notice and plan of the lands which it has been found necessary to take possession of for the purposes of this School, and this seems to be in accordance with what was arranged between the Commandant, myself, on behalf of Lord Ailsa and the tenants, Messrs Marshall and Forbes. Claims for compensation will at present lie with the tenant, but should any emerge for the Proprietor, communication will be made to the Command Land Agent, Headquarters, Scottish Command, Edinburgh.

Over 250,000 civilian Belgian Refugees arrived in Britain between 1914-1918 and of these Glasgow received close to 19,000. Dispersed to towns and villages throughout Scotland, Belgian families were given accommodation and supported throughout the war. The small Dam House at the Kirkoswald end of Blawearie Road was used to house some of these exiles. Louise Van Hee (Madame Soete) died here on 29 January 1917 aged 64 and is buried in Kirkoswald churchyard. A memorial cross was erected by Florent Loosvelt.

With the influx of so many service personnel, complaints soon arose, one such from the engineer of the Glasgow & South Western

Turnberry Airfield and aircraft after the storm of 24/25 October 1917

Belgian Refugees at Girvan

Railway Company regarding the deposit of rubbish adjoining the company's filter line. The factor immediately wrote to the CO at Turnberry saying he had not had an opportunity of inspecting the alleged nuisance, but as this would seem to arise from the occupation of the ground by the Royal Flying Corps, he would be obliged if the matter was looked into, and if there was such a nuisance, take steps to have it removed.

On 1 April 1918 the Royal Flying Corps became the Royal Air Force and became independent of Army or Navy influence. The Turnberry Aerodrome training facility was absorbed into the RAF.

One month later, on 11 May 1918, Commandant Lt.-Col. Bell Irving left the staff at Turnberry and was replaced by Lt.-Col. Lionel W. B. Rees, VC, who had been Commanding Officer of 1 School of Aerial Fighting and Gunnery, which moved down the coast from Ayr Racecourse to Turnberry. It was amalgamated with 2 (Aux) School of Aerial Gunnery, becoming 1 Fighting School, the training establishment being expanded to provide courses for two-seater fighter pilots, gunners, and bomber crews. L.A.C. Arthur Charles 'Pop' Lewcock wrote in his diary, 'Everybody well completely fed up, 200 applications for transfer put in, nothing doing.'

The Assistant Commander of 1 School of Fighting and Gunnery wrote to Girvan Town Council asking for their support of, 'Lectures to civilian audiences for the purpose of stimulating interest in Air Force Work.'

The farmers were still suffering the inconvenience of the aerodrome, although perhaps 'inconvenience' is slightly understating the fact, as shown in the following letter:

Lieut.-Col. Rees VC
Commandant
No. 1 School of Aerial Gunnery
TURNBERRY
29 August 1918

Dear Sir,
The following is a copy of a letter which I have received from the tenant of Turnberry Lodge Farm, viz:
'The above Corps are shooting down

on the byres, stable and other houses, a great many of the slates are shattered. The policeman took up 2 bullets to the Adjutant, 1 off the roof and 1 lying in the courtyard. Windows are shattered and slates are getting loose and broken with the extra vibration. Would you kindly see the Commanding Officer because it is very dangerous to the liege's resident here.'

I visited the farm last week but could see no trace of extensive damage, and unfortunately the tenant was from home and I did not see him. I understand however, from a telephone message from Mr Bone that he had handed to the Policeman a machine gun bullet picked up in the yard and one extracted from the roof of the byre.

Instructions are no doubt issued to pilots and observers to use every precaution whilst practising with machine guns, and I feel sure it is only necessary to draw your attention to cases such as the above in order that instructions may be emphasised to prevent an occurrence of such incidents.

Yours faithfully,
Tom Smith
Factor

A reply from Lieut.-Col. L. Rees VC, MC, explaining the cause of machine gun bullets 'falling' on Turnberry Lodge farm steading was received, but sadly we do not have access to this correspondence, the reasons given must have satisfied the factor because he answers, 'While I felt sure this had happened through inadvertence, having received the complaint, it seemed necessary to draw your attention to it to prevent a recurrence. Mishaps will occur in such dangerous operations as your men engage in, and inhabitants must put up with the annoyance caused by the noise of aeroplanes and Machine Gun Fire. Meantime my purpose has been served and I am obliged for your courteous reply to my letter.'

Louise Van Hee

In conclusion, when writing to Mr Robert Bone, tenant of Turnberry Lodge farm, the factor mentions that he called at the farm yesterday, 'I looked around the steading but could see no trace of bullets striking any of the houses but agree with you that any such practice must be of considerable danger.' He continues, 'I have been served with a notice for the taking of a piece of land adjoining the Hospital erected on Turnberry Lodge farm, for a small annexe; the extent of the ground occupied being 288 sq yds or nearly 13 poles … Under the Defence of the Realm Act the Proprietor has no claim until the occupancy of the land ceases, but the tenant has a claim for any loss he may suffer through the occupation of the land, and I would suggest that you make up a claim, and I enclose a form which you can fill out and forward to the Directorate of Lands.'

Once more the staff at Turnberry Aerodrome were called upon to help with the sawmill. Just before the end of the war, and almost a year after the fire that destroyed it, owing to the shortage of labour and materials, the sawmill had still not been re-erected. The situation, however, was becoming somewhat urgent, there was a considerable stock of felled timber readily available. The sawmill had been almost entirely used producing timber for estate purposes, farm buildings and fencing material and for pit sleepers etc., it was felt the sawmill could be usefully employed during the winter months in the production of this essential requirement.

The lack of fencing material was being felt all over the estate, and it was vital that stobs and paling should be made available for their maintenance. As both the above-named requirements were considered works of national importance, and it was practically impossible to obtain labour to carry out the proposed reconstruction, an application was submitted to Sir John Hunter, Administrator of Works and Buildings for the Air Board, on behalf of the Marquess of Ailsa, that the project might be favourably considered, and the necessary permit issued to Sir Robert McAlpine & Sons, who were then executing contracts at Turnberry School of Aerial Fighting. It was proposed to obtain the material for the re-erection from a large shed formerly used as a joiners' shop, moulding loft etc., in connection with the shipbuilding yard which his lordship had established at Maidens, but which had been disused for many years, the material of which was eminently suitable for re-erecting the sawmill.

Messrs. McAlpine & Sons were quite willing to favourably consider the extra work, providing that the necessary permit could be obtained. They estimated the time required would be two to four weeks. Soon McAlpine & Sons had authorisation to undertake the dismantling of the present buildings in Maidens Shipyard and their re-erection as a sawmill near Glenside and a permit was issued to that effect.

It was immediately apparent that the question of the transportation of the materials from Maidens to the new site, about a quarter of a mile from Glenside, presented unforeseen difficulty, Mr Barclay, the manager for McAlpine & Sons mentioned the matter to Lieut.-Col. Rees, VC, MC, who very kindly agreed to place the matter before the RAF Area Command with the view of obtaining permission to place at the disposal of

Messrs McAlpine & Sons sufficient Government transport to enable them to carry through the contract.

The Marquess of Ailsa was most anxious that the sawmill should be re-constructed and put in working order as soon as possible, but without facilities for transporting the material, it would have to be postponed.

The Factor also took time to write to Lieut.-Col. Rees, VC, MC, on the matter, 'I therefore trust that you will be able to place the matter before the Command in the most favourable light, in order that work may be proceeded with. I understand that Messrs McAlpine will be ready to commence operations very shortly.'

Sporting activities were encouraged among the men of the Royal Flying Corps, a Gun Club was formed and permission to shoot on the estate lands, Factor Tom Smith writes:

I duly received your letter of the 15th instant with relative copy of the Rules of the Turnberry Gun Club, which I have submitted to the Marquess of Ailsa for his approval. His lordship approves of the rules of the club, and has instructed me to prepare and forward to you for approval the conditions upon which he is prepared that you should lease the shootings on Turnberry Lands and will be glad to have the draft returned approved at your early convenience in order that it may be written out for signature … Please fill in the full names of Colonel Bell-Irving and yourself, and also give the full names and designations of your members of Committee … His Lordship has agreed to the suggestion in your letter that the Northern Boundary of the Shootings should be the Ardlochan, Morriston and Balvaird road. He desires me to point out that shooting tenancies usually terminate at Whitsunday (15 May) and commence on 11 August, beginning of the game season, but that as rabbits may be shot during the whole year, he is willing that the club should have this privilege up till the beginning of next game season, and I have accordingly made the terms of the first let to run from 11 November to 1 August, and (needs) thereafter. This will make the let coincide with the usual shooting terms. I am obtaining copies of the Ordnance Survey on which the boundary of the shootings will be shown and will have these in time to be signed along with the agreement. Should your committee wish to discuss any points with me, I shall be very glad to call at Turnberry at an hour convenient to you.'

By the autumn of 1918 Turnberry was one of the largest training stations in Britain with 1,215 officers and other ranks, including 205 women, and 204 pilots under instruction.

The establishment of the School of Aerial Gunnery at Turnberry required a substantial staff of varying trades and ranks, apart from the instructors and students, the nominal roll of clerical, maintenance and other sundry personnel was immense.

Included in this list were a Flight Sergeant (Disciplinary) who was required to command a unit of ten Regimental Policemen in their task of safeguarding the valuable Government property and buildings, which were widely scattered over an area of three miles and continually exposed, liable to heavy storms and difficult to protect. The Regimental Police were responsible for the safety of the Station and Aerodrome with an establishment of about 1,300 comprising Officers, other ranks and WRAF.

The Master Clerk of the Station was accountable for the fourteen administrative offices of various sizes scattered throughout the aerodrome. Due to the floating population of officers under instruction passing through a highly technical and condensed course, at any one time numbering over 180, combined with the permanent staff of over 1,300 other ranks and WRAF, considerable clerical work was required. The records of pay and allowances of officers particularly required the most careful organisation and control, as the officers could

be posted out at short notice. His main duty was to ensure proper and prompt record keeping and manage co-operation between the sub-divisions of the clerical staff.

In the different administrative sections, General Clerks were responsible for arranging and recording all movements of officers such as clearance certificates, travel warrants and notification to other departments, managing a card index system recording school records, next of kin details, officer's nominal rolls and records. The Pay Office dealt with all correspondence, the supervision of Pay and Mess Books and the Ration Index, also the Officer's Allowance certificates. These were for advanced field allowances and orders for overseas, which had to be signed three times by an average of 300 officers turned out per month.

A separate Flight Clerk was required to keep the Fighting and Gunnery Records, which were very complicated and called for considerable clerical work. With twelve pupils expected to be passed out of gunnery each day, all tests, marks, etc., had to be recorded. Reports on each pupil's ability as he arrived at and left the station had to be made. Separate clerical officers were employed on each Flight and at Headquarters for the reports connected with the Aircraft and Engine Log Books.

On the technical site numerous different trades were engaged, ranging from mechanics and engine fitters, to general labourers. The Chief Mechanic supervised the work of the two Flights at the school, assisting the Group Commander. Two Chief Mechanics (Armourers) and four Sergeant Mechanics (Armourers) took general charge and responsibility for the guns and ammunition, supervising the aerial practices and all necessary arrangements for aerial firing. Gear Fitters, both men and boys, were tasked with the maintenance and repair of gears, which required constant attention and adjustment to keep the machines in the air. Likewise, the Instrument Repairers whose duty it was to make 'on the spot' corrections to the Camera Guns on the Fighting Machines. The course of Aerial Gunnery was so condensed that it was vital that the equipment was always serviceable if the output of pupils was to be maintained. The Armourers were responsible for keeping the guns maintained and scrupulously clean in view of the continual firing they were subjected to. A tinsmith was on charge to renew the tin 'checks' which had to be constantly replaced due to crashes, etc. All 'empties' (used shell cases) from aerial firing had to be accounted for to Ordnance and were collected in a 'check' or box below the guns. A Draughtsman (Mechanic) was employed in connection with the experimental work in the Gunnery Instructional Section.

Twenty Engine Fitters were employed at Turnberry, ten on Rotary Engines, seven on Stationary Engines and three on test benches. There were over 143 engines at the school needing constant maintenance and repair.

On the ranges 24 men worked the Roller targets during firing, with two men to keep the targets in repair. On the 25 yd and 50 yd Deflection Range, one Corporal, eight Markers and one Flag Man were employed on each, the same on the 25 yd Stoppage Range. The 100 yd Deflection Range required one Corporal, ten Markers and one Flag Man, and on the Harmonisation Range two Markers. Five Orderlies were required for duties on all ranges and at Gunnery Headquarters, four Target Repairers and seven men on the Ammunition Dump. Five men were employed filling Ammunition Belts on the Aerodrome and four men manned Observation Posts, watching for machines falling into the sea. To this end the Motor Boat Section had three Drivers, three Deckhands and three Coxswains, with a Sergeant to supervise the care, maintenance and proper handling of the craft.

Other technical trades at Turnberry in the Aeroplane Repair Section were Blacksmiths, Acetylene Welders (2 WRAF), Carpenters to repair propellers, Vulcanizers and an Electrician who had the task of maintaining the very complicated range of telephones at the school. There were Painters and Fabric Workers, mostly WRAF, it was noted that an extra NCO was needed to control the Fabric Workers in the Fuselage Repair Shop and Assembling Shop, to produce the greatest possible amount of work.

A squad of Plumbers, Bricklayers and Plasterers kept the station buildings in good order.

A Corporal was allocated to oversee the work of the 47 Batmen (one to every eight officers) in the hotel, to allocate them to their duties and to be responsible for their cleanliness and discipline. When the Batmen were not looking after their officers, they would be placed on fuel duties at the boiler house or as night watchmen. Ten female General Domestic Workers cared for the 100 bedrooms in the hotel, corridors, bathrooms and offices.

Hordes of other support personnel such as drivers, medical staff, butchers (who were kept busy dividing over 1,300 rations daily), bakers, cooks, mess orderlies and labourers completed the contingent of Turnberry Airfield.

After the signing of the Armistice on 11 November 1918, Turnberry School of Aerial Fighting reduced its training throughout December, finally closing in January 1919.

3
WWI Airmen

The average age of the students who came to training establishments such as Turnberry was about eighteen years old, just out of school. A few others were quite elderly, being almost twenty-one.

Many valiant young men were drawn to the bohemian, glamorous image of the fighter pilot which had been feted by much public attention and adulation, to be part of the 'great adventure, and to get paid to do it as well!' Once admitted into the RFC, pilot cadets were usually sent first to a ground school, for as little time as four weeks. Here they learnt basic drill and mess room etiquette, they also undertook courses on aerial observation and wireless telegraphy, basic mechanics and 'sail-making', which taught how to properly cover aircraft. Astonishingly, there were no lessons in the theory or mechanics of flight. After this basic training the pupil was sent to an EFTS (elementary flying training squadron), where he would finally take to the air, receiving only a few hours of basic instruction before taking his first solo flight. Training at Turnberry was undertaken in two parts, first being an intensive ground gunnery course then combat exercises in the air. The School was now considered to be a single-engine operational conversion unit, the last stage in the training of aircrews before they were sent into action with a front-line squadron.

After leaving Turnberry, those who made it through training went on to do great things, during and after the war. Some sadly did not survive more than six weeks over the front, in

Chief of French Air Force (Centre) with Lieutenant Colonel Richard Bell-Irving and Captains R.T. Leather and S.H. Harris

one case lasting only 45 minutes on his first patrol. Many of the famous airmen of WWI passed through Turnberry, either whilst training or as instructors - a few were to reappear during WWII. Here are their stories, where possible in their own words, through letters, diaries and memoirs. Names have been lifted from old photographs taken at Turnberry and service records searched, in an attempt to 'write a portrait' of the spirit of the men. Ranks given are those that were held whilst at Turnberry.

Lieut.-Col. Louis Arbon Strange DSO, MC, DFC and Bar, OBE

The man tasked with the establishment of 2 (Aux) School of Aerial Gunnery at Turnberry, Lieut.-Col. Louis Arbon Strange, was born in Dorset in 1891 into a farming family. In 1909 he enlisted in the Dorset Yeomanry. His family had ridden with the yeomanry since the regiment's formation during Napoleonic Wars. A year later his mounted troop was detailed to help with the crowd control at the first ever flying display at Bournemouth. He witnessed Britain's first aviation fatality when the Wright Flyer flown by Charles Stewart Rolls (co-founder of Rolls Royce) broke up in mid-air, killing the pilot. Despite this, Louis Strange was totally entranced by aviation and within a year had completed his flying training at Hendon. On 8 October 1913 he was commissioned as a 2nd Lieutenant (on probation) in the Royal Flying Corps (Special Reserve). Much taken with all aspects of flying he took to the skies whenever he could, participating in numerous air races of the day, and instructing trainee pilots at Hendon.

At the outbreak of war in July 1914, Strange obtained a commission in the British Army as a 2nd Lieutenant in the Dorsetshire Regiment, but continued on attachment to the Royal Flying Corps, with the rank of Flying Officer. In August 1914 he joined 5 Squadron, based at Gosport in Hampshire. Between 15/16 August the squadron was sent to France - he flew his Farman aircraft via Dover and across the English Channel to Amiens and then on to the unit's new base at Maubeuge, near the Belgian Border, despite

severe weather and a defective airframe.

Always an inventive 'ideas man', Louis Strange was forever trying to improve the performance and armament of his aircraft, such as inventing home-made petrol bombs, a machine gun mounting (the observer in two-seat aircraft carried a single mounted Lewis Gun in the front cockpit mounted on metal tubes designed by Louis Strange. These became known as 'Strange Mounts') and a bomb chute, consisting of a steel tube set in the floor of the aircraft to drop 7-pound shrapnel bombs through.

On 16 February 1915 he was appointed Flight Commander with the acting rank of Captain and posted to 6 Squadron. He instantly became friends with air ace Lanoe Hawker VC. With his ideas and Hawker's engineering skills, they came up with the concept of a mounted Lewis machine gun on the top wing of a Martinsyde scout aircraft to enable it to fire forward, clearing the arc of the propeller. This 'idea' nearly cost Strange his life.

Later, in combat, while under enemy fire, Louis Strange stood up on his seat to change the jammed ammunition drum on the Lewis gun, which he had mounted on the top wing of his Martinsyde. While trying to pry it loose, the aircraft stalled, rolled onto its back, throwing him from the cockpit and went into a flat spin, upside down from 10,000 ft. Left hanging for dear life to the very drum he was trying to loosen, he managed with a lot of swinging and kicking to get back into the cockpit, smashing his instruments and controls in the process. He succeeded in righting the aircraft within 500 ft of the ground and returned to the aerodrome, landing safely. After all the excitement he fell into an exhausted sleep for over 24 hours and told no one of his adventures. Hoping to forget the whole incident and keep it a secret, Strange was dismayed to find that the German pilot he had been firing at had witnessed the whole incident and had assumed that their courageous assailant had been killed. As was the custom between air combatants, they dropped a wreath and a note describing how Strange had met his death. An embarrassed Captain Strange found he had become the stuff of aviation history. His next invention was a pilot's safety harness!

Ill health in the form of appendicitis returned Louis to Britain, where after the operation it was discovered that a swab had been left inside him. This extended his convalescence and, not quite fit for active service, he was promoted to Lieut.-Col. and tasked with setting up the first air gunnery schools such as Turnberry, he then served at the Central Flying School before taking command of the 80th Wing.

On 1 August 1919 Strange was granted a permanent commission in the Royal Air Force and on 1 November 1919 he was promoted from Squadron Leader to Wing Commander. Forced to retire from the service through further ill health in 1921, he didn't forget his love of aviation, and in the late 1920s became a director and chief pilot of Simmonds Aircraft Limited and continued when it became the Spartan Aircraft Company. He was also a director of the Whitney Straight Corporation. He flew the company aircraft in several competitions, including the Simmonds Spartan G-AAGN in the 1929 King's Cup Air Race, and the Simmonds Spartan G-AAMG in a 45-mile race from Woodley to Hanworth and back. At the Reading Air Fete in June 1930, he came in 2nd with an average speed of 99 miles per hour.

During World War II, 50-year-old Louis Arbon Strange joined the Royal Air Force Volunteer Reserve as a Pilot Officer, eventually retiring from service in June 1945. He was awarded the Order of the British Empire for his wartime contribution.

Louis Strange's military awards are as follows:

Military Cross – Second Lieutenant (temporary Captain) L. A. Strange, Dorsetshire Regiment and Royal Flying Corps. 'For gallantry and ability on reconnaissance and other duties on numerous occasions, especially on the occasion when he dropped three bombs from a height of only 200 feet on the railway junction at Courtrai; whilst being assailed by heavy rifle fire'. [*London Gazette*, 27 March 1915]

Distinguished Flying Cross - Lieut.-Col. Louis Arbon Strange, MC (Dorset R). 'To this officer must be given the main credit of the complete success attained in two recent bombing raids on important enemy aerodromes. In organising these raids his careful attention to detail and well-thought-out plans were most creditable. During the operations themselves his gallantry in attack and fine leadership inspired all those taking part'. [*London Gazette*, 2 November 1918]

Distinguished Service Order - Lieut.-Col. Louis Arbon Strange, MC, DFC. 'For his exceptional services in organising his wing and his brilliant leadership on low bombing raids this officer was awarded the Distinguished Flying Cross not long ago. Since then, by his fine example and inspiring personal influence, he has raised his wing to still higher efficiency and morale, the enthusiasm displayed by the various squadrons for low-flying raids being most marked. On 30th October he accompanied one of these raids against an aerodrome; watching the work of his machines, he waited until they had finished and then dropped his bombs from one hundred feet altitude on hangars that were undamaged; he then attacked troops and transport in the vicinity of the aerodrome. While thus engaged he saw eight Fokkers flying above him; at once he climbed and attacked them single-handed; having driven one down out of control he was fiercely engaged by the other seven, but he maintained the combat until rescued by a patrol of our scouts'. [*London Gazette*, 7 February 1919]

Bar to the Distinguished Flying Cross - Pilot Officer Louis Arbon Strange, DSO, MC, DFC (78522), RAF Volunteer Reserve. 'Pilot Officer Strange was detailed to proceed from Hendon to Merville to act as ground control officer during the arrival and departure of various aircraft carrying food supplies. He displayed great skill and determination whilst under heavy bombing attacks and machine-gun fire at Merville, where he was responsible for the repair and successful despatch of two aircraft to England. In the last remaining aircraft, which was repaired under his supervision, he returned to Hendon, in spite of being repeatedly attacked by Messerschmitt's until well out to sea. He had no guns in action and had never flown this type of aircraft previously, but his brilliant piloting enabled him to return with this much needed aircraft'. [*London Gazette*, 21 June 1940]

Wing Commander Louis Arbon Strange died in 1966.

Gilbert Strange, a seven-kill ace, younger brother of Louis, was killed during the war and is remembered on the war memorial at Tarrant Keynston.

Lieutenant Colonel Richard Bell-Irving OBE
The first Commander of 2 (Auxiliary) School of Aerial Gunnery was Lieut.-Col. Richard Bell-Irving. He was the son of a Scottish immigrant to Canada, Henry Ogle Bell-Irving. Richard was one of six sons and four daughters and was educated at the Loretto School, near Edinburgh. Henry Ogle Bell-Irving established the Anglo-British Columbia Canning Company, and became a prominent businessman in Vancouver, and ever proud of his Scottish heritage helped to raise the

Seaforth Highlanders of Canada. Richard was married to Kathleen Hume Morris on June 27, 1913, in Vancouver.

Richard Bell-Irving enlisted in the 29th Battalion before transferring from the Canadian Expeditionary Force to the RFC in January 1916 and becoming a Wing Instructor in gunnery. He was posted to 2 (Auxiliary) School of Aerial Gunnery as Chief Instructor and was then promoted to Commandant in March 1917. Whilst at Turnberry he wrote a book titled *Notes on Guns, Gears and Sights* which was published in 1918 at the *Observer* Office, Union Arcade, Ayr.

He was awarded the Order of the British Empire as noted in the *London Gazette*, dated 7 January 1918.

All of Richard's brothers, Henry, Roderick, Malcolm, Duncan and Aeneas, served during the First World War, and of these, four brothers were awarded ten decorations for bravery. Duncan and Malcolm served with the Royal Flying Corps as fighter pilots, in addition, a sister Isabel was nurse at Lady Ridley's hospital for wounded officers in London. Lieutenant Colonel Richard Bell-Irving left Turnberry in May 1918.

Lieutenant Colonel Richard Bell-Irving and office staff

Lieutenant Colonel Richard Bell-Irving

Richard Bell-Irving and Harry Butler

Lieutenant Colonel Richard Bell-Irving with Major Frank Steel and concert party at Turnberry Hotel

Captain Henry John 'Harry' Butler

The first aircraft of 2 (Aux) School of Aerial Gunnery, an FE2b 4981, was flown into Turnberry Airfield on 22 January 1917 by Captain Harry Butler.

Harry Butler was born in 1889 into an isolated farming community in Koolywurtie, South Australia and from an early age he developed an obsession for all things mechanical, especially anything to do with flying. He had no interest in schoolwork, having elementary tutoring at the Koolywurtie schoolhouse and no secondary education. Harry could usually be found building model machines, reading anything he could find on the subject of aviation, or tormenting his mother's chickens by measuring their wingspans and testing their aeronautic properties.

Harry Butler

As with most boys, when he grew older, his mechanical interest turned to motorbikes and he built his own using an engine imported from the United Kingdom.

About 200 km from Koolywurtie in the town of Smithfield, was a Mr Carl Wittber, who was

attempting to build an aeroplane from scratch, manufacturing every part himself, including the engine. Hearing of this, Harry would rush to get his work on the farm done and travel to Smithfield on his motorbike every weekend, where he worked alongside Wittber on the machine, and managed to make short flights in the homebuilt aircraft.

Captain Harry Butler with FE2b 4981

In February 1915 Harry joined the Australian Flying Corps as an engineer, achieving very high marks in his preliminary exams, but he was impatient to get into the air, and thinking his chances of flying training were better abroad, he resigned, withdrawing all his savings and borrowing what he could for the fare to England to join the Royal Flying Corps.

Enlisting as an Air Mechanic in the early part of 1916, his considerable skills and ability was immediately recognised, and after just a few weeks he was gazetted 2nd Lieutenant. By 29 September 1916 he had obtained his Royal Aero Club Aviators' Certificate, although he was graded not fit for operational service above 5,000 feet due to severe asthma. After serving for a while at Hythe with Colonel Louis Strange, he moved to Turnberry with him and was posted as Captain, Flight Commander and Fighting Instructor to 2 (Aux) School of Aerial Gunnery.

Prior to leaving Hythe, he wrote to his aunt on 23 December 1916, 'Two of us are going to Scotland and take charge of a school up there in a couple of weeks' time, so I will be able to tell you a bit about the Scotch lassies in their kilts.'

Harry Butler with Carden Cyclecar 'Toothpick'

The other officer he mentions was probably Louis Strange or his 'pal' Richard Bell-Irving.

Butler's trip in the FE2b from Hythe to Turnberry, via Glasgow, a distance of just under 500 miles, was gruelling, and he arrived totally exhausted after flying through some stormy January weather. He was also credited with bringing the first airmail from Glasgow to Turnberry. His flights in Scotland were not without incident, as he recalled after the war. 'In January 1917, flying between Ayrshire and Dumfries, in one of the pusher type machines,

Harry Butler and BE2e 3677

a mechanic inadvertently left a spanner on the engine behind the pilot. The implement worked its way to the propellor and broke it. Fortunately, he was flying at about 11,000 feet and was able to glide to the ground.'

On 20 June 1917 Harry Butler was promoted and made temporary Captain and Flight Commander at Turnberry. He wrote home enclosing a photograph of his latest aircraft. 'This is a photograph of one of England's leading battle planes which I have been given. It was presented to the RFC by the Colony of Mauritius, my Gunner/Observer sitting in the seat (with the gun) is Captain Anderson, and he comes from Mauritius, and with this bus on the morning of the 26th of this month he and I played a leading part in capturing two German submarines, on the strength of it I've been Mentioned in Despatches and I think he has also been complimented as an Aerial Gunner, that is the gun sticking out over the nose of the machine. I notice you address my letters as Flight Captain, that is wrong, and the letters might go astray if not properly addressed.'

The leading battle plane Butler refers to was FE2d 6527, 'Colony of Mauritius', donated by Mauritius to the war effort. Gunner Albert John Gordon Anderson, born in Mauritius, a solicitor's clerk pre-war, was a gunnery instructor at Turnberry from February to December 1917, He had also been stationed at Hythe with Butler.

While an instructor Butler would fly to France and attach himself to an operational squadron for a few weeks at a time to keep up to date with new aircraft and fighting methods. Taking part in patrols he was able to study the German aerial tactics at first hand, returning to Britain to pass on his experience. In his time as Fighting Instructor he trained over 2,700 pilots.

In September 1917 he wrote to his father, 'I am still going strong and I can and do fly anything with wings on. I have looped the loop nearly 300 times. Well, I will probably be leaving Scotland shortly, I have applied for a command of a flight of DH4 aeroplanes in a fighting squadron in Egypt, so I expect I will be off there very soon. These DH4 machines are the fastest buses in the world. They do over 130mph. I have

one of them here for doing anti-submarine patrol work and it is a beauty.'

In 1946 an Adelaide picture framer discovered a photograph, on the back of which was written, 'It was taken by the school photographer as I was on a Bristol Fighter about sunset to do a submarine patrol. You notice the Christmas card looking island in the distance. That is known as Ailsa Craig, and it is about halfway between here and Belfast.'

After completing almost 600 flying hours, in November, Butler wrote to the Commanding Officer at Turnberry requesting a transfer to Egypt, citing his chronic asthma as the reason. The Lieutenant Colonel at Turnberry wrote to the Headquarters of the training division in London in support of Butler's application for transfer, stating that, 'this Officer suffers from asthma and is considered only able to fly at low altitudes in a warm climate.' At a previous medical examination it had been noted that, 'He is quite unfit for general service owing to the conditions of his lungs. He suffers from chronic Bronchial Catarrh and Emphysema which leads to asthmatic attacks and shortness of breath. Both lungs are affected. While fit for service at home, he is unfit for flying at high altitude, and for the exposure and strain of conditions of general service.' It was also noted that he smoked five cigarettes and six pipes per day!

In spite of the Medical Board recommending that Harry was 'fit for general service in dry warmer climates or fit for half service with no flying above 5,000 ft', he was never posted to Egypt. He remained at Turnberry and rose through the ranks to become Chief Flight Instructor.

While at Turnberry he fed his passion of all things mechanical, having with him a Carden Cyclecar which he called 'toothpick', driven by a motorcycle engine, and a motorcycle and sidecar on which he enjoyed trips around the countryside with a 'Scotch' lady friend. He made the most of his stay in Ayrshire, enjoying sightseeing in the area and taking lots of photographs.

Posted to France, Harry was wounded on active service whilst over Douai during February

1918. Upon recovery he was posted to 2 School of Aerial Fighting at Marske in April 1918, and was promoted to Fighting Instructor and Group Commander, and in December of that year was awarded the Air Force Cross. Captain Butler served with the Royal Flying Corps and then the Royal Air Force, until Armistice Day in 1918 and was transferred to the unemployed list in June 1919.

Harry Butler had no plans to give up flying though - he bought two aircraft that the RAF had retired and put up for disposal before his return to Australia. These were Bristol M1c C5001, which he had flown at Marske, and an Avro 504K, together with three spare 110 hp Le Rhone engines which could be fitted to either airframe.

Butler was keen to share his passion of flying and saw the future potential of civil aviation. Arriving back in July 1919, followed soon after by the arrival of his aircraft and his team of engineers, Lieut. Kauper (inventor of a gear that enabled airmen to shoot between the revolving propeller blades of an aeroplane), Sgt Major Crawford and Jack Lewis. By the end of the month Harry had his M1c painted red and named the 'Red Devil', and began giving aerial displays, demonstrations of aerobatics and stunt flying to his fellow Australians. He operated as the Captain Harry J. Butler & Kauper Aviation Co. Ltd., in partnership with Harry Kauper. Together they established the first airport and the first passenger flight business in South Australia.

On 6 August 1919 Harry Butler became the first man to fly across the gulf to Yorke Peninsula, located north-west and west of Adelaide in South Australia, in turbulent conditions and delivered the first air mail to be carried over water. The wind gusting up to 110 km/h did not put him off - he was determined to return to his hometown by aeroplane. His only grudging concession to the inclement conditions was the wearing of an inflated inner tube around his neck, in case he had to 'ditch', and shipping in the Gulf had been placed on alert and were looking out for him. Approaching his home at Minlaton, Harry gave a spectacular display of stunting and aerobatics to the waiting crowd of over 6,000 people. On

landing he was met by an official delegation of local dignitaries who welcomed home their famous son, 'Butler of South Australia'. After being honoured and feted in various ceremonies for four days on Southern Yorke Peninsula, Harry made ready for the return trip to Adelaide on 11 August, loading two bags of mail, and one very special message he had prepared himself. Flying low over Koolywurtie schoolhouse, he weighted his personal airmail, and dropped it from his plane. The message read, - 'My old school and its scholars. I sincerely hope that this message from the air will bring to you all the very best of luck'. This was later framed and hung on the wall of the remote schoolhouse for many years.

The Red Devil with Harry at the controls was to become a famous sight across South Australia, with aerial displays of aerobatics, and he was known to fly past the Adelaide Children's Hospital, to give an impromptu display for the patients. He also took part in Australia's first Aerial Derby, which was held in Adelaide in 1920, and won the 30 km race and as a grand finale thrilled the crowds with his breath-taking aerobatic stunts.

In 1919 the Australian government offered a prize of £10,000 (Australian) for the first Australians in a British aircraft to fly from Great Britain to Australia. Successfully arriving in Darwin from England in March 1920, the Smith brothers completed the challenge, with their crew in their great Vickers-Vimy, in which they then flew on to Adelaide. Harry Butler and the Red Devil was waiting to escort them in.

Harry wound up the company of Captain Harry J. Butler & Kauper Aviation Co. Ltd., in September 1921, finding it difficult for it to work as a commercial venture, although Harry still believed in the future of aviation and was as enthusiastic as ever, buying the two aircraft from the business for his own use.

The Avro had been having persistent engine problems for a while, but even with all his mechanical knowledge Harry couldn't identify the problem. The biplane had been worked hard in the preceding months on a tour of the Yorke Peninsula, giving flying displays. Harry performed a complete engine overhaul but was unable to find the cause of the misfire.

In early January 1922, he took a flight in the Avro, near Minlaton, with mechanic Rex Miles, again trying to diagnose the unusual sound of the faulty engine. He performed all the necessary checks but still couldn't find the cause of the misfire. After taking off, having reached a height of about 400 m, the engine seized, and Harry could do nothing to avoid the subsequent crash. Harry Butler was grievously injured, suffering serious head injuries and his face was almost destroyed. Rex Miles escaped unhurt.

Harry's flying days were over - he was left badly disfigured, undergoing a series of surgical procedures to rebuild his forehead, nose and jaw, and he also suffered from severe headaches, dizziness and problems with concentration.

Harry Butler died suddenly of an unexpected cerebral abscess on 30 July 1924, an unseen injury from the earlier Avro accident. He was just 34 years old. He was buried in Adelaide with full military honours, and his funeral was attended by thousands of people who came to pay their respects to the Australian aviation hero.

A memorial to 'Butler of South Australia' was raised by the local community, a grand columned headstone with a bronze model plane (later stolen) on the top.

After his death, his plane, the 'Red Devil' was hung from the rafters of a shed in Adelaide until it was purchased by Mr C. Miller who, after restoring the aircraft, competed in races and exhibitions the following years. The 'Red Devil' continued to fly until 1945 when it was officially retired.

Today, Harry Butler's beloved M1c, 'Red Devil' can be seen at the Butler Memorial building in Minlaton, South Australia.

Major Frank Steel OBE

Frank Steel remembers: 'In 1914 Kaiser Bill broke out. I joined up at once getting an immediate commission with the Essex Regiment as I had always been keen on the OTC at school (Charterhouse). I was posted to Parkeston near Harwich and my company headquarters were located at Ray farm on the side of the local golf course where I was able to play regularly. Having reported for duty in mufti I footslogged over much of Essex in this garb until a Dovercourt tailor could make me some uniform.

A very pleasant contact at this period was with the RN Submarines based at Parkston Quay. I had the luck to be taken out for target practices in E5 and H11 when three or four submarines took it in turns to submerge and attack a destroyer. This was a real thrill.

Major F. Steel and NCOs

I went to France in January 1915 and had the luck to be attached to the 2nd Gloucestershire's who sent me to a regular Battalion recently returned from the hotter parts of the earth. I therefore had the luck to command a battalion of the old regular army. During my six weeks of survival we went through the first gas attack of the second battle of Ypres and on 9 May I lost 50 out of 60 of my men in half an hour.

After being wounded in June 1915 (an improvised mills bomb went off in my face and caused me lose an eye) I went home, but being somewhat sensitive to walking about with a heavily bandaged eye, I introduced myself to the Essex bowling club where I was welcomed by the members who were mostly about 80.They could not have been more kindly and I put my knowledge of Bowls to good effect when I later found myself at Turnberry.

Going back to my Regiment I got tired of sitting at Felixstowe as machine gun officer waiting to knock down Zeppelins with ordinary .303 ammunition. Having a brother in the regular army I learned that that the Royal Flying Corps were anxious to take on "earthworms" who would not be tempted to interfere with their wrestling with the mysteries of the pay and mess books by going up for a "flip", killing themselves and leaving the administration of the unit in a mess.

In October 1916 I was seconded to the RFC and found myself posted to Edinburgh where a new elementary training squadron was being formed on a farm called Gogar Mains at Turnhouse. I feel that I was the first administrative officer at what is now the main airport at Edinburgh. The squadron was flying Maurice Farman aircraft. The main points of interest during my stay were that the aerodrome was adjacent to a golf course and that we were near to Rosyth and found ourselves being called on to provide services for the Royal Navy crews recalled from the delights of the Capital whenever there was an alarm.

We were administered from Newcastle and I found myself on very good terms with the Adjutant although I never saw him. At the end of 1916 I found that I had been chosen as Adjutant to one of the plum jobs of the service – 2 (Auxiliary) School of Aerial Gunnery at Turnberry. I served there for the rest of the war until I was demobilized in Sept 1919'.

Second Lieutenant David Ogilvie Duthie
David Ogilvie Duthie was born in 1894 in Coupar Angus, Perthshire. A clerk in civilian life, he enlisted in the army on 1 May 1917, gaining an appointment to a temporary commission in the RAF in August 1917.

Posted to 35 Squadron in September 1917 on Observer Duties 2nd-Lieutenant David Ogilvie Duthie was wounded on 23 November when flying with Lieutenant L. G. Paling in A.W. FK8 B5767 on reconnaissance, the aircraft was badly damaged by machine-gun fire and landed at Longavesnes aerodrome.

On recovery from his wounds he was posted to 2 Squadron and was reported missing on 23 August 1918 when A.W. FK8 F4264 flown by Lieutenant E. O. Drinkwater was seen to dive straight to earth from 500 feet between Givenchy and La Bassée while engaged on a photography mission. It was reported that they had been hit by anti-aircraft fire.

The *Blairgowrie Advertiser* of 31 August 1918 reported:

> Second Lieut. David O. Duthie, RAF, son of Mr A. Duthie, Causewayend, and who was wounded some time ago, is now reported missing since 23rd August. This officer was formerly engaged in the railway offices at Coupar Angus, Alyth, and Crieff. It was not until 29 May 1919 that word was received from the Red Cross that 2nd Lieutenant David Ogilvie Duthie had been killed on the day he went missing and was buried 'At the fence of the canal, near Auchey, By Copenhagen'. [sic].

2nd Lieutenant David Ogilvie Duthie is commemorated on the Arras Flying Services Memorial.

Captain Sydney Herbert Bywater Harris

Born in 1881, Sydney Herbert Bywater Harris left home in Ilford at the age of 17, bound for the Klondike gold rush. Realising that he had arrived too late to make his fortune, he joined the US Army instead, where he saw action in the Boxer Rebellion in China and the Philippines Insurrection. Returning to England, Sydney joined the King's Colonial Imperial Yeomanry, later renamed the King Edward's Horse in 1905, transferring to the Royal Flying Corps in August 1914. Awarded his flying certificate on 18 August 1915 at Military School, Gosport, flying a Maurice Farman Biplane, he was then posted to France with 23 Squadron flying the FE2b, and on 30 April 1916 was badly wounded by anti-aircraft fire while gun-

Sydney Herbert Bywater Harris

spotting over enemy lines. Invalided home, after recovering, Sydney was posted to the School of Aerial Gunnery in October, as an Instructor (on probation), arriving at 2 (Aux) School of Aerial Gunnery, Turnberry on 17 February 1917. Appointed Chief Instructor, Captain Sydney Harris held that post until December 1917 when he was posted to command 4 (Aux) School of Aerial Gunnery, Marske with the rank of Lieutenant Colonel. He was awarded the Air Force Cross in June 1919. He was transferred to the Unemployed List in August 1919.

Ever the adventurer, Sydney Harris became involved with the International Brigades in the Spanish Civil War in 1936, which was also a useful way of fleeing his imminent bankruptcy. On his return to Britain, he joined the RAFVR, and at the outbreak of WWII was posted to Turnhouse, near Edinburgh, as Section Controller. Always the man of action, Sydney, at the age of 58 managed to arrange a posting to France as Adjutant with 1 Squadron where his duties included liaising with the French Air Force, later transferring to 1 ATS near Perpignan, and managed to retreat from France at the last

minute with the German Blitzkrieg only a few hours away.

Although suffering ill-health, Sydney managed to serve throughout the 2nd World War and was made Commander of Marchwood Park at East Grinstead in 1947, looking after the recovering members of the 'Guinea Pig' club, which was formed in July 1941 to support aircrew who were undergoing reconstructive plastic surgery after receiving burn injuries in the Second World War. He was in Command for a year before retiring from the RAFVR, at the age of 65.

After spending much of his retirement living in Spain, Sydney Herbert Bywater Harris returned to England and died in Bath, Somerset in 1960.

Captain Ralph Towlerton Leather AFC

Ralph Towlerton Leather was born in Dover in 1892, the son of Colonel Gerard F. T. Leather JP of Middleton Hall, Belford, Northumberland. Educated at Eton and Durham University, Ralph was Commissioned Lieutenant in the Warwickshire Yeomanry (TF). Seconded to the RFC, he obtained his Royal Aero Club Aviators Certificate in March 1915, flying a Maurice Farman Biplane. He was posted to France with 2 Squadron in June 1915 and then to 21RS Abbassia, Egypt, in October 1916. He was awarded the AFC in the 1919 Birthday Honours list. He received a permanent commission as Flight Lieutenant in 1921 and served in India 1921-1925. He was promoted as per the *London Gazette*, 1 January 1923, Flight Lieutenant to be Squadron Leader, and again in 1937 Wing Commander to be Group Captain.

During World War II, Group Captain R. T. Leather AFC officially opened and commanded RAF Benson, Oxfordshire, in April 1939. This was 12 Operational Training Unit, established to train pilots, observers and air gunners in Fairey Battles and Avro Ansons for operations against enemy forces.

Captain Ralph Towlerton Leather AFC died at Montreux Palace Hotel, Montreux, Vaud, Switzerland in 1958.

STUDENTS

The first training course at Turnberry began on 16 February 1917 and included the following students:

Lieutenant Frank William Balls *Course 1, Squad 6 Vickers*

Born in 1894, the son of William George and Emily Mary Balls of Stowmarket, Suffolk, Frank William Balls was commissioned as a Lieutenant in the 3rd Battalion, Suffolk Regiment, and was attached to the RFC in August 1916. After initial training, finishing at Turnberry in March 1917,

2nd Lieutenant Matthew Brown (Bunty) Frew
Course 1, Squad 6 Vickers

Born Glasgow in 1894, Matthew Frew was educated at Hutcheson's Grammar School and prior to the Great War he was employed with the City of Glasgow Life Insurance Company. He enlisted with the 9th Battalion Highland Light Infantry on 9 September 1914 and served with the regiment in France. In August 1916 Frew was posted from the battalion back to Britain where he volunteered for service in the Royal Flying Corps and after going through the preliminary training, he was gazetted Temporary 2nd

Lieutenants Hone, Wood and Bristow

Lieutenant Balls was posted to 84 Squadron but did not appear to go with them when they moved to France in September. He was subsequently posted to 109 Squadron, where he spent just over a month, before being transferred to 23 TS (Training Wing HQ) Aboukir, Egypt, on 22 January 1918.

Lieutenant Frank William Balls died age 24, on 1 July 1918 of Pyaemia (septicaemia) and is buried in the Alexandria (Hadra) War Memorial Cemetery Egypt.

Lieutenant (on probation) on 20 October 1916 and posted to 24 Reserve Squadron, Netheravon, for elementary flying training.

Completing his training, some of which was at Turnberry, where he appears amongst Squad 6, Vickers course, his first operational posting was to 'B' Flight 45 Squadron, St. Marie Cappel, France, on 28 April 1917. In October 1917, Frew was appointed 'B' Flight Commander and Temporary Captain. He was awarded the Military Cross as per Supplement to the *London Gazette*, 7 March 1918:

T./2nd Lt. Matthew Brown Frew, Gen. List and RFC. For conspicuous gallantly and devotion to duty on patrol, showing a fine offensive spirit in many combats. He has shot down five enemy aeroplanes, on one occasion leading his formation to attack twenty-two Albatros Scouts, and himself shooting one down

This was quickly followed by the Military Cross Bar, Supplement to the *London Gazette*, 23 April 1918:

T./2nd Lt. Matthew Brown Frew, MC, Gen. List and RFC. For conspicuous gallantry and devotion to duty in shooting down three enemy machines in two days. He has destroyed eight enemy machines and driven down many others out of control.

In August 1918 he was given the Distinguished Service Order; Supplement to the *London Gazette*, 16 August 1918:

T./Capt. Matthew Brown Frew, MC, Gen. List and RFC. For conspicuous gallantry and devotion to duty. On one occasion when leader of a patrol he shot down an enemy aeroplane, two others being also accounted for in the same fight. On a later occasion he destroyed three enemy machines in one combat, all of which were seen to crash to the ground. Immediately after this combat he had to switch off his engine and make an attempt to glide towards our lines five miles away on account of his machine having received a direct hit. Owing to the great skill and courage he displayed in the handling of his damaged machine, he succeeded in bringing it safely to our lines. He has destroyed twenty-two enemy machines up to date.

Captain Matthew Frew was credited with 23 victories, but in 1918 he was injured when his

Sopwith Camel was hit by anti-aircraft fire. He returned to Britain and served as an instructor for the rest of the war. Frew remained in the Royal Air Force and in 1922 he was posted as Flight Commander 6 Squadron to Baghdad and flew with the squadron in operations directed against the Kurdish Chief Sheik Mahmoud. Returning to Britain he was an Instructor at the Armament and Gunnery School, Eastchurch, and was promoted to Squadron Leader in 1927. He was then Chief Fighting Instructor at 1 F.T. School before being posted to Iraq on Air Staff, Operations H.Q. in June 1931, taking part in the Northern Kurdistan campaign. Frew was awarded a Bar to his DSO for this campaign. He then served as an Officer Commanding 111 Squadron at Hornchurch, from May 1933 until August 1934; he served in the same capacity with 10 Squadron, being promoted Wing Commander in 1934. With the outbreak of the Second World War Frew was Group Captain, Senior Air Staff Officer 23 (Training) Group at Grantham; after a brief period in command of the RAF Station at North Weald, he was appointed Head of the RAF Mission and Director of Air Training in South Africa on 19 September 1940; Frew´s main responsibility was the administration of the Commonwealth Air Training Scheme in South Africa, which trained aircrew not only for the RAF and the SAAF but also for the Air Forces of other Allied countries.

Matthew Brown (Bunty) Frew KBE, CB, DSO & Bar, MC & Bar, AFC was knighted and retired with the rank of Air Vice-Marshal in 1948. He was also awarded the Silver Medal of Military Valour (Italy), Commander of the Royal Order of George I (Greece) and Military Cross, 1st class (Belgium).

Air Vice-Marshal Matthew Brown Frew died in Pretoria, South Africa in 1974.

Lieutenant John William Fraser Neill *Course 1, Squad 6 Vickers*
Lieutenant John William Fraser Neill was born in Edinburgh in 1895. At the outbreak of World War I he was commissioned as a Lieutenant in the 3rd Battalion Royal Scots. Attached to the Royal Flying Corps on 25 September 1916, he obtained

his aviator's certificate number 3859 flying a Maurice Farman biplane at Thetford on 10 November 1916. After further training, finishing at Turnberry in March 1917, Lieutenant Neill was posted to 49 then 55 Squadron on 2 June 1917.

On 5 September 1917, Lieutenant J. W. F. Neill & 2nd Lieutenant T. M. Webster, flying D.H.4 A7530, took off on a bombing raid to Melle. The formation was attacked by ten enemy aircraft near Gitsberg and Lieutenant Neill was last seen between Thielt and Litchterve going down apparently under control with smoke billowing from the fuselage. Rudolf Berthold of Jasta 18 claimed D.H.4 A7530 as his sixteenth victory. Lieutenant Neill & 2nd Lieutenant Webster were both taken Prisoner of War. Repatriated on 16 October 1918, after a spell in hospital, Lieutenant John Neill was transferred to the unemployed list on 28 February 1919.

After the war, Lieutenant John William Fraser Neill became a Chartered Accountant and a partner of Price Waterhouse Accountants. He died in Palm Beach, Florida, USA, in 1980.

2nd Lieutenant Leslie George Paling *Course 1, Squad 6 Vickers*
Leslie George Paling was born in 1895 in Nottingham, and prior to the outbreak of war was a civil engineer. Joining the 4th Battalion Notts. and Derby Regiment (Sherwood Foresters) Special Reserve, he was attached to the Royal Flying Corps in September 1916. After some time at Wendover on disciplinary duties, training at Oxford and then Turnberry, Lieutenant Paling was posted to France with 35 Squadron on 9 June 1917.

On 23 November with observer 2nd Lieutenant David Ogilvie Duthie, while flying Armstrong Whitworth FK8 B5767, he took off on a reconnaissance mission, when after the aircraft was damaged by machine-gun fire and Lieutenant Duthie wounded, he was forced to land at Longavesnes aerodrome. Promoted to T/Captain on 23 March 1918, Paling, with observer Capt. J. R. Fasson were shot down in Armstrong Whitworth FK8 C3631 near Mons while on patrol. For his bravery in this action

he was awarded the Military Cross. On 26 June 1918 the *London Gazette* announced the award of the Military Cross for conspicuous gallantry to Lieut. (Temporary Captain) Leslie George Paling. Notts, and Derbyshire Regiment, Special Reserve and RFC:

> For conspicuous gallantry and devotion to duty. While on contact patrol work, he engaged a hostile battalion moving to the attack with bombs and machine-gun fire, inflicting heavy casualties. On his return he was attacked by ten enemy scouts, all of which he succeeded driving off after hard fight but was forced land in front of our lines. Having destroyed his machine, he gave information to the dispositions of the enemy and the nearest hostile field batteries. His exceptional courage, dash, and capacity for hard work have proved invaluable.

This was followed shortly afterwards on 1 April by force landing in a wood north of Belloy after the propeller of his FK8 was smashed in a hailstorm. Both Capt. L. G. Paling and Capt. J. R. Fasson were unhurt.

Returning to Home Establishment on 17 April, Captain Paling spent the remainder of the war at various Training Stations in England before being placed on the unemployed list at the end of October 1919. Leslie George Paling died in 1973.

(2nd Lieutenant David Ogilvie Duthie, after recovering from wounds received in the action of 23 November 1917, was posted to 2 Squadron. He was reported killed on 23 August 1918 between Givenchy and La Bassée and is remembered on the Arras Memorial. Word later came through from the German Red Cross that he is buried by the canal fence north of Auchy les Mines)

Lieutenant Arthur Percival Foley Rhys-Davids *Course 1, Squad 6 Vickers*
Perhaps one of Britain's most famous airmen

to appear at Turnberry, Arthur was born on 26 September 1897 in London, son of Thomas William Rhys-Davids and Caroline Augusta Foley Rhys-Davids. He gained his School Certificate in July 1913 with higher marks than any other student and at the age of 14 joined Eton College as a King's Scholar and subsequently received a scholarship to Oxford. In mid-1916 Rhys-Davids applied for a commission in the Royal Flying Corps, and on 28 August 1916 he reported for training at the CFS Oxford. Completing his training with a course at Turnberry he was assigned to 56 Squadron RFC on 7 March 1917.

It was said that fearing he might be shot down and captured, he always carried a book of Blake's poetry into combat. In his first dogfight on 7 May 1917, his flight Commander Albert Ball was shot down whilst Rhys-Davids survived the attack by Kurt Wolff of Jasta 11. On 23 September 1917, during one of the most famous dogfights of the war, he shot down a Fokker Triplane piloted by Werner Voss (see also 2nd Lieutenant Samuel Leslie John Bramley). When Carl Menckhoff arrived on the scene in an Albatros Scout and attempted to help Voss, Rhys-Davids shot him down too. His final tally was 27 enemy aircraft; one shot down in flames, one 'destroyed in flames', one 'driven down', two 'forced to land', 15 'out of control' and seven destroyed. His commanding officer James McCudden was full of admiration of Rhys-Davids fighting spirit.

In the Supplement to the *London Gazette*, 18 July 1917, it is noted that he was awarded the Military Cross:

2nd Lt. Arthur Percival Foley Rhys-Davids, RFC, Spec. Res. For conspicuous gallantry and devotion. On many occasions he has shot down hostile machines and put others out of action, frequently pursuing to low altitudes. On all occasions his fearlessness and dash have been most marked.

Again, in the Supplement to the *London Gazette*, 17 September 1917, he was awarded the Military Cross Bar:

2nd Lt. Arthur Percival Foley Rhys-Davids, MC, RFC, Spec. Res. For conspicuous gallantry and devotion to duty whilst on offensive patrols. He has in all destroyed four enemy aircraft and driven down many others out of control. In all his combats his gallantry and skill have been most marked, and on one occasion he shot down an enemy pilot who had accounted for twenty-nine Allied machines. His offensive spirit and initiative have set a magnificent example to all.

On 1 October 1917 he was awarded the Distinguished Service Order. It was announced in the Supplement to the *London Gazette* on 18 March 1918:

2nd Lt. Arthur Percival Foley Rhys-Davids, MC, RFC, Spec. Res. For conspicuous gallantry and devotion to duty in bringing down nine enemy aircraft in nine weeks. He is a magnificent fighter, never failing to locate enemy aircraft and invariably attacking regardless of the numbers against him.

On 27 October 1917 Arthur Rhys-Davids DSO, MC and Bar was reported as missing in action. He was last seen chasing an Albatros east of Roulers, and then pursuing a group of German aircraft over Roeselare, Belgium. He was never seen or heard from again. His remains have never been found.

On 29 December 1917 it was reported that a German aircraft had dropped a note to inform the RFC of Rhys-Davids's death. Rhys-Davids has no known grave and is remembered on the Arras Flying Services Memorial.

2nd Lieutenant James John Scaramanga
Course 1, Squad 6 Vickers

James John Scaramanga hailed from a wealthy Greek shipping family and was born in Redhill, Surrey, on 25 July 1898, the son of John and Louisa Scaramanga. The Scaramanga name was made famous by author Ian Fleming, who named the villain in his novel, *The Man with The Golden Gun*, after James's cousin, George Ambrose Scaramanga. George Ambrose Scaramanga had been schoolmates at Eton College with Ian Fleming, the two did not get on and had a major falling out, and as perverse revenge, Fleming made the name synonymous with evildoers.

James John Scaramanga was commissioned into the Special Reserve and joined the RFC. After training at Turnberry in February 1917 he was posted to 2 Training Squadron where he spent some time as an assistant instructor before being sent to France as an Observer with 20 Squadron on 28 December 1917. Lieutenant Scaramanga scored his first aerial victory on 9 March 1918 when, with pilot Lieutenant Douglas Graham Cooke in Bristol F2b C4605, sent an Albatros D.V out of control south of Comines. His next victory was with pilot Lieutenant Dennis Latimer in Bristol F2b C4615, when on 13 March 1918 they sent down another Pfalz D.III out of control near the Belgian border.

On 11 April 1918, both he and his pilot, Canadian Major J. A. Dennistoun, were forced to land near Neuve-Église, Bas-Rhin, France, under heavy machine-gun ground fire. Their aircraft Bristol F2b B1275 was destroyed and both men were seriously wounded. On 6 June 1918, after he had recovered, Lieutenant Scaramanga was posted to 22 Squadron.

On 19 June 1918, Lieutenant James Scaramanga with Lieutenant John Gurdon as his pilot, shot down three Fokker D.VII aircraft southeast of Armentières. Two were destroyed, and a third sent out of control, followed on 27 June when he sent a Fokker Dr. I out of control, again southeast of Armentières.

On 1 July 1918 he sent a Pfalz D.III out of control over Armentières and shot down his eighth enemy plane on 4 July when a Fokker D.VII was destroyed over Noyelles. The next dogfight involved a D.F.W. (Deutsche Flugzeug-Werke) which he destroyed north of La Bassée. Lieutenant Scaramanga's last three victories all took place on the same day, 10 July 1918, with Lieutenant John Gurdon flying Bristol F2b C1003. They engaged three Pfalz D.III aircraft, all over Armentières-Lille. One was destroyed and the other two sent down out of control.

Scaramanga was severely wounded in this dogfight and briefly lost consciousness. On coming round and with his one good arm, he shot down the enemy Pfalz D.III that had been attacking them. He died later that same day, soon after he and his pilot, Lieutenant John Everard Gurdon, landed, just short of his twentieth birthday. His pilot, Lieutenant Gurdon, survived.

Although being credited with twelve aerial victories and seriously wounded twice, the second time fatally, he does not appear to have been decorated.

Lieutenant James John Scaramanga is buried at Aire Communal Cemetery in Aire-sur-la-Lys, Pas-de-Calais, France.

2nd Lieutenant James Scott Williams *Course 1, Squad 6 Vickers*

James Scott Williams was born at Goldenville, Nova Scotia, Canada, in 1893. In January 1915 he was employed as a chauffeur/mechanic when he joined the Canadian Army, 2nd Divisional Supply Column as a lorry driver with the rank of private. The unit proceeded to England in May and embarked for France in September. After spending eight months at the front, he was granted a Temporary Commission in the Royal Flying Corps in May 1916. No record can be found regarding his early flying training, although contemporary accounts state that Williams learned to fly in 27 minutes. Lieutenant James Scott Williams was subsequently posted in September 1916 to 22 Squadron, a reconnaissance unit whose mission was the location and surveillance of enemy supply dumps, railheads and troop movements. Williams flew almost 500 hours over enemy lines in France and was awarded the Military Cross, Air Force Cross and was Mentioned in Despatches for conspicuous gallantry.

Major J.S. Williams (Centre) CO of School of Instructors, Ayr 1918 was a pupil at Turnberry 1917

Returning to the UK on 12 February 1917, three days later he arrived at Turnberry on a Vickers Gun Course. On leaving Turnberry he was posted to Gosport, where he was one of the originators of the Gosport School of Special Flying and he remained there as a Flight Commander until July 1918. From Gosport Williams was sent to Ayr, to organise and command the Flying Instructors' School for Scotland and Ireland. After the war he commanded the two Royal Air Force Aircraft Exhibitions which toured England for six months.

On demobilization he returned to Canada where he started one of the first commercial air services, initiating routes between the Noranda goldfields and Haileybury in Ontario and he was considered one of Canada's outstanding pioneers in north country flying.

In 1920, he was given the commission of organizing the new Canadian Air Force, gaining the rank of Wing Commander, and in 1921 he handed over command to Air Marshal Lindsay

Gordon and took up a position as a test pilot in the United States. In a memorandum issued by the Department of Militia and Defence it was noted: 'Wing Commander J. S. Williams has the reputation of being one of the best pilots in the world, and one of the best flying instructors. He has character and can command men but is himself somewhat impatient of control. Office routine is irksome to him and (his whole soul is in flying) he begrudges time devoted to administration'.

Prior to World War II, Williams managed several gold mines in Nova Scotia, and was Managing Director of Laurentide Air Service Corporation, a civil aviation firm which was established to provide services for mining enterprises. His record states that he had logged over 2,000 flying hours and had flown most types of aircraft, land and water.

James Scott Williams re-joined the RCAF in an administrative role in 1940. He served at Calgary and Picton, Ontario. He was posted to the accident investigation branch at Ottawa before 5 Manning Depot at Lachine in 1942. He took temporary command at Lachine and was then posted to 2 Initial Training School, Regina, as Commanding Officer, relinquishing his command on 1 November 1943.

Wing-Commander Williams, MC, AFC, died on 1 January 1944, age 51, after a long illness, and is buried in Montreal (Mount Royal) Cemetery.

Captain Claude Manley Gibson
Claude Manley Gibson was born in London in 1895, son of Walter Matthew Gibson (who had been Secretary to the Privy Purse through the reigns of Queen Victoria, King Edward VII and King George V) and Katherine M. Gibson.

Transferring to the RFC from the 10th Battalion East Surrey Regiment, 2nd Lieutenant Claude Manley Gibson gained his Royal Aero Club Aviators' Certificate on 3 July 1915 flying a Maurice Farman Biplane at the Military School Birmingham.

Posted to 12 Squadron over the front in December 1915, he suffered a couple of mishaps when flying Bristol Scout 5313, the first on 26 April when his aircraft overturned on landing from a patrol, and then on 16 May he was hit by ground fire on return from a patrol and he ended up crashing into a ploughed field next to the aerodrome but escaped with minor injuries.

Moving to 11 Squadron in early June 1916, this time flying Bristol Scout 5312, he experienced engine failure on take-off for a special patrol. While trying to make an emergency landing, he tried to turn to avoid a sunken road and side slipped into the ground. On 25 June during a special mission flying Nieuport Scout 5173 the throttle cable snapped and in attempting to make a forced landing the aircraft overturned at Haute Avesnes.

Returning to Home Establishment in August 1916, 2nd Lieutenant Gibson spent some time at the Machine Gun School at Hythe before being posted for duty at Turnberry in March 1917 until July 1918, as Flying Officer, then Flight Commander of the D.H.4 flight with the rank

of Captain (Temp.) Captain Gibson was further posted to 2, and 3 FS, until he was transferred to the unemployed list in January 1919.

Claude Manley Gibson died in 1947 in Hove, Sussex, England. His elder brother, Lieutenant Malcolm Reginald Gibson, 7th East Surrey Regiment, was killed in action at Hulloch, France, on 8 Oct 1915. 2nd Lieutenant Edgar Darnell Gibson, 2 Squadron, RFC, Claude Manley Gibson's younger brother, was killed on active service near Bethune, France, on 9 Oct 1917, aged 19.

Lieutenant Stanley Frank Allabarton
Born in Olney, Buckinghamshire, Stanley Frank Allabarton was educated at Bedford Modern School, and was then an Engineer Apprentice. On 15 August 1914 he enlisted in the Royal Fusiliers 19th Battalion (City of London Regiment). In September that year he entered the theatre of war in France until the 4 of August 1916 when he transferred to the Royal Flying Corps and was commissioned 2nd Lieutenant. After initial flying training at Hendon, where he obtained his 'ticket' on 22 December 1916 flying a Caudron Biplane, he arrived for a course at Turnberry in April, after which he was posted to 19 Squadron in France on 12 May 1917 with the rank of Flying Officer. Just one-week later Allabarton took off on an offensive patrol flying SPAD B1627 when he was forced to land behind enemy lines near Harnes due to anti-aircraft fire damage. He was taken prisoner and sent to the POW camp at Karlsruhe, in Germany. He was repatriated on 2 January 1919 and transferred to the unemployed list in March 1919.

Stanley then worked for the War Graves Commission, and subsequently for a petroleum company which led to living in Iran, Egypt and other countries in the middle east. He next moved to Canada and was an 'unofficial' pilot to Prime Minister Pierre Trudeau and flew him around the many islands that make up the country of Canada. After a few years he moved back to Britain and became an antiques dealer in Brighton where he died in 1957.

2nd Lieutenant Brian St. John Harvey Atteridge

Born in Paddington, London, in 1898, and educated at Merchant Taylors School in London, Brian St. John Harvey Atteridge transferred to the Royal Flying Corps from the Royal Garrison Artillery in December 1916. After pilot training at Oxford and Netheravon, where he flew M.F. Shorthorns and BE2e's, Atteridge came to Turnberry in April 1917. Graduating from the Lewis Gun Course at Turnberry he was posted to D Squadron CFS where he was injured on 12 June flying BE2e A8638 which stalled and nose-dived. Subsequently classed permanently unfit for any further military service he was discharged on 29 March 1918.

In civilian life, Atteridge was a Tea Planter in India arriving in Bombay in 1919 and is found on the 1939 Register returned to Britain and living in Shoreham-by-Sea, with his wife Lilian, both of whom give their occupation as 'Theatrical Artist' (Travelling). Brian St. John Harvey Atteridge died in 1995 in Brighton.

2nd Lieutenant Raymond James Brownell, CBE, MC, MM

Born in Hobart, Tasmania, in 1894, Brownell attended Scotch College, Melbourne. On graduation he was apprenticed as an accounts clerk; he enlisted in the Australian Imperial Force on the outbreak of the First World War and was posted to the 41st Battery, Australian Field Artillery.

He served during the Gallipoli Campaign before transferring to the Western Front. He was awarded the Military Medal for his actions during the Battle of Pozières. On 21–22 July 1916, Brownell was in action with his battery at Pozières, during which the unit was under attack by severe German shellfire. Throughout the battle, Brownell established and maintained communications between the battery and firing line, despite fatigue or personal risk to himself. Commended for his 'particularly meritorious service and gallantry in this work', Brownell was subsequently recommended for the Military Medal. The notification for the award was published in a supplement to the *London Gazette* on 16 November 1916.

Transferred to the Royal Flying Corps in 1917, Brownell undertook pilot training in England and was posted to 3 School of Military Aeronautics at Oxford prior to arriving at Turnberry for instruction in gunnery at the beginning of April 1917. Commissioned as a 2nd Lieutenant in September 1917, he was posted for operational service over the Western Front with 45 Squadron, flying Sopwith Camels. On 10 September, he took part in his first patrol, during which he shot down a two-seater German aircraft. In his time flying with 45 Squadron, Brownell accumulated a total of five victories and achieved 'ace' status before the squadron was transferred to Italy in December. Later that month, Brownell and his observer, Lieutenant Henry Moody, shot down German ace Alwin Thurm over Asoloin. He was awarded the Military Cross and credited with shooting down 12 enemy aircraft by the end of the war.

Records of Brownell's awards note that these were, 'For conspicuous gallantry and devotion to duty. Within the last three months he has brought down six enemy aeroplanes, four of which were seen to come down in flames, the other two falling completely out of control. The dash, gallantry and offensive spirit displayed on all occasions by this officer are worthy of the highest praise'.

Brownell was discharged from the RFC in 1919 and returned to Australia, once more taking up accountancy in Melbourne. In 1921 he was commissioned in the Royal Australian Air Force and by the beginning of the Second World War had risen to the rank of Group Captain. In August 1940, Brownell was posted to Singapore to establish and command an RAAF station on the island, as well as administer the RAAF squadrons located in Malaya. Promoted to acting Air Commodore in 1941, he returned to Australia and was selected to lead 1 Training Group. He was Air Officer Commanding Western Area from January 1943 until July 1945. Brownell was present at the Japanese surrenders in Manila, Tokyo, and on Morotai.

On 3 January 1947 Brownell was presented with the CBE. Retiring from the air force on medical grounds in March of the same year, Brownell then became a partner in the stockbroking firm of S. G. Brearley & Co., located in Perth.

Raymond James Brownell, CBE, MC, MM, died in Western Australia in 1974 at the age of 79 and was buried with full Air Force honours. His autobiography, *From Khaki to Blue*, was published posthumously.

2nd Lieutenant Randall George Henry Davis

Randall George Henry Davis was born in Farnham, Hampshire in 1894, the son of George Davis, Queen's Messenger (a courier employed by the British Foreign and Commonwealth Office).

At the outbreak of war, Randall was employed as an insurance clerk before being commissioned temporary 2nd Lieutenant in The Queen's Own (Royal West Kent Regiment) in December 1915. Transferring to the RFC in August 1916, 2nd Lieutenant Davis completed his elementary flying training at Oxford and then to the CFS at Netheravon where he was injured on 22 December 1916 when he stalled BE2c 2659 on his first solo flight.

On recovery, he completed the Lewis Gun Course at Turnberry in April 1917 and in October was posted to 28 Squadron in Italy flying Sopwith Camels. Transferring to 45 Squadron in 1918 he was sent to the Western Front, returning to the UK in October for training at the South West Area Flying Instructors School.

In February 1919 he fell victim to the influenza pandemic that was sweeping the world

Davis, Gibson and Markham

and was hospitalised, succumbing to pneumonia he was severely ill for months and was finally discharged from hospital in mid-November 1919 and was demobbed at the end of the month.

Randall Davis, now a professional singer, embarked at Southampton on the SS *Royal George* bound for Los Angeles on 8 January 1920.

Lieutenant Randall George Henry Davis died on 29 May 1920 of Tuberculosis Meningitis and is buried in the Hollywood Cemetery, California.

2nd Lieutenant Curtis Matthew De Rochie

French Canadian Curtis Matthew De Rochie was born on 15 Nov 1895 in Cornwall, Ontario, Canada. He attended Cornwall High School and in 1914 entered the Faculty of Medicine. Joining the RFC in October 1916, he completed his flying training at Oxford and proceeded to Turnberry in April 1917. On graduating from Turnberry, he was offered a position as a flying instructor, but chose instead to go to France 'and do his bit' on 2 June 1917 with 27 Squadron.

2nd Lieutenant De Rochie took part in a number of successful patrols, one of which was described by another airman. He recalled a raid in June 1917, in which their homeward bound formation was attacked by four German scouts. He himself was attacked by two enemy fighters, one of which fired at him: 'His stream of lead carrying away a piece of my flying cap and making a mess of the instrument panel'. Left straggling, the rest of the formation being over a mile away he was 'left, stunting desperately, a mere 400 ft up, 60 miles from the line and two painted devils doing their utmost to down me ... when suddenly, I saw one of our machines diving by me with its Lewis Gun going in one long burst. I noted the pilot was bareheaded and knew at once that he was a Canadian named De Rochie. He had left the formation to turn back to what might well have been certain death in order to give me a hand. His spectacular dive had scared off the enemy machine he had been firing at and though the other hung on until we reached our lines, he didn't attempt anything serious'.

'De Rochie was killed a week or so later. His action in coming to my rescue as he did was more than unusual for, when over enemy territory and with enemy scouts about, all that every Martinsyde pilot was instructed to aim at, was above all to keep in formation and to follow the leader home as fast as possible.'

On 14 July 2nd Lieutenant Curtis De Rochie was last seen flying his Martinsyde A6266 over Zarren whilst on a bombing raid. He was seen going down in flames, and he jumped out at 10,000 ft, his burning aircraft falling in the Yser canal.

The German ace Julius Buckler, who shot De Rochie down, wrote about it in his book *Malaula*. Buckler describes how he was flying at 3,200 metres above the clouds in search of a reported enemy airplane:

> In this solitude I met an enemy airman. He came from the direction of Bruges and was apparently about to fly home .... He was flying 200 metres higher than I, and I followed him and pulled up and shot at him from below.
>
> He immediately turned around, which was just what I wanted. I pretended to make a run for it, and he dived after me. Then I made a sudden turn and he raced past me and, sitting on his neck, I fired a few well-aimed rounds from 150 to 290 metres range.
>
> His machine emitted smoke and then burst into flames. Now came the most horrifying thing I have ever experienced in my flying career. I saw the pilot stand up - the brave man did not want to burn - preferring a leap to his death from 3,000 metres rather than endure a death by fire. I cannot describe my emotions as I watched this person plunging into the depths before my eyes. I cursed the war.... First my fellow aviator and then the burning machine plunged into the sea of clouds, with me following. When I again appeared from out of the clouds, I just managed to see the burning motor with the spark-scattering remains of the airframe whirling around it strike the earth.

Buckler landed at his airfield and jumped into a car and drove to the crash site: 'The English flyer lay there as though he were sleeping. Someone had fished his corpse out of the Yser (canal). Everything was crushed internally. His trunk and thigh were torn up by bullets that had passed right through him. What tremendous energy this person must have possessed to still find the strength to jump out of the aeroplane with these frightful wounds! I could endure everything again if I had to, but I would not want to experience my thirteenth victory a second time.'

2nd Lieutenant Curtis Matthew De Rochie, who died age 22, is buried at Ostende New Communal Cemetery, Belgium.

Captain Wilfred Allan Fleming MC

Born in Thayetmo, Bengal, in 1890, the son of Allan Stopford Fleming of the Civil Service, Wilfrid Fleming joined the 1st Battalion Devonshire Regiment and was stationed in Jersey at the outbreak of World War I. Transferring to the Royal Flying Corps in 1917, he completed his training at Turnberry and was then posted to 56 Squadron flying SE5.a's.

We find some information regarding Fleming in *Harrow School War Memorials Volume V*:

> Captain Fleming was gazetted to the 1st Devons in 1910, and on the outbreak of the War was at once sent to France. He was mentioned in Despatches and awarded the Military Cross for distinguished services in the Machine Gun Section, in June 1915. He was then attached to the Royal Flying Corps, and after a course of training at Tidworth returned to France in July 1917. On 10 August he was reported missing. 'On that day he was sent out with a patrol of four machines over the Menin-Roulers road, east of Ypres. It was a bad day, with a lot of cloud and a forty-mile wind blowing from the west, and it was his first flight on an SE5 aeroplane. At about 1 p.m. his patrol was attacked by some eight German Albatros Scouts, and a sharp fight began,

in the course of which our formation got scattered. On emerging from a cloud, the patrol-leader saw Fleming heavily engaged with three enemy planes far to the east. He was putting up a splendid fight, firing at close range on one of the enemy, while another of the Germans was close behind firing at him. The patrol-leader went to his assistance, and together they so settled the Germans that they brought down one and drove the others off. Then the leader and Fleming started to return home in the teeth of the wind, and in a bank of cloud the leader lost sight of Fleming but from our aerodrome his machine was seen to land behind the German line, and it was thought he must have been forced down by lack of petrol. However, in January 1918, definite information came through that Fleming was killed in the action of 10 August and was buried in the cemetery of Ledeghan.

Captain Wilfred Allan Fleming MC, flying SE5a A8923, was last seen east of the Roulers-Menin road. Leutnant Walter Stock of Jasta 6 claimed A8923 as his third victory. He is buried in Harlebeke New British Cemetery, Belgium.

Lieutenant Francis Behrens Best
Born in Handsworth, Birmingham, in 1893, son of Robert Hall Best (Chandelier Manufacturer) and Emma Maud Best, Frank was educated at Bedales, near Petersfield in Hampshire.

Although he had passed the Oxford and Cambridge Board Higher Certificate in Mathematics, English and Mechanics but failed in German, his father was not keen for him to attend university but instead serve his time in the various departments of the family business. This was followed in 1913 by six months in Dusseldorf, studying at the School of Industrial Design, and then in Paris returning home in July 1914. Joining the Army Service Corps, Frank embarked with his division on SS *Maidan* in March 1915, bound for France. Returning to England in November 1916, he was then posted

in April to The Curragh in Ireland, just after the Easter Rising. Totally unhappy with this posting, he and his brother Robert applied for transfer to the Royal Flying Corps. They both received orders to report to the RFC School of Instruction in Oxford. Following graduation from ground training, Frank was posted to Dover for flying instruction. It is reported that after going solo he had a few near misses. Early in May 1917 he arrived at Turnberry and was delighted to find his brother Robert on the same course. Both brothers graduated on 21 May and Frank was posted to Hertfordshire for further flying training before being posted to 19 Squadron in June with a total of 58 hours 10 minutes flying time.

The following is an account of his time at Turnberry, based on his diaries and letters, later published by his brother Robert Dudley Best, *A Short Life and a Gay One, Diary of Frank Best, 1893-1917*:

On arriving at Turnberry, one entered the hotel direct from the railway station. We were at once struck by the opulent atmosphere, with its faint aroma of good food, wine and cigars. The place had been taken over by the authorities just as it was - Head Waiter, chefs, staff, cellars and all. The bedrooms, bathrooms and appointments were what one would expect in a first-class golfing hotel. The bars were attractive, and it was in one of them that we found ourselves on that first night. As we sipped our gin-and-Italians I experienced, once again, that joyous feeling of companionship. After the second we began to get slightly hilarious, more with pleasure and high spirits than with the stimulant. On the way into the dining room we had to control an inclination to laugh uproariously. We sat down, side by side and were once again hard at it, our heads together comparing notes, making exiting plans for the future, discussing ideas for songs and dramatic sketches, and projects for the business.

Next morning, we lit our pipes as we strolled about on the asphalt terrace in front of the Hotel which, we discovered, stood on a low wooded ridge of hills. Below us we saw the Firth of Clyde, with Ailsa Craig on the left, the Isle of Arran to the right and, beyond, the Mull of Kintyre. We learnt that on a clear day the Irish coast was plainly visible. A drive to the Hotel wound up the side of the hill from a narrow coast road. Between it and the sea we noticed sand-dunes and two golf courses. A wide flight of steps led direct from the terrace to some tennis courts at the bottom.

A lighthouse near the ruins of Turnberry Castle completed the landscape. We looked at each other and nodded approval of the sporting and recreational prospects.

Immediately to the North, that is to say on our right as we faced the sea, aeroplanes could already be seen taking off and landing. The low wooded ridge on which the Hotel stood, continued North up the landward side of the aerodrome; it was surrounded by Bessonneaux Hangars. The flying speed of aeroplanes was then such, that it was essential to land into the wind. But with a light breeze blowing off the sea the pilot would be forced almost to touch the treetops of the wooded ridge in order to avoid overrunning the narrow aerodrome and crashing into the fences bordering the coast road. Further north, a large potato field provided what golfers call a 'natural hazard' for pilots with engine trouble trying to scrape home. Beyond, and a mile or so away, was a small fishing village called the Maidens.

Founded earlier that year, the Turnberry School of Aerial Gunnery, like the Oxford and Reading Schools, formed part of the general expansion initiated by Douglas Haig the previous year. It was designed to give final gunnery instruction to pilots. As part of the training we were taken up in a pusher, sitting out in front in a sort of balcony, and encouraged to loose off ammunition at a large flag towed behind an aeroplane at the end of some fifty yards of wire. Sometimes a careless pupil would mistake the towing machine for the target. At least one pilot, who had just landed, was seen to be pointing in anger to a number of bullet holes in his machine.

We would also dive at targets placed amongst the sand-dunes, or at rafts floating out at sea. Tractor machines, with a machine-gun firing through the propeller, would be flown dual control. Once, as I remember, a misunderstanding arose. The instructor and the pupil each thought that other was in control, until they ended in the sea! This was what befell Sir Robert Lorraine, 'the actor airman'. The consequences were not serious, except for a wigging from the CO, Colonel Bell-Irving.

We saw some sensational flying. I have a vivid recollection of Casey Callaghan, well known in those days as a stunt pilot, flying at about a thousand feet above the aerodrome. He would suddenly turn his machine so that the wings appear to be almost perpendicular to the earth, and then sideslip right down to within a few feet of the aerodrome, righting himself at the last moment.

The types of machines in use were limited to about four. But from time to time one or two German machines, with British markings of course, would appear in the sky in order to help the pupils recognise them in flight and as targets for camera gun.

Much time was given to work on the machine-gun range and with the camera gun. we also attended a number of lectures. As I write these lines, I have open before me my book of notes. Some of them recall the portentous and parrot-like accents of the Sergeant Instructor. They were most detailed and thorough:

"The remainder of the gasses follow down the barrel and hit against the cone-shaped barrel mouth-piece, where they are spread out evenly and strike against the walls of the front radiator casing, driving the air out and so causing a suction, which draws in air from the rear along the radiator pin".

But there was one aspect of the machine-gun lore on which our lives were to depend. For these lethal implements were inclined to jam, lock and suddenly cease to function. Nothing could be more exasperating than to get the enemy machine nicely in the ring-sight, press the trigger - and then - after perhaps a short burst - silence. Or to be attacked by enemy machines and suddenly to find oneself deprived of effective defence.

We were shown again and again, how to free these 'stoppages' practising and rehearsing 'immediate action' i.e. 'crank handle onto roller' – 'belt, left front', 'release crank handle', 're-lay', and 'carry on'.

The lecturer, Captain Stokes of 60 Squadron, emphasized that tactics of fighting were constantly evolving, and this was certainly confirmed later in one of Frank's letters home: 'Spinning is getting rather hackneyed now...'

We were told that 'stalling or flying low, when the Hun was on your tail, had its uses'. Looping and the 'corkscrew or Immelmann turn' (A half loop followed by a vertical bank reversing the direction of flight), perplexed him. Success in fighting depended more on science and skill than on mere luck.

We were taught to take notice of the 'dead-fields' or angle through which the pilot was prevented, by the structure of his own machine, from firing. If you could manoeuvre yourself into one of them, one might approach unseen and take him unawares. Clouds should be used as a cover for ambushes and surprises. On seeing another machine, a pilot should always approach it as if it were hostile until absolutely certain of its identity. It was important to be able to identify machines in the air. The atmosphere and light made this deceptive.

Flying straight was dangerous. One should swing about to uncover blind spots. It was important to keep one's height and beware of decoy machines, to say nothing of 'Archie Traps', when the unsuspecting pilot might be lured into a barrage of anti-aircraft fire. If, in a formation flight, your engine went 'dud' one should not straggle but either go home or keep up your speed losing height.

The procedure for getting the formation into position was explained to us in detail. The leader, having reached the required height, made wide circles round the aerodrome, other machines falling behind in position and cutting off corners if late. When the officer in charge on the ground was satisfied that all were in correct positions, he would put out a letter 'K' in white strips on the ground and away they went.

While in some machines Morse messages could be sent to the ground by wireless there was no telephony between the machines. Communication was made by certain signals, rocking the wings, and so forth. If your gun jammed, you were to catch up with the leader and rock the machine fore and aft. Coloured 'Verey' lights (fireworks discharged from a pistol) were supposed to indicate that the pilot was going home, in cases of engine failure. A rendezvous was fixed to which the pilots should repair after a scrap.

Scouts escorting slow machines made 'S' turns above them, always keeping in pairs. 'Although', the lecturer concluded, 'there are so many things to think about in formation flying, you soon get used to it, and it becomes almost second nature to keep your position.'

During what was to be the last fortnight of our partnership the atmosphere suggested a holiday. The sun shone brightly. We played a round of golf nearly every day, using our own clubs which had been sent on to us from home. On the first Saturday we got away to Ayr after lunch and met Jack Marshall at a hotel, he being my future brother-in-law, the husband of Beryl's Sister.

Golf and work continued pleasantly enough during the second week. We were getting to know some of the other officers at golf or at one of the bars, over a game of billiards, or through Frank's playing of the piano. On the Friday we were asked to arrange a smoking concert. He recorded that he 'spent the evening routing around for material, which eventually turned out a big success.' 'We were in high spirits, delighted to be doing some of our old numbers again - Rag Maninoff and Rag-time Folk Song and others. His playing of the piano seemed exceptionally easy and effective. Our audience was responsive, joining in the choruses, and generous in its applause'.

Time was running short. We felt that somehow, we had still not fully exploited the sporting possibilities of the Ayrshire County; so, thinking that the sea trout might be running, we got a lift to the Girvan Water and fished with a wet fly for an hour or two. We had no spectacular success. But that night it rained hard. We were reminded of Albert Wood. There might be some colour on the water. But would the local fishing folk stand for a brandling fished upstream? We felt doubtful.

Sunday 20/5. Exam 10-12. Work washed out at 4.30, so with rod borrowed from corporal set off for Girvan Water again. Funny incident - we thought we were being sleuthed by two figures, because we were fishing with worm instead of fly. But the two figures turned

out to be the President and Secretary of the local fishing club, and far from objecting they eventually supplied us with information and tackle for fishing worm. Dined at King's Arms and returned by car.

The next day saw our final separation, though we were subsequently to speak to each other twice on the telephone. It was another wet day. The officers who had just taken the course were assembled in the lounge and were first addressed by the instructors. Later, Col. Bell-Irving gave us some final words of advice and encouragement. We left Turnberry at 5.10 and dined at Jack Marshall's house in Glasgow - then caught the North Western Express at 9.45 - changed at Crewe - 3.40 for Birmingham.

This time there were no emotional twinges such as I had experienced on leaving Thetford. My mood, like his, was ebullient. Optimistic as ever, he was anticipating further pleasurable experiences on faster and more efficient machines, while I had in mind the girl I was to meet the next day and whom, though not aware of it at that time, I was to marry within a little over two months. For I had been posted to Turnberry as a pilot instructor and with the prospect of my being settled for a few months at the same station the two families agreed that the marriage should take place without delay.

Posted to active service, on 29 July 1917, Lieutenant Francis Behrens Best flying Spad B3531 with 2nd Lt. B. A. Powers brought down an Albatros east of Ypres, and Spad B3531 was itself shot down shortly after, Vizefeldwebel Misch, Jasta 29 claiming it as his first victory.

The family received a letter from the CO of 19 Squadron who wrote to say that Frank was missing, adding 'I sincerely hope that we shall hear that he is safe and a prisoner....'

On 24 September the Best family received official notification of their loss which included the following information published in the German newspaper *Deutscher Allgemeiner Zeitung*, 'List of English flying losses during the month of July 1917 … B3531 Occupant dead.' A cautionary note was added, 'This has been identified as referring to Lieut. F. B. Best. However, this report has not been accepted as official, so should be treated with reserve.'

The War Office officially confirmed the death of Lieutenant Francis Behrens Best on 26 August 1918. He has no known grave and is remembered on the Arras Memorial.

His brother, Robert Dudley Best, married Beryl Gladys Smith on 28 July 1917. He was wounded in action flying an RE8 with 53 Squadron. He survived the war and died in 1984.

Sergeant Christopher William Henry Bowers 932

Born in 1892 in Birmingham, Christopher Bowers was an apprentice engineer when he was an early entrant into the Royal Flying Corps as an Air Mechanic in 1913. He served two years in France with 2 Squadron prior to obtaining his Royal Aero Club Aviators' Certificate 3960 on a Maurice Farman Biplane at the Military School, Brooklands in December 1916.

Sergeant Bowers was the pilot of FE2b 6975 who died at Turnberry. His passenger, 2nd Lieutenant James Stevenson, was also killed.

Sergeant Christopher William Henry Bowers is buried at Girvan (Doune) Cemetery and is remembered on the Turnberry War Memorial.

2nd Lieutenant James Stevenson

Born in 1897 at Evie, Orkney, the son of Thomas Stevenson and Ann Reid, prior to joining the RFC, James was an art student. 2nd Lieutenant James Stevenson was killed, age 20, on 1 May 1917 whilst flying with Sergeant Christopher Bowers, when aircraft FE2b 6975 crashed adjacent to the Warren, near Turnberry Lighthouse. Buried at Evie, Orkney, 2nd Lieutenant James Stevenson is remembered on the Turnberry War Memorial.

L.A.C. Arthur Charles 'Pop' Lewcock 8743

We can read a few details on Lewcock from a diary donated to the Shuttleworth Collection. After serving a year in France with 20 Squadron, 'Pop' Lewcock was posted to 2 (Aux) School Aerial Gunnery, Turnberry as a Motor Driver/Mechanic. Whilst stationed there he kept a diary/notebook in which he recorded events, poetry and autographs. Some of the entries are listed below:

May	FE2b Crashed	2 Killed
July	FE2b Crashed	2 Killed
Aug	FE2b Crashed	2 Killed
Sept	BE2c Crashed	1 Killed
24/25 Oct		Hangars Blown Down, Umpteen Machines Smashed
28 Nov 1917 First Flight.		Hangars Rebuild BE2c's &??
1 Feb		BE2c Crashed: Capt. Davis Unhurt
17 March 1918		Sopwith Camel No 9204 Crashed Pilot Killed
17 March 1918		Sopwith Camel No 9222 Crashed Pilot Killed
1 April 1918		Sopwith Camel No 9210 Crashed Pilot Killed
11 May 1918		Lt.-Col. Bell Irving O.B.E. Left Staff
12 May 1918		No 1 School Aerial

Fighting and No 2 School Aerial Gunnery amalgamated. Everybody well completely fed up, 200 applications for transfer put in, nothing doing.

26 May 1918	Crashed	Killed
27 May 1918		Crashed Killed
28 May 1918		Bristol Fighter Crashed. Right Wing crumbled up in air. 2 Killed Lt. Makepeace and Observer.

2 June 1918	Crashed Killed
5 June 1918	DH9 Crashed 2 Killed
21 June 1918	DH9 Crashed 2 Killed Capt. Henderson & Lt. Redler
19 Aug 1918	Sopwith Camel Crashed
23 Aug 1918	Sopwith Dolphin collided with DH9 in the air. 2 Killed. The Dolphin Pilot climbed out on the tail & Lt ? saved himself. Motor accident. Capt. Childs Killed. Collided with Crossley tender when returning from the crash.
5 Sept 1918	DH9 Crashed & caught fire. 2 Killed
12 Dec 1918	Sopwith Camel Crashed. Pilot Killed
31 Dec 1918	DH9 Crashed & Caught fire. Deaths 2.

2nd Lieutenant Augustus Orlando Balaam
Squad 5
Augustus Orlando Balaam was born in Hackney in October 1892. Prior to his enlistment, he was employed as a tailor's cutter. Augustus was commissioned as a 2nd Lieutenant in the 4th (Territorial Force) Battalion, Suffolk Regiment, on 18 November 1915. Attached to the Royal Flying Corps on 12 May 1917, he trained at the Central Flying School and then Turnberry before being posted to 16 Squadron on 4 August 1917.

On 16 September 1917, '2nd Lieutenant Balaam and Lieutenant Irwin Brooke Wallis, 16 Squadron, were on a photographic mission when they were attacked by two enemy scouts which

they drove away. They were then attacked by a large Nieuport-type machine which they engaged and hit, and it was last seen falling into mist'.

Lieutenant Augustus Orlando Balaam and Lieutenant Donald St. Patrick Prince-Smith were killed flying RE8 B5896 on 24 October 1917. They were last seen going down out of control in a spin until 600 feet then nose-dived with engine full on and burst into flames on artillery observation at Méricourt, Pas-de-Calais. The loss was claimed by Vzfw. Julius Buckler of Jasta 17 as his nineteenth victory. Lieutenant Augustus Orlando Balaam is buried at Aubigny Communal Cemetery Extension, Aubigny-en-Artois, Pas de Calais, France.

Captain Gerald Dixon-Spain *Squad 5*
Gerald was born in Bath, Somerset, in 1889, son of Rev Thomas Dixon-Spain. Between 1909 and 1911 he was an architect's assistant with the firm of J. J. S. Naylor Esq. in Hanover Square, London, giving up that position to become a schoolmaster at St. Cross, Walton on the Hill, Surrey. At the outbreak of WWI, Gerald Dixon-Spain enlisted in the 1st Sportsman's Battalion, 23rd Royal Fusiliers. He was the Battalion Musketry Officer in 1915 and Captain, commanding a Company on active service in France, 1915-1916, seeing action at the Battles of Albert, Bazentin, High Wood and Boritska, all part of the Somme Campaign.

Attached to the Royal Flying Corps, Gerald Dixon-Spain served as a Flying Officer Observer with 20, 43 and 65 squadrons between March and December 1916. A notice appeared in the Supplement to the *London Gazette*, 26 September 1916:

Military Cross: 'For conspicuous skill and gallantry on many occasions. Capt. Dixon-Spain, with 2nd Lt. Reid as pilot, attacked and drove back a hostile machine. A few minutes later four hostile machines were seen, three of which were attacked one after another and driven back, the fourth being accounted for by another patrol. Another time they attacked two hostile machines, shot down one and drove the other back. Two days later they attacked two more machines, of which one is believed to have been destroyed, the other being pursued back to its aerodrome.

Gerald Dixon-Spain was injured on 8 December 1916 while flying in Henry Farman A1185 which overturned while he was attempting a force landing in a ploughed field. On recovery, in July 1917 he was graded permanently unfit for flying as pilot or observer and was posted to Turnberry on an aerial gunnery course before taking an instructor's course at Perivale, then on to Hythe as Assistant Instructor of Gunnery, where he was promoted to Chief Instructor and Squadron Commander by 20 August 1917. He was the Commanding Officer at Hythe for two months in 1918 during the absence of a Commandant. On 15 May 1919 Major Gerald Dixon-Spain was transferred to the unemployed list. He was awarded an OBE (Military) on 3 June 1919.

In the 1930s, Major Gerald Dixon-Spain was Clerk of the T.T. Races, and in September 1936 was Officer in Charge of the Control Office at the International Tourist Trophy Race held at Ards Circuit, Belfast. He was also the Deputy Manager of The Competitions Department of The Royal Automobile Club. Major Gerald Dixon-Spain died in Suffolk in 1968.

Lieutenant William Geoffrey Meggitt *Squad 5*
Meggitt was born in 1894 in Nottingham. A warehouseman's assistant, he was commissioned as a 2nd Lieutenant (on probation) in the 3rd Battalion, Welsh Regiment, on 31 March 1915. He was later seconded to the Royal Flying Corps and was appointed Flying Officer (Observer) on 28 October 1916.

He was posted to 25 Squadron, as an observer in FE2b's, and gained his first victories on 22 October 1916, destroying two Type D aircraft, the first with pilot Sergeant William Drummond Matheson, one action in the morning, south-west of Seclin, and another in the late afternoon north-west of Lille, with pilot 2nd Lieutenant D. S. Johnson. On 17 November 1916 he was part of

a force which saw further fighting and destroyed an enemy aircraft over Vitry, in an aircraft piloted by Captain R. Chadwick. On 26 December 1916 Lieutenant A. P. Maurice and Lieutenant William Meggitt were both injured when their FE2b A5451 spun in and was wrecked after engine failure during bombing over Lozinghem, Pas-de-Calais.

On 15 February 1917, this time flying with Captain Lancelot Richardson, he drove down out of control another enemy aircraft over Avion.

Meggitt was awarded the Military Cross, which was gazetted on 17 April 1917. His citation reads:

> 2nd Lieutenant William Geoffrey Meggitt, Welsh Regiment, Special Reserve and Royal Flying Corps.
> 'For conspicuous gallantry and devotion to duty whilst one of a patrol engaging five hostile machines. He drove down one enemy machine and then attacked another, which was seen to go down vertically. He has previously brought down three hostile machines.

He then trained as a pilot and was appointed Flying Officer on 8 June 1917. Promoted to Lieutenant on 1 July, and after training at Turnberry in August, he was posted to 22 Squadron, flying the Bristol Fighter. He gained his fifth aerial victory on 10 October, destroying an Albatros D.V over Moorslede. with Observer/Air Mechanic 1st Class Arch Whitehouse. The next day he brought down another enemy Albatros, with Captain F. A. Durrad as his observer, while taking part in fighting against six enemy aircraft.

Lieutenant W. G. Meggitt MC was shot down during an offensive patrol, and his observer Captain F. A. Durrad was killed on 8 November 1917 whilst flying Bristol F2B B1123. Meggitt was initially listed as missing in action, but was in due course reported as wounded and captured, serving as a prisoner of war at Karlsruhe in early 1918.

He was repatriated following the Armistice of

11 November 1918 and, on 1 August 1919 was granted a permanent commission in the Royal Air Force with the rank of Lieutenant. Promoted from Flying Officer to Flight Lieutenant in the 1922 New Year Honours, Meggitt served for some time overseas before returning to fly with 41 Squadron at RAF Northolt. On 28 January 1927 flying Armstrong Whitworth Siskin J7171, Flight Lieutenant William Geoffrey Meggitt MC was killed when his single-seat fighter crashed into the garden of a house in Norbury, London.

Lieutenant Walter Cross Moore *Squad 5*
Born in Glasgow in 1895, Walter Cross Moor was the son of a varnish manufacturer. Walter joined the Army Service Corps (Territorial Force) on 19 March 1914. He was seconded to the Royal Flying Corps on 5 May 1917, and after training at Turnberry was posted to 45 Squadron.

Information is found in RFC Communiques – '22 July 1917, Captain Geoffrey Hornblower Cock and Lieutenant Walter Cross Moore of 45 Squadron shot down an Albatros Scout in flames over the Commines-Warneton region in Belgium.' And further on 3 September 1917 from RFC Communique 104:

> While on an Offensive Patrol over Ledeghem – Dadizeele, Sopwith Camels of 45 Squadron engaged 14 Albatros scouts flown by pilots who appeared to be more experienced than the majority. Captain Arthur Travers Harris (later to be known as Bomber Harris) dived at one enemy aircraft which was attacking the rear Camel and shot it down in flames. Another enemy aircraft, however, followed Captain Harris, so Lt. Walter Cross Moore dived at it and shot it down out of control. Lieutenant Oscar Lennox McMaking also drove one enemy aircraft down out of control. After the fight three enemy aircraft were seen wrecked and burning on the ground. Captain Harris fired 150 rounds at his enemy aircraft which went down vertically. Lieutenant Moore fired off 200 rounds and Lieutenant McMaking 100 rounds.

Lieutenant Walter Cross Moore was restored to the Army Service Corps Establishment on 26 November 1917. He was promoted to Captain on 22 January 1918, and resigned his commission in 1922.

Lieutenant Kenneth Mann Rodger *Squad 5*
Kenneth Mann Rodger was born in 1894 in Bothwell, Lanarkshire. He was educated at Uddingston Grammar School and the University of Glasgow. Studying medicine before the war, he joined the forces as a Lieutenant with the Argyll and Sutherland Highlanders. Kenneth was attached to the Royal Flying Corps on 16 October 1916 and proceeded to train at Reading and finally Turnberry before being posted to 43 and then 70 Squadron, flying Sopwith Camels on 29 November 1917.

On 29 January 1918, Lieutenant Kenneth Mann Rodger, flying Sopwith Camel B3890 was last seen in combat with 10 Albatros Scouts between Moorslede and Roulers, east of Poelcapelle. It was later reported that he had been shot down by Ltn. Karl Bolle of Jasta 28 and taken prisoner.

Held as a prisoner of war for almost a year at Rastatt Camp, Baden, Germany, Lieutenant Kenneth Mann Rodger was repatriated on 13 December 1918, arriving back to hospital in Britain on 23 January 1919. Leaving the Royal Flying Corps in April 1919, he gained his diploma in Psychological Medicine in 1922 and practised as a Physician in mental asylums such as Horton near Epsom, Menston in Shipley, Yorkshire, Bristol and Salop.

In 1948 Kenneth Mann Rodger MD, MBChB, DPM, and RCPS, was the Medical Superintendent of Nervous Disorders at the Ingutsheni Mental Hospital, Bulawayo, Rhodesia, which was British Central Africa's first and largest mental institution, founded in 1908.

Lieutenant Colonel Lionel Wilmot Brabazon Rees VC, OBE, MC, AFC
Lieutenant Colonel Lionel Wilmot Brabazon Rees took over command of Turnberry from Lieutenant Colonel Bell-Irving when 1 School of Aerial Fighting and Gunnery from Ayr merged

with 2 (Auxiliary) School of Aerial Gunnery in May 1918.

Born in Wales in 1884, Lionel was the son of an army officer. He was educated at Eastbourne College before entering the military academy at Woolwich, and thence obtained a commission in the Royal Garrison Artillery in 1903. Rees served with the West African Frontier Force, until transferring to the RFC in August 1914. He had learned to fly, at his own expense, receiving Aviator's Certificate 392 in January 1913. He was posted to Upavon as an instructor, and then to 11 Squadron where he saw action flying the Vickers FB5. Lionel Rees assumed Command of 32 Squadron in January 1916, until July when he was badly wounded in the leg.

Lionel Rees

'Rees encountered ten enemy bombers while on patrol in an Airco D.H.2. Attacking alone, he drove down two of the enemy aircraft. While attacking another, his Lewis gun ran dry. Before mounting a fresh drum of ammunition, he pulled a revolver only to lose it somewhere in the cockpit. Badly wounded in the leg, Rees turned away as the two-seater retreated back across the lines'.

For this action he was awarded the Victoria Cross. He was considered an exceptional pilot, and was accredited with eight confirmed aerial victories, he was a superb marksman and was one of the first aces. He retired from the Royal Air Force as a Group Captain in 1931. Lieutenant Colonel Lionel Wilmot Brabazon Rees died in the Bahamas in 1955.

2nd Lieutenant Walter Horace Carlyle Buntine MC

Walter ('Lyle') Horace Carlyle Buntine was an Australian, born in Toorak, Victoria, in 1895, the son of the headmaster of the Caulfield Grammar School in Melbourne. He was a medical student at Melbourne University when he enlisted in the Australian Army Medical Corps on 15 May 1915, leaving Melbourne for overseas on 17 July. He applied for a commission in the British Army and became a 2nd Lieutenant in the Nottinghamshire and Derbyshire Regiment (Sherwood Foresters). Transferring to the RFC, he qualified as a pilot and was awarded Royal Aero Club Certificate 2982 on 16 May 1916. In July he was posted to 11 Squadron, flying FE2bs, who were based at Savy in France, and who saw heavy combat in the skies over the Somme Battlefield. On 9 September 2nd Lt. Buntine was flying FE2b 6988, with Sgt G. J. Morton as observer, on a bombing escort mission to Iries. They were engaged in combat with enemy aircraft over Pommier and were credited with destroying an LVG C type near Achiet-le-Grande. Walter was wounded by a bullet in the arm in this action, for which he was awarded the Military Cross, the citation reads:

> In recognition of his gallantry and devotion to duty in the Field; —2nd Lt. Walter Horace Carlyle Buntine, Notts. & Derby. R. and RFC. For conspicuous gallantry and skill. As escort to a bombing raid he attacked several hostile machines, one of which fell to the ground nose first. Later he was attacked by three enemy machines, his own machine being damaged and himself severely wounded. With great skill he managed to land in our lines, though most of his propeller was shot away and his machine otherwise much damaged.

He returned to Australia to convalesce and was back in the UK by late April 1917 and was posted as an Instructor to 2 (Aux) School of Aerial Gunnery at Turnberry.

Walter Buntine was killed at Turnberry on 19 June 1917 while employed as an observer in FE2b A817, Pilot Sgt. S. C. Appleton, when the aeroplane stalled on a turn after take-off, crashed at Dalquhat farm, and then caught fire after hitting the ground; both airmen were killed in the accident. A Court of Inquiry report recorded:

> Accdt. due to stalling on turn at a height insufficient for recovery to be made from ensuing nosedive. At the time of the accdt. the machine had not reached the spot apptd. for practice to be carried out in - were gaining their height while on journey out. Machine caught fire on striking ground.

2nd Lieutenant Walter Horace Carlyle Buntine MC is buried at Girvan (Doune) Cemetery, and is remembered on the Turnberry War Memorial.

Sergeant Stanley Chalmers Appleton 3661

Stanley Chalmers Appleton was born in 1895 in Warrington, Lancashire, the son of James and Mary Appleton. Stanley was a plumber when he joined the Royal Flying Corps in February 1915 at Farnborough. He was attested Air Mechanic 2nd Class and had been promoted to Sergeant by April 1916, obtaining his Royal Aero Club Certificate in November on a Maurice Farman Biplane at the Military School, Brooklands. Posted to 19 RS, Sergeant Appleton was injured whilst flying as a passenger in FE2b 4930 flown by Sergeant G. Finerman, when after engine failure, he tried to land in field but struck a tree. Graded a 1st Class Pilot by February 1917, Sergeant Appleton was posted to Turnberry for gunnery training when he was killed flying Royal Aircraft Factory FE2b A817 on 19 June, west of Dalquhat farm, Kirkoswald. 2nd Lieutenant Walter Buntine MC, Observer, was also killed. Sergeant Stanley Chalmers Appleton is buried at Warrington Cemetery and is remembered on the Turnberry War Memorial.

2nd Lieutenant George Barclay Buxton

Born in October 1892, the son of missionary, Rev Barclay Buxton, who worked in Japan. George returned to England at the turn of the century and was educated at Repton, subsequently working in a London bank, before becoming assistant manager of the East African Industries agricultural estate at Maseno in British East Africa in 1912. He was a deeply religious man and decided to become a missionary and join his elder brother Alfred at a mission in the Belgian Congo in early 1914.

War intervened, and cancelling his trip to the Congo, Buxton volunteered for service in the King's African Rifles. He spent from mid-August until October on scouting duties, taking part in several expeditions along the frontiers of British East Africa. In July 1915 he returned to England and volunteered for military service there.

In the late summer of 1915 Buxton was granted a commission in the British Army and trained at Tring in Hertfordshire. On completion of his training in January 1916 he was posted to Egypt to join the 5th Norfolk Regiment. He only served with this unit for a few weeks and in April he was appointed ADC to the divisional commander, Major-General Sir Steuart Hare. Feeling he wanted to do more for the war effort and finding his post was 'not exacting', he volunteered for service in the Royal Flying Corps in 1917.

Completing a three-week course in the theory of flying at the Flying School at Abbassia, Egypt, in March, he was then posted to Ismailia for flying training. He returned to England in May and by July had qualified as a pilot of scout aircraft, after further training in Lincolnshire and at the School of Aerial Gunnery at Turnberry. He was posted to 1 Squadron in France stationed at Bailleul.

On one of his first patrols on 25 July 1917, his aircraft was 'absolutely riddled with bullets' although he was not wounded, but on 28 July he was shot down and killed flying Nieuport 17 A6680. He was last seen 'being driven down over Zonnebeke with 4 enemy Albatros scouts on his tail.'

2nd Lieutenant George Barclay Buxton has no known grave and is commemorated on the Arras Flying Services Memorial, Pas de Calais, France.

2nd Lieutenant Cyril Ashley Cooper

Born in 1893, the eldest son of Ashley Cooper and Henrietta Sharp of Winchmore Hill, London, Cyril was educated at Owens School, Islington. Leaving his employment as an insurance clerk, he enlisted in the Army Service Corps on 14 August 1914 and served with the Expeditionary Force in France and Flanders. He was commissioned 2nd Lieutenant in the East Yorkshire Regiment in France and upon returning to England he was attached to the Royal Flying Corps in August 1916, obtaining his wings in April 1917.

He was killed on 29 June 1917 in a flying accident at Maidens whilst acting as an instructor. He left a widow, Kathleen A. Cooper (nee McIntosh), who he had married in January 1917. On the reverse of the Casualty Card the name and address of his next of kin is given as Mrs K. Cooper, Maidens, Ayrshire, which would suggest that she was living there at the time of his accident.

2nd Lieutenant Cyril Ashley Cooper is buried in Doune Cemetery, Girvan, and is remembered on the Turnberry War Memorial.

Air Mechanic 2nd Class Harry Towlson 35101

Harry Towlson was born in 1887 at Basford, Nottinghamshire, the son of William Towlson and Sarah Clarke. He married Florence Chilton in 1914 and joined the RFC as an Air Mechanic 2nd Class in 1916 stating his occupation as Wheelwright.

He died at Turnberry in an aeroplane accident on 29 June 1917 when flying as a passenger in FE2b 4926 flown by 2nd Lieutenant Cyril Cooper. He is buried at Basford Cemetery, Nottinghamshire, and is remembered on the Turnberry War Memorial.

Major James 'Stuffy' William Gordon

Born in 1888 in Cawnpore, India, the son of Glasgow engineer, James T. Gordon, James William Gordon had achieved a BSc in Engineering and was working as a civil engineer when he joined the RFC in 1915. Completing his flying training he was awarded his Royal Aero Club Aviators' Certificate on 15 August at the Hall School, Hendon. He was posted to 47 Squadron in Salonika in July 1916 but was invalided in October, suffering from malaria and dysentery. James William Gordon was awarded the Croix de Guerre on 1 May 1917, and arriving back in Britain in that month he was graded P.U. (permanently unfit) for General Service and as suitable for H.S. (Home Service). He arrived at Turnberry as Wing Instructor on parole in June 1917 and was promoted to Chief Gunnery Instructor by mid-December, staying at Turnberry until March 1919.

Major Gordon was mentioned in the Air Ministry List of 22 January 1919 for 'valuable services rendered in connection with the war'. He was transferred to the unemployed list on 6 April 1919, and was awarded the OBE (Military Division) in June 1919.

After WWI, James William Gordon, OBE, became the Manager of the Jodhpur State

Dr. Gardener and Major Gordon

J.W. Gordon

Railway, Rajputana in India, and he was made a Companion of the Indian Empire (CIE) in the 1936 New Year Honours.

2nd Lieutenant Henry Samson Wolff

Henry Samson Wolff was born in 1898 in London and in 1916 joined the Inns of Court Officers Training Corps as soon as he left school, transferring to the Royal Flying Corps a few weeks later. After initial training at Reading and Thetford where he received dual instruction on the Maurice Farman Shorthorn, going solo after four hours. He was then posted to Sedgeford for higher training advancing on to the Avro 504K including aerobatics and cross country flying. Wolff then completed his aerial gunnery course at Turnberry before being posted with a total of eight hours solo flying time, to 40 Squadron arriving at the end of September 1917.

Henry Wolff was said to be only 5-foot-tall, and he required his seat to be packed with cushions to enable him to see over the cockpit of his aircraft. On the squadron he was nicknamed 'The Mighty Atom' and 'Little Samson'. He started flying line patrols in October and was flying his first SE5a operational patrol with Flight Commander Lt. W. MacLanachan and Lt. R. E.

Boin when they spotted an enemy 2-seater. Not having the experience to realise that the 2-seater was a decoy and that there were 12 enemy aircraft higher up, Wolff attacked the 2-seater, at which point the Albatros scouts dived to attack, engaging Lt. W. MacLanachan and Lt. R. E. Boin. Busy concentrating on the decoy, Wolff pushed home his attack and sent it down in a spin after a burst from his Lewis gun. He then joined the fight with Lt. MacLanachan and Lt. Boin. After landing, Wolff discovered several bullet holes in his top wings, some only inches from his reserve fuel tank. He admitted later that he had been so intent on bringing down the decoy that he hadn't seen any other enemy aircraft.

On another patrol he managed to shoot his own propeller off, due to his Constantinesco interrupter gear not being synchronised properly. He had to shut down his engine and gliding through heavy A.A. fire managed to crash land in the British trenches and made his way back to the Squadron (where he had been posted as missing), courtesy of a lift in a tank.

On 28 January 1918 2nd Lieutenant Wolff, while on patrol, spotted a formation of six enemy scouts. He attacked a two-seater over Gondecourt, firing several bursts at point-blank range and tracers were seen to enter the fuselage. The enemy aircraft was seen to turn completely over and dive very steeply of control, he closely followed it down but was forced to break off the combat, owing to being by attacked by another two enemy scouts.

Lieutenant Wolff claimed another Albatros Scout as sent down out of control at Pont-à-Vendin on the 9 March.

By 28 May, after completing nine months with 40 Squadron and officially credited with five victories, he was transferred back to England diagnosed and hospitalised suffering from 'Flying Sickness' (thought to be caused by differences in air pressure and temperature, which led symptoms such as severe headache, pounding ears, and breathing difficulties). This could also be described as Neurasthenia which was a nervous condition brought on by fatigue and mental strain.

After four weeks leave, he was posted to 49 TDS at Doncaster as an instructor. He spent the next few months here until he was scheduled to return active duty in France just as the war ended.

2nd Lieutenant Henry Samson Wolff was discharged from the RAF in March 1919 and worked in advertising, enlisting again in the RAF at the outbreak of WWII. He served as a staff officer at 22 Group H.Q. and was promoted to Wing Commander. Henry Samson Wolff died in 1972 in Henley Oxfordshire.

Captain Edmund Leonard Zink

Edmund Leonard Zink was born in Maida Vale, London, in 1899, the son of artist George Frederick Zink. He was commissioned temporary 2nd Lieutenant in the Suffolk Regiment in 1915. Giving his year of birth as 1897, to enable him to qualify for acceptance into the RFC in 1916, he received his pilot's certificate on 27 September, and was posted to 18 Squadron on 16 November. 2nd Lieutenant Zink claimed four victories whilst flying an FE2b.

On the evening of 23 April 1917, Leutnant Hermann Goring of Jasta 26, flying with other Albatros Scouts, came upon four FE2b's of 18 Squadron, employed on a bombing raid. He wrote, 'Aerial fight with British two-seater, from a squadron of four English. I set on fire an opponent after a short fight with PH ammunition. Burning crash north east of Arras. Shortly afterwards two more aerial fights with a two-seater and a single seater without result'.

The 'burning crash' Goring refers to was FE2b A823 flown by 2nd Lieutenant Zink with observer 2nd Lieutenant George Bate. With the engine on fire the wounded Zink managed to keep flying and heading for the safety of their lines while Bate fought of Goring and tried to prevent him finishing them off. 2nd Lieutenant Zink made it back safely and was thereafter sent back to Britain to recover from his wounds. 2nd Lieutenant George Bate was unharmed, although a bullet slicing through his leather jacket and uniform. (He was later killed in action on 29 Apr 1917 and is buried in the Queant Road Cemetery, Buissy, France.)

From June 1917, Edmund Zink then served as an Instructor at Turnberry and then Ayr, before joining 32 Squadron as a Flight Commander in August 1918, shortly becoming an ace, scoring his fifth and final victory whilst flying a SE5a North of Equancourt.

He was granted a short service commission in the RAF with the rank of Flying Officer, effective 24 October 1919 which he resigned in November 1920. Edmund Leonard Zink died in1921 Paddington, London, England of influenza.

Air Mechanic 2nd Class Edward Ernest Hall 63054

Edward Ernest Hall was born in 1887 at Cramlington, Northumberland, the son of Edward and Jane Hall. He was a driver of a motor lorry prior to the war and he joined up in December 1915. He served with the Army Reserve and was attached to the RFC in March 1917.

Air Mechanic 2nd Class Hall was posted missing from 2 (Aux) School of Aerial Gunnery, Turnberry, on 20 July 1917. His body was found five days later on the rocks south of Turnberry Lighthouse. The cause of death was 'Accidental Drowning'. He is buried at Buried at Cramlington Old Burial Ground, Northumberland, and is remembered on the Turnberry War Memorial.

Stuart Charles Sillem

2nd Lieutenant Stuart Charles Sillem

Stuart Charles Sillem was born in Kensington, London, in 1898. Leaving Eton College, Sillem joined the RFC in October 1916. After training at Turnberry, he was posted to 27 Squadron. On 12 August 1917, while flying Martinsyde A1573 on a bombing mission over Mouveaux aerodrome, France, he was brought down in flames north of Lille during combat with Hauptmann Otto Hartmann, a German Ace of Jasta 28. 2nd Lieutenant Sillem, killed in action, aged 19, has no known grave and is listed on the Arras Memorial.

Captain Frederick Lawrence Bristow *Gunnery Instructor*

Frederick Lawrence Bristow was born in 1889 in Surrey, the son of a Flour Miller. Transferring from the Royal Garrison Artillery, Captain Bristow was employed as a Gunnery Instructor on the Vickers and Lewis Guns at Turnberry from May until November 1917. He was then posted to the School of Aerial Gunnery and Fighting at Leuchars as Chief Gunnery Instructor until the end of the war when he was transferred to the unemployed list in March 1919.

Captain John 'Jack' Alfred Hone *Gunnery Instructor*

Born on 28 Feb 1895 at Tewkesbury in Gloucester, the son of George Hone and Sarah Edgell, John Alfred Hone was educated at Warminster Grammar School. He was commissioned as a Lieutenant in the 9th Battalion, Gloucestershire Regiment, and transferred to the RFC in May 1917. After receiving instruction as an observer and attending a course at 2 (Aux) School of Aerial Gunnery at Turnberry, Lieutenant Jack Hone was posted to 20 Squadron as an observer in July 1917.

On 17 August, patrols of 20 Squadron engaged several enemy aircraft formations, during which Lieutenant O. H. D. Vickers & Lieutenant J. A. Hone in FE2d A6376, drove down four Albatros Scouts out of control over Halluin within the space of five minutes. Their own aircraft engine and tanks was badly shot through and they had to force land.

Returning to England in February 1918, Jack Hone was posted to Uxbridge for a course for Gunnery Instructors and was then posted to Turnberry in April 1918 for the rest of the war. He was transferred to the Unemployed List in April 1919. Returning to civilian life he was a Stockbroker on the London Stock Exchange. John 'Jack' Alfred Hone died in 1978 in London.

Lieutenant Joseph Wesley Lightbourne Birkbeck *Squad 7 July 1917*

Joseph Wesley Lightbourne Birkbeck was born in 1895 in Shipston on Stour, Warwickshire. Prior to the war he was a student at Birmingham University before being commissioned as 2nd Lieutenant in the 15th battalion of the Durham Light Infantry. Seconded to the Royal Flying Corps in July 1916, Birkbeck trained at Reading before completing the Turnberry Course in July 1917 and was then posted to 27 Squadron who were flying missions in support of the British offensive at Cambrai.

Returning to Home Establishment on 6 February 1918, he was graded as suitable as an instructor in day bombing and returned to Turnberry with the rank of Flight Captain until he was transferred to the Unemployed List on 28 March 1919, having been awarded the Air Force Cross in the New Year's Honours of 1919.

After World War I, Joseph Wesley Lightbourne Birkbeck was a Master at Queen Elizabeth's Grammar School, Hartlebury, Worcestershire. During World War II Birkbeck served in the RAFVR in Administrative and Special Duties Branch (Intelligence).

Lieutenant Charles Ernest Ottiwell 'Otto' Cowell *Squad 7 July 1917*

Born in 1891, Charles Ernest Ottiwell Cowell, an estate agent's clerk, transferred to the RFC on 16 October 1916, from the 4th Battalion East Lancashire Regiment. After initial training at Reading, he was appointed 2nd Lieutenant on 12 June 1917 and posted to Squad 7 at Turnberry in July.

Lieutenant Cowell was found unfit for General Service for one month from 5 November 1917, then gradually allowed limited flying, 'no stunting' and then served as a ferry pilot, again 'no stunting' allowed.

Since his entry into the Royal Flying Corps he had flown numerous types, such as Maurice Farman Shorthorn, Avros, BEs, RE8s, Bristol Scouts, Sopwith Scouts, Martinsydes and SPADs.

From his service record Ernest spent most of his time after initial training with 11, 15 and 56 Training Squadrons, it wasn't unusual for talented individuals to be 'creamed off' to instruct.

After the war, Cowell continued flying when on 19 March 1920 he was injured in an accident at Twickenham.

A notice of an accident involving him appeared in the *Birmingham Dailly Gazette* on 20 March 1920:

An aeroplane from Northolt was passing over Twickenham yesterday afternoon about half-past three when the engine stopped, and the machine crashed into the trees at Marble Hill Park. The pilot, Otto Cowell, who belongs to Manchester, was injured and was taken to Twickenham Hospital, and did not recover consciousness for some hours. John Edge who was travelling with him, was not so seriously injured, and he was taken to Richmond Hospital.

Followed by the *Portsmouth Evening News*, 23 March 1920:

The two airmen, Otto Cowell and Joseph [*sic*] Edge, who were injured in an aeroplane accident at Twickenham last Friday, continue to make good progress.

His leaving hospital was reported in the *Edinburgh Evening News*, 31 March 1920:

Otto Cowell, the Airman whose machine fell in Marble Hill Park, has recovered

sufficiently to leave the Twickenham Hospital. When over Twickenham his engine failed and seeing Marble Hill Park, he steered for it. As he approached, he saw that there were a number of children playing in the large open space, and that if he came down there, he would injure some of them. He therefore turned the machine into the trees, realising as he did so that it meant almost certain death. Fortunately, neither he nor his observer was hurt as much as at one time feared. Cowell has been flying for seven years, and this was his first accident.

Between the wars Charles Cowell worked as a film agent for Warner Brothers. Further military service followed in the RAFVR (Training Branch) and he was for many years Commandant of the ATC, Christchurch, Bournemouth. Charles Ernest Ottiwell 'Otto' Cowell died on 28 Sep 1955 in Bournemouth, Hampshire.

2nd Lieutenant Geoffrey Clapham

Born in Keighley, Yorkshire, in 1896, son of John Clapham and Marian Weatherhead, Geoffrey Clapham was employed by the Blackburn Aeroplane Company of Leeds before enlisting in the RFC as a 3rd Class Air Mechanic. He was discharged in May having been selected for appointment to a temporary commission as 2nd Lieutenant. After training, appearing in Squad 7 at Turnberry in July 1917, 2nd Lieutenant Clapham was posted to 54 Squadron in France.

In August, flying Sopwith Scout B1783 on landing from offensive patrol, his aircraft overturned, he was unhurt, a month later he is again mentioned in Communiqués when his Sopwith Scout B1799 was shot through by enemy aircraft while on offensive patrol.

On Monday, 28 January 1918, 2nd Lieutenant Clapham dived on one of three enemy scouts, the Albatros went down out of control and was later seen to crash near Hancourt.

Transferring to 78 Squadron in February 1918, Geoffrey was promoted to Captain in September. He was then posted to 2 School of

Navigation and Bomb Dropping at Andover. He was awarded the Air Force Cross in February 1919, and was transferred to the Unemployed List in July 1919.

By 1922 Geoffrey Clapham was racing motorcycles, entering the Senior T.T. 500cc Race on the Isle of Man, a race of 6 laps (226.5 miles). A cycling report appeared in the *Yorkshire Post* (Motor Cycling News) in April 1922:

> The Ilkley Motorcycle Club has again nominated Geoffrey Clapham as a rider of a Scott in the senior event and the Scott Co. has entered three Squirrels.
> 1. Alec Bennett (Sunbeam), 2. Walter Brandish (Triumph), 3. Harry Langman (Scott), 4. C. P. Wood (Scott), 5. Graham Walker (Norton), 6. Tommy de la Hay (Sunbeam), 7. A. H. Alexander (Douglas), 8. V. Olsson (Sunbeam), 9. Geoffrey Clapham (Scott), 10. J. W. Adamson (Norton)

In 1924, after a spell as a civilian pilot in Brazil, Geoffrey joined the RAF Reserve, taking a re-qualifying course at the William Beardmore Civilian Flying School, where he qualified with the comments, 'A very keen and capable pilot.'

Geoffrey Clapham married Christine Winifred Booth in 1928, shortly before taking up a post as an aviation instructor in Bolivia. Arriving back in Britain, having spent two years flying civilian aircraft in South America in and around the Amazon. He became Chief Flying Instructor of Hooton (Liverpool & District Aero Club) Flying Club a position he held for many years.

In June 1932, flying a Comper Swift, Geoffrey attained second place in the air race around the Isle of Man, recording the fastest speed in the race of 126mph. In June the next year, a 'Great Air Pageant' was held at Speke near Liverpool with over two hundred aircraft taking part, including Lord Malcolm Douglas-Hamilton (who was later tragically killed with his 21-year-old son, Niall, in an airplane crash in Cameroon, West Africa in 1964) with his 'Monster Amphibian Plane' in which he was

giving pleasure flights, landing on the Mersey. Flight Lieutenant Geoffrey Clapham, 'a well-known Merseyside pilot gave a "crazy-flying" display.' That same month Geoffrey became a father with the birth of his son Crispin.

With war approaching, a Civil Air Guard was established in 1938, with the intention of providing pilots to assist the Royal Air Force in a time of emergency. Liverpool had its own C.A.G. Clubhouse where Geoffrey lectured on 'The Sequence of Flying Instruction.'

In 1939 Geoffrey was a 'Flying Officer (RAF) Awaiting Posting', however no further service records can be found.

Geoffrey was the landlord of the 'Ye Old Vicarage', Holmes Chapel, in Cheshire in 1954, losing his 20-year-old son Crispin in a motorcycle accident that same year. Flying Officer Geoffrey Clapham died in Cheshire in 1967.

Lieutenant Ronald Hibbert Cross *Squad 7 July 1917*

Ronald Hibbert Cross was born in 1896, and in September 1914 became an officer in the Duke of Lancaster's Own Yeomanry, before transferring to the RFC with the rank of 2nd Lieutenant, then Temporary Lieutenant as of 3 December 1915. He appears at Turnberry on Squad 7, July 1917, before being posted to 23 Squadron on 18 August, flying DH2's and SPADs both on offensive fighter patrols over the front and low-level strafing attacks against German troops.

According to the Squadron diary:

> Lt. Cross, 23 Squadron, attacked an enemy balloon south of Houthulst Forest. He was then attacked by EA so dived into a cloud, but the balloon was seen by another pilot to burst into flames.

After being admitted to hospital on 31 October, Ronald Hibbert Cross was struck off charge with 23 Squadron and posted to 6 Brigade Headquarters.

After the war he became a merchant banker and was elected Unionist M.P. for Rossendale in Lancashire, in 1931. By 1938 he had risen to become Parliamentary Secretary of the Board of Trade, and in May 1939 was appointed the Minister of Economic Warfare. This was followed by Minister of Shipping in May 1940, then in 1941 he was appointed as High Commissioner in the Commonwealth of Australia for H.M.'s Government in the United Kingdom, which post he held until 1945. He was defeated at Rossendale in the Election of June 1945 but was again elected to the House of Commons by being elected the Conservative M.P. for Ormakirk, Lancashire, 1950, which he resigned a year later when he was chosen to be the Governor of Tasmania. He was appointed KCVO (1954) and KCMG (1955). He retired and returned to England in 1958. Mount Ronald Cross in western Tasmania is named after him. Sir Ronald Hibbert Cross, 1st Baronet, KCMG, KCVO, PC, died in London in 1968.

Lieutenant Oswald Douglas Hay *Squad 7 July 1917*

Born on 31 July 1894 at Peterhead, Aberdeenshire, Oswald Douglas Hay was commissioned 2nd Lieutenant in the Gordon Highlanders in 1915 and was seconded to the Royal Flying Corps in March 1917. A short article appeared on 13 June 1916 in the *East Aberdeenshire Advertiser*:

> Four Brothers Serving - All four sons of Mr William Hay, coal merchant. Maiden Street, are with the Colours. In the 'London Gazette' of Friday last, his youngest son, 2nd Lieutenant Oswald Douglas Hay, 3/5th Gordon Highlanders, was announced as having been promoted Lieutenant April 24th last. Robert Stephen Hay, an elder son. has been transferred from the 3/6th Seaforth Highlanders and appointed a machine gun officer; Stephen Hay has received his commission dated 3rd June, as a Lieutenant in the Royal Flying Corps; and the eldest son William, who has been in business in the Straits Settlements, is in training with the

Volunteers at Kuala-Lumpur. A creditable family record!

Completing the Gunnery Training at Turnberry, in which he was graded 1st Class, Lieutenant Hay was posted to 66 Squadron which deployed over the Somme in June 1917, his special duties listed as Top Scout.

On 15 August, flying Sopwith Pup A6212, he was wounded, and force landed at 11 Sqn aerodrome after a longeron was shot through on offensive patrol. Struck off strength of the Expeditionary Force, Lieutenant Hay after recovery was posted as Assistant Instructor at 34 Training Squadron and was placed on the Unemployed list in April 1919.

After demobilization Oswald Douglas Hay returned to Aberdeen where he entered the fish curing business. He died on 6 Jun 1922 at Bucksburn, Aberdeen in a motor car accident.

2nd Lieutenant Ernest Barnes-Hedley *August 1917*

Ernest Barnes-Hedley was born in 1886 in Southwark, Surrey. After serving an engineering apprenticeship in Ordnance with Vickers Ltd., he emigrated to Australia. Ernest joined the Australian Imperial Force on 22 August 1914 at Sydney and was appointed to the Machine Gun Section of the 1st Battalion, 1st Infantry Brigade. Posted to Ismailia in defence of the Suez Canal, by April 1916 he had been promoted to Company Sergeant Major. In November 1915 he served at Gallipoli, and by March 1916 was in France. He was mentioned in dispatches by Sir Douglas Haig, 'For distinguished and gallant services and devotion to duty'. After serving two years and 226 days with the Machine Gun Corps, on 30 October 1916 Ernest Barnes-Hedley joined the Royal Flying Corps and was posted to Harlaxton in Lincolnshire for flying training before arriving at Turnberry. Posted to France with 6 Squadron in July 1918, that same month he was hospitalised in The Duchess of Westminster's Hospital for Wounded Servicemen, in the casino at Le Touquet. He returned to England classed unfit for general service and sent on leave for four weeks before resuming flying duties with 6 Squadron.

Ernest Barnes-Hedley died in September 1948 in Sydney, New South Wales.

2nd Lieutenant Samuel Leslie John Bramley

Born in 1895 in Romsey, Victoria, Australia, son of Harry and Elizabeth McAlpine, Samuel was an electrical fitter when he joined the 3rd Australian Pioneer Battalion, AIF, in February 1916 as a Signaller. He was discharged in January 1917, having been appointed to a Commission in the RFC, and was posted to Exeter College of Instruction, the School of Military Aeronautics, Oxford, and finally 2 (Aux) School of Aerial Gunnery at Turnberry in August 1917.

2nd Lieutenant Bramley was posted to 57 Squadron in France on 1 September and was reported missing on 23 September, over Roulers, flying DH4 A7643, while returning from a bombing raid on Hooglede. His observer, 2nd Lieutenant John Matthew de Lacey, was also posted as missing. German Ace, Leutnant Werner Voss from Jasta 10 was credited with the victory, his last, before he was shot down later that afternoon by Lieutenant Arthur Rhys Davids.

2nd Lieutenant Samuel Leslie John Bramley is buried in Harlebeke New British Cemetery, Belgium.

2nd Lieutenant Edward Watson (Ted) Powell

Edward Watson Powell was born in 1892 in Melbourne, Australia. A clerk in civilian life, Powell enlisted in August 1914 and served with the 1st Australian Division at Gallipoli and was attached to the Royal Flying Corps in March 1917. Upon completing training at Turnberry, Powell was posted to 84 Squadron flying SE5a's and was sent to France on 23 September 1917.

He was killed in action a little over a month later, on 31 October 1917, flying SE5a B4874, last seen between Roulers and Menin.

2nd Lieutenant Edward Watson Powell has no known grave and is commemorated on the Arras Flying Services Memorial.

2nd Lieutenant Frank Cecil Ransley *August 1917*

Born in 1897 in Caversham, Oxfordshire (now in Berkshire), Frank joined the Royal Garrison Artillery as a gunner in 1914. Transferring to the Royal Flying Corps in May 1917 he was commissioned as 2nd Lieutenant. After training at Turnberry, he was appointed Flying Officer and confirmed in his rank on 29 September. Posted to 48 Squadron flying the Bristol F2B two-seater fighter, by 15 May 1918 he was promoted to Flight Commander with the temporary rank of Captain. Captain Frank Ransley became an Ace with nine victories by June 1918 and was awarded the Distinguished Flying Cross:

> This officer displays conspicuous gallantry and skill. On a recent occasion, while on patrol he was attacked by seven enemy scouts; he and his observer drove down two, and by skilful manoeuvre and dash he rallied his formation, which were being driven down, and succeeded in driving off the remaining enemy scouts. He has, in all, destroyed three hostile machines and driven down three others completely out of control.

Captain Frank Ransley was also awarded the Croix de Guerre with Bronze Star for his service in France.

Leaving the Royal Flying Corps in April 1919, Frank Ransley joined the Prison Service and was Governor of various institutions such as Cardiff, Wakefield and Wandsworth prisons. He was made an OBE in 1958.

Captain Frank Ransley died at the age of 95 in 1992, at which time he was one of the oldest surviving World War I aces.

Lieutenant Francis James Scott *August 1917*

Born in Korumburra, Australia, in 1893. Prior to joining up with 2 Australian Flying Squadron in September 1916, Francis was a Grazier on Goree Station, New South Wales.

On 4 October 1916 Francis was appointed L/Cpl for active service abroad. On arrival in Plymouth just before Christmas he was posted to 69 Australian Squadron, Royal Flying Corps.

After training at Northolt and Reading, and in August 1917, Turnberry, Francis Scott was promoted to 2nd Lieutenant after which he was sent to 4 Squadron, Australian Flying Corps, with the rank of Lieutenant and proceeded to France on 16 December 1917.

After four months of intense fighting, Lieutenant Scott was hospitalised at Etaples, suffering from Neurasthenia (nervous exhaustion) and was transferred back to Britain and after a spell in Shirley RAF Hospital, where he spent months convalescing before returning to Australia in August 1918.

He was awarded the Military Cross, the citation reads:

For conspicuous gallantry and devotion to duty after carrying out a bombing attack, he observed an enemy plane – into which he fired 80 rounds. The hostile plane turned on its back and crashed to earth. Four days previous to this he had destroyed an enemy two-seater machine and had sent down out of control a hostile scout. In addition to these, he has destroyed another two-seater machine. He has displayed marked courage and determination.

Lieutenant George Guy Barry Downing

Born in 1893, George Guy Barry Downing was the third and youngest son of Mr and Mrs G. C. Downing, of Beverley, Llanishen, Glamorgan. He was schooled at Waynflete, near Reading, and at Charterhouse before studying art at the Slade School in London. He was a very talented artist and a number of his drawings had been reproduced in the illustrated magazine *Colour*.

In August 1914, he joined the Old Public School Boys' Training Camp at Tidworth Pennings in Wiltshire, was recommended for a commission, and posted to the Welsh Regiment.

George applied to be transferred to the RFC, but before this was approved, he was posted with his regiment to France, and was wounded on 25 September 1915, at the Battle of Loos. Having recovered from his wounds he was subsequently transferred to the RFC, and after completing training, in July 1916 was sent to the front. In September his aircraft was seriously damaged by German anti-aircraft gun fire while flying over the German lines, swiftly followed by the attack of two enemy aircraft, but he managed to fend them off and to land safely within the Allied lines. He was sent home to recover from his wounds but was then incapacitated with a severe attack of diphtheria.

By March 1917 he was certified fit for light duty and was employed as an instructor at various aerodromes in the UK. He received orders to return to the front, but these were countermanded, and he was posted to the training school at Turnberry as an instructor. He had only been there a few days when he was killed, in an accident whilst flying BE2c 5416. 2nd Lieut. E. Allman, Observer, was injured. He left a widow and baby son.

Lieutenant George Guy Barry Downing was buried with full military honours in St Isan cemetery, Llanishen, Glamorgan, and 'a very large congregation from the village and the neighbourhood gave expression to the true sympathy that goes out to Mr & Mrs Downing and Mrs Barry Downing in their great sorrow.' He is remembered on the Turnberry War Memorial.

2nd Lieutenant Indra (Laddie) Lal Roy DFC

Born in Calcutta on 2 December 1898, the son of Piera Lal Roy, eminent barrister and director of public prosecutions in Calcutta, Indra Lal Roy moved to London in 1901 to better his English education at St. Paul's School for Boys in Hammersmith. Indra was fifteen when war was declared and immediately signed up for the school cadet force at St Paul's. He had won a scholarship to Oxford, but so desperate was he to become a fighter pilot 'Laddie' applied to the Royal Flying Corps but failed the medical examination because of 'defective eyesight'. Undaunted, he sold his beloved motorcycle to raise funds to pay for a second opinion from a prominent eye specialist. This time he passed the sight test, and the initial decision was reversed.

Immediately thereafter, on 4 April 1917, he signed up for the Royal Flying Corps. By July, the eighteen-year-old was commissioned as a 2nd Lieutenant and within a few days Indra Lal Roy was training at Vendome in France. This was followed by completing a course at Turnberry before being posted to 56 Squadron at the end of October 1917.

In early December 1917, 2nd Lieutenant Indra Lal Roy was shot down by a German fighter and crashed in no-man's land. His body was removed from the wreckage and taken to the local hospital where he was pronounced dead.

But not happy with this verdict, Indra regained consciousness, and attracted the attention of the horrified hospital staff by banging on the door of the mortuary. He was immediately sent back to Britain for treatment. On recovering from his injuries, Indra was declared no longer fit to fly. But again, he disagreed, he was determined to return to flying. Laddie continued to plead his case to be allowed back in the air until his officers finally gave in. In June 1918 he passed a further medical test and was posted to 40 Squadron as Temporary Lieutenant and sent back to France. By early July he had brought down his first enemy aircraft over Arras in Northern France, quickly followed over the next few days by another nine German machines.

Towards the end of July 1918, 'Laddie' flew a daring mission over the trenches in Carvin in France. Carvin at that time was occupied by the Germans as a garrison town for both German military and civilians situated at the rear of the front line. He suddenly found himself surrounded by Four German Fokker aircraft who attacked, and a 'dog fight' followed. Putting up a brave fight, Laddie shot down two German planes one after the other, before his own aircraft was hit by gunfire, bursting into flames and coming down over German-held territory. At the age of nineteen, it was to be his final flight.

In September 1918, the RAF officially confirmed his death and he was posthumously awarded the DFC. The citation reads:

A very gallant and determined officer, who

in thirteen days accounted for nine enemy machines. In these several engagements he has displayed remarkable skill and daring, on more than one occasion accounting for two machines in one patrol. (Supplement to the *London Gazette*, 21 September 1918)

Flight Magazine published his obituary: 'He was one of a band of young Indians studying here who, precluded until recently from any chance of obtaining commissions in the Army, found scope for striking a blow for the Empire in the new arm of our forces.'

The famous Red Baron, the German flying ace, even paid tribute to the brave young Indian and dropped a wreath from the skies on the spot where 'Laddie' fell.

2nd Lieutenant Indra Lal Roy was India's only officially accredited air ace of WWI, achieving ten kills prior to his death. He is buried in Estevelles Communal Cemetery in France.

Lieutenant Evelyn Alexander Wilson 'Pappy' Ffrench *Squad 10 September 1917*

Born in Ballarat, Australia, in 1878, the only son to Acheson Evelyn and Marion Wilson Ffrench, Evelyn Ffrench joined the Legion of Frontiersmen just before the Boer war, and later the 4th Victoria Imperial Bushmen in 1900. This unit departed from Melbourne on 1 May 1900 in the troopship *Victoria*, with an establishment of 629 officers and men, 778 horses and 11 wagons. They saw action at Ottoshoop, Malopo Oog, Wonderfontein, Hartebeestfontein, Uitral's Kop, Doornbult, Wolmaranstad, Harteb, Hoopstad, Zeerust, Matjesfontein, Philipstown and Read's Drift. Commissioned as a 2nd Lieutenant in the Royal Artillery on 20 November 1900, he was wounded on 8 April 1902 when he suffered two bullets through his right arm, and another grazed his foot. He resigned his commission on 2 May 1903.

At the end of the Boer War he arrived in Kensington, London. His occupations were listed as Australian Bush Poet and 'Gentleman'. He also owned the 'Imperial School of Colonial

Instruction' at Shepperton, outside London, in partnership with Mr. C. E. Morgan, for the schooling of rich immigrants bound for Australia. All aspects of 'Colonial Craft' were taught, including riding, roping, stock whip, carpentering, stock management, blacksmithing etc. Morgan and Ffrench were also the patentees of the Morgan Packsaddle and Packboat, a contraption which, 'Fits any horse, donkey, mule, camel or ox, and converts into a boat in a few minutes capable of carrying 1,000lbs over any river.'

Being a skilled rider, he was at one time called 'The World's Greatest Horseman', and had a travelling horse show. There is also reference that he produced and appeared in a film called *A Texas Elopement*. Captain Evelyn Ffrench was a rather bohemian character, six feet six inches tall, sporting a magnificent moustache and looking every inch the showman. Not only an accomplished equestrian, but a proficient author as well, he published works on horses, the frontier and poetry, using the pen name 'Jeffrey Silant', and had a collection of stories published in *The Pall Mall Magazine*, 1913, *The Badminton Magazine*, 1916, and *The Royal Magazine*, 1914-18.

Evelyn Ffrench

On the outbreak of World War I, Ffrench joined the Royal Field Artillery, serving in France with 83 Battery. He lost his right leg during the second battle of Ypres (22 April - 25 May 1915). After recovery, he applied for and was accepted into the Royal Flying Corps and was posted to 1 School of Instruction at Reading in February 1917. He received his Royal Aero Club certificate 4722 on 23 May 1917 at the British Flying School, Vendome in France. Just a couple of weeks later, on 9 June 1917, he was badly injured when the Caudron, 3087, he was flying suffered engine failure and crashed. Recovering from this, he was posted to 42 Training Squadron, based at Hounslow. Awarded his Central Flying School 'A' pilot's certificate, he joined 56 Training Squadron at London Colney. He appears in Squad 10 at Turnberry on a Bombing and Gunnery Course in July, then to the Pilots Pool at 1 Aircraft Depot at St Omer in France, where he was found by his friend Major Gregory, the Squadron Commander of 66 Squadron. When Gregory discovered that Ffrench was waiting to be posted to a squadron he requested him from the depot and sent Airman Frederick L. Burns to collect him. Once on the Squadron flying Sopwith Pups and Camels, he became affectionately known as 'Pappy' by the enlisted men, as he was the oldest active pilot to serve and fly with the unit, he became well-known and liked, wearing spurs on his boots, and talking about his aeroplanes as if they were horses! Unfortunately, he was a better horseman than pilot and he was famous on the airfield for his exceedingly bad landings, thought to be caused by his wooden leg slipping off the rudder bar when he touched down!

Evelyn Ffrench

In late 1917, 66 Squadron moved to Verona in Italy and it was here Ffrench was injured again when he made one of his poor landings on 5 December. During the following months Evelyn Ffrench did not fly many operational patrols with 66 Squadron, but he was injured again on 9 February 1918, admitted to 37 Casualty Clearing Station, and on 8 March he was embarked for 57 General Hospital (Western General, England), but disembarked on 17 June in France, graded unfit for flying. He was sent to Headquarters RAF at St Omer for administration duties.

On 17 June 1918 he was gazetted Temporary Captain and on 3 August Ffrench went before a medical board and was graded 'A' fit to resume flying. He was posted back to Home Establishment on 2 September and returned to 42 Training Depot Squadron at Hounslow.

On 23 December 1918, Captain Evelyn Alexander Wilson 'Pappy' Ffrench was killed in a flying accident at Hounslow whilst flying Sopwith Snipe E8179. He was cremated at Golders Green in London.

Lieutenant William Whitefield McConnachie
Squad 10 September 1917
Born in 1893 in Wilton, Roxburghshire, the son of schoolmaster David McConnachie, William Whitefield McConnachie joined the Royal Navy on 2 November 1914 as a Petty Officer, prior to which he was a law student. Transferring to the RNAS Armoured Car Division (RNACD), McConnachie was posted to France and later the Dardanelles. In the summer of 1915, the RNACD was disbanded and control of armoured cars was taken under the wing of the army, coming under the command of the Motor Branch of the Machine Gun Corps. McConnachie was discharged from the RNAS to allow him to take up a Commission with the Heavy Section of the Machine Gun Corps.

Transferring to the Royal Flying Corps in September 1917 as per *London Gazette* Supplement, 8 October, 'Temp. 2nd Lieut. W. W. McConnachie, MG Corps, to be transferred to RFC.'

After training at Turnberry, he was posted to 70 Squadron flying Sopwith Camels in a fighter and ground attack role over Pas-de-Calais in France.

While with 70 Squadron Lieutenant McConnachie scored 4 Victories. On 15 May 1918 flying Camel C1825, he shot down an enemy Albatros near Ervillers firing 800 rounds into it, followed by sending another down out of control over Aveluy Wood. On 21 May he shot down another Albatros Scout out of control at Henin and on 27 May, flying Camel D1825, he brought down an enemy two-seater which crashed at Hamelincourt. Although suffering damage to his own aircraft, longeron and plane shot through by enemy fire, he was uninjured.

He was posted home to 143 (Home Defence) Squadron in Detling, and then 56 Squadron at Aboukir, in Egypt. In September 1926 now Flight Sergeant William Whitefield McConnachie was transferred to the Reserve list and relinquished his Commission in 1930 on completion of service.

2nd Lieutenant Allan Adolphus Veale
Squad 10 September 1917
Allan Veale was born in 1893 in Cambridge, New Zealand. Prior to the war he was a civil engineer when he enlisted as a Private in 3rd Auckland Infantry Regiment, at the YMCA centre in Auckland, on 5 August 1914. He took

Allan Adolphus Veale

part in New Zealand's first action of the First World War as part of the Samoan Expeditionary Force in 1914, being part of the advance party that invaded the German territorial islands of Samoa, arriving at Apia, Samoa, on 29 of August 1914, only to find the German garrison deserted.

Joining the Canterbury Infantry Battalion as part of the 5th Reinforcements, Allan then left for Suez, Egypt, on 13 June 1915. He was wounded in September 1915, at Gallipoli, and after recovering in England, he returned to his battalion in France, where he was again wounded in June 1916. On recovering he was attached to corps headquarters in France, and was then nominated for a commission in the RFC

Transferring to the RFC on 1 February 1917. Allan trained at Oxford and Hendon, was promoted to 2nd Lieutenant on 5 April, before being posted to 20 Reserve Squadron at Wye where he was injured in an aeroplane accident on 26 May. Reported in *The New Zealand Herald* of 28 of August 1917:

> Lieut. Allan A. Veale, RFC, son of Mr W. Veale, Cambridge, has been declared fit for duty again after recovering from very serious injuries sustained in action in May. The injuries from which he has just recovered were very severe, his machine being crippled by anti-aircraft fire, and falling more than 600 ft out of control.

In July he continued his training, ending at Turnberry with Squad 11 in November 1917, 2nd Lieutenant Veale was posted on 30 December 1917 to 19 Squadron stationed at Ste-Marie-Cappel, France, flying the Sopwith Dolphin.

2nd Lieutenant Veale was killed in action on 22 January 1918, flying Sopwith Dolphin C3826, when the aircraft dived into the ground after the wings folded up following completion of a loop, near the Abeele-Watou road by the Belgian border. According to the *New Zealand Herald* of 2 February 1918:

> VEALE. On January 22, killed in action. Lieutenant Allan A. Veale, of RFC, Imperial Forces, only son of William and M. Veale, Hamilton Road, Cambridge; aged 24 years and 6 months.

2nd Lieutenant Allan Adolphus Veale is buried at Bailleul Communal Cemetery Extension, Nord, France.

2nd Lieutenant Eric Waterlow *Squad 10 September 1917*

Eric Waterlow was born on 4 November 1893 in Maida Vale, London. He emigrated to Regina, Saskatchewan, Canada, as a farmer in March 1913. He enlisted in the 16th Light Horse, Canadian Cavalry, and then 10th Regiment Canadian Mounted Rifles on December 29, 1914 and served as a Trooper.

Returning to Britain he joined the RFC on 19 May 1917, obtaining his Royal Aero Club Aviator's Certificate 4776 on 8 June at the Military School at Ruislip on a Maurice Farman biplane. After training at Turnberry with Squad 10 in September 1917, he was posted to 25 Squadron on 22 September.

Promoted to Temporary Captain on 1 April 1918, he was Mentioned in Dispatches on 20 May by Sir Douglas Haig. Eric Waterlow was awarded the Military Cross the next month. Recorded in the *London Gazette* of 22 June 1918:

> For conspicuous gallantry and devotion to duty. He carried out two long range reconnaissances, flying at a very low altitude, and brought back most valuable information. During one of these flights he was attacked by a hostile scout, which he destroyed. He has carried out four exceptionally long flights, during each of which he took a great number of photographs. He has always undertaken himself the longest and most arduous operations given to his flight, and by his skill, gallantry and determination has on each occasion completed his task with the greatest success.

On 16 July 1918 he was flying with fellow Canadian Lt. James Matthew Mackie DFC (originally from Muirkirk in Ayrshire) on a photographic mission to Tournai in DH4 D8380, when they were shot down over Pilckem near Ypres. Lt. Mackie was killed and Captain Waterlow was listed as missing and later confirmed dead.

He was awarded the Distinguished Flying Cross posthumously. The *London Gazette* of 21

September 1918 makes reference to this:

> This officer has carried out thirty-three bombings raids and over forty solo photographic and long distance reconnaissances far over the enemy lines. In one flight he took no less than 108 photographs. In these services he has proved himself an exceptionally skilful and resolute pilot; his railway reconnaissances have been markedly successful.

Captain Eric Waterlow, MC, DFC, is buried at the Ypres Reservoir Cemetery in Belgium as is Lieutenant James Mackie, DFC.

Lieutenant Cyril Whelan *Squad 10 September 1917*
Cyril, also known as Cyril Waxman, born in 1899 in Melbourne, Australia, received his Royal Aero Club Aviators Certificate 4824 on 16 June 1917 at Edgeware. His father was Albert Whelan, a very famous Australian music hall entertainer of the early 1900s known

Cyril Whelan

by his whistling signature tune. Cyril had worked as an actor before the war, and after obtaining his 'ticket' was posted to Turnberry in September 1917 as part of Squad 10.

Lieutenant Cyril Whelan died in a flying accident on 25 April 1918 at Kennington in Kent while acting as assistant instructor with 42 Training Squadron RAF at Wye Aerodrome.

Cyril was the pilot of an Avro 504, with Lieutenant Edmund Marrable as observer, when his machine was involved in a mid-air collision with a Sopwith Pup, also from Wye being flown

by 2nd Lieutenant Alwyne Gordon Levy. It is thought that the pilots, who were first cousins, attempted to approach and signal each other with tragic results, both aircraft came down near the Golden Ball Public House. All three officers were killed outright.

A report appeared in *Table Talk*, Melbourne, Victoria, Australia on 9 May 1918 and in the *Jewish Herald*, Victoria, Australia on 17 May 1918:

AVIATORS FALL
TWO COUSINS COLLIDE

Flight-Lieutenant Cyril Whelan (Waxman) and Flight Lieutenant Levy, killed in a collision in England, were sons of Australians well known in the literary and theatrical world – Mr Albert Whelan, now residing in London, and Mr Bert Levy, now in America. Both fathers held leading places in literary circles in Melbourne some years ago, and both have since been successful on the vaudeville stage. Mr Levy revisited Australia a few years ago, and his entertaining caricature act proved a big hit on the Tivoli circuit.

The cables give only vague details of the deaths, but it is surmised that the boys, who were first cousins, were manoeuvring at the base of the Royal Flying Corps, when they endeavoured to approach and signal each other, with disastrous results; or they may have been flying in the same machine as pilot and observer.

Much sympathy is felt for the parents, but more especially for Mr and Mrs Levy, as their son was an only child, and just turned eighteen. Lieutenant Whelan was twenty-two years of age. The boys were nephews of Councillor Joseph Waxman, the late Mr Louis Waxman, Mrs Aaron Blashki (Sydney), and Mrs Hyman White. The Levy family have had a double bereavement, for Sergeant Albert Levy, MM, who was killed on 25th March, another only son, being Mr and Mrs Jack

Levy's boy, was a cousin. Both Lieutenant and Sergeant Levy were grandsons of Mrs S. Levy, St. Kilda.

Flight Lieutenant Cyril Whelan is buried at Kensal Green Roman Catholic Cemetery, Kensal Green, London.

2nd Lieutenant Herbert Allen Yeo *Squad 10 September 1917*
Born in 1897, Herbert Allen Yeo served in the Somerset Light Infantry before transferring to the RFC in April 1917. After training at Oxford, Denham and Turnberry in September, 2nd Lieutenant Yeo was posted to 19 Squadron.

On 7 December 1917, 2nd Lieutenant Yeo was shot down whilst flying on offensive patrol in SPAD B3559, by Leutnant Max Ritter von Müller from Jasta 2. He crashed behind enemy lines at Moorslede, was taken prisoner and sent to the Landshut compound in Bavaria. He was repatriated to Britain on 14 December 1918. Herbert Allen Yeo died in December 1971 in Tiverton, Cheshire.

Major Joseph Cruess Callaghan MC
Born in 1893 in Dublin, the son of Joseph Patrick and Croasdella Cruess-Callaghan, Joseph Callaghan was living in Texas at the outbreak of World War I. He returned to Britain and was commissioned as a 2nd Lieutenant in the 7th (Service) Battalion, Royal Munster Fusiliers, in January 1915, transferring to the RFC in September 1915. He received the Royal Aero Club Aviator's Certificate 1829 on a Maurice Farman biplane at the Military School, Norwich, on 4 October 1915. Posted to 18 Squadron, in 1916 he scored his first victory flying an FE2b, forced to crash-land near Château de la Haie because of damaged controls. He discovered his observer was dead, shot through the head. Callaghan himself was wounded in action on 31 July 1916. Promoted to Major and Squadron Commander, Callaghan was at Turnberry from February to October 1917, employed as an aerial gunnery instructor, where his flying stunts earned him the nickname the 'Mad Major'. It was said

that he flew under the Old Brig of Ayr!

According to the *London Gazette* Supplement of 13 February 1917, 2nd Lt. (temp. Capt.) Joseph Cruess Callaghan, R. Munster Fusiliers and RFC was awarded the Military Cross:

> For conspicuous gallantry in action. He displayed marked courage and skill on several occasions in carrying out night bombing operations. On one occasion he extinguished a hostile searchlight.

Returning to combat in April 1918 as a Sopwith Dolphin pilot and Commanding Officer of 87 Squadron, he had scored four more victories to become an ace by the end of June. On 2 July 1918, Callaghan was killed when his Dolphin was shot down in flames by Franz Buchner of Jasta 13. He had single-handedly attacked a group of as many as 25 German enemy aircraft.

He is buried in Contay British Cemetery, Contay, Somme, France.
2nd Lieutenant Eugene Owen Cruess Callaghan, brother of Major Joseph Callaghan, was killed

Joseph Cruess Callaghan

J.C. Callaghan and Unknown

on 27 August 1916, flying BE12 6545 with 19
Squadron over the Western Front. He has no
known grave and is listed on the Arras Memorial.
Another brother, Captain Stanislaus Cruess
Callaghan, was killed in a flying accident in
Canada on 27 June 1917 and is Buried at Barrie
(St. Mary's) Roman Catholic Cemetery, Ontario,
Canada.

2nd Lieutenant Herbert Anderson *Squad 11
November 1917*
Pictured at Turnberry with Squad XI, in
November 1917, is Herbert Norman Scott
Anderson, who was the son of the Provost
of Kinross. Having completed his course at
Turnberry, Lieutenant Anderson was posted to 80
Squadron at Beverley in Yorkshire before being
sent to the front.

On Christmas Eve, 1917, Herbert Anderson
was sent up within the boundaries of Beverley
aerodrome to get some flying practice in the
type he was to fly in active service. He had been
flying and practicing manoeuvres successfully
for almost an hour when, spinning down from

1,500 to 400 feet, he suffered engine failure. He
put the aircraft into a dive, in an attempt to gain
speed to get his engine restarted, but did not have
enough height to regain control and crashed.
At the subsequent inquest it was found that had
Lieutenant Anderson had more height he would
have managed to recover control of the aircraft
and save himself.

The body of 2nd Lieutenant Herbert Anderson
was returned to Kinross for burial and was
conveyed from St Paul's Episcopal Church to the
cemetery on a gun carriage escorted by pipers
from 4th Battalion Royal Scots Fusiliers, who
also provided a firing party of 40 men. Herbert
Anderson was just 18 years old.

Flight Lieutenant Richard Josiah Gosse *Squad
11 November 1917*
Richard Josiah Gosse was born in 1894 in British
Columbia, Canada. He enlisted on 9 December
1916 at Vancouver and was posted as a Sapper
to the Inland Water Transport Section of the
Royal Engineers, based at Sandwich in Kent.
Discharged on 28 February 1917, he joined

the Royal Flying Corps on 5 March 1917 and was graded 2nd Lieutenant on 23 July. Initial training at Upavon was followed by appointment as Flying Officer on 26 September, arriving at Turnberry on 23 November. Completing the course, he was posted to 20 Squadron.

Flying Officer Richard Gosse was hospitalised on 20 March 1918 and from there to Netheravon for dispersal on 29 April.

After the war Richard Gosse was involved in the fishing and cold storage business with Gosse-Millard Packing Co., becoming managing director when the firm amalgamated with BC Packers in 1928. He resigned this position in 1931 and became Managing Director of BC Ice and Cold Storage Co. Ltd., then President of Gosse Consultants Ltd. and Trans Canada Freezers Ltd. Richard Gosse died in Vancouver, Canada in 1956.

Lieutenant William Stanley Martin *Squad 11 November 1917*

Born in Geelong, Australia, in 1893, motor mechanic William Martin enlisted in July 1915 and was posted to the 10th Field Ambulance of the Australian Imperial Force. He proceeded to France in November 1916. where he served as an ambulance driver. Transferring to the Australian Flying Corps in April 1917 he reported to the School of Aeronautics at Oxford. Promoted to 2nd Lieutenant in October, he was sent to the School of Aerial Gunnery at Turnberry on 15 December, leaving on 2 January 1918, when he was classed as unfit for flying due to illness. On recovery in early May he joined 4 Squadron in France.

Lieutenant Martin participated in the bombing of La Gorgue aerodrome (Beaupre-sur-la-Lys) when on 18 May he and Lieutenant R. C. Nelson broke away from their formation of eleven aircraft to attack four enemy triplanes near Bailleul, managing to shoot down two of them, one of which was totally destroyed.

Lieutenant William Martin, flying a Sopwith Camel of 4 Squadron, was shot down and killed by a Pfalz enemy fighter on 12 June 1918 (flown by Ltn. Raven Freiherr von Barnekow of Jasta 20)

He is buried in the military cemetery of Ebblinghem, not far from Clairmarais aerodrome

Lieutenant George Nowland *Squad 11 November 1917*

George Nowland was born in Melbourne Australia in 1892, the son of James Frederick and Janet Nowland. Prior to joining the Australian Flying Corps, he was a tent maker in North Fitzroy Melbourne.

After training at Turnberry, George Nowland was sent to 4 Squadron at Clairmarais (North) aerodrome Pas-de-Calais in France.

On 22 May 1918, while flying Sopwith Camel D1909, diving and firing on enemy balloons, he collided in mid-air with Sopwith Camel D1924 flown by fellow pilot Lieutenant Alex Finnie. Both aircraft broke up and fell from about 10,000 ft, crashing to the ground five miles over enemy lines at Neuf Berquin, both men were killed. Lieutenant George Nowland is buried at Pont-Du-Hem Military Cemetery, La Gorgue.

2nd Lieutenant Thomas Franklin Rigby *Squad 11 November 1917*

Thomas Franklin Rigby was born in 1895 in Lanark, Ontario Canada. Leaving Turnberry, he was posted to 3 Squadron Royal Flying Corps in France. On 27 March 1918 2nd Lieutenant Rigby was flying Sopwith Camel C1570 on a low-level mission over Fricourt, approximately 5 kilometres east of Albert, when he was shot down and killed by German Ace Ltn. Hans Kirschstein of Jasta 6.

He is commemorated on the Arras Flying Services Memorial, Pas de Calais, France.

2nd Lieutenant Harry Estcourt 'Courty' Robinson *Squad 11 November 1917*

Harry Estcourt Robinson was born in 1898 in Ealing, London, with twin brother William Stanley Robinson, who also later served with the Royal Flying Corps. Harry enlisted on 14 March 1917 and served with the colours for 99 days before transferring to the RFC. He completed his flying training at Oxford prior to training at Turnberry in November with Squad 11. Completing his course, he was posted to 72

Training Depot Station at Beverley as an assistant instructor on 26 January 1918.

On Sunday 21 April, three airmen were killed in a single accident when two aeroplanes, Avro A2636 and D194, collided near Cherry Burton. One of those killed was 2nd Lieutenant Harry Robinson. At the subsequent inquest it was stated that Harry Robinson, who was the instructor, and pilot pupil 2nd Lieutenant John Alfred Clayton, were in one machine, and Lieutenant Evan Idres Howell, in another.

An officer who witnessed the collision in the air at a height below 2,000 feet, stated that he did not see what they were doing immediately before the accident.

The commanding officer also stated that he saw the machines falling to earth and at once flew across to them, where he found the aircraft on the ground absolutely wrecked, and the three airmen dead in the wreckage.

Another witness who saw what happened, a local farm worker, said the collision was due to one aircraft attempting a turn while too close to the other, hitting the other aeroplane's wings. One machine crashed into a field and the other fell by the roadside. The verdict was 'accidental death'.

2nd Lieutenant Harry Estcourt "Courty" Robinson is buried in Twickenham Cemetery.

Harry's twin brother, Lieutenant William Stanley Robinson, survived WWI but died in a guerrilla camp in Perak, Malaya during WW2, when he was serving in the Special Operations Executive (Oriental Mission SOE) based in Malaya supporting the guerrilla movements.

Harry and William Robinson had a very entertaining stepbrother, Stan Laurel of 'Laurel and Hardy' fame.

Lieutenant James Pomeroy Cavers

James Pomeroy Cavers was born in 1891 in Ontario, Canada, son of William Andrew Cavers and Estelle Pomeroy. He was educated at the Galt Collegiate Institute, Upper Canada College and Toronto University.

Enlisting with the Eaton Machine Gun Battery in 1915, he embarked on HMT *Metagama*, which sailed from Montreal on 4 June, arriving in England two weeks later. He was sent to France in February 1916 with the now Eaton Motor Machine-Gun Battery and served with that unit at the front until June 1916 when he was wounded by a gunshot wound to the leg. He returned to hospital in England, and had recovered by mid-September 1916 and was posted to the Canadian Engineers' Training Depot for a course on Military Engineering after which he was taken on strength of the 1st Battalion Canadian Calibration Troop. Struck off charge on 4 July 1917 on receiving a Commission in the RFC, he proceeded to Oxford for training and obtained his Royal Aero Club Aviator's Certificate at Edgeware on 17 July flying a London and Provincial Fuselage Biplane. Arriving at Turnberry for a gunnery course on 10 December 1917, he left for an Aerial Fighting course at Ayr on 22 December. Posted to 47 Squadron in February 1918 where he was wounded slightly on 17 August but 'remained at duty', he was subsequently posted 150 Squadron fighting the Salonika Campaign on 22 August.

Lieutenant James Cavers had two claims for enemy aircraft, on 1 September while flying Bristol M1c 4907 he brought down an LVG enemy aircraft over Serres, Greece, after firing 200 rounds. It was seen to turn steeply and went into a sideslip at around 2,000 feet. Cavers couldn't stay around to confirm his victory as his engine was running rough from battle damage.

On 2 September 1918, again flying M1c 4907 he brought down another LVG, after firing 100 rounds. The enemy aircraft was seen to be out of control and one wing was seen to come off, before bursting into flames.

The next day, flying Bristol M1c C4907 on a photographic reconnaissance escort mission, he he was attacked by six enemy Albatros fighters. Although outnumbered he stayed and fought in an attempt to allow the slower reconnaissance aircraft to escape, but his plane was hit, and he crashed into Lake Dojran. He survived the crash but was killed by strafing fire from the enemy aircraft while swimming to shore. The killing was seen by a patrol from 150 Squadron, who immediately engaged the enemy and were credited with bringing down four of them.

Ltn. Fritz Thiede, the Staffelfuhrer of Jasta 38, claimed Lieutenant James Pomeroy Cavers as his sixth victory.

The following two letters were written by Cavers' superior officers:

Headquarters,
Royal Air Force,
Salonika.
7 September 1918

Dear Madam,
You will doubtless have heard of your son's death, both officially and from his Squadron Commander (Capt. Goulding) so I will not go into details. I write in the name of the Wing Commander (Col. Todd) Officers, NCOs and men of the 16th Wing, to offer our heartfelt sympathies in the sad loss you have sustained of so brave and good a son. He was absolutely fearless, and his high moral character set a good example to all those who came in contact with him. He was one of our best pilots and had brought down two enemy machines in flames the two days previous to the date on which he himself was unfortunately brought down by 5 or 6 enemy machines, which he bravely attacked.

In his death the British Empire and the RAF are heavy losers, and we all of us here cannot express our sympathies with you too greatly.

If there is anything I can do, please ask. His personal effects, etc., are all sent to you through the usual Military Channels and will arrive someday.
Yours truly,
C. Hodgkinson Smith,
Capt. S.O., 16th Wing RAF

Dear Major Dyas,
You will have heard through official sources of the loss of your brother, and it is with the deepest sympathy I endeavour to portray the most gallant way in which he carried out his duty and so added his name to that glorious list of those who have given their utmost.

He had been doing the most excellent work, and three days previous to his death, had been successful in sending two enemy machines to earth, and his loss is deeply regretted by all. He was on escort duty, protecting a reconnaissance machine, when he was attacked by six enemy machines. His engine and controls were hit, and he was forced to come down followed by the enemy. The combat had been over Lake Doiran into which he was forced to descend. A formation of his squadron saw the fighting and went to help as fast as possible, but unfortunately arrived too late to save your brother, but succeeded in shooting down four of the six 'Hun' machines.

Cavers was an excellent swimmer and I thought he probably had managed to reach shore and sent out patrols at night in an endeavour to find him, but in this I was disappointed. He was an excellent fellow and most popular in the squadron. His death though quickly avenged is keenly felt, and all officers deeply sympathize.

The General Officer commanding the army in Salonika sent the following telegram to the Squadron which I think you will be glad to read.

"I greatly regret the loss sustained in Lieut. Cavers' death; he was a splendid example."
Your brother's kit is being forwarded. If there is anything further I can do to assist you I shall be only too glad to do so.
Sincerely,
A. G. Goulding

Lieutenant James Pomeroy Cavers was Mentioned in Dispatches *'For Gallant Conduct and Distinguished Services Rendered.'* He has no known grave and is remembered on the Doiran Memorial, Greece.

Lieutenant Jack Henry Weingarth

Jack Henry Weingarth, from Sydney, New South Wales, was a civil engineer in civilian life. He enlisted in the Australian Flying Corps on 11 January 1917. His unit embarked on the *Shropshire* from Melbourne, Victoria, on 11 May 1917, to Britain. Jack flew his first Sopwith two-seater aircraft on 11 November at the Central Flying School, Upavon and, by 22 November, had completed twenty hours of solo flying. In a letter home he complains that he, 'nearly perished with the cold whilst flying an Avro', the school being situated on Salisbury plain, described as *'jolly nippy'*.

From Upavon he wrote on 12 January 1918, 'I am expecting to go overseas soon. I'm a full Flying Officer now and can do anything at all in a bus. Fly upside-down, side loop, spin, dive vertically and over etc. It is all fun'.

Jack had logged a total of 42 hours and 45 minutes of flying and had also received an 85% mark in ground gunnery with the Vickers gun when he left Upavon, being posted for further training to 2 (Auxiliary) School of Aerial Gunnery, Turnberry, Jack wrote home on 27 January:

I arrived at the above address, which is in Scotland, on the West Coast, about a week ago. We left St Pancras station in London at about 8.50am and arrived here about 9 o'clock the same night. This place was in peace time, a hotel, one of the most luxurious places I have ever been in. The tariff is 6/- a day, but it will cost us 4/- a day. Of course, it has been commandeered by the RFC. This place is a finishing off school in gunnery, before you go to Ayr about 18 miles away, which is the School of Aerial Fighting. Service pilots are all sent here before going overseas. I should be a week at Ayr. Passed my exam just a while ago, so I've only a few days here now.

Having completed his course at Turnberry, Jack wrote from Middleton in Ayr:

As you see I've changed my address again. I stayed at the School of Gunnery at Turnberry for two weeks then got shifted to here. This place is a School of Aerial Fighting and is a finishing off school before proceeding overseas. They do the most reckless and amazing things here, such as looping straight off the ground and follow the leader through trees and between houses and up and down streets etc. To see a dozen machines following one another, diving and zooming, all flying within feet of the ground is very funny. I have finished the course on aerial fighting and am now in the pool awaiting orders for overseas. They have put me temporarily on the staff, fighting pupils, leading formations and patrols and when I'm not doing that, I take a machine and fly for miles around about ten feet from the ground. We call it hedge-hopping or contour-chasing. Sometimes we fly out to sea for a change.

On 19 March 1918, Lieutenant Weingarth was posted to 4 Australian Flying Corps, which was then based near Bruay, France flying Sopwith Camels over the front.

After the war Jack served as an instructor at 6 Training Squadron based at Minchinhampton, Gloucestershire, where he was killed on 4 February 1919 whilst acting as a flying instructor with Lieutenant W. H. Millard in an Avro 504. Lieutenant Jack Henry Weingarth is buried in Leighterton Church Cemetery, Gloucestershire, England.

During World War I, young men came from all corners of the Commonwealth to volunteer with the RFC, over one thousand Americans flew with the Royal Flying Corps and RAF, some joined via Canada before America entered the conflict.

In August and September 1917, two groups of over 200 cadets who had graduated from ground school at the Schools of Military Aeronautics, established at several US Universities, arrived in Britain for flight training. The first group of 53 arrived on the *Aurania* and the second, larger, group on the *Carmania*. Both contingents had initially been posted to Italy, but upon docking in Liverpool the cadets were informed that there was a change of plan and that they were to remain and train in England, to repeat ground school at the School of Military Aeronautics at Oxford University and were billeted in Christ College.

The first group of fifty-three cadets who had arrived on the *Aurania* in August became known as the 'first Oxford detachment,' and those who landed from the *Carmania* in September came to be called the 'second Oxford detachment.'

Of the two contingents to pass through Oxford, forty-nine were later to be killed in action and training accidents, or die of their injuries, a further thirty-one would be seriously injured and fifteen would be taken prisoner of war.

1st Lieutenant Harriss Percy Alderman

Harriss Percy Alderman

1st Lieutenant Harriss Percy Alderman was born in 1894 in Wilmington, North Carolina, USA. Arriving at Turnberry with Lieutenant Robert Miles Todd, Alderman completed his higher training and was posted as a Camel pilot to the 17th Aero Squadron.

On 7 August 1918 1st Lieutenant Alderman flying Sopwith Camel D9499 overshot the aerodrome and hit the bank of the road and overturned on landing from an operational patrol. He was unhurt. Five days later he was wounded in action when his Camel was shot up in combat and he had to force land near Ramskapelle, Nieuwpoort. Ltn. Theodor Osterkamp claimed this as his 21st victory.

Lieutenant Rodney Williams, also flying that day in Sopwith Camel D6595, gives details of the flight:

Shortly after passing Ostend on our way to the lines eight Fokker Biplanes came along from the south east on the same level with the five highest machines. One of these five had 'gun trouble' and could not fight, so he climbed up out of danger (so he thought) to watch the fight. The odds—two to one in their favour—just suited the Huns so they attacked the remaining four Camels.

The minute I perceived that there was going to be a scrap I turned around and climbed for all the machine was worth in an attempt to get into the scrap and one of the other three followed me. The other two having engine trouble (as I found out later) had to continue toward home. I got to the higher level just in time to see a Hun come down on the tail of Gracie's machine. From the way Gracie was flying I judged that he must have been wounded for he made no effort to shake the Hun off.

Though I was over 500 yards away I opened fire at the Hun to warn Gracie and try to scare off the Hun, but to no avail. When he had gotten within 75 yards of the Camel, he opened fire on it, and I saw it hesitate for a second, then it shuddered, one wing collapsed and it fell in a dizzy spin into the sea eight or nine thousand feet below. I did not stop to watch it go down but flew head on at that Hun firing both guns all the while. I very nearly ran into him in my rage, but his machine nosed down on my last burst & I passed over him. Then jerking my machine around to get on his tail, I found that he had disappeared in a spin.

I was now just a few hundred feet below the main 'dog fight' out of which two Huns came diving on my tail. It was useless trying to out manoeuvre both of them, so I just threw my machine around until they were right on top of me and then by a tight right-hand climbing turn I came up behind them & they both dropped into spins. A moment later I forced another Hun into a spin to escape my gun fire.

Taking a look around I saw but one of

our machines left and that one surrounded by three Huns. He was some distance above and behind me and again I had to climb to his level before I could be of any assistance. From the letter on his machine 'C' (D9495) I could make out that Snoke was the pilot and the way he went after those three Huns was a sight to behold. When the Huns discovered that I had come to his assistance they quickly beat it for home, apparently even three to two were not odds enough to suit them.

On crossing the lines, I saw one of our machines land on the Belgian coast just a mile or so beyond. Flying down close I could make out that it was Alderman and he was apparently wounded, for several Belgian soldiers were helping remove his flying clothes. He was standing up all right, so I returned to the aerodrome where I found that Armstrong was wounded in both the arm & the back.

A few minutes later Alderman arrived at the aerodrome. He was only slightly wounded in the back, but he thought that he was bleeding to death because the petrol was running out of the tank onto his back and that was the reason he landed up in Belgium. He was the one whose guns had jammed & though he thought he was out of the fight because of his altitude evidently one of the Huns had taken a long distance shot at him with the consequent results.

Both wounded men were taken to a nearby hospital where they were quickly operated upon for the removal of the bullets.

Surviving the war, 1st Lieutenant Harriss Alderman returned to the USA and died age 27 in his hometown of Wilmington, on 3 February 1922, of tuberculosis.

Lieutenant Lloyd Andrews Hamilton DFC, DSC

Born in Troy, New York, in 1894, the son of Rev John A. and Jennie B. (Andrews) Hamilton,

Lloyd Andrews Hamilton graduated from Syracuse University in 1916, then studied at Harvard's Graduate School of Business Administration. Enlisting in early 1917, he was selected for aviation training in July. After completing ground school class on 25 August 1917, he embarked on the *Carmania* for England and attended 2 School of Military Aeronautics at Oxford. By mid-November, he was posted to flying schools in Tadcaster and South Carlton. From here he arrived at Turnberry in February 1918.

One of the other cadets of his detachment, Perley Melbourne Stoughton, wrote of him:

Hamilton is very much of a success. Luckily, he was posted to another aerodrome [*sic*] not so crowded and has put in a long time in the air. He is very much talked of, not only by the Americans but also by the English and is especially well thought of by his instructors. He is more than clever with a machine and is the best pupil at his aerodrome [*sic*] at present and probably one of the best in the RFC. I understand that he was offered the job of instructor in this country but passed it up.... Flying is natural with him and he puts the fastest 'bus' through all her stunts with all kinds of assurance.

Lieutenant Lloyd Hamilton had moved to Ayr on 2 March 1918 to finish his training, and by 14 March received notification that he was to be posted to operational RFC squadrons in France.

On 16 March he embarked at Folkestone for Boulogne en route to 3 Squadron (Sopwith Camels) at Warloy-Baillon, about twelve miles northeast of Amiens.

It was usual for newly arrived pilots at the front-line squadrons to spend at least three weeks getting to know the area before becoming operational and going over the lines. It would seem Hamilton's first sorties were around 11 April. The next day Hamilton was credited with destroying an Albatros DV, over Ovillers-la-Boisselle.

From an article taken from a letter to his father and published in the *Burlington Free Press* on 7 May 1918, we read:

Young Hamilton on the day he wrote ... had brought down his first individual Hun, although previous to that he had participated in fights when the Huns had gone down but his entire 'flight' of six machines had chipped in to do the trick. The fight was on April 12 . . . and in this fight Hamilton's 'flight' of six machines accounted for at least four Hun machines in the space of two hours and all returned safely. In the letter to his father Hamilton wrote from time to time during the day. At 5:15 he tells that he is going out at 5:30 and hopes that he can get his individual Hun. At 9:30 he resumes writing and has won his victory. The flight . . . went far over the German lines. When they were returning, they pounced on five German Albatrosses. . .. Hamilton was flying below the rest of the formation when he saw what he describes as "a beautiful black and white machine with spots on it. The Hun tried flight, but Hamilton was 'on his tail. . . . The American poured 400 rounds of ammunition as fast as he could work his machine guns. He saw flames burst out in the cock pit of the German's machine and saw him dive for the earth. They were then up in the air 9,000 feet and Hamilton followed him down a mile to about 4,000 feet and saw him crash to the earth in a burst of flames, which covered him. The Hun was later declared officially dead and was credited to Hamilton. The rest of the flight was meanwhile engaged in battle with other members of the Hun flight and Hamilton turned to the right to climb back up to their assistance when another Hun machine flashed by him and dashed to the ground. The machine turned out to be the victim of the Captain of the English flight. Hamilton was then so near the ground

that he was plainly visible, and four Hun machines started after him while the anti-aircraft guns opened up a hot fire. He got away, however, and with the rest of the flight returned safely. They were sure they had accounted for four and possibly five machines.

In June 1918, Hamilton transferred from 3 Squadron to the US 17th Aero Squadron, many more combat patrols followed. In all Lieutenant Lloyd Andrews Hamilton was credited with 10 enemy aircraft.

On 24 August when returning from a bombing mission with another Camel, they flew towards a German observation balloon spotted floating at about 1,000 feet from the ground about halfway between Beugny and Haplincourt. Both aircraft fired on the balloon, and it burst into flames and went down. The pilot of the other Camel, Lieutenant Jesse Campbell, stated in his combat report, 'I saw Lieutenant Hamilton firing all the way down at close range on it, he also stated that, 'Lt Hamilton did not return from patrol and was seen apparently out of control near the balloon. Lieutenant Lloyd Andrews Hamilton had been killed in action.

Hamilton's parents received a letter from the medical officer of the 17th Aero Squadron, Dr Jacob Johnson Ross, on 12 September 1918, informing them that their son had, 'gone down in his machine behind the German lines and was probably killed.' He continued:

He was without doubt the best pilot we had, fearless and heady. I have seen him come in during the push with his wings literally shot to pieces. At the time he was due back I was standing with Lieut. Wells and had just remarked that the strain had been terrific, as we wanted to know who failed to return, especially so in the case of 'Ham,' as I knew his parents and we had many things in common. When we went to Jesse Campbell of Detroit, who was up with him, he said he saw him go down apparently out of control. He did not go down far back of the lines and if this push continues, they may pass the place where he was seen to go down. If we are still here, I will make a trip to that spot and see what I can learn.

In early November, Hamilton's parents and his fiancée received word through the Red Cross that Hamilton had been killed.

After much searching the grave of an 'unidentified airman' was located near Lagnicourt (Pas de Calais). On further investigation by the Graves Registration Service it was found that, 'On cross erected by the Germans is the inscription: "Here lies American Flyer, died August 24, 1918" and on burnt wood of propeller is the inscription "Mr Lord Hamlington [*sic*]"'. The grave was discovered, 'in a machine gun pit forty paces from a completely demolished steel frame Scout 9-cylinder rotary plane.' The engine number of 52820 was still discernible. This confirmed that it was Hamilton's Camel D1940.

Lieutenant Lloyd Andrews Hamilton was buried in the American Military Cemetery at Bony. His body was exhumed in 1921, and reburied in Pittsfield Cemetery in Pittsfield, Massachusetts.

Hamilton Field at Novato, California was named in his honour in 1932.

For his bravery he was awarded the Distinguished Flying Cross:

On 13 August 1918, Lt. Hamilton led his flight on a special mission against Varssenaere aerodrome. He dropped four bombs from 200 feet on some aeroplane hangars, making two direct hits and causing a large amount of damage. He then machine gunned the German officers' billets and made four circuits of the aerodrome, shooting up various targets. On the first circuit, he destroyed one EA on the ground which burst into flames when he shot it up. On the third circuit he repeated this performance, setting afire another Fokker biplane. His dash and skill very materially helped in the

success of the operation. In addition, this officer destroyed a Fokker biplane over Armentières on 7 August 1918. On 12 July he brought down two EA in flames and on two other occasions has driven down out of control enemy machines. He is an excellent patrol leader.

Similarly, he received the Distinguished Service Cross:

The Distinguished Service Cross is presented to Lloyd A. Hamilton, First Lieutenant (Air Service), US Army, for extraordinary heroism in action at Varssenaere, Belgium, August 13, 1918. Leading a low bombing attack on a German aerodrome, 30 miles behind the line, Lieutenant Hamilton destroyed the hangars on the north side of the aerodrome and then attacked a row of enemy machines, flying as low as 20 feet from the ground despite intense machine-gun fire, and setting fire to three of the German planes. He then turned and fired bursts through the windows of the chateau in which the German pilots were quartered, 26 of whom were afterwards reported killed.

Cadet George Atherton Brader
George Atherton Brader was born in 1893 at Kingston, Luzerne, Pennsylvania, USA, the son of George G. Brader and Elizabeth Atherton. Graduating from the Wharton School of Finance in Wilkes-Barre in June 1916, George enlisted in the Aviation Service in May 1917, training at the Maddison Barracks, New York, and the Air Service Ground School at Ithaca, before embarking on the *Carmania* bound for Britain and further training. He graduated from CFS in February 1918 prior to being posted to Turnberry, where he was killed on 5 April 1918 whilst flying SE5a C1762.

Cadet George Atherton Brader is buried in Girvan (Doune) Cemetery and is remembered on the Turnberry War Memorial.

CO Richard Bell-Irving with Majors Gordon, Steele and NCOs

Lieutenant Gustav Hermann Kissel

Born in 1895 in Maryland, USA, Gustav Kissel was the son of prominent banker, Rudolph Hermann Kissel. Gustav graduated from Harvard in 1917, just as America was entering the Great War. Commissioned into the United States Air Service and attached to the Royal Flying Corps, he was sent for training in Britain, arriving at Turnberry in February 1918.

Gustav Kissel kept a daily diary throughout his service. The following is his account of his time at Turnberry and Ayr:

Saturday 2 February 1918
I got up early and took the 8.50 train out of St Pancras Station. I wrote letters all morning in the train and talked with two Scotchmen. The trip was very long and tiring. After changing trains twice, I eventually got here to Turnberry at 9.30 in the evening and was given a room. This is a wonderful large golfing hotel with the links on the dunes right on the sea front. I

have a splendid room and all the comforts of home, linen sheets, hot baths etc.

Sunday 3 February 1918
We got up early for breakfast and reported at 8.30. We were then all given instruction as to what to do and were numbered. My number here is 153. All morning and afternoon we went to lectures in the gunnery huts and I must say I was bored and glad when the day was over.

Monday 4 February 1918
Today was just as bad as yesterday and worse because everything was repeated all over again today, added to that we were given a lot of notes to copy in the bargain and I am beginning to feel as if I were back again in ground school. To complete our day of misery we weren't finished at 4.30 but had to also attend a lecture for an hour at 5.15 on the C.C. gear which was terribly boring.

Tuesday 5 February 1918
Today wasn't half bad. We spent the whole time shooting on the range and it was a lot of fun. We first did stoppages, then C.C. gear and then deflection practice. We had another lecture again tonight.

Wednesday 6 February 1918
This day was rather unfortunate for us because we did all of yesterday's work over again. We were supposed to go on the 100yd range, but it was out of commission, so we had to content ourselves with reviewing. This evening I did a little writing and played bridge with some pretty poor players.

Thursday 7 February 1918
Absolutely nothing doing worth mentioning. Shooting is good enough fun but after a while you get tired of it. I went on the 200yd range and the moving fuselage and got fairly good scores, but that doesn't necessarily mean that I can shoot. At lunch time I went to a lecture on the Kauper Interrupter Gear which was quite good.

Friday 8 February 1918
More shooting today as usual and some of it practises on stoppages. Lord knows I ought to be pretty good at them by now.

Saturday 9 February 1918
Today we worked on the range all morning and in the afternoon, being Saturday, we were free to do as we liked. I had just received several letters in the morning, so I decided to answer them first and then spent the rest of the time enjoying myself.

Sunday 10 February 1918
Our work was down on the range and in the huts as usual this morning, but we had the afternoon in the hotel for the purpose of writing up our notes. I had already done some of my notes already so after I finished, I had a very pleasant afternoon playing bridge.

Monday 11 February 1918
A terribly boring day on the range and in the huts again, thank goodness it is my last in X group for I should certainly never be able to stand another. In all my life I have never been so awfully fed up before.

Tuesday 12 February 1918
This morning we had our oral examination on guns, gears and fighting. Before I went in several of us made up a pool of two shillings each for the person who got the best mark. The exam was great fun and I enjoyed it a lot. I always did like oral exams a great deal. Incidentally I got 99%, the highest mark of the lot, and therefore won 12 shillings, which didn't hurt my feelings!

Wednesday 13 February 1918
Nothing doing all morning on account of the mist, so I stayed indoors, wrote letters and played bridge. In the afternoon. Things didn't look very promising and I managed to get someone to play golf with me. No sooner had we started than the hooter blew, and we had to go to the aerodrome for flying, but when we got there it was washed out.

Thursday 14 February 1918
Today was quite a busy day. We went down to the aerodrome early and spent all morning doing our tests. We were taken up in BE2c's and BE2e's and personally I have never flown with such terrible pilots. The afternoon was just as bad and I'm thankful to say that I'm almost finished.

Friday 15 February 1918
No flying today either in the morning or

afternoon so I spent the whole day playing golf.

Saturday 16 February 1918
Our last day at Turnberry. We took the five o'clock train for Ayr and arrived here about half an hour later and were taken to our billet in a tender. Our quarters would be alright but there is no furniture except for a bed and a lot of us are in one room. The food is good though and the living room fairly comfortable, so I dare say we will manage well enough.

Sunday 17 February 1918
We went up to the aerodrome this morning in a bus and found that we have to wait a while in the pool owing to the number of students ahead of us. Several machines were up, and I am very keen to get flying again. This afternoon I wrote letters and read.

Monday 18 February 1918
Owing to the rain and bad weather no one went to the aerodrome, so I made good use of the time by going down to the town and getting my hair cut with Dowling.

Tuesday 19 February 1918
When we went to the aerodrome this morning, they had nothing for us to do so we were given the day off. I walked around the town and was much impressed by the neatness and prosperous appearance of the many storefronts. One never sees a wooden house in this country. All afternoon I read and did practically nothing.

Wednesday 20 February 1918
Today I started flying. I was first taken up for 40 minutes in an Avro and put through stunts and I almost got sick. I don't mind stunting myself, but I hate it when anyone else does it to me. A little later I went up in a Camel and tossed it around to my heart's delight. In the afternoon I went up on a poor formation and not much fun.

Thursday 21 February 1918
Terrible wind and rain all day. I spent the better part of the morning reading and writing letters. In the evening Allen and I walked downtown and took in the movies, but they weren't any good.

Friday 22 February 1918
We got up late this a.m., so rather than go without breakfast we went down to the Station Hotel and had some there. In the afternoon, I left my tunic to be pressed and then went to the hotel again and indulged in a bath.

Gustav spent Saturday playing golf at Troon and visiting local homes. Sunday was a very lazy day, 'I got up very late this morning and then spent the entire day indoors owing to the wind and rain. In the afternoon I wrote letters and read a book.

Monday 25 February 1918
Fine weather and flying. In the morning we did a sham battle. Two formations of Camels went as escort to a formation of Avros which was theoretically supposed to bomb Troon. We were met by hostile aircraft and our duty was to fight them off and keep them from bothering the bombers. There were so few of them though that they had no chance.

Wednesday 27 February 1918
We had flying this a.m. and I did two hours in all and had a fine time. I first did a bit of fighting against Hamilton and fared pretty well. I then flew over the Watson's house [an Ayr family who had befriended him] and successfully put the wind up them all, I next flew over to Turnberry with Hamilton and together we raised all the hell we could by diving on everything and everybody and chasing them to cover.

Friday 1 March 1918
Flying all day today, but this morning I was informed that I had finished the course, so didn't get a chance to fly in the afternoon. It seems that most of my course consisted in fighting students for I never had a single scrap with an instructor and hence didn't learn much.

Saturday 2 March 1918
I flew this morning on one of the pool Camels and had a splendid time contour chasing up the beach.

Sunday 3 March 1918
This morning was a half holiday, so we all took it easy, but this p.m. there was flying, and I took a short flip early in the afternoon. At about 3 o'clock the Watsons came up to look around the aerodrome and I got them a pass. Unfortunately, there was a very bad crash in a Camel of one of our Naval fliers, but luckily, he wasn't killed.

Monday 4 March 1918
This a.m. I flew for an hour and 10 minutes and in the first part of the time I scrapped Colonel Rees in a DH2 and it was great fun. I then flew to Turnberry and looked around.

Tuesday 5 March 1918
Played golf at Troon all day.

Wednesday 6 March 1918
This morning Davidson and I went out to Old Prestwick and had a round of golf. This p.m. it was too windy for flying so I went out with several other fellows to see Burns' Cottage and Monument and its surroundings were very beautiful.

Thursday 7 March 1918
I went up to the aerodrome today but owing to the lack of machines in the pool I couldn't get a flip. I was standing by the shed when a Camel spun into the ground. I was one of the first out to the crash and helped get the man out of the wreckage, but he was hopelessly

Gunnery Staff outside golf clubhouse

mashed up and dead before we even got him out. Ortmeyer [sic] - 1st Lieutenant Andrew Carl Ortmayar, Aviation Section, Signal Corps. US Army. Buried in Ayr Cemetery], an American officer also spun into the ground and was killed.

Friday 8 March 1918
Played golf at Prestwick all day. When I got back, I heard that Velie [Ensign Harry Glenn Velie, US Navy. Buried in Ayr Cemetery], an American Naval Flier had also spun himself into the ground to his death. We seem to have had a rather unfortunate week.

Saturday 9 March 1918
Played golf at Troon all day. It certainly is great to get so much golf and I'm not a bit anxious to leave here.

Sunday 10 March 1918
This morning I slept rather late and then walked down to the Station Hotel with Bisset to get a good hot bath, something quite impossible at Westfield. This afternoon I borrowed a Camel from G Flight and flew for a half an hour. When I came in there was about 200 feet of telegraph wire wound around my propeller shaft, I must have struck it head on when flying up the streets of Troon and consider myself very lucky not to have smashed my prop or to have been upset by it.

Monday 11 March 1918
This morning I went out to Prestwick and played a round with Hunter the professional.

HQ Office Staff

Tuesday 12 March 1918
More golf at Troon.

Wednesday 13 March
More golf at Troon.

Thursday 14 March 1918
I went up to the aerodrome this a.m. and was informed that five other American 'Camel' pilots and myself were to leave for overseas tonight. In the afternoon I managed to get a short flip and then went and packed up my things. I dined with the Abbotts (of Midton Road, Ayr) and the Watsons came down to the station to see me off. When I got to Troon, Miss Laird and Miss Gilmore were at the station to say goodbye to me on the way through, so all in all I had a fine send-off.

The next few days were spent at CFS where, I had nothing to do all day but enjoy myself. I was given a Camel to fly around in all day and I had a great time. In the morning I took a short flip, but I was gone for three hours in the afternoon and visited several aerodromes to look up friends.

Sunday 17 March 1918
Today was very misty but Oliver and I decided to go to Hamble to see fiends. We managed to get there successfully and landed the Avro in a field right next to their house. We stayed for lunch and afterward gave them an exhibition battle for amusement.

Monday 18 March 1918
I got all my kit packed up this morning and arranged for my departure. I then had an hour to spare and took my last flip while in England. Waters and I left at 1.30 for Larkhill where we went through the gas chamber and afterward, we left Salisbury for London.

Tuesday 19 March 1918
I reported at Masons Yard again this morning and received orders to leave on the 7.35 tomorrow morning for Boulogne. I then ran around town getting money and doing various other errands.

Arriving at Boulogne Lieutenant Gustav Kissel was ordered to the pilot's pool at Candas. Arriving on 22 March he writes: 'All day with absolutely nothing to do but rest and look around. This is a huge ASD and there are all sorts of imaginable machines here under repair. Many Hun prisoners are being used around the camp and troops go by every minute, so the place has quite a war like appearance. I have met several fellows that I knew in England and was very glad to see them'.

Saturday 23 March 1918
Today Clay and I were posted to 43 Squadron, but no transport came for us as they are moving.

After being transported to 43 Squadron at Avesnes, he was put in C Flight with Captain Woollett as his Commander. 'At present I will fly when I get a chance behind the lines in order to learn the country and in about two or three weeks I'll start in as a war pilot and then the excitement will begin. As far as I can make out there is at least a 50/50 chance of getting away with it'.

The next couple of weeks were spent on Orderly Duty, and occasionally collecting new machines from St Andre. Easter Sunday was an unpleasant day in which Kissel and Clay had to go through the kits of six officers who had been posted as missing over the front and make a detailed inventory of their things. 'It took us all day and I certainly am glad to be finished. It's the first time I ever spent an Easter Sunday like that'. The following days allowed for settling in and a few flights around the countryside getting his bearings, before being sent to St. Andre to bring back a new machine, 'to be mine henceforward'.

Wednesday 10 April 1918
I was up early this morning standing by to go on my first patrol at eight o'clock, but execution orders never came through all day. Tested my new machine and she flies very well although the engine doesn't rev quite as high as I might wish. Tonight, a regiment marched by, each battalion playing its band and the men singing as they went up to the trenches. It was a most impressive thing to hear and filled one's mind with the wonders of the war.

This was the last diary entry Lieutenant Kissel would make. He took off on his first mission on 12 April flying Camel D6558. Taking off at 09.35, his patrol clashed with a large number of enemy aircraft and Kissel's plane was shot down over Flanders at 10.20. Vizefeldwebel Friedrich Ehmann of Jasta 47, claimed Kissel as his fourth victory. He is buried in Pont du Hem Cemetery near Armentieres, France.

Lieutenant Kenneth McLeish
Born in 1894 in Illinois, the son of a Scots immigrant, Andrew McLeish, Kenneth was educated at Yale University. Enlisting in the United States Naval Reserve in 1917, he spent time as an instructor before being posted on assignment to Britain, where he continued his training with the Royal Flying Corps at Gosport and arriving at Turnberry in February 1918.

There is a short account of McLeish in *Wings of Honor: American Airmen in WWI*:

When ordered to report to the RFC gunnery school at Turnberry, Kenneth felt this was almost like a homecoming. His entire family had visited Scotland, the homeland of his father, ten years before the war. Situated on the Firth of Clyde southwest of Glasgow, Turnberry School offered concentrated instruction in the use, mechanics, and maintenance of automatic weapons. Days were filled with lectures and demonstrations, with the evenings devoted to copying notes and studying.

Very little flying was attempted. The nature of the course and the perpetual fog and drizzle precluded active operations. As Kenneth noted, Scottish weather was 'fit for ducks and sea monsters only!'

Difficulties caused by poor flying conditions and the grueling, exacting instruction schedule paled before the ire aroused by the arrival of the 'roughnecks' or 'Hard Guys', a compliment of navy enlisted men trained at flight schools in France and now ordered to Scotland to complete their work. McLeish ridiculed their supposed lack of breeding and education, and their unwillingness to conform to the British system.

With the work at Turnberry completed in a few weeks, McLeish and his companions moved on to the nearby School of Aerial Fighting at Ayr. Here the task of producing combat flyers was completed. Among the many students at this RFC finishing school could be found pilots from England, Ireland, Canada, Australia and South Africa. In addition to the small Navy contingent, dozens of Army Air Service pilots were there, part of the 'Lost Battalion' of American Aviators dispatched to England in the fall of 1917. Future aces Elliot White Springs, Laurence Callahan, George Vaughn, Reed Landis and many others trained in Scotland.

At Ayr the men flew front-line aircraft, mostly Camels and SE5s. They engaged in frequent aerial battles and often employed camera guns to determine the winner. As the days passed individual encounters gave way to group operations, mimicking the tactics employed at the front. After only two weeks of advanced work at Ayr, the Navy pilots were judged proficient and ready for combat.

Straining to graduate pilots as quickly as possible, English flight schools rushed the men through their course, with the inevitable result that scores were killed

HQ Office Staff

or injured in training mishaps. While stationed in Scotland, McLeish learned that Yale friends Fred Stillman and Lyman Cunningham had both died in recent accidents. On one occasion he counted twenty-three crashes within a four-day period.

Posted to 213 Squadron on 12 October 1918, Lieutenant Kenneth McLeish was listed missing two days later. Reports state that, 'he was found unwounded yet dead in a sleeping repose beside his undamaged Sopwith Camel. The thought was that he landed with some mechanical problem

and succumbed to poison gas of British origin. Lieutenant Kenneth McLeish is buried in the Flanders Field American Cemetery, Waregem, Belgium.

2nd Lieutenant Robert Miles Todd
Born in Cincinnati, Ohio, in 1897, Robert Miles Todd was studying engineering at the University of Cincinnati and Ohio State University, before joining the Signal Enlisted Reserve Corps on 6 August 1917. Prior to this he had raised almost $200,000 as a Liberty Bond salesman. He graduated from the School of Aeronautics in October and after completing his training

at Turnberry, he was posted to the 17th Aero Squadron in France, flying Sopwith Camels.

During August 1918, Todd destroyed four enemy aeroplanes and an observation balloon. On 26 August 1918, Todd and seven other Camels of the 17th Aero Squadron found themselves in a fight with 40 Fokkers of Germany's Jagdgeschwader 3.

After shooting down his fifth enemy aircraft and qualifying as an 'Ace', he himself was shot down by German ace, Lt. Rudolf Klimke of Jasta 27. 2nd Lieutenant Todd was wounded and the German soldiers rescuing him from the overturned wreckage of his plane broke his foot. He was heckled and abused by German housewives on the way to prisoner of war camp and was held captive until December 1918.

In his memoirs, which were published in 1978, entitled *Sopwith Camel Fighter Ace*, he recounts his memories of Aerial Gunnery at Turnberry:

After two weeks of flying the Camel, I had enough time to be sent to the School of Aerial Gunnery at Turnberry, Scotland. Aldy [Harriss P. Alderman] was about a week behind me in his training, so I told him I'd wait for him in London so that we could go up to Turnberry together. On our 'leaves' we had gone to London and had some good times. We had been told of the American Officers Red Cross Hostel, run by a Mrs Nichols. She was from Dixie and Aldy's Southern charm had her eating out of his hand. We had a very comfortable room; the beds and bathrooms were excellent. The rate we paid for the room also included a very fine breakfast as the food came from the Red Cross, and nothing was omitted.

While in London, we became members of Murray's Club. There were several clubs in London at that time. Most had a bar and dining room where we could get drinks after hours and occasionally something to eat. There was a small group of Negro musicians there, to play for the

dancing. When Vernon and Irene Castle were dancing on the stage before the War, they had a large band of musicians with their act. When Vernon joined the RFC, these musicians were abandoned but stayed on in London, breaking up into small groups to play at different clubs. We were successful in meeting a group of girls who were members of the club and came every day to dance. Irene Castle was one of them and I dared Aldy to dance with her which he did without hesitation. Aldy, besides having his Southern charm, was an excellent dancer, and she really told everyone about it.

We became aquatinted with most of the girls, but we centred in on two sisters, Sonia and Valia Sovorava. They were white Russians and had escaped from the Reds and reached England with their worldly goods. Their father was a very successful businessman and they lived in a very fine home in Kensington. Sonia was married to an English Naval Officer who was at sea. She had a flat not far from the club, and both girls usually stayed there. Aldy and Sonia hit it off from the start, as did Valia and I. We had some good time with them - dancing, visiting other clubs and hotels, also going to the theatre.

Aldy finally joined me, and after two more days in London, we boarded the train for Turnberry, Scotland. After reporting in we were assigned sleeping quarters in the Turnberry Hotel, a very fashionable vacation resort. The adjoining golf course had been turned into an aerodrome and the hotel sat halfway up the side of a mountain, looking out over the aerodrome and into the bay, a most beautiful spot. The CO wanted to know where I had been. I told him, 'waiting for Aldy to finish up'. He didn't think much of my reason for the delay in reporting and thought that punishment was necessary.

He said the standing order was for

all lights in bedrooms to be out at 10 p.m. each night, and that for the next two weeks I was to check the three floors where the cadets were quartered and report to him the following morning any violations of this order.

The Turnberry Hotel was a very exclusive resort and catered for the aristocratic society of Britain. The room Aldy and I were assigned, was in the front overlooking the aerodrome (golf course) and across the bay to Ailsa Craig, a round mass of rock sticking out of the water several miles away. We had very nice comfortable twin beds, lounge chairs, a cupboard and a dresser. We also had our own bathroom. We were on the 2nd floor which was nice because the elevators were not running. All of our meals were served in the main dining room and we were introduced to several Scottish dishes immediately; namely, kippered herrings with our eggs and treacle for our cereal. We liked both and the service was very good. The English enlisted men make good servants. They did the maid service around the hotel as well as waiting on tables. The dining room was known as the 'Officers Mess - No.2 Auxiliary School of Aerial Gunnery, RFC.' We were billed weekly for our meals and other services that we had.

This is a good time to write about money as we were paying mess bills, and on our trips away from the bases we had hotel, meals and drinks to pay for. Aldy and I went to the theatre as often as possible, and we saw many good shows in London as well as the Palladium. The British Army Officers were paid through the bank of 'Cox & Co.'; they were paid for three months after death or capture so it was general practice to overdraw your account. What US Officers did that opened accounts with this bank was to deposit or mail our pay vouchers to the bank which, in turn, collected from the US Paymaster and deposited the pay in our accounts. Most of us were experiencing a bank account for the first time and keeping a cheque book. The bank assumed that our army paid off like the British, so we were able to overdraw if we needed it.

The aerodrome was the golf course complete with bunkers, sand traps and a dandy water hazard that circled through the whole field. This was a brook; a typical Scots hazard on a golf course. The prevailing wind came off the sea, which was not bad for take-offs; but to land we had to come down the mountain where the hotel was and sideslip for a very short landing strip to avoid the brook and other hazards. The planes were not maintained as well as the ones we flew at the training field in England, and Aldy was assigned to a Camel that was not rigged right; it pulled to the left and was hard to land. The planes were equipped with two Vickers machine-guns, firing through the propeller, and we fired on ground targets - plane's silhouettes laid out on the beach away from the field. We also had camera guns that we could fire when fighting with each other. The films from these showed how close you came to making a kill.

We played billiards and cards for recreation. The Aerial Gunnery School No. 1 was at Ayr, Scotland. One night when we were in Ayr, we met 'Andy' Anderson from Hawaii. Andy knew a civilian family that wanted to entertain Yanks - so later that week, I joined Andy and we went to their home and we spent the afternoon playing tennis and drinking tea.

After serving us a nice dinner they drove me back to Turnberry. Andy Anderson played the ukulele quite well and later in life wrote and published many of the famous Hawaiian tunes we all know today.

Turnberry pupils, August 1917

During World War II, Robert Miles Todd returned to active duty and served in England but due to poor eyesight he did not fly. He served in logistics for both the Eighth and the Ninth Air Force until 1944, and also served as a Commander of a staging area for gliders. He retired from military service as a Lieutenant Colonel in 1965.

In 1981 he was one of a group of aces honored by France, at an Armistice Day celebration, for their contribution to the war effort during the Great War, and in 1982, he belatedly received the Purple Heart for the wound he received on 26 August 1918.

There appeared a feature in *The Journal Tribune* of Wednesday, 11 November 1981:

FORMER FLYING ACE HONOURED IN PARIS

CINCINNATI (AP) - Robert Todd, an 84-year-old, World War I flying ace, says today's astronauts have it easy compared with dangers he and other early pilots faced.

'Those astronaut fellows are in clover. The planes we had were nothing but kites. No heat, no brakes, no parachutes,' the author of Sopwith Camel Fighter Ace,

published in 1978, said this week.

He was to be among a group of heroes honoured in Paris, France, today as the world's first air war was commemorated.

Todd, a native of Cincinnati who has lived in San Diego, California, since 1952, spent time in Germany as a prisoner of war after he crashed in France.

He was an engineering student at the University of Cincinnati when President Woodrow Wilson visited the campus in 1917 and he enlisted in the service shortly after the United States declared war.

He was assigned to the Signal Reserve Corps, later learned to fly and was credited with shooting down five German airplanes before crashing near Bapaume.

German soldiers broke his foot pulling him out of his up-side-down craft.

He said German pilots taunted the captured American pilots, 'But the worst treatment was from German women. They would crowd past the guards, spit on us and try to kick and punch us. The soldiers themselves had their bellyful of war.

'They were just like us — a bunch of kids trying to fight a war we didn't know anything about,' he said.

Todd was one of seven other American aces along with their counterparts from

Germany, Britain, France, Belgium, Italy and Hungary gathered together by a coalition of aviation groups for Armistice ceremonies.

Todd went back to Europe in World War II to command a 5,000-man air-depot group but didn't fly because of his eyesight.

He said that after the war he supervised disposal of 'combat material' overseas but resigned in disgust. 'They were even disposing of typewriters and filing cabinets as "combat material."' he said. 'US manufacturers had lobbied so those goods wouldn't be dumped on the market back home.'

Robert Miles Todd died on 20 January 1988.

Lieutenant Alexander Robert 'Andy' Anderson

Alexander Robert 'Andy' Anderson was born in Honolulu, Hawaii, in 1894. He was the son of Robert Willis Anderson and Susan Alice Young, whose father was a mechanical engineer from Lanarkshire in Scotland who had gone to Hawaii in the 1860s. He had established an iron works that created machinery used in the cane sugar industry. 'Andy' Anderson's father, also of Scottish ancestry, was born in New York, trained as a dentist, and moved to Hawaii. Both of Alexander Anderson's parents were musical, and he himself later became famous as a writer of Hawaiian music, one of his compositions was 'Lovely Hula Hands.'

Graduating from Cornell with a degree in mechanical engineering in 1916 he enlisted just after the US entered the war and was sent to Fort Niagara, where he volunteered for aviation training. He was sent to ground school at Cornell, graduating on August 25, 1917. Embarking on the *Carmania* he sailed to England to train with the RFC and attended ground school (again) at Oxford.

By late March 1918 he had done enough flying to qualify for his commission and Anderson went on to higher training on SE5a's at Turnberry and then Ayr in April. In July 1918 he was posted to the pilot's pool at 1 ASD at Marquise in France and then assigned to 40 Squadron stationed at Brias.

Lieutenant Alexander Robert Anderson flew his first offensive patrol on the morning of 7 August, being part of a large formation dropping bombs east of Arras and Lens. By 26 August he had taken part in no less than 17 patrols and the next day when preparing to set off on another, his engine refused to start. He managed to find another aircraft and took off with three other pilots, two of which had to turn back with engine trouble.

Flying over the German lines at about 5,000ft, Anderson got into a fight with five Fokker biplanes, shot in the back and knee he was forced to land in enemy territory about ten miles southeast of Arras. He managed to climb out of his aircraft, but because of his wounds was unable to evade capture. Initially held at a POW hospital in Mons, he was then transferred to a prison camp at Fresnes-sur-Escaut in German-occupied France. Late in September he and four other POW's managed to escape and, although two were recaptured, Anderson and the remaining escapees, John Owen Donaldson and Theose Elwyn Tillinghast, made their way through Belgium and Holland, finally reaching the American Consul in Rotterdam by the end of October. Arriving back in the UK, Alexander Anderson spent the remainder of the war as an instructor at Ayr.

Along with other men of the 2nd Oxford detachment, he left Liverpool on the *Mauretania*, disembarking at New York on 2 December 1918. Alexander 'Andy' Anderson died in 1995 in Honolulu, Hawaii, just short of his 101st birthday.

Lieutenant John McGavock Grider

Born in 1892 in Sans Souci, Arkansas, the son of William Henry Grider and Susan John McGavock. Like his father before him. he was a farmer. John was married to Marguerite Samuels and had two sons. Grider enlisted as a cadet in the Aviation Section of the United States Army Signal Corps in June 1917 and was sent to University of Illinois's School of Military Aeronautics, where he had his first flight. He graduated on 25 August and set sail from New York on the *Carmania* in September arriving in Liverpool on 2 October.

There was a letter from him published in the *Osceola Times*, 5 October 1917:

John McGavock Grider

Dear Papa

I am writing on a tablet that came in a Red Cross comfort kit. The ladies have given us all sweaters and mufflers and a little kit containing everything imaginable.

I have written so many farewell letters that I am almost ashamed to write, but you know we never know until 10 minutes before we go on board ship when we sail. For the last few days we have been kept in constant expectation of sailing. Our things are all packed, and we are not allowed to go out of calling distance of the barracks. I don't think we will be here long.

We have the best equipment of any squadrons that has left this country, our flying clothes alone, for the 150 men, cost the government $37,000. Everything is of the best and is fur lined--pants and shoes and all. I never saw such fine things. We will be warm this winter anyway.

I don't want you all to worry about me, papa. I am big and ugly enough to look after myself anywhere, and I have a dandy bunch of friends here. I don't know whether I will ever leave the service or not. It's a great life. We are going over on a passenger ship, the *Carmania*, and we are going first class, as cadet officers. I guess there is class to us.

Papa I want you to write often and I want you to buy me a pair of First Lieutenant's bars to wear on my shoulders when I get my commission. I could easily buy them myself, but I want to wear a pair give me by my father. I think they would bring me luck. Heaven knows, I have been lucky so far, if it only holds. Our officers are the finest men I ever met and treat us like a gentleman expects to be treated.

Goodbye, papa, and good luck. I hope to bring honor to your name, and if I don't, I'll pass out trying. This is a wonderful adventure and a wonderful opportunity, and I am making the most of it.

Your affectionate son,
McGavock

Grider and his two friends, Lawrence Callahan and Elliott Springs, were sent to 2 School of Military Aeronautics at Oxford University. While there, Grider was offered a non-flying commission, which he turned down. From Oxford he was posted to machine gun school at Harrowby Camp near Grantham in Lincolnshire. In November he undertook further training at 12 TS at Thetford and got down to some serious flying. He wrote to his sister, 'It's beautiful to see a sunset from about five thousand feet, you cannot imagine anything like it. You wouldn't be surprised to see God any time. I absolutely can't describe it.'

From Thetford he went to London Colney in December and arrived at Turnberry Aerial Gunnery School in March 1917.

From Turnberry, Lieutenant Grider wrote to his sister, Josephine:

Your little bud is now a graduate pilot. I can fly any old thing they build. I can't fly very well yet, but I can fly safely. I have passed the crashing stage. I think I am pretty lucky, too; ten of our boys went west in training, and not over seven or eight of the two hundred are through yet. I expect my commission to arrive in about two weeks. I was recommended five weeks ago so it should be coming along any day now."

This is a wonderful place! The flying

corps has taken over this wonderful summer hotel and we are billeted in it. My window overlooks the sad sea waves on one side, and on the other, one of the most famous golf courses in the world. Famous I suppose on account of the very high, thick grass that grows all over it. They say it sometimes takes hours to find your ball; it must be a wonder.

This is a gunnery school. I leave before long and go to a fighting school and learn to use all the tactics I studied at St Albans. It's like teaching a boxer all sorts of funny side steps and counters and then teaching them how to use them. I will get my final polish there and then I will be the finished product. I am glad, too. It's been a long, hard old grind, lots of fun, but you know you kind of get fed up with being a 'Hun' as they call the pupils, and want to be a real, honest-to-God pilot after a while.

I have put in about twelve hours on the meanest thing with wings, known as a Spad. They are fast and have no windshield on the training busses, so you sit there with a playful 120-mile wind in your face, and shout to see if you can hear it; you can't! They make you as deaf as a post, and the wind pressure on my goggles rubbed the skin off my nose. If you can get away with a Spad, though, you fly anything.

I only broke one tiny wing tip landing too slow and pan caked. Well, you know, just lost flying speed and at six feet instead of six inches and sat down rather suddenly. Perfectly safe, but hard on the machine. I have rather a good record about busting. I have not strained a wire except that one wing tip.

Sis, I have glorious prospects of going over the lines with the fastest company in France. I am pulling every wire at my command and doing everything possible. If I make the grade, my reputation is

made; if I don't, why then I stand a good a chance as any of the other boys. I hope I make it; you will be proud of me if I do.

The days are getting longer, and I believe that before I go to France, the weather will be quite pleasant. If I go out with the Squadron I want, I will have a wonderful time. This Squadron is being run on social lines. I always told you that a thirst and a sense of humour were good assets. They are invaluable in the army, especially the British Army, which is the finest there is next to our own WCTU organisation. Our boys are the stuff though; I tell you they show up well anywhere you put them. They are keen, too, and dead anxious to be in France, seeing the show. One of the Americans has six Huns to his credit already. I hope I do as well.

My only objection to this business is that you might get shot or hurt some way.

After training at Turnberry and Ayr John was posted to Hounslow, from here he wrote: 'I got commissioned First Lieutenant on 1 April; I got a telegram and, of course, considering the date, paid not the slightest attention to it. It was genuine, however, and I am now a real honest-to-God Officer.'

John Grider was posted to 85 Squadron flying SE5a's, serving under Major Billy Bishop. By May 1918, Grider was flying patrols over the front lines in France, and always when going into combat, he carried a doll given to him by British actress Billie Carlton.

On 18 June 1918, barely a month after his arrival at the front, returning to base flying SE5a C1883 after a successful patrol with Lieutenant Elliott Springs near Armentieres, France, Lieutenant Grider never made it back. According to Springs, 'Grider was following him back towards the lines and then disappeared from view,' and that he was, 'Last seen over enemy lines near Menin.'

Three days later, Grider's father received a telegram stating, 'Lieutenant Grider reported missing in action June 18, 1918.'

An article appeared in the *Osceola Times* on 12 July 1918:

MCGAVOCK GRIDER IS MISSING

W. H. Grider yesterday received a letter advising him that his son McGavock Grider, who has been with the allied fliers in France, has been missing since June 18. The following extract is taken from the letter: 'Your son did not return yesterday after a flight over enemy lines and having no news of him we are forced to give him up as missing. He and another pilot were chasing an enemy machine far over the lines. Together they brought the enemy down and started to return to our lines. A strong wind was blowing from the west and the other pilot saw McGavock follow in the direction of our lines. On looking a moment later, he was not to be seen and the best that could have befallen him is that he might be a prisoner. It is usually a month or six weeks before we hear from prisoners captured and will notify you at once upon receiving and further information. CAPT. G. H. BAKER, Royal Aviation Corps, France.' This letter was dated June 19 and was received by Mr Grider on July 11. McGavock Grider was one of our best known and best liked young men. He enlisted about a year ago and was anxious to get to the fighting front. He was well liked by his comrades in France and Captain Baker spoke of him in the most complimentary manner. We sympathize with the father and sisters in this hour of suspense and trust that they will soon receive news that McGavock is still alive and well.

A letter to Grider's father from the CO of 85 Squadron, Captain George Brindley Aufrere Baker dated 14 August 1918 states: 'you will have learned that the worst, I fear, has befallen your son. . . . We had hoped with hopes we thought well founded that he was alive, but it seems from a message dropped over the lines, that although the name was misspelled, the name referred to was his.

Credited with shooting down four enemy planes, Lieutenant John McGavock Grider was lost between Houplines and Armentières. His body was never recovered, and he is remembered on the Tablets of the Missing at the Flanders Field American Cemetery, at Waregem, Belgium.

In 1926 Elliot Springs published *War Birds: Diary of an Unknown Aviator*, which was also published serially in *Liberty Magazine*, apparently based on, at least in part, Grider's own war diary. Springs initially made no mention of this, but acknowledged in the second edition, that Grider had given him the diary before his death and expressed a desire that the story be told. In 1927, Grider's sister sued Springs, receiving a $12,500 settlement. In 1988, Texas A. & M. University Press published a new edition of *War Birds* listing Grider as author.

On 22 March 1941, Grider Army Airfield, named in Grider's honour, became operational near Pine Bluff (Jefferson County); it is now part of Pine Bluff's municipal airport.

1st Lieutenant Thomas John Herbert

Born in Cleveland, Ohio, in 1894, Herbert graduated from Western Reserve University in 1915 and stayed on to study law. Enlisting in 1917, he undertook ground school at Ohio State University and graduated on 1 September 1917.

Departing Halifax on the *Carmania* with another 150 cadets for further training in Britain, Tommy Herbert completed the course at Turnberry in the spring of 1918. While awaiting posting to a front-line squadron, he served as a ferry pilot, crashing a Bristol F2b, breaking a couple of ribs and spraining his back.

In early June, Lieutenant Herbert with others from the 2nd Oxford detachment were posted to the pilot's pool at Rang-du-Fliers in France, finally receiving notification that they had been assigned to 60 Squadron. Another of the airmen Lieutenant Paul Winslow wrote in his diary:

At last, after three weeks here (Rang du Fliers No. 2 Pool, France) Tommy and I were posted to 60 Squadron. At eleven o'clock 60's tender came and picked us up, bag and baggage. We stopped at Abbeville and then carried on to 60, arriving at 2:30, and were told that we did not belong there but at 56, and were shipped off without any lunch. Arriving here the change of atmosphere was very apparent. Captain Gerald Joseph Constable Maxwell, the CO—Major Euan James Leslie Warren Gilchrist—and everybody turned out to meet us. It seems they had us transferred when we were on route. I think we landed on both feet, as this is the top squadron in the RAF. We are south of Doullens and patrol from Albert to Arras.

After the usual three weeks of familiarisation to the squadron and the geography of the area, 1st Lieutenant Thomas John Herbert began taking part in patrols over the lines. On 4 August 1918 RAF Communiqué mentions that, 'Lieut. T. J. Herbert, 56 Squadron, shot down a Pfalz Scout which broke up south-east of Bray'.

Lieutenants Read and Herbert

In his combat report Lieutenant Herbert wrote:

The formation dived on eight enemy aircraft, Pfalz Scouts at 9,000 ft south east of Bray at 7.50 I opened fire on rear machine of formation. This machine went into a steep spiral. Then I saw leader (Lieut. William Otway Boger) being attacked by three enemy aircraft, so I dived and attacked the nearest one. Opening fire at 100 yds range, I fired a

long burst from very close range with both guns and seemed to avoid colliding with enemy aircraft. I then saw enemy aircraft turn on its back and go into a spin. It was seen to break up by other members of the patrol. Leader fired red light as there were 7 enemy aircraft (Fokker Biplanes) coming from above. Leader and I came home at about 5,000 ft.

On 8 August, again from RAF Communiqué, 'Lieut. T.J. Herbert, 56 Sqn, brought down a Fokker DVII in flames north-east Chaulnes'. Twenty-five minutes later Herbert was last seen in combat with large formation of enemy aircraft north-east of Chaulnes. Lieutenant Paul Winslow wrote: 'Tommy Herbert and I engaged, and Tommy got on one's tail, while I shot another off Tommy's tail.... Tommy and I did right hand climbing turns to the west and then Tommy turned again, and that was the last I saw of him.' On landing back at Valheureux, Winslow continued, 'Tommy did not turn up, and I am awfully worried. It might be that he landed on the French front and could not get word through—at least, that is what I try to think. My best friend, too.'

Later the next evening it was discovered that Herbert was in hospital and had been wounded in the leg and face. He gave his combat report from his hospital bed:

On patrol led by Capt. Boger, Lieutenant Herbert attacked a large formation of enemy aircraft (Fokker Biplanes) at about 7.05 p.m. N.E. of Chaulnes. Saw Captain Boger dive into midst of formation and lost sight of him. Attacked one detached enemy aircraft and put four long bursts into him. Saw him go on his back and burst into flames. Was attacked by two enemy aircraft from behind and did a climbing turn to the right. An unknown SE5 attacked one and the other shot me through the knee, and the petrol tank. Switched to emergency tank and headed west with two enemy aircraft firing continuously. Managed to get about 10 feet from the ground and crashed due to faintness and loss of blood, somewhere S.E. of Amiens. Was picked up by Canadian soldiers and taken to No.1 Field Ambulance at Amiens.

Ltn. Rudolf Klimke of Jasta 27 claimed 1st Lieutenant Thomas John Herbert as his eleventh victory.

Transferred from the field hospital Herbert was taken initially to No. 8 General Hospital at Rouen and then on to England.

In September 1st Lieutenant Thomas John Herbert was awarded the DFC. The citation reads:

For skill and gallantry. On August 8th, 1918, when with a formation of six machines, this officer attacked from between 18-20 Fokker biplanes N.E. of Chaulnes and shot one down in flames. During the fighting Lt. Herbert was hit in the leg and his machine was shot through the petrol tank. It was only by skilful manoeuvring that he managed to recross the line. As he was landing, he fainted from the loss of blood and his machine crashed. On August 4th, 1918, he destroyed a Pfalz Scout which broke up at 9,000 ft over its own aerodrome at Cappy. This machine was at the time on the tail of the patrol leader and Lt. Herbert's dash and clever manoeuvring undoubtedly saved his patrol leader from a very dangerous position. This officer took a very prominent part in a low bombing attack on Epinoy aerodrome on August 1st from a height of 200 ft. On this occasion he killed 3 mechanics by machine gun fire and shot up hangars and billets. The influence of this officer's keenness, determination and gallantry has made itself felt throughout the squadron.

He spent some time recovering in the US Army Base Hospital at Dartford in Kent, before embarking on HMS *Celtic* with other sick and wounded servicemen. Leaving Liverpool on 23 January he arrived back in the US on 2 February 1919.

After the war, Tommy Herbert returned to Western Reserve University to complete his studies and subsequently had a distinguished legal career. He was a onetime Governor of Ohio as well as serving a term on the Supreme Court of Ohio.

1st Lieutenant Parr Hooper
Parr Hooper was born in 1892 in Pikesville, Baltimore, Maryland, USA, son of Herbert Hooper of the Oakland Manufacturing Company. Parr trained as a mechanical engineer at Baltimore Polytechnic, graduating in 1910. Later he entered Cornell's Sibley College of Engineering, graduating with a BSc in mechanical engineering in 1913. He was then employed in Philadelphia at the Lanston Monotype Company before moving to the New York Shipbuilding Corporation at Camden, New Jersey. In the spring of 1917 Parr volunteered for the Aviation Section of the US Signal Corps and entered ground school at the Ohio State University School of Military Aeronautics on 25 June 25 1917.

After ground school, Parr Hooper embarked on the SS *Carmania*, bound for England. Arriving at Liverpool in October, he was sent to Oxford and over the following months to Grantham, Lincolnshire, for a course of instruction on the Vickers machine gun, then on to Northolt for flying training, going solo on 4 December 1917. Moving to London Colney for further instruction, he finished the required twenty hours solo by mid-February 1918. He graduated from elementary training and was recommended for a commission.

After a spell of leave Parr Hooper was posted to Turnberry for higher training. Throughout the war he wrote a series of letters home, some excerpts from these follows:

Turnberry, Scotland
March 20, 1918

Dear Mother and Father,

They shipped us out of London Colney Monday afternoon. Tommy Herbert, Frank Read, and myself were all posted here together. We were allowed Monday afternoon and evening in London and took the train at 8.50 Tuesday morning for here. We arrived at 9.30 p.m.

We had a pretty tiresome trip up here. It was a foggy day so we could not see much country. There was a diner car on, and we made a good use of it by getting lunch, tea, and dinner.

This is very luxurious living for war times. Turnberry is a famous golf course, right on the coast. It was developed by the railroad. We all live in the large swell hotel. We have beds, sheets, fine bathrooms, hot water etc. Herb, Frank, myself, and an RFC chap are in one room. Our window looks right out over the Firth of Clyde and Irish Sea. There is a big army of WAACs here waiting on table and fixing up our rooms. The meals so far have been wonderful. The sea air is so bracing that I feel like eating all the time.

Turnberry, Scotland
March 22,1918

Dear Mother and Father,

Well, as has been the case for the last three letters I have a lot to write about but no time to write. They keep us very busy here all day and we have a big bunch of notes to write up at night. Also, this bracing sea air and wonderful windy weather makes me want to pound my ears for 9 hours every night. And still some more—my roommates, Tommy Herbert and Frank Read, are getting so blooming congenial that when we get together here in the room to write we spend most of our time laughing and talking. So, this life of 'war deluxe' is not conducive to letter writing.

This afternoon I had a fine trip. We were given the afternoon for holiday.

Frank and Tommy had resolved to stay home and write notes. I wanted to go to Ayr to get measured for a suit, so I piked off on the train immediately after lunch. Here at the Aerial Gunnery School we have all the Scout and Bristol fighter pilots for two weeks in their training, so I ran across a lot of old friends. Ritter is just finishing up here now, and my old Oxford swimming mate Scotie Henderson is here now also. He and I went into Ayr together.

Ayr is about 15 miles up the coast from here. The railroad follows the coast very closely. It is a most picturesque ride. There are several ruins of old castles on bluff points, and valleys and coves, and beaches. The hills rise right from the rocky shore and are being tilled with great diligence. At one place a high ridge hill runs out to the water terminating in a high bluff point known as the Heads of Ayr. To the north of this point the coast makes a big cove and the town of Ayr is at its northern point. From one point on the railroad south of the Heads of Ayr you can look north and on the right is the sea then the wonderful rocky bluff of the Heads of Ayr and then to the left of the bluff, seen over the ridge of the hill which makes the bluff, you can see the big cove with the city and its spires in the distance. The coloring was beautiful. Green hills with brown plowed fields, hedges, purple mist around the rocks and the sheen of the sun on the water.

Coming home was a real treat. The sun was just getting low when I started, and when I finished the ride, an hour later, it was in the horizon mist over Ireland. So, I got all stages of sunset combined with the changing foreground of the hills and coast as the train meandered on.

I had a very successful visit in the town. Scotie and I did all our shopping and I got measured for a new uniform. I also met several of our fellows, who have finished the fighting school at Ayr, that I have not seen since we were at Grantham. Scotie took a 3:45 train to another town, and I took a walk around the byways of Ayr.

It is a very quaint old town. Lots of stone houses with the floor lower than the street, and thatch roofs. I strolled down to the docks. Saw some very interesting ships, cranes, coaling arrangements, and even found a little steel ship building yard.

The course here is very well arranged, and the lectures and practices are very fine. We spend about 7 hours a day on the machine gun range. We have about 3 men using one gun and keep it pretty busy. The guns are arranged in aeroplane fuselages which are pivoted on a ball and socket joint. The regular aeroplane controls are all rigged to balance and point this fuselage just as though you were flying the bus in the air. We practice correcting gun stoppages and sighting on model aeroplanes set at various angles.

But the best part of this place is the life in a swell hotel. Wonderful air, view, baths, beds, sheets, food, service, and everything that makes life pleasant.

Tonight, the clock gets put forward an hour to start its summer day light saving career, so I have to call this off. Tomorrow is Sunday but it is going to be like a regular working day for us.

Turnberry Hotel
Ayrshire Scotland
March 25, 1918

Dear Mother and Father,
I just want to pass a few remarks about the immediate past. In a couple of days, I will be in the flying classes and if it is windy, I hope to get a chance to tell the back dope.

Last night we put the clock ahead one hour. It really is a great idea. Herb and Frank and I joke a lot about it whenever we consider the time but as a matter of fact, we have gained one hour of time, and one hour of daylight each day, without any mix-up or loss.

Today the wind was N.W. and blew the mist away from the land on the opposite side of the Firth of Clyde. The Isle of Arran with its rugged peaks became very plain and we could see the long peninsula west and south of it also. The old sentinel Ailsa Craig which stands 12 miles S.W. was very plain. It is a small island that rises in the form of a rocky peak 1,040 feet out of the Firth. Today we could see the details of the ravines and cliffs, and bluffs on Ailsa and Arran, and Frank and I have been raving about coming back here after the war and cruising about this coast for a summer with an auxiliary sloop. I suppose Lake Champlain and lots of places in the States are just as pretty, but somehow this seems to have the stronger appeal. It seems so different; its people and associations are so different and (as I know really so little about its history I will say) I imagine its past glory and romance is of the kind that appeals strongly to me. I know the Vikings sallied forth from away across the North Sea, but this shore looks like a fitting place for such a race.

Of course, this high school stuff does not count for much. A fellow may be a shark here and get potted by the first Hun that gets at him in France, but just to ease your mind with the idea that, so far, I seem to be going to be able to take care of myself, I will tell you that I have been pointing the old gun pretty well. I made one perfect score on the 100 yd range with the model flying right angles to my line of sight, and the staff sergeant seemed to think it was quite exceptional, saying that very few had ever been made. That was twenty shots all in the bull's eye. Since then I have put all twenty in the bull's eye twice on the 50 yd. range when judging and allowing for the enemy machine approaching or retreating at different angles. I do not hear of any other perfect scores being made or having been made recently.

Yesterday they unexpectedly sent a bunch of the English fellows from here overseas. Some friends of mine who have not yet finished all the training.

Moving to Ayr Parr writes:

From Turnberry we used to see some interesting convoys. Ships and destroyers and dirigibles.

I am in the non-flying pool here and will be for as long as the weather is dud. Six of our fellows left here for France last night. I received a commission as First Lieut. in Aviation Section Signal Officers Reserve Corps Easter morning March 30. I do not get Lieut's pay and do not wear the uniform until I get my active service orders which will come in perhaps a week. I will cable you then. I am getting a new uniform made and will get a picture taken for you. In my next I will tell you something about this place. It is fine.

Ayr, Scotland
April 3, 1918

Dear Mother and Father,
I surely have had a wonderful day. Frank Read and I have been seeing Ayrshire on bikes. When we returned tonight, we saw a lot of soldiers about town. They surprised us and we remarked, 'There must be a war on.' I believe I edited Sherman's remark about war once before, now I want to revise it again. 'War is H—, sometimes.' Today was not one of the times. We went up to the airdrome [sic] this a.m. and learned that we would not get out of the pool, so we decided to take a walk. Coming thru Ayr we hired two bikes and started out nowhere in particular but south in general. I had my map and we thought we would go to certain hills and lakes. We followed along fine roads thru

very picturesque valleys and over hills. The little towns were very interesting. We stopped at a little inn at Kirkmichael and got lunch. Rations limited us to one and tuppence each, but the landlady treated us very well. Then we went to the little store and bought some oat crackers and chocolate candy (28¢ ¼ pound). While riding to the village we had seen a monument on the highest nearby hill. Some of the old men we talked with said there was a fine view to be seen from it, so we decided to eat our dessert up there. We rode a little way towards it, left our wheels and climbed up. This is hilly sheep raising country. Very steep and rugged hills covered with grass, and moss and a matty sort of short bushy stuff. Stone walls dividing steepest and wildest parts and a wonderful old castle and estate in every section of each valley.

We were blessed with a wonderful day. Quite cool, rather clear, and wonderful white clouds in a blue sky to make sky scenery and hide the sun every now and then.

Climbing this young mountain was hot work so we left our coats and belts on the first stone wall. The view from the top was a wonder. We nestled in the sun in a bit of a pocket in the hilltop to keep off the wind and eat our dessert while taking in the country. The moss is made up of a sort of soft bushy vine that beats grass all hollow for a mattress. The little village of Straiton, that we had come through, looked like one big stone barn. We could see one handsome castle and the estate around it and several villages tucked between the hills.

This fellow Frank is a fine mate. He is very refined and genteel and a sport from A to Z. He enjoyed it all as much or more than I did and was singing and raving all the time about living up there with some wonderful girl after the war.

After our fest and a carroos [sic] over

the map we decided to have a look at Loch Doon. So, we clambered down, got our wheels and rode further back into the hills. Finally, we came to the end of the road and the last farm house. Mr and Mrs Farmer Scotsman and their really pretty daughter were very sociable and gave us detailed directions to follow over the trackless hills. They invited us in to have a glass of milk, but we declined and started to wheel our bikes over the sheep hills. I cannot describe the grass or matty covering of the hills. It seemed to be of indefinite thickness, tuff [sic] grass, heather, matty moss, bushy moss, etc. They cut it out in slabs about a foot thick and burn it. I suppose it is peat when it is pressed and dried. We dragged our bikes along. There were flocks of sheep, shepherds, dogs, individuals walking, all around in the landscape but each group miles and miles away and apart. We passed Loch Derclach and Loch Finlas and then came over the ridge to where we could see 'Ye banks and braes O bonny Doon'. The moor and the hills, and the smaller lakes were very interesting and picturesque, but I was disappointed with Loch Doon. Its shores were so bare and barren it looked anything but bonnie. We rode home along the valley of River Doon. It surely was a fine trip and gave us a good idea of the Burns country. We covered 47½ miles. Such sights and pleasures are going to make it harder for Mr Hun to get at us, because, as Frank said, we will have to see this war thru.

It is lucky for you all that we were not over the lines for this push. Sometimes we wished we were in it, and at other times we were satisfied with the position in which providence has put us. There surely has been some doings over there. Probably we will be there in time for our push when the Hun is tired out by his own.

Ayr is the best town next to Oxford that I have been in over here. It is on

a sandy point. Half of its shoreline is developed as a park with a good street and stone promenade just where the beach and land join. It is evidently quite a popular bathing beach in summer. Back of this park is the best residential section. The houses are very handsome. They are built of grey sandstone with towers, gables and chimneys like little castles. Each place has a pretty garden and plenty of grounds around it, and all surrounded by tremendous stone walls. The streets are very wide, well paved and kept in good condition. I do not know anybody who lives here but the people in general look very nice.

The fighting school is under the command of Col. Rees, VC. He visited the USA. last fall. The aerodrome is at the racecourse. There are quite a lot of all varieties of aeroplanes and from averaging the various 'for and against' remarks they are well rigged and cared for busses. I spent my first mornings looking the place over, examining the machines, and watching the flying. I was very much interested in a Hun Albatros and an English made monoplane. They are both several years old. The Albatros is not in flying condition now. The monoplanes are wonderful little things. Their performance in the air looks so good Frank and I wanted to go to France on them. If I get a day or so in the graduate pool here, I may get a chance to fly one of them. You can see everything from them that is in the hemisphere from your horizon up, but you cannot see at all well below. That makes it difficult to land, so I hope I don't smash it for them.

The other morning they had an 'air day'. A flock of Avros (20 or 30) went out on a reconnaissance patrol, protected by SE5s flying above them. Then the Camels and Bristol Fighters went out in formations and scrapped them—air sham battle. We could not see them when they got far away, but there was a lot of excitement on and over the aerodrome when they were taking off and trying to get in their proper groups in the air. I wish I would get out of this blooming pool and get to flitting again. I have not flown for so long it will seem like a first solo when I take off again. I think we will get started tomorrow.

I have not gotten my active service orders yet, and I am really not a 1st Lieut. until they come, because my Lieut's. pay does not start until I am put on active duty. I put up all the regalia when I got my commission because even as cadets we are instructed to wear about half of it and there are so many American air mechanics in town saluting us that it is much better to be a full Lieut. when they do it. Easter morning, I started wearing the entire outfit. I have not put any wings up yet because I cannot buy any here. When I get my active service orders, I will cable you. This Lieut. dope has been drawn out and held up so long that it does not thrill a bit. I remember how tickled I was to wear those little wings on my shirt sleeve when I came home from ground school.

I wish you could see these English WAAC's, pronounced wack. They are great sports. They are mostly recruited from the ordinary class, about the same social grade as the Tommies. The ones who do not work as cooks, waitresses, maids, etc., namely those who drive cars, lurries [sic], clerk and pay office and messenger boys around the airdrome [sic], drill every morning the same as the men. It seems very funny. English drill is a scream anyway. The severity and bull voice of the Sergeant Major drill master makes the Tommies act like energetic galley slaves. The drill master puts 'the fear of God' up them and their faces look very drawn on parade. But the girls are a scream. The Sergeant Major can bellow his head off and it has no effect on them,

except perhaps it makes them smile the more. And with all their happy and talkative manner in ranks they drill pretty well. When they are given double time, the stout ones drop out and the rest jog along in broken rank as though it is the funniest thing they do.

We get good food here, but English people do not know the first thing about preparing food. It is due to too much servant [sic].

Tommy (Herbert) and I took the bikes out this morning. We went down the coast south of the Heads of Ayr and rode down to the beach and played around on it with the bikes. Then we climbed up the rocks of the cliff of the Heads of Ayr and took some pictures. Coming home we cut into a little by road, coasted down a winding and steep cow path. After several hair breadth escapes my bus finally chucked me down a ten-foot bank into a newly plowed field. No casualties. Then we discovered we both had punctured our tires and had to walk them 3 miles home. We just got back in time to get the last end of lunch.

Carleton House, Ayr.
April 7, 1918

Dear Mother and Father,
I am feeling pretty good this evening. Have had a wonderful ride on a Spad, have learned of the possibility of getting a week's leave before going overseas, and got my pay check this afternoon. Pretty good cause for delight, eh!

While it is on my mind, I will tell you about my Spad ride. This morning being Sunday morning it was a holiday. Flying started at 2.30. It has been very windy all day, but fairly clear, chunky white clouds coming and going at about 3,000. It was too windy for Avros, so I got permission to take up the Spad. Frank tried to get the use of our other Spad, which belongs to the graduate pool, so as we could joy ride together, but something was reported to be wrong with it, so I went up alone. The bus I was taking did not have an air speed meter, altimeter, or engine temperature indicator. At London Colney we thought that the temperature indicator was very important. Everybody said the engine was O.K. and that if any change at all occurred with the circulating water it would heat up. I knew I could detect heating better than I could chilling so felt satisfied. I ran her up well on the ground and gave her some good tests over the airdrome. Everything seemed O.K. so I started to take on the fun.

The rigging of the bus was fine, and I practiced some stunts. She half rolled beautifully. I never used to be able to tell just where I was or how I was doing the half roll and did not come out of it correctly very often. But with this bus today everything seemed natural and easy. From all the flying I read about that is going on in the States I presume you know just what the ordinary stunts are.

After playing about a little while I started down the coast keeping over the land. There were just enough clouds to make things interesting. As they flew by, they would hide portions of the land that you wanted to see, then you would be completely enveloped in them, then they would be above you. For the first time I really played with them. I climbed up over certain banks and kept in the sun, then I would go down through a certain opening to see the ground. The landscape was the most interesting I have ever flown over. The hills around Loch Doon rise to 2,600 feet. The three streams, River Ayr, River Doon, and Water of Girvan have well defined valleys with naturally situated towns. Then there is the coast with its coves, beaches, cliffs, and coastal hills. Then there was the Firth of Clyde well decorated today by waves and white caps

and punctured by the good old landmark Ailsa Craig. Across the firth, at times, I could see Arran island and its mountains. I was very much interested in the hills and moors that Frank and I became acquainted with on our bike trip.

I went south some miles below Girvan, came back and then struck west for Ailsa Craig at about 5,000 feet. I had been looking through hazy distance and admiring Ailsa for some weeks and I always wanted to go out and see it. I went close and low, circled around it, and gave it a real good look over. It surely is an interesting old sentinel. It is about ¼ -mile dia. and 1,080 feet high. It is 12 miles from shore standing in deep water with a narrow white fringe of beach all around its granite cliffs. The cliffs mostly rise straight up, with sharp vertical flutes, making the surface rough, to about 200 feet. Then the rest is like a rugged mountain that has been smoothed and rounded up by the wind. It has creases, and ravines and ledges. Some of the surface is bare rock and some is covered with a straggly green stuff. I suppose mossy bushes. One ledge has the ruin of an old stone tower and house on it. It was great sport frisking about the ancient old dome. Watching the wind drift you into it when to windward, and then riding the terrific bumps in its shadow to leeward where the wind was joining together after passing around it. Then I thought I had been down low long enough so climbed towards the mainland.

I was quite thrilled at the thought that this was my first flight over the sea. It was very nice to hear and feel that the engine was doing its job O.K. I saw a couple of steamers. They looked like motorboats, so small I could hardly keep my eye on them. I slipped down to two that were pretty close and played around one. It seemed to be standing still. I could dive by it as fast forward or aft and fly across its bow as easily as its stern. I rolled, half rolled, and fooled around in hope that I was amusing some old salts.

Then I went ashore and came low to inspect Culyean [sic] Castle and estate. It looked fine from a close aerial view. I hedge hopped all the way home. It is a very amusing way to travel. Everybody comes out to look and the horses and sheep gallop out of your path. I dived on some of the people to let them know I saw them and then exchanged hand waves as I departed.

One of the prettiest places was the River Doon by the old Brig and Burns Monument, there being two very pretty estates and castles close by. Then I returned home and found everybody, mechanics, instructor, ambulance, etc., very glad to see me as I was the last bus out, and they all wanted to go home. I don't care if it blows hard for a week if I can get that Spad. Frank and I are hoping to get two of them up together. I want to go over to Arran and to Glasgow.

This morning I walked out to Alloway and saw Burns's birthplace, monument, and the Old Brig o' Doon.

Carleton House,
Ayr.
April 9, 1918

Dear Mother and Father,
I have had a right disappointing day.

This afternoon I was scheduled to fly over Glasgow and Troon in a stunting formation on a Spad and drop bushels of the enclosed leaflets and snowflakes on the populace while they were gathered at a patriotic open-air meeting. Then do stunts for them and help thrill them to the point of further diggings into their purses. Well the formation went, saw Glasgow and had their fun but I remained on the airdrome [sic] with Mr Spad who refused to work properly. We fiddled and

fooled all afternoon but could not get the engine to perform properly. I learned considerable about the bus and think it will go tomorrow.

The flight that Frank Read is in got a new Spad today so we have hopes of taking some trips together. I'll see Glasgow yet.

Won't it be nice to fly over John Brown's Yard and look it all over from a position where Mr Luke cannot throw you out (assuming that they have no anti-aircraft guns stowed in the yard).

Carleton House,
Ayr.
April 13, 1918.

Dear Mother and Father,
This week has been rather uneventful. The flying has been rather informal. We have gone up about as we liked when we could get a bus to ride in. My Spad has been a bit balky on starting but have had several more good rides in it. The weather has been very thick and hedge hopping usually was the order of the day. Thursday, I took up my first passenger. I was starting off in an Avro. The mechanic, a very young (and foolish) chap asked me to take him up. He strapped himself in the front seat and away we went. I say 'foolish' because a fellow with a good safe ground job should never trust his neck with a novice like me. It was very hazy, and we had to stay close to the ground. I guess he thought I gave him a rather tame ride as he looked pretty cool and unperturbed all the time. It seemed awfully funny to fly an Avro with a load of two people in it after buzzing around in a Spad. She loops so slowly you think you are on a worn out ferris wheel, and you feel like you wanted to do something like smoke a cigarette or bat flies off the wall to pass the time away while she is turning over and sliding out of a half roll. A spin felt like we had

anchored our tail to a parachute.

I nearly got lost again yesterday. I went out to practise flying in the fog. Thought maybe I could develop a good sense of direction by exercising it. It is some sport flitting about in the darkness of fog without any instruments. One of our fellows humorously remarked about my adopted bus 'That is the kind of a machine I like, because it has nothing in it to worry about. It has no stagger, no dihedral, no inherent stability, no air speed indicator, no engine temperature indicator, no radiator shutters, no altimeter, and no compass, but it's got the gliding angle of a brick, and therefore is as much fun to fly as a shingle.'

The great Capt. McCudden VC, MC, DSO, etc. ad infinitum has just been stationed here as assistant CO. He has given us a couple of lectures. He is a very decent chap. He tries to get us very bloodthirsty and conveys the idea that it is the easiest thing going to get a string of 50 or so Huns. I would love to hear Baron von Ristoffen [*sic*] (or whatever the Hun ace's name is) talk to the Hun pupils.

Carleton House,
Ayr.
Scotland
April 17, 1918.

Dear Mother and Father,
Yesterday I took my first ride on an SE5. The type of scout that is doing so well in France now. It was not a particularly good one, not as tight and smooth mechanically as a Spad but very easy to fly although every time I tried to half roll it to the right, she would full roll. The propeller torque is very strong, and you have to account for it all the time. It makes a lot of difference whether your engine is off or on. She will hardly spin to the left but goes around as fast as a Spad when nose spinning to the right.

Carleton House,
Ayr.
Friday April 19, 1918.

Dear Mother and Father,
Frank and I had been wishing to get a bus off together and have a joy ride. Wednesday afternoon we wangled it. I sat in front (of Avro) for the first time and flew the bus. It feels a bit strange in the front seat as there were no instruments and the position of the 'stick' is close into you. Also, the view and the lack of any length of fuselage ahead of you is quite strange.

We started out in great glee. I took my camera. We were going to try to take each other's pictures, I was going to try to put the wind up Frank, stunting and hedge hopping, and we were going to see some mutually interesting country. But we were very muchly bemoaning the fact that although we had everything it seemed possible for our hearts to desire, we did not have a girl to go see. Just think of it—an aeroplane, a mate along making it possible to land and get started again, wonderful romantic country full of beautiful castles and estates, the time to make the trip, but no girl. It was very tough luck.

Well we went up and I found it (naturally) impossible to worry Frank with any of my contortions and ravine scuttling so we went up and took our pictures (not developed yet). While doing this we got lost, it being quite misty as usually. So, we explored some unknown hills and moors and made west for the coast. (I could see where the sun was without climbing up above the clouds.)

Just before we got to the coast our engine started missing on one cylinder. When a rotary misses it goes pretty dud. I tried to nurse it by gliding and starving and diving and trying various combinations to clean her out but miss she would. I could not see the revs, speed, or petrol pressure, but as Frank was not trying to communicate any alarming news to me, and as we seemed to be limping along fairly well, we continued. We got the coast below Girvan and flew back to Turnberry where we landed on the aerodrome. I gave the engine a looking over. It seemed to be intact mechanically, so we started her up and came home, keeping an eye out for field to land in in case she let us down. We got home all right. One spark plug porcelain had cracked and prevented that cylinder from firing.

I took a ride on a Hun Albatros the other day. It was a lot of fun. It flew so differently from our busses and was so roomy in the office. The 6 cyl. Mercedes engine sounds very nice and chug-chuggy. She is rather old, so we do not strain her by tossing her about, I had just a nice ladylike ride, gentle curves and climbs and dives. It was very interesting to see where the Hun couldn't see you.

My idea of fun now is to fly strange machines. I wish I could get a chance to fly a Camel, a Bristol Fighter and a Bristol Monoplane. As you know now, I have flown six different kinds of busses, Rumpties (Maurice Farmans), Avros, Pups, Spads, SE5s and Albatroses.

Now I will begin the main burden of this note. I went to a wonderful party Wednesday night. It seemed they needed some more dancers and as Americans are considered, over here, to be dancers I was invited along with three other Americans, and the main personnel of British Flight Commanders and Captains. We were the only rank so low as 1st lefts., and everybody else had up a couple of decorations. Captain McCudden was in the party.

We got in automobiles and rolled about five miles out into the country to a castle. It was a wonderful house. The kind of architecture that has fine big halls and staircases and yet extensive and ambling sort of inviting you to get lost, and when you do you find yourself in some very cosy room. The place was beautifully furnished. I don't believe there is anything in the States that can touch these British castles. Our multimillionaires may build grand homes, but they cannot reproduce the attractive and hospitable atmosphere that results from the care and taste of many generations. You would love these places.

We were relieved of our duds and went upstairs to the music room and met the host and the guests. O my! How it tickled me to cast an eye on those girls. I realized in an instant that there was none of that disappointing London beauty in the place. This was real Scotland and with a vengeance. Well to make a long story longer, Catonsville had nothing on that gathering.

We danced downstairs in the main hall on a floor that looked like they had been polishing it since Cromwell. It was so dark and polished. At the start the music was a gramophone, but pretty soon we brought down one of the numerous pianos, somebody produced a drum and cymbals and we had very clever music by several of the present talents.

One of the first things I did after picking out my favourites, I told them of the tough luck Frank and I had experienced and made arrangements never to let it happen again. Not the engine trouble mind you, the girl trouble. It seems that I have been beating around the bush quite long enough so I might as well 'fess up I discovered a very interesting friend. Some party, and it all lasted until 3.15 a.m.

You should have seen the supper. The swellest wedding imaginable in peace time could not touch it. There was a most marvellous planked fish about 3 feet long with exquisite trimmings that did not get touched. Then there were all kinds of things like oyster patties, jellied meat, whipped cream and cake and fruit. I was not blotto' from the 'pre-war liquor' but I only have a very hazy impression of the great display of fancy eats.

There was champagne punch made up fresh every ten minutes that foamed up in your glass like you were taking it right out of the bottle. And the goblets. They would have made good wash bowls. They were cut glass but reminded me more of things I have seen made of stone and bronze, namely, their bases looked like the pedestal of a monument and the cup like an inverted church bell. And all of this was spread in the most entrancing room. Beautiful dark woodwork and panelling, private little niches and circular coves where the towers were formed, raised floor to them, and marvellous old furniture. It's lucky for me that she had me a bit blotto, or these other things would have.

Without going into the multitude of lovely details— The next morning I grabbed an SE5, and went on a hunt up the river Ayr for Sorn Castle. It was my second ride on an SE and this particular bus was much better than the first one I took up. Doing 110 mi/hr it did not take me but an instant to get there (13 miles). Everything about it checked up with my study of the maps. I knew she would not be at home because Sunday afternoon is the first time I could see her, but I thought I might as well get a look at the place. So, I staged a bit of a show for her maid and the countryside. I am not much of a circus artist but with that bus anybody could put on a right good show. The most fun was to dive steeply at the house, pull her up, push on engine and zoom up about 1,000 feet practically vertically then instead of putting her nose down forward pull it back and go off in a loop. I managed to half roll fairly well this time, but the best-looking

stunt must have been the roll and a half to the right that she did beautifully every time.

Frank did not go on the party. I told him all about the girls. There are two of them that live within four miles of each other and we planned to go call on them. The other girl, Miss Latta, did not say that she was going to be away, so we decided to take a chance this morning. Frank did not turn up at the regular time. I got rather excited waiting for him while the Avros were leaving one by one, so finally I decided to buzz off taking a boy mechanic with me to start the bus again. I located the place as marked on the map just after my engine started to run on eight cylinders instead of all nine. I ventured down and circled around but could not see anything that looked like the house. There were two collections of what I thought were stone barns, and an inn. I decided that that locality was Failford but that the castle itself must be someplace else and I would have to gain more information about it. So, I tried to make Sorn Castle. The bus would not climb but the engine gave just enough revs to keep us poking along. We got to Sorn Castle and I put her down in a nearby field, making a bum cross wind landing. I left the youngster to watch the bus and I beat it for the castle. A constable and a couple of farmers came up. They directed me to the castle and added that I would be very welcome there because an aviator had given a wonderful exhibition over it yesterday morning.

A respectful and pleased old butler let me in. No, Miss McIntyre was not in. She had taken the early train to Glasgow, but I could come in and he would be very pleased to show me to the phone. I called up my Flight Commander and told him about the engine and its location. He said he would send somebody out, and then I could buzz off and enjoy myself, a knowing and sporty Scotsman (Flight

Commander Gerald Joseph Constable Maxwell), he realized that I did not put her down at a castle for nothing. So, then I called up Miss Latta, got the location of her house, and said I would try and get to see her after the mechanics arrived. While I was waiting to get these calls through, I was ushered by the butler into a charming big room. It was about 25′ × 100′ with a break of two steps across the floor about 1/3 way from the end and pillars and arches up to the high ceiling at that section, sort of making two rooms of it. There was a fine fire going in a cosy diagonal fireplace near this divisional section. One end of the room had a big church organ built into it.

Mrs McIntyre came in and charmed me with her hospitality and kind interest and walked out on the grounds a little way towards the bus with me.

When I got back to my bus, I did a few of the things I should have done immediately on landing. Turned her into the wind and made a more thorough examination of the engine. Although some small part was rolling around inside of it, all the tappet rods and valves seemed to be working O.K. In a few minutes we heard a machine coming in the air and an instructor with a mechanic landed. We changed the bad plugs and after a little fiddling got her running again. By this time the wind had increased considerable and a storm was just starting. The engine ran up so well that I decided to fly her home.

Just before I took off, I saw Mr and Mrs McIntyre standing at the edge of the field looking on. I hope I can feel as pleased every time I take off on a dud engine as I did that trip. Because I thought to myself, she is bound to go enough to get off and if she conks out, I can plant her down near Miss Latta's house. But she gave pretty good revs. and not much shaking so we buzzed on home through the sleet storm.

The weather has been bad all this afternoon. Frank and I are going calling tomorrow morning if we can get a bus.

Carleton House,
Ayr.
April 21, 1918

Dear Mother and Father,
Today has been a prize. I stayed in bed until 11:30, and this evening has been spent sitting around a fire in our anteroom. Ha-Ha! You wonder how I can call this a prize day. It is because of this afternoon. Just one little afternoon.

I had an engagement to take tea at Sorn Castle with the McIntyres. The only way to get there was via aeroplane, so I was very anxious about the weather. It was rather windy all morning and I was afraid Avros would be 'washed out'. When I arrived at the aerodrome to my joy, I saw them wheeling out the Avros. I consulted my Flight Commander about what I could take up. He said to flip an Avro. I suggested that if I was going to fly an SE5 I would like to do it early in the afternoon as I had a fine chance to take an Avro to tea. He got me right away and said, 'O well then you might as well wash out the SE, as there are others I want to fly it today and buzz off.' Two o'clock seemed very early to leave for tea, but there was a storm brewing, a little rain (almost sleet) was falling, so we decided we had better get off before we were made to stay home. Frank sat in front and I piloted from the rear. The storm passed with a little rain and pretty good bumps.

We went first to Failford House and played around to see if we could attract Miss Latta's attention. They all came out and we could see Miss Latta among them. We staged a few simple stunts. An Avro with two people is a pretty slow and clumsy bus. Then we saw their message LAND laid on the lawn in letters about 6 feet high. We had no intentions of landing there, but in the face of that invitation we could not resist. I could not find a field. All the land is cut up in little ridges due to the streams and winding river. From 500 feet almost everything looks level. We dropped down to a couple of fields but were always disappointed and had to go up again. Then I decided on one and resolved to plant her and plant her we did—right on a hillside. We did not run along the ground 15 feet, and our tail skid sticking into the ground kept us from rolling down the hill backwards.

The usual crowed gathered, and in a few minutes the Latta party arrived. We met Mr and Mrs, her brother, a Captain in infantry, and two guests. They invited us to stay, but of course we could not.

Our take off was quite thrilling. We did about 20 mi/hr up the hill, zoomed a hedge fence before we had flying speed and then started to fly off the other side of the hill and got away. We went immediately to Sorn and on our second trial got a field where the ridges ran near enough into the wind to stall her down.

A tremendous mob, Sunday afternoon, gathered from all points of the compass. The constable was among them, all dressed up in civvies and we left our bus in his care and beat it to the castle. Miss McIntyre soon came to us and then Mr and Mrs and two guests a Mrs Somebody and a Mr Weir.

We had a wonderful visit. Walked over the wonderful picturesque estate, Frank and I with Miss McIntyre Then Mr took me over part of the castle. It surely is a wonder. 600 years old and kept up to top notch. All modern luxuries, yet all the old beauties preserved and brought out. He has one bedroom of beautiful oak panels. The only furniture is a magnificent old carved 4 poster (and roofed over) bedstead. Every panel opens like a door and behind them were concealed the slickest furniture you ever

saw. Desks, wardrobes, chest of draws, mirrors, modern bathroom, etc. There is no hardware (latches, handles, or hinges) visible from the outside. Even the door by which we entered was a panel and I could not find which it was. The view from the windows was wonderful. The cliffs and banks of the river are all gardened with vines, rhododendron and flowers.

I cannot begin to describe even the small part of the place that I saw. The spiral stone steps in a tower with look out windows, loopholes, and signal bells, his two studies, one down in the foundations with stone arched roof.

We had a fine tea. Afterwards Mrs played the organ. Then we got out the Victrola and had a dance.

It surely was a wonderful place and wonderful people. Frank and I had the time of our lives.

Our take off was again very exciting. The field was pretty large. It was bordered to windward and the two adjacent sides by large pine trees. There seemed to be considerable eddy down from them, because we got off the ground O.K., but could not climb over the trees. I held her down to keep flying speed, and it looked about like we would strike them about their middle (up and down). Just as we got to them the bus seemed to get a grip on the air and responded to a zoom which took our nose over their tops and apparently got us into good air because we flew over them very nicely. I turned half a circle. The engine was running regularly but did not seem to have sufficient pep. The rev. counter was out of whack. Of course, we were supposed to put on a show for the folks below. We barely had time to get home before the mechanics were due to leave (6 p.m.) so I thought I would consult Frank. We were up 200 feet. I shut the engine off (so as to be able to talk to Frank) and asked him what he thought of the engine. Before he could turn his head around the prop stopped dead. I struck her nose straight down. To my surprise she did not take. Down we went, luckily just clearing an edge of woods. Still the prop stood still. I dove her right to the ground and as I pulled up the prop started turning and took the throttle and pulled us up over the opposite trees. I cannot imagine what made it stop. You can always shut off the engine and the motion of your glide keeps the prop. slowly spinning around so that you can turn on the petrol and start her pulling any time. Also, a slight dive ought to start a dead prop on an Avro. I decided for sure then that the engine did not suit me so buzzed off home.

I have been in lot worse situations when by myself and they only affect me as being interesting, but with Frank in the front seat with no belt, it was not particularly funny. We got back just in time to go in the hangars with the last machine to come in.

Carleton House,
Ayr.
April 26, 1918

Dear Mother and Father,
I have finished all my training now and am ready to try and do something. I have flown an SE5 enough to be able to play around with it, but I still cannot land it properly. I have done in two undercarriages since I have been here.

Instead of going into the pool here and hanging about until I am assigned to some duty, I am going away on 7 days embarkation leave. So that while they are getting me assigned, I can be having a trip. The Whiting's have written to some friends in Glasgow and they have invited me to call. I expect to get into the highlands.

Frank Read has not finished his time yet so I will have to go alone.

I had a fine party Wednesday evening. I went to dinner at McIntyre's. She had a Red Cross class (as pupil) on until 8 p.m. I hired a bike and abided by the bike-man's advice—allowed myself plenty of time for the hills. It was a very good wheel, three speeds. I left here at 5:50 and got there—Sorn Castle at 7:10. The wheel worked so nicely, the afternoon was so fine, and the country so pretty that the 15 miles seemed very easy. Perhaps the goal had something to do with it. I pedalled about the immediate country until 7:50.

I wish you could have seen the dining room, and service. Absolutely class. Also, an elegant meal, not to mention the company.

Mother and Father let me talk to Alison for about ¾ of an hour after dinner, and then joined our party. I had a lot of fun dancing to the Victrola. Then Mrs played the organ and Father and daughter and I talked.

I left at 10:30 and had a beautiful bright moonlight ride home, practically all downhill.

After a few days leave, Parr Hooper was posted to Lympne, near Folkstone in Kent, where he was billeted in Lympne Castle. From here he undertook a couple of Ferry Flights before writing from France:

May 12, 1918

Dear Mother and Father,
I have been assigned to No. 32 Squadron, 9th Wing, Royal Air Force (RAF) of the British Expeditionary Forces (BEF) and will go there to-morrow.

They are using the type of machines I like the best.

I had a lovely cold swim in the surf of the English Channel all by myself this afternoon.

There is nothing much to tell about today. We landed here about 10:30 after a slow train ride and a long tender ride. We are bunking in frame huts. Meals O.K. I expected to go to my squadron after dinner today, but the transportation did not come. My baggage is all packed so I must hurry and unpack it so as to get out my cot and go to bed.

On 15 May, Lieutenant Hooper took part in his first Offensive Patrol over the Hun Trenches at Albert, Bapaume and Douai.

On 10 June 1918 flying SE5a C9625, 1st Lieutenant Parr Hooper was leading a low altitude patrol, strafing German troops over No Man's Land when he was seen to fall slowly from the air and crashed near Château de Sorel. It is thought that Gefreiter Wilhelm Laabs of Jadgstaffel 13 shot him down.

32 Squadron
BEF
France

11th June 1918

Dear Mr Hooper:
Your son did not return from patrol on the morning of the 10th of June 1918.

He was leading a patrol and the pilots who were with him have no definite information. He was seen to go down very slowly and might only be wounded.

The Germans were making an attack and the patrol were shooting up the troops from a very low altitude and doing most excellent work under his leadership.

The short time that he was in this Squadron he proved himself to be exceedingly brave and a good leader. He will be a great loss to the flying corps, US Flying Corps and especially this Squadron at the present time.

He would have been with me only a few weeks longer, as I should have sent him, as a Flight Commander to the US Flying Corps.

In the mess he was one of the most popular officers and will be greatly missed and I only wish that I had more like him. This whole squadron and myself wish to send you their deepest sympathy and only hope that he is a prisoner in German hands being well cared for.
Yours Very sincerely,
J. C. Russell, Major.
D.C. 32 Squadron.

1st Lieutenant Parr Hooper is remembered at the Aisne-Marne American Cemetery and Memorial, Belleau, Departement de l'Aisne, Picardie, France

Lieutenant Edward Frank Hollander

Edward Frank Hollander was born at Philadelphia, Pennsylvania, in 1897. In 1917 while at the Officers' Training Camp at Plattsburg, New York, Hollander was selected for aviation training, graduating from ground school in August. He was then posted for flight training overseas and sailed on the *Carmania* to Britain as part of the 2nd Oxford detachment.

On arrival, Hollander was sent to 2 School of Military Aeronautics at Oxford and, in November, on to Harrowby Camp, near Grantham, for training on Lewis and Vickers machine guns. Further training followed at Northolt and London Colney, before he arrived at Turnberry in May 1918 for a course on aerial gunnery, where he was one of the pallbearers at the funeral of Lieutenant George Squires. From Turnberry he moved to Ayr to train in aerial fighting. It was at Ayr that he met and married local girl Helene McGregor Lochhead on 6 September 1918.

Lieutenant Edward Hollander was seemingly not assigned to an operational squadron. The only movement recorded on his RAF service record states at '1 Stores Depot' in October and November 1918.

Lieutenant Edward Frank Hollander and his Scots wife sailed for America on the USS *Plattsburg* in August 1919. They settled in Rhode Island, and he was employed as a salesman for a tool manufacturer until his death in 1962.

Lieutenant Rueben Lee Paskill

Rueben Lee Paskill

Son of a carpenter, born in Ohio in 1893, Paskill graduated from Hastings High School in 1911. Prior to enlisting he had been a student at Chicago's Armour Institute of Technology, then becoming a Civil Engineer with the Virginia State Highway Commission.

Serving with the USAS, Lieutenant Rueben Lee Paskill was attached to the RAF and trained at Turnberry in May 1918, leaving in June, being posted to 32 Squadron on the Western Front.

He died on 9 August while flying SE5a E1327 on offensive patrol and was last seen at 5,000 feet over Misery, some 30 miles east of Amiens. Ltn. Rudolf Klimke, Jasta 27 claimed him as his thirteenth victory.

He is remembered at the Somme American Cemetery and Memorial, Bony, Departement de l'Aisne, Picardie, France

Lieutenant Alexander Miguel Roberts

Alexander Miguel Roberts was born in 1895 in Mexico City, Mexico, the son of Hiram Ridgway Taliaferro Roberts and Concepcion Galvez. Enlisting in May 1917, he was a student of Electrical Engineering at Mississippi State University.

Arriving in Britain with the 2nd Oxford Detachment, his flying training was not without incident. In December 1917, he was injured when the Maurice Farman Shorthorn B1970 he was flying stalled at 50 ft, nosedived and was written off at 25 TS. After training at Turnberry and Ayr he was posted to 74 Squadron, where on 6 June 1918 he had another prang, this time in a SE5a C8845. When landing from a special mission, the tail skid buried itself in a deep furrow and broke a longeron. Lieutenant Roberts was unhurt. The next day he was slightly injured flying SE5a C1779 when he had to force land due to engine failure during an operational patrol.

On 19 July while flying SE5a E5948 on his first action against enemy aircraft, Roberts successfully shot down the first plane he engaged. As he watched the enemy plane go down, he failed to see the German ace pilot, Josef Jakobs, flying a Triplane, who caught him off guard and shot through and disabled his aircraft. Josef Jakobs claimed him as his 24th victory.

Lieutenant Roberts was posted as missing by the RFC, he was last seen over Menin. In September word was received that he had been taken prisoner of war, notified by a message dropped by German Aviators.

Having survived the crash landing, he spent many weeks in captivity, and made several attempts to escape. In one bid for freedom, he jumped off a moving train while being transferred between prison camps in Liege, Belgium, and Aachen in Germany. He tried to reach allied lines, but the German Army captured him again. In another attempt he escaped and hid in the Black Forest for nearly a week before being forced to return to the camp due to hunger.

While Roberts was in prison camp at Karlsruhe, he received a visit from the German ace who shot him down. Jakobs told Roberts that he felt relieved to learn that he had survived the crash and informed him that the pilot of the aircraft that Roberts had shot down also survived.

In 1919 Lieutenant Roberts was repatriated to the United States where he continued his career as a pilot. He flew in air shows and cross-country air races becoming well-known nationwide in newspaper and radio articles of the time. Alexander Roberts was keen to use his celebrity status to promote the aviation industry and urged his hometown and other communities in Mississippi to build more airfields ready for the new 'Aerial Age.'

In 1942 Roberts joined up to serve in World War II. He was promoted to the rank of Lieutenant Colonel and served as an official aide and aviation advisor to the US Army.

Lieutenant Colonel Alexander Miguel Roberts died age 92 in 1988 in Tampa, Florida, and is buried in the National Cemetery in Bushnell.

Lieutenant John Arnold Roth

John Arnold Roth was born in 1892 in St Louis,

Missouri, the son of John Henry Roth and Alice Stiensmeyer. A graduate of the University of Missouri, he was working for the US Department of Agriculture in Washington D.C. when he enlisted in June 1917. Arriving in Britain with the 2nd Oxford Detachment for further training at Turnberry and Ayr, John Roth was posted to Central Despatch Pool where he was injured while ferrying Bristol Fighter D8021 on 1 June 1918. The Casualty Card for the incident has the comment, 'Is he alive?' written across it in pencil.

Posted to 60 Squadron on 21 October 1918, Lieutenant Roth flying SE5a F545, 'Landed rather fast and broke wheel in force landing at Savy aerodrome after losing his way while delivering message to 87 Squadron.' Lieutenant John Roth only served a few weeks with 60 Squadron before the Armistice.

Sailing from Brest in France in February 1919, with other returning Military Personnel, on the SS *Northland*, John Roth disembarked at Philadelphia, Pennsylvania, and returned to civilian life working in the animal feed business.

John Arnold Roth served with the USAF from May 1942 until November 1945, retiring with the rank of Lieutenant Colonel. Lieutenant Colonel John Arnold Roth died in 1972 at Cape Girardeau, Missouri, USA.

His son, John Rodes Roth, Combat Infantryman, who had been awarded the Purple Heart whilst serving with General Patton's Third Army was killed in killed in action at Neustadt, Germany, in 1945.

1st Lieutenant Harold Goodman Shoemaker

Harold Shoemaker was born in Bridgeton, New Jersey, USA, in 1892. He volunteered for the United States Air Service in the summer of 1917 and on 14 July he was sent for training with the Royal Flying Corps in Britain. After training he joined 74 Squadron RAF on 3 July 1918, where he scored five victories as an SE5a pilot, before he was reassigned to the 17th Aero Squadron on 29 August 1918.

In October while flying a Sopwith Camel on a mission to bomb German ammunition dumps over Wambais and at Esnes in France, his plane

was in mid-air collision with that of Lieutenant Glenn Dickinson Wicks. It was reported that his aircraft 'was seen going down behind German lines in a tailspin'. Initially listed as missing, the International Red Cross later reported that Harold Goodman Shoemaker died in a prisoner of war camp on 23 October 1918 in Germany. He is buried in the Somme American Cemetery at Bony, Picardie, France.

1st Lieutenant George Clarke Squires

George Squires, born in St Paul's, Minnesota, in 1896, was a student when he joined the officers' training camp at Fort Myer, Washington D.C., in 1917. In July, George volunteered for the flying service and was posted to Camp Borden, Canada for training. On 17 August he was involved in a mid-air collision which resulted in the death of the pilot of the other aircraft. Squires was entirely exonerated. After completing his training in Canada, George was sent to Fort Worth, and on 19 January 1918 with the 17th Aero Squadron he sailed for England. After training at 65 TS, now 1st Lieutenant George Squires arrived at Turnberry in May for a course in aerial gunnery.

1st Lieutenant George Squires, aged 22, was killed in a flying accident at Kirkoswald on 18 May. The following letters were sent to his mother:

My dear Mrs Squires,
I take the liberty of writing to you because I had the privilege of being with George at his last station in Scotland and because I know somewhat of the 'void' that is in your heart after receiving the more or less cold-blooded official communications of his 'going up.'

You may perhaps remember me as a Flying Officer introduced to you on the station platform at Chillicothe and I also talked to you from Garden City Camp.

I happened to be in the air myself at the time of George's accident and when I came down and learned the horrible news, it was almost as hard to believe as that the sun would not shine again.

George had done about thirty hours flying in 'Camels', the type of machine he was up in. He really liked to fly them and could fly them better than any of our other fliers that have watched. I am flying 'Camels' myself and know to what extent he surpassed me in handling what is really one of the most difficult machines to fly. He was feeling his best that morning I know, as I had sat beside him at the table.

He had asked a number of times for a machine that morning and had made several flights besides the necessary ones for the course in fighting, which we were engaged in.

You have probably been informed of the details of the matter, but I can perhaps add a few.

He came down some distance from the Drome and the only actual witness was a labourer, who was questioned at the 'inquiry'. He had heard the sound of an aeroplane overhead and told of a sound which the flying officers present interpreted as a stall on a turn and a spin into the ground, from what height we have no very definite idea.

George had been working particularly at rolls, half rolls and the 'falling leaf stunt' and the machine, as I said before, being a most tricky and treacherous one, it does not take long to come to grief. He was a splendid pilot and apparently physically fit, and what happened is what is liable to happen to anyone, no matter how long his experience at this game. There is, of course, the possibility of something wrong with the machine, but at that station they are much better and better maintained, and any that I had up were a pleasure to fly compared to the average.

The funeral took place on a warm pleasant afternoon and the procession which formed in front of the Officers' Mess was made up of Royal Air Force tenders and touring cars. The first car carried the body in an oaken coffin of Mission style, bearing a brass plate, with the inscription, 'Lieut. G. Squires, A.S.S.C., USA. Age 22 years, Died May 18, 1918.'

It was covered with Old Glory and two beautiful wreaths, largely made up of sweet peas, from the Royal Air Force and the American Flying Officers of the station. There was also a great bouquet of many coloured tulips tied with a knot of red, white and blue.

The pallbearers were all American Flying Officers: Lieutenants Roberts, Roth, Hollander, Smith, and Shoemaker and myself of the old 17th.

The procession moved along the edge of the Firth of Clyde to the cemetery in a nearby town, where the last rites were said in the impressive and inspiring ceremony of the Church of England, by an Army Chaplain. And so, we stood with sad hearts and came to the salute at the sound of the volleys of the firing squad. And I shall never forget the marvellously beautiful silver-toned bugle notes that rang out as we paid our last tribute to the remains of 'gorgeous George' as we liked to call him for his princely qualities. And surely, Mrs Squires, no mother ever had a son more loved and respected by his fellows than was George. His was a heart of gold and I am sure that America has not contributed to this maelstrom, for the 'democratization of the world', a finer gentleman.

I am getting up close to the big show now and I shall have a supreme incentive to do not my bit - but my best - against the despicable Hun, with this fresh memory of the young manhood he has sacrificed. We have now lost four of the fliers who came over with our squadron to train in England.

I consider myself the representative of Minnesota in particular in this scrap, as my home is Bemidji, although I have spent the last four years in. St. Paul and

Minneapolis, and I shall fight hard to justify her expectations of her sons. My mother passed away a few years ago but I shall try to have my father meet you if he is ever in the Cities.

My heart goes out to you for the supreme sacrifice you have paid, but no mother has more cause for pride.

Sincerely,

Lt. Ralph D. Gracie [France, 26 May]

Here follows an extract from letter by L. S. Baikie to Mrs Squires:

The Schoolhouse, Kirkoswald, Maybole, dated 23 May 1918.

I happened to be about half a mile away when your son came down. He was circling slowly down from a fair height when, as it appeared to me, he lost flying speed and nosedived. I heard the crash and ran for the place where it appeared to me, he had fallen. Poor fellow! when we got there, he was past help. A farm worker, Robert Lawrie, was there a couple of minutes before me and had your son removed from his seat; he cut the fastenings. We had him on the roadside and made a pillow for his head. We despatched a phone message for a doctor and ambulance. He was still breathing and as we did not then quite realize the extent of his injuries. A crowd was there by that time. I got the men there to take off their coats which we put over and under him. As I said before we were more than half a mile from anywhere, but we ran and got tepid water and bathed his wounds. He, however, never recovered consciousness. When the doctor arrived, he pronounced life extinct.

Although he received multiple injuries, the only noticeable wounds were one on the left cheek and another gash above his left knee. I was assured medically he suffered no pain; also, that he could not have lived. This worried me greatly.

I mean it worried me in respect that the man was lying here, too far away from a house to be removed to one and we could do so little. What little we could do we did. After the arrival of the ambulance he was gently lifted on to a stretcher and put inside.

He was flying in single-seated Sopwith Camel machine, I think of about 130 h.p. The military will know better than I do what went wrong, but it appeared to me he had engine trouble and probably miscalculated his distance in trying to land to have things put right.

It was a pathetic sight. Your son's face will haunt me as long as I live. It is however a consolation to me to have been there; we did what in us lay to soothe his last moments.

He died for us; he is your son and no doubt hard to lose and he is dead before his time; but he has died a hero's death, as much so as if he had fallen in the stricken field.

With deepest sympathy,

Yours sincerely,

L. S. Baikie.

Lieutenant Edward Russell Moore of the Air Service, Signal Corps, AEF, wrote to his mother from Turnberry, (as published in the *Evening Missourian*, 15 July 1918):

We paid our respects to one of our fellows last Thursday. I had to act as a pallbearer. It was a fellow who entered the ground school at Champaign III shortly after I did. An unlucky crash, of course – a very depressing incident which is among all too many now.

Another letter regarding the crash was written by the Commandant:

1 School of Fighting & Gunnery, Royal Air Force,

Turnberry, Ayrshire,

27 May 1918

Dear Madam:

It is with the very deepest regret and sympathy that I have to confirm the news, which you have no doubt received by telegraph, of the death of your son, 1st Lieut. George Squires, who was at the time attached to this school for a course of instruction in Aerial Fighting and Gunnery. I beg to give you the following brief particulars of the fatal accident and of the funeral arrangements.

On Saturday May 18th, at 10:30 in the morning, the machine of 1/Lt. G. Squires, who was carrying out practice in certain fighting evolutions, was seen to circle slowly and begin descending, the descent taking place apparently normally until the machine dived at a steep angle and crashed to the ground. Your son was in sole control of the machine, no passenger being carried on the type he was flying.

A Court of Inquiry investigated the circumstances on the following day, and it was not established from all the available evidence what was the definite cause of the accident. There was no structural defect suspected in the machine, as the latter had been reported in perfect condition and OK' for flying on its descent from a flight undertaken by one of the instructors immediately before the fatal one.

In view of the evidence given by a farmhand who witnessed the accident, it is considered almost certain that your son was endeavouring to make a forced landing owing to engine failure, and that he lost flying speed, (i.e., the speed sufficient to maintain a machine in the air) at a height which did not permit him to recover from the ensuing dive. The fall took place about 1,000 yards south of Kirkoswald village, Ayrshire, at a point almost exactly two miles eastward of and inland from this aerodrome.

Immediately upon the report of the accident happening, the motor ambulance with Medical Officer proceeded to the scene, and found 1/Lt. Squires dead, having been extricated from the wreckage by a local resident who was first on the spot.

The cause of death was fracture of the base of the skull, both thighs had been fractured and there were other injuries, and it is considered that death was practically instantaneous upon the machine striking the ground.

Units of the Royal Air Force are not permitted to telegraph or cable direct to the next-of-kin in cases of fatalities occurring where the nearest relatives reside out of the country, but you no doubt heard as soon as possible, after our reporting the accident to the proper British and American authorities.

In regard to funeral arrangements, the interment took place at Doune Cemetery, Girvan, at 2 p.m. on Tuesday, May 21st. Girvan is the nearest town of any size and is about 5 to 6 miles distant from this station.

The burial service was conducted by the Rev. G. B. Allen, the Church of England Chaplain to the Forces Stationed here, and the following American brother officers of the Aviation Service acted as pallbearers: - 1/Lt J. A. Roth, 1/Lt. E. Hollander, 1/Lt. A. M. Roberts, 1/Lt. R. L. Paskill, 2/Lt, H. R. Smith, 2/Lt. L. T. Wyly.

A large number of other American officers and British officers of the school attended, the latter including the Commandant, Lieut. Col. L. W. B. Rees, VC, MC.

A firing party accompanied the cortege, firing three volleys over the grave at the conclusion of the service, after which the 'Last Post' was sounded by the bugler. The coffin was covered with the American flag and three floral tributes from the American brother officers of your son, the officers of the staff of the school and the Camp YMCA authorities.

The position of the grave is No. 6, Section M, First Division, Doune Cemetery, Girvan, Ayrshire, and the arrangements for title deed etc, also for disposal, according to US Regulations, of your late son's effects, are in the hands of Lieut. G. J. Dwyer, United States Aviation, Headquarters, 35 Eaton Place, London SW, to whom all his property at this school has been handed. As a temporary measure a plain white wooden cross of the pattern prescribed for a British Officer or man, is being placed above the grave, pending later arrangements for memorial.

The interment took place in brilliantly fine weather, which, indeed, prevailed upon the day of the accident. If there is any further information which I can furnish, will you please let me know, and be assured that any assistance which it is, in my power to afford will be most willingly given.

With the deepest sympathy of all ranks at this School with yourself, and other relatives of the late Lt. Squires in the bereavement you have suffered.
Believe me to remain
Yours very faithfully,
Rees, Lieut, Colonel Commandant,
No. 1 School of Aerial Fighting & Gunnery
Royal Air Force.

As explained, 1st Lieutenant George Clarke Squires is buried in Girvan (Doune) Cemetery and is remembered on the Turnberry War Memorial.

Lieutenant Ralph D. Gracie

On 2 April 1917, American Lieutenant Ralph D. Gracie of Bemidji, Minnesota, applied to join the Air Service. He completed his initial flight training at an air base of the Royal Flying Corps, near Toronto, Canada, and in early 1918, he and the rest of the squadron arrived for training at Turnberry.

A few months later on 12 August 1918, in a letter to his father, Daniel Gracie, Samuel B. Eckert, Commanding Officer of the 17th Aero Squadron wrote, 'Your son, Lieutenant Ralph D. Gracie, left this aerodrome with the other members of the squadron. While about 12 miles over the line they encountered a strong enemy formation and, in the encounter which ensued, the wings of one of our planes was seen to give way and the machine fell into the sea, the wing spars no doubt having been splintered by bullets. When the squadron returned, Lieutenant Gracie was missing, and I fear it must have been he who fell into the sea. It will, I am sure, be a comfort to you to know that your son was loved by us all and universally admired for his character as a man and his oft proved bravery'.

Lieutenant Ralph Gracie's body was not recovered. Eventually, after the war, Gracie's grave was found in a German cemetery where he had been buried on 13 August 1918. His remains were taken home to Bemidji and he is buried in the family plot in Greenwood Cemetery.

2nd Lieutenant Lawrence Theodore Wyly

Lawrence Theodore Wyly was born in Cardington, Ohio, USA, in 1893. In 1910, at the age of 16 he was working as a teamster for a timber company, before receiving his BA from Oberlin College in 1916. Following his graduation, he enlisted in the United States Aviation Service, training at camps in Toronto and Texas before arriving in Britain on the SS *Adriatic* for further training at Turnberry and Ayr. Completing his training in July 1918, he was posted to 148th American Squadron, flying Sopwith Camels on combat operations attached to the Royal Air Force.

On 13 August 2nd Lieutenant Wyly scored the first of his four victories, bringing down a Halberstadt CL which crashed north-east of Roye. Suffering engine trouble he was forced to make a dead-stick landing behind allied lines. Sending a signal back to his squadron requesting help, Lieutenant F. E. Kindley drove out to collect him and found Wyly at an AA battery, 'having the time of his life popping away at Hun planes as they came into range'.

Two days later flying Camel E1409, Wyly brought down a Fokker D.VII out of control east of Chaulnes, again he didn't make it back to base having to force land east of Rosieres-en-Santerre after his petrol tank was shot through by anti-aircraft fire and he was wounded.

Recovered from the gunshot wound to his left forearm, Lieutenant Wyly scored his final two victories towards the end of September. On the 24th he brought down a Fokker D.VII which crashed north-east of Cambrai, and on the 26th drove down another Fokker D.VII out of control over Épinoy.

Flying Camel F6191 on operational patrol, 29 October, Wyly crashed south-west of Sepmeries after his petrol tank was shot through by enemy aircraft fire.

Belatedly, 2nd Lieutenant Lawrence Theodore Wyly was awarded the Distinguished Service Cross in 1923.

Returning to the USA, he was then employed in civil engineering, primarily for the Northern Pacific Railway Company, designing and supervising construction of railway and highway bridges. He joined the civil engineering faculty of Northwestern University in 1938 as an assistant professor of civil engineering. Wyly retired from the university in 1961 as professor of civil engineering and as director of the model truss bridge research project.

Lawrence Theodore Wyly died at Wilmette, Illinois, USA in 1981.

Lieutenant Edward Russell Moore

Edward Russell Moore was born in Audrain, Missouri, USA, in 1895. On his World War I Draft Registration Card he gives his occupation as 'Horseman'.

The following letter to his mother from Turnberry was published in the *Evening Missourian*, on 15 July 1918:

I will start in about 21 May; I had finished my course at my last station and was hanging around waiting to be posted here. One of the fellows asked me to fly down to Bournemouth and bring a friend of his back. Being a kind-hearted lad, I agreed to do the job. I had engine trouble before I started. The mechanic finally fixed me up. I took off and circled the 'drome a few times. Everything seemed ok, so I pushed off. I reached Bournemouth alright, picked up my passenger and started for home. I had enjoyed a delightful trip down; the weather was fine; naturally I felt venturesome. But of course, the same old story. The engine cut out. I picked out a lovely field but failed to turn into the wind until we were about seventy-five feet from the ground. I did not have time to pull the bus up before the wheels struck the ground. Crash! Bang! Whack! The passenger came out with a disfigured face and I nearly broke a leg and got an ugly cut across my kneecap. The machine was worthless. It was my first crash after practically a hundred hours of flying. It was the last time I flew at this other station. I hated to dirty my clean slate, but no one was hurt seriously, so I am still happy. But I have not come to the remarkable part of my story. The owner of the farm took me to his home and the mother of the family bathed my knee and bandaged me up. I was all day getting away. The mechanics came to the place late in the evening. I had dinner with the family, a boy my age, six girls and the parents; a homey party, a mother's care, a most delightful evening. When they learned my home address they asked if I knew Fayette, Well, did I? The Dean of the Girls' College in Fayette is a brother of Mr Warren's. They are going to write to him very soon. I am invited to spend my seven days embarkation leave with them. I want to stop there a couple of days, but I hate to spend too much time with people here. Not that I do not appreciate and enjoy their hospitality, but everything is rationed, you know, and it means a lot to these people to entertain visitors.

Well, back to the camp that evening. My leg was so stiff I could not fly. I did not expect to be posted until the twenty-seventh, so I pushed off to Bournemouth for a couple of days on the seaside. Our Commanding Officer is a real sport. We had nothing to do and he remembered he was a boy once himself; besides all of the Americans have tried to deliver the goods whenever on duty. I had averaged fourteen hours of flying a week after I had finished the regular course there. Quite likely Major Harcourt realised our happy training days are probably about over.

I never hope to be under a better command than while I was down there on the 'plain' [Salisbury Plain]. The staff always dropped their dignity after working hours and met us as man to man, a thing one cannot forget, especially in a foreign land away from the association of your own people.

This has been a delightful change, both from the standpoint of military training and personal enjoyment. From tents to what in peacetimes is a flourishing pleasure resort, an immense hotel which faces the sea, situated on a fairly high hill, terraces dotted with flower beds and shrubbery, tennis-courts, croquet grounds on down to a very famous golf course, then on to the beach which is about a mile from the hotel. Every modern convenience is at our service. The meals are very good for war rations – a most excellent place to get into condition for our work at the front. The work and discipline are very stiff, but that is what makes soldiers. I have practically finished a week's course on advanced ground gunnery, six hours a day on the range and two hours for lectures. Following this comes a course in aerial fighting and aerobatics under the most expert instructors, men who have spotted Huns by the dozen. It is up to us to get all we can out of this course. I'll be flying again in a few days.

The Scotch are the most interesting and hospitable people I have had the pleasure to be associated with since I have been globetrotting. Everyone volunteers to assist you; they seem to really enjoy your company. They are not as reserved and cold as the other people of this big, little island. There is not much enterprise or at least not much in evidence in the cities, but the farmers are making things hum. Acres and acres of spuds – all kinds of products. Everything is expensive – that will hold good everywhere until long after the war, I guess. I want to visit the old Scotch hills and hollows in peacetime.

We paid our respects to one of our fellows last Thursday. I had to act as a pallbearer. It was a fellow who entered the ground school at Champaign III shortly after I did. An unlucky crash, of course – a very depressing incident which is among all too many now.

Bevin and I went into Ayr last Friday, our day off here. Sunday is just another day in army life. This is near the birthplace of Bobby Burns. The natives are all very proud of Bobby and they have every reason to be too. We saw a show in the evening, two of the liveliest acts were put on by Americans. You cannot imagine how these things appeal to us.

I often sit here in the window in the evenings and dream away as I look over the craggy coast; off in the distance is a fringe of the Emerald Isle; then the sun drops down in the west. I wonder just what part of the good old states Old Sol is beaming on then. Well, I just wish I could hook onto him some evening, ride along, drop off where I cared to, then ride back the following evening.

Just three months after completing his training at Turnberry, Lieutenant Russell Moore as pilot took part in action near Thiaucourt in France with the 8th Aero Squadron on 9 October 1918. With 1st Lieutenant Gardner Philip Allen as observer, they undertook a photographic mission behind German lines. Along with two defensive aeroplanes they had just started their mission when they were attacked by upwards of eight enemy aircraft which followed them throughout their flight, all the time firing at them. Lieutenant Moore managed to stay airborne with both flying wires cut by bullets, a landing wire shot away and his elevators riddled with bullet holes. His wings were also badly damaged by gunfire, despite this, and being constantly under attack, he kept on course enabling 1st Lieutenant Allen to take pictures of the exact enemy location required. At no time did 1st Lieutenant Allen make any attempt to defend their aircraft with his machine guns, concentrating solely on the task in hand displaying disregard for their personal safety and showing a courageous devotion to duty both officers successfully completed their mission. They were both awarded the DSC for this action.

The citation reads:

The President of the United States of America, authorized by Act of Congress, July 9, 1918, takes pleasure in presenting the Distinguished Service Cross to First Lieutenant (Air Service) Edward Russell Moore, United States Army Air Service, for extraordinary heroism in action while serving with 8th Aero Squadron, US Army Air Service, AEF, near Thiaucourt, France, 9 October 1918. Lieutenant Moore, with First Lieutenant Gardner Philip Allen, observer, took advantage of a short period of fair weather during generally unfavourable atmospheric conditions to undertake a photographic mission behind the German lines. Accompanied by two protecting planes, they had just commenced their mission when they were attacked by eight enemy planes, which followed them throughout their course, firing at the photographic plane. Lieutenant Moore, pilot, with both firing wires cut by bullets, a landing wire shot away, his elevators riddled with bullets, and both wings punctured, continued on

his prescribed course, although it made him an easy target. Lieutenant Allen was thus enabled in the midst of the attack to take pictures of the exact territory assigned, and he made no attempt to protect the plane with his machine guns. Displaying entire disregard of his personal danger and steadfast devotion to duty, the two officers successfully accomplished their mission.

Returning to the USA after the war, Edward Moore was an aviator working for, among others, the Polaris Flight Academy at War Eagle Field in California. He died in California in 1969.

2nd Lieutenant Howard Rea Smith

Born in 1888 in Indiana, USA, Howard Rea Smith was educated at the University of Michigan, receiving a BA degree in Forestry and Landscape Gardening. Enlisting in the aviation service of US Army in June 1917, he passed out of ground school at Illinois. Completing further training in Canada, and Texas he was Commissioned 2nd Lieutenant, arriving in England in January 1918, with 28th Aero Squadron and he was posted to Turnberry for Aerial Gunnery training.

On 27 May 1918 2nd Lieutenant Howard Rea Smith was killed in a flying accident at Turnberry. His body was returned to the United States and buried in South Mound Cemetery, Newcastle, Indiana. He is remembered on the Turnberry War Memorial.

2nd Lieutenant Robert Schermerhorn McNair

Robert Schermerhorn McNair was born in 1895, Connecticut, USA, the son of Robert H. McNair, MD, and Antoinette B. Schermerhorn. In June 1917 he was employed as a salesman for the Goodyear Tyre and Rubber Company when he joined the RFC in Toronto, arriving in Britain on the *Carmania* in August. In November he was selected for appointment to temporary Commission and completed his flying training before arriving at 2 (Aux) School of Aerial Gunnery.

2nd Lieutenant Robert Schermerhorn McNair died at Turnberry on 17 March 1918 flying Sopwith Camel B9204, which crashed on the shore 400 yds from the Brest Rocks.

The Official Casualty Card reads:

Court of Inquiry 20538/1918 2/Lt. R. S. McNair

The court having considered the evidence are of the opinion that the cause of the fatal accident to 2/Lt. R. S. McNair, was due to lack of judgement: that the machine got into a slow spinning nose dive at a low altitude, after completing a dive at the target, and the pilot through apparent lack of judgement, failed to recover control: that he had been stunting in contravention of distinct orders to the contrary. That he was killed instantaneously on the impact with the water, and that death was not due to drowning.

He is buried at Girvan (Doune) Cemetery and is remembered on the Turnberry War Memorial.

2nd Lieutenant Edwin Charles Hull

Edwin Charles Hull was born in 1895 at Kempston in Bedfordshire, England, the son of carter and contractor, Edwin Hull and Eliza Wolfe. Edwin was living in Canada and employed as a clerk (stenographer) when he enlisted in the Canadian Mounted Rifles, Machine Gun Section, at Ottawa in July 1915. He arrived in England on the SS *Missanabie* in October that same year. Transferring to the Royal Flying Corps in May 1917 and, after training at Oxford, he was posted to Turnberry for higher training.

2nd Lieutenant Edwin Charles Hull died at Turnberry on 17 March 1918, flying Sopwith Camel B9222 which spun into the ground. He is buried at Bedford Cemetery, Bedfordshire, and remembered on the Turnberry War Memorial.

2nd Lieutenant Charles William (Bob) Janes

Charles William Janes was born in 1895 in Islington, London, son of William Janes, chemist,

and Louisa Simmonds. Before joining the RFC in December 1917, he served as a Sergeant with 2nd County of London Yeomanry (Westminster Dragoons) in Egypt and Gallipoli.

Posted to Turnberry for training, 2nd Lieutenant Charles William Janes died on 11 April 1918 flying Sopwith Camel B9210 when he lost control and crashed into the sea. He is buried in Abney Park Cemetery, Stoke Newington, London, and is remembered on the Turnberry War Memorial.

Sergeant John Stabb Tuckett 100034

John Stabb Tuckett was born in 1899 in Birkenhead, Cheshire, the son of John George Stabb Tuckett, ship's rigger, and Eva Jane Wakeham.

Joining the RFC as a Sergeant Mechanic in October 1917, John Tuckett had graduated as a pilot on 27 February 1918. Posted to 1 School of Aerial Fighting and Gunnery, Sergeant Tuckett was killed on 26 May when he fell out of Sopwith Camel B5560 about 1,000yds south of Kirkoswald.

He is buried in Birkenhead (Flaybrick Hill) Cemetery and is remembered on the Turnberry War Memorial.

Lieutenant Reginald Milburn Makepeace MC

Reginald Milburn Makepeace was born in Darlington, County Durham, in 1890, the son of John P. Makepeace, printer and compositor, and his wife Mary A. Milburn. The Makepeace family emigrated to Montreal in Canada, where Reginald worked for the Canadian Pacific Railway.

Commissioned as a 2nd Lieutenant (on probation) in the Royal Flying Corps on 17 November 1916. Makepeace was posted to 20 Squadron in June 1917, as a pilot, flying FE2d's. He scored eight victories. The squadron was re-equipped with the Bristol F2 Fighter and he brought down a further eight enemy aircraft. His seventeenth victory, in January 1918, was as an observer/gunner flying with 2nd Lieutenant John Stanley Chick of 11 Squadron.

Makepeace was awarded the Military Cross

on 26 September 1917, which was gazetted on 9 January 1918. The citation read:

> Second Lieutenant Reginald Milburn Makepeace, Royal Flying Corps, Special Reserve.
> For conspicuous gallantry and devotion to duty whilst on an offensive patrol. He and his gunner shot down three enemy aircraft in quick succession, having attacked a large hostile formation, about twenty in number, with great dash and determination.

Posted to Home Establishment in March 1918, Lieutenant Reginald Makepeace MC was assigned to 1 School of Aerial Fighting and Gunnery at Turnberry as a Fighting Instructor. He was killed on 28 May, flying Bristol F2b B1178 as an instructor with student 2nd Lieutenant Thomas McClure, when a wing folded back during flight, resulting in a crash.

The Official Casualty Card contains the details:

> Court of Enquiry 23804/1918 Lt. R. M. Makepeace and 2/Lt. T. A. McClure
> The Court having viewed the wreckage, taken evidence and duly considered it, are of the opinion that the accident was caused by one or two things, either a main spar or a drift wire breaking, allowing the planes to fold up. The evidence goes to show that this breakage was probably caused by the pilot opening up his engine too suddenly, while on a dive, thus putting a sudden strain on some part of the fittings of the right hand planes, probably weakened by the stunting which had occurred before when fighting with SE and afterwards.

Lieutenant Reginald Milburn Makepeace MC is buried at Anfield Cemetery, Liverpool and is remembered on the Turnberry War Memorial.

2nd Lieutenant Thomas Albert McClure

Thomas Albert McClure was born in 1898, a native of Dun Laoghaire, Co. Dublin. He served with the 1/2nd Battalion, Connaught Rangers, before transferring to the RFC and qualifying as an Observer in March 1918.

2nd Lieutenant Thomas Albert McClure was killed on 28 May at 1 School of Aerial Fighting and Gunnery, Turnberry, whilst flying with Second Lieutenant Reginald Milburn Makepeace. He is buried at Toberclare, Co. Westmeath, and is remembered on the Turnberry War Memorial.

2nd Lieutenant Howard Richmond Henry Butler

Howard Richmond Henry Butler was born in 1897 at Toorak, Victoria, son of Walter Richmond Butler and Emilie Millicent Butler (nee Howard).

Giving his occupation as 'architect', Howard joined the Royal Australian Engineers as a Sapper in Melbourne, Victoria, on 11 November 1915. He was discharged from RAE on 31 May 1916 at his own request and enlisted on 15 June with the Australian Imperial Force (AIF).

Now with the rank of Gunner, Howard was posted for recruit training at Maribyrnong on 19 June 1916, after completion of such he embarked from Melbourne, Victoria on the *Ascanius* on 11 May 1917 with the Field Artillery Brigade Reinforcements. He was admitted to the ship's hospital on 9 July 1917 suffering from Parotitis (Mumps) and on arrival in England was admitted to Devonport Military Hospital. He was discharged to duty on 24 July to 1 Command Depot at Perham Downs, Wiltshire, where he continued his training in and around Salisbury Plain.

Gunner Howard Butler transferred to the Australian Flying Corps; he joined the School of Military Aeronautics at Reading in early October and was taken on strength at Wendover as 2nd Air Mechanic. By mid-January 1918, he had been assigned Elementary Instruction in Aviation and was sent to 5 Training Squadron at Shawbury. Butler graduated as a Flying Officer (Pilot) and was appointed 2nd Lieutenant on 8 April 1918. Completing the Lewis Gun and Vickers Gun course 2nd Lieutenant Howard Butler arrived at Turnberry on 31 May.

On 2 June, while flying Sopwith Camel

B9262 as part of a formation flight, 2nd Lieutenant Howard Butler was killed. A Court of Enquiry was held at the next day at Turnberry by order of Lieutenant Colonel L. W. B. Rees, VC, MC, Commandant, 1 School of Aerial Fighting and Gunnery, for the purpose of investigating the circumstances of the accident.

Evidence to the Enquiry:

1st Witness. 85617. 3rd A.M. Waller F.W., states:
'I am the rigger on Sopwith Camel No. B9262 and prior to its being flown by Lieut. Butler on June 2nd, 1918 on its last flight, I examined the rigging and found it to be satisfactory.'
(Sgd.) F.W. Waller 3rd A.M. 85617.
2nd Witness. 63019. 2nd A.M. Hirst R. states:
'I am fitter on Sopwith Camel No. B9262 WD 43775 and prior to its being flown by Lieut. Butler on June 2nd, 1918, I examined the engine and found it to be satisfactory.'
(Sgd.) 63019. 2nd A.M. Hirst R.
3rd Witness. Lieut. J.H.R. Bryant RAF, G.L. states:
'I flew Sopwith Camel 9262 on May 31st 1918, which was the last flight previous to its being flown by Lieut. Butler. I then found the engine and rigging to be quite satisfactory.'
(Sgd.) J.H.R. Bryant
4th Witness. Capt. A.D.C. Browne, RAF, G.L. states: -
'On June 2nd 1918, I sent off a formation of 4 Sopwith Camels, one of which was flown by Lieut. Butler. Whilst taking off Lieut. Butler flew about 40 yds. to the left of another machine. He evidently thought he was much closer than he really was as he 'zoomed' his machine steeply to stalling point and essayed a left-hand turn. I saw his machine begin to spin. It turned twice and then hit the ground from about 200 feet and was wrecked.'
(Sgd.) A.D.C. Browne. Capt. RAF

5th Witness. Capt. I.H.D. Henderson, MC Argyll & Sutherland Highlanders and RAF states:
'On June 2nd 1918, I observed two Sopwith Camels take off one thirty yards in front of the other and about 50 yards apart. The first machine started to do a slow climbing turn to the left. The second machine also started to turn. Then the first machine pulled his nose up into a steep left-hand zoom. He did a flat turn at the top, stalled and spun into the ground from about 200 feet.'
(Sgd.) I. Henderson. Capt. RAF
6th Witness. Capt. J. Gardner, RAMC states:
'A telephone message was received at the hospital at 10.23 a.m., June 2nd 1918, to effect that a crash had occurred on the aerodrome. I proceeded there at once by motor ambulance and found the body of the pilot, Lt. Butler, had already been removed from the wreckage of a Sopwith Camel. Life was extinct. Examination revealed the following injuries to which death was due. Several small scalp wounds. Fracture of several ribs with rupture of right lung. Fracture of right thigh also severe fractures of both legs. Death occurred within 5 minutes of the crash.'
(Sgd.) J. Gardner. Capt. RAMC.
Lt. H.R.H. Butler's logbook was produced. 'He has flown 42 hours 40 minutes solo and 57 hours 45 minutes in all. He has flown 15 hours 40 minutes on Sopwith Camels.'
Finding:
The Court having examined the above evidence finds that: 'Lieutenant H. R. H. Butler met his death though turning on the top of a zoom too close to the ground. The accident committee was not required.'

The deceased officer was accorded a military funeral, firing party, buglers and pallbearers being in attendance. Major J. W. Gordon,

Chief Instructor in Gunnery, represented the Commandant at the funeral and a large number of officers, both of the staff and pupils of the school were present. Chaplain, Rev. G. B. Allen, conducted the service, and the 'Last Post' was sounded at the graveside. 2nd Lieutenant Howard Butler was buried on 5 June 1918 in Doune Cemetery, Girvan.

The obituary notice in the *Melbourne Grammar School Magazine – Old Melburnians*, gives some of his story:

> Howard Richmond Henry Butler, who was killed in an aeroplane accident on 2nd June 1918 at Turnberry, Scotland, was the only son of Mr Walter R. Butler. He was born in 1898 and entered the Preparatory School in 1909. He came up to the Senior School in 1911 and left in 1913. He was an architectural student, articled to his father, when he enlisted. He left Australia with reinforcements for the field artillery in April 1917, and obtained his transfer to the Australian Flying Corps. He qualified for wings and was appointed 2nd Lieutenant on 8th April 1918. The fatal accident occurred at the aerodrome of No. 1 School of Fighting and Gunnery of the Royal Air Force. The machine was seen to 'stall', and began to spin downwards, finally crashing from a height of 200 ft. He was buried at Girvan.

The Age, Melbourne, Victoria (5 June 1918) reported:

FLIGHT LIEUT. BUTLER KILLED

Information has been received by cable that Flight Lieut. H. R. H. BUTLER, only son of Mr and Mrs Walter R. Butler, of 'Studley', Toorak, has been killed in an aeroplane accident. The late Flight Lieut. H. R. H. Butler left Australia with Field Artillery reinforcements in April 1917 and obtained his transfer to the Australian Flying Corps in England, qualifying for his wings only last month. He gave up his profession in which he was training as an architect to join the colours and had just passed his 21st birthday.

In *Flight Global* magazine of 20 June 1918, we find:

2nd Lieutenant HOWARD RICHMOND HENRY BUTLER, Australian Flying Corps, who was killed in an accident while flying in Ayrshire, on June 2nd, aged 21, was the only son of Mr W. R. Butler, architect, of Melbourne, Victoria. He was educated at the Melbourne Grammar School, and, after serving some months in Australia with the Royal Australian Engineers at Swan Island, he enlisted in the Australian Artillery. He was in camp some time at Maribyrnong, Victoria, and left Melbourne in May 1917, with artillery reinforcements. He served in the Australian Artillery until October 4th, 1917, and then joined the Australian Flying Corps, being gazetted 2nd Lieutenant on April 8th last. He was completing his training before going out to the front.

2nd Lieutenant Howard Richmond is remembered on the Turnberry War Memorial.

Captain Hugh William Elliott

The son of William Underwood Elliot and Ann Elliot, Hugh was born in Biggleswade, Bedfordshire, in 1897. He served as a pilot with 48 Squadron on the Western Front between July 1917 and February 1918 flying Bristol F2b's, and he was credited with shooting down five enemy aircraft; so is technically an ace.

On returning to the UK he served as a Staff Officer (Instructor) at 1 Fighting School, Ayr and then Turnberry. Captain Hugh William Elliott was killed at Turnberry while instructing on 5 June 1918, flying a DH9 which crashed into the sea. His pupil, Lt. Richard Brumbach Reed, US Air Service, was also drowned.

Captain Hugh William Elliott is buried in St Peter Churchyard, Boxworth, Cambridgeshire, and is remembered on the Turnberry War Memorial.

Lieutenant Richard Brumbach Reed
Richard Brumbach Reed was born in 1891 in Ohio, the son of John Perry Reed and Estella (Brumbach) Reed. Enlisting in the Reserve Corps in July 1917, he attended the School of Military Aeronautics at Ithaca, New York, before arriving in Britain for further training in May 1918.

Lieutenant Richard Brumbach Reed was killed at Turnberry on 5 June flying as a passenger under instruction in DH9 C1231, with pilot Lieutenant Elliot, when 'part of the fuselage broke away and interfered with the working of the empennage thus throwing the machine into an uncontrollable spinning nose-dive into the sea.' His body was not found until nineteen days later and, as was policy at the time, was laid to rest in Scotland. His remains were repatriated at the end of the war and he was buried in Woodland Union Cemetery, Van Wert, Ohio. Lieutenant Richard Brumbach Reed is remembered on the Turnberry War Memorial.

Captain Ian Henry David Henderson MC

Sir David Henderson

Born in 1896, Ian Henry David Henderson was the only son of Lieutenant General Sir David Henderson KCB, KCVO, DSO, Commanding Officer of the RFC, and Dame Henrietta Caroline (née Dundas). After graduating from the Royal Military College, Sandhurst, in January 1915, he was commissioned as a 2nd Lieutenant in the Argyll and Sutherland Highlanders. Ian Henderson was seconded to the RFC in August 1915 and after flying training was promoted to Lieutenant in June 1916, then appointed Flight Commander with the acting rank of Captain.

He was posted to 19 Squadron flying the BE12, where he scored his first three victories and was awarded the Military Cross, gazetted in October 1916:

2nd Lieutenant (Temporary Captain) Ian Henry David Henderson, Argyll &

Sutherland Highlanders.
For conspicuous gallantry and skill on several occasions. He drove down a machine out of control, and two days later dispersed six enemy machines which were attacking his formation. A few days later again he brought down an enemy biplane, the observer being apparently killed. A week after this he attacked and drove down another machine which had wounded his leader. He has also carried out several excellent contact patrols and attacked retiring artillery and a kite balloon.

In February 1917 Henderson was reassigned to 56 Squadron, shooting down a further four Albatros DVs., and he represented his father at the memorial services for British Ace, Albert Ball, who was killed in action on 7 May 1917.

Returning to Britain, Ian Henderson was appointed to the General Staff, until March 1918 when he arrived at 1 School of Aerial Fighting, Ayr, and then Turnberry. Captain Ian Henderson was killed on 21 June 1918, while flying in a DH9, testing the Lewis gun with Lieutenant Redler. It is stated on the official Casualty Card that, 'The cause of the accident is in our opinion due from the evidence that there is nothing to state definitely and the cause must be left open pending the report from the Accidents Committee whom are investigating the case.'

The findings of the Accidents Committee report states that the aircraft stalled and spun in. Captain Henderson was killed instantly, and Lieutenant Redler died later of his wounds.

Captain Ian Henderson is buried at Girvan (Doune) Cemetery (along with his father Sir David Henderson) and is remembered on the Turnberry War Memorial.

Lieutenant Harold 'Herbert' Bolton Redler MC
Harold Redler was born in 1897, a citizen of Moorreesburg, South Africa, and West Monkton, Somerset.

On 15 March 1918, Lieutenant Redler MC, whilst serving with 24 Squadron, flying an SE5a, shot down Adolf von Tutschek of 'Jagdstaffel 12' near Brancourt. Von Tutschek was second in value to the German Air Service, only to the infamous 'Red Baron', Von Richthofen. Lieutenant Redler had already completed a tour of duty with 40 Squadron and had scored ten victories. On returning to the UK Harold was killed in an accident at Turnberry flying a DH9 with Captain Ian Henderson MC.

The official report of their fatal accident at Turnberry states that the aircraft stalled and spun in. Captain Henderson was killed instantly, and Lieutenant Redler died later of his wounds. Even after the passage of many years, rumour hangs in the air. It had been suggested that both men were 'fooling around' at the time of the accident.

Gerald Maxwell and Ian Henderson

I received a visit from the niece of 'Herbert' Redler, who was tracing her family tree. She was able to confirm the rumour that, yes, the airmen were engaged in horseplay. A photograph, signed on the back by a witness, Mrs Telfer of Glenhead farm, stated that both men were on the 'outside' of the aircraft when it went out of control and crashed. It seems it was a favourite prank to swap seats in mid-air. Mrs Telfer tended Harold Redler until further help arrived at the scene.

The notification of his award of the Military Cross appeared in the *London Gazette* the day after he died:

T./2nd Lt. Harold Bolton Redler, Gen. List, and RFC

For conspicuous gallantry and devotion to duty. He encountered four enemy two-seater machines and attacking the lowest drove it to the ground with its engine damaged. Later, he attacked one of five enemy two-seater machines, and drove it down out of control. He has destroyed in all three enemy machines and driven three others down out of control. He continually attacked enemy troops and transport from a low altitude during operations and showed splendid qualities of courage and determination throughout.

Supplement to the *London Gazette*, 22 June 1918.

Lieutenant Harold 'Herbert' Bolton Redler MC is buried in St. Augustine Churchyard, West Monkton, Somerset, and remembered on the Turnberry War Memorial.

2nd Lieutenant John David Dunbar

John David Dunbar was born in 1897 at North Portal, Saskatchewan, Canada, son of Alexander Dunbar, farmer, and Hannah Hopkins. John joined the Royal Flying Corps as a Cadet in Toronto in August 1917 and received a temporary Commission in January 1918. Arriving in Britain for further training, 2nd Lieutenant John David Dunbar was posted to 1 Fighting School Turnberry.

On 25 July 1918, carrying out aerial gunnery practice, flying Sopwith Dolphin E4437, 2nd Lieutenant John Dunbar was killed when the wings of the aircraft folded up while diving at raft target. 2nd Lieutenant John David Dunbar is buried in Girvan (Doune) Cemetery and is remembered on the Turnberry War Memorial.

2nd Lieutenant Raymond Hinton Grove

Raymond Hinton Grove was born in 1892 at Yarraville, Melbourne, Victoria, the son of Joseph and Margaret Ann Grove (nee Lodwick). He enlisted at Keswick, South Australia, on 4 August 1915, with the Australian Imperial Force.

Raymond was an Engineering Surveyor, 23 years old, and newly married to Myra Irene.

With the rank of Acting Corporal, Raymond Hinton Grove was posted to 'A' Company, 2nd Depot Battalion, for recruit training. He was transferred to 7th Reinforcements of 27th Battalion on 16 September 1915.

Raymond embarked from Adelaide, South Australia on the *Medic* on 12 January 1916 then embarked from Alexandria on SS *Oriana* on 21 March 1916 for Marseilles, France. His service records document his 'failing to comply with an order' while based at Etaples and he was reverted to the ranks.

Now Private Raymond Grove, he was attached to 7th Brigade Machine Gun Company on 22 July 1916 and was transferred to 7th Australian Machine Gun Company in mid-November 1916 and then to the 1st Anzac Light Railway Unit with the rank of Sapper in February 1917. He was again transferred this time to the 17th Anzac Light Railway Operating Company in June 1917 and was promoted to Temporary Corporal by August.

He was transferred to AFC (Australian Flying Corps) in October 1917 and joined 1 School of Aeronautics at Reading as a Cadet on 2 November 1917. By March 1918, he had been posted to 34 Training Squadron at Tern Hill for Higher Instruction in Aviation and in June graduated, and was commissioned 2nd Lieutenant. He was then posted to 1 Fighting School, Turnberry for a course on gunnery.

On 19 August 1918 2nd Lieutenant Grove was killed in training, as a result of an aeroplane accident when the aeroplane he was flying nose-dived into the sea two miles south of Turnberry Aerodrome. His body was not recovered at the time of the accident.

The Royal Air Force report on the incident records, 'After completing dive at target the pilot commenced the climb steeply. The engine cut out and the machine stalled and spun into the sea. Body not yet recovered.'

A Court of Enquiry was held at Turnberry on 20 August 1918 by order of Lieut. Col. L. W. B. Rees, VC, MC, Commandant, 1 Fighting School,

for the purpose of investigating the circumstances attending the accident of Sopwith Camel B9212 as a result of which 2nd Lieutenant R. H. Grove was missing and presumed dead:

Finding:
1st Witness. Lieut. H.T.J. Jagger. RAF states:
'I flew Sopwith Camel No. B9212 for about an hour on 17th August 1918. The engine and machine were then in perfect order.'
(sgd.) Herbert T.J. Jagger. Lt. RAF
2nd Witness. 2/Lt. T.C. Cox. A.F.C. states:
'On August 19th I was doing gunnery practice on a Sopwith Camel along with 2/Lt. Grove, diving on Raft No. 3. I had just completed a dive and was watching 2/Lt. Gove diving in his turn. 2nd/Lt. Grove dived down on to the target to within about 200 feet of the sea and then flattened out. He appeared to have some difficulty in getting his engine again, but nevertheless he did a climbing turn to the left apparently without his engine and stalled the machine which spun into the sea.'
(sgd.) 2/Lt. T.C. Cox A.F.C.
3rd Witness. Captain. A.D.C. Brown. RAF states:
'On August 19th 1918, I sent 2/Lt. Grove up on Camel B9212 to do gunnery practice. This machine was inspected on the morning of the 19th August and was in perfect order. This machine was flown by 2/Lt. Jagger on August 17th and since that day has only been flown by 2/Lt. Grove.'
(sgd.) A.D.C. Brown. Capt. RAF
4th Witness. 41578. 1/AM Entwhistle. RAF states:
'I was the rigger in charge of Sopwith Camel B9212. I examined the machine before it was flown by 2/Lt. Grove on August 19th and found it in perfect order.'
(sgd.) 41578. 1/AM Entwhistle. H.
5th Witness 58419 1/AM Woolgar G.F. RAF states:

'I was the fitter on Sopwith Camel B9212 I examined the engine before it was flown by 2/Lt. Grove on August 19th and found it in perfect order.'
(sgd.) 58419 1/AM Woolgar G.F.
The Court having considered the evidence are of the opinion that the accident to Sopwith Camel B9212 was due to an error of judgment on the part of the pilot, 2nd Lt. Grove.

His body was recovered from the sea on the 28 August 1918.

2nd Lieutenant Raymond Hinton Grove was buried at 2 p.m. on 30 August in Doune Cemetery, Girvan. From the burial report of 2nd Lieutenant Raymond Hinton Grove – 'Coffin was good polished Oak. The deceased officer was accorded a military funeral, firing party, buglers and pallbearers being in attendance. The coffin was draped with the 'Union Jack' and surmounted by a beautiful wreath presented by officers of the school. The service was conducted by the Rev. G. Gibson, and during the ceremony three 'Camel planes', driven by comrades of deceased, circled round the last resting place. The 'Last Post' was sounded by buglers in attendance. An oak cross will be erected on the grave. Administrative Headquarters, A.I.F. London were represented at the funeral.

2nd Lieutenant Raymond Hinton Grove is remembered on the Turnberry War Memorial. His best friend Lieutenant Will Parkes, of Creswick, Victoria was killed on 1 September 1918 in an aeroplane accident at Leighterton, Gloucestershire, when RE8 D4975 stalled and spun into the ground.

2nd Lieutenant Archibald McFarlan

Archibald McFarlan was born in 1893 in Mozufferpore, Bengal, India, son of Scotsman, Archibald McFarlan, an indigo planter, and Eleanor Henrietta Nicholson. Prior to the war Archibald was a surveyor, living in Canada, when he joined the 31st Canadian Infantry Battalion (Alberta Regiment) CEF at Calgary in February 1915. He married Alice Penzer in Canada in April, just before sailing for active service in France. In September 1917, after serving two years and suffering continual painful bouts of Metatarsalgia, for which it was recommended that he wear special boots, he was transferred to England for posting to the RFC Cadet Wing. His wife Alice moved to the Isle of Man during the war, and a son, also Archibald, was born in 1918.

After initial flying training, now 2nd Lieutenant Archibald McFarlan was posted to 1 School of Fighting and Gunnery at Turnberry on 3 August 1918. He was killed in the mid-air collision of DH9 C1334 and a Sopwith Dolphin on 23 August. His body was found in a field on Minnybae farm.

2nd Lieutenant Archibald McFarlan is buried in Girvan (Doune) Cemetery and is remembered on the Turnberry War Memorial.

Flight Cadet Andrew Anderson Hepburn

Andrew Anderson Hepburn was born in Dunfermline, Fife, in 1900, son of William Hepburn schoolmaster, and Elizabeth Anderson. Joining the RAF, he was attached to 2 Cadet Wing. Hepburn was posted to 1 Fighting School, Turnberry, on 17 August 1918, for training. Whilst employed as an Observer on fighting practice in De Havilland DH9 C1334, Flight Cadet Andrew Hepburn was killed on 23 August, when the aircraft was involved in a mid-air collision. His body was later found in Ballochneil Wood. Flight Cadet Andrew Hepburn is buried in Dunfermline Abbey Churchyard and is remembered on the Turnberry War Memorial.

Captain James Martin Child MC

James Martin Child was born in Leytonstone, North London, in 1893, the son of Tylney Harris and Constance Octavio (Oxley) Child. He emigrated to Canada, working in banking and mineral prospecting. At the beginning of the war, he enlisted in the Canadian militia but was unable to deploy with the Canadian forces and returned to England at his own expense in February 1915 on the SS *Adriatic*. He was commissioned into the Durham Light Infantry, transferred to the Manchester Regiment and was then seconded to the Royal Flying Corps. 2nd Lieutenant Child received Royal Aero Club Aviator's Certificate 2377 on 31 January 1916. After serving with 4 Squadron, he was posted to 19 Squadron flying SPADs in July 1916. Child scored three victories with 19 Squadron in 1917. He was Mentioned in Despatches and on 28 September 1917, the King of Belgium invested him as a Chevalier in the Order of Leopold II. Later that year, he moved to 84 Squadron as a Captain and Flight Commander and he claimed five more victories flying the SE5a. In February 1918, he was relieved from combat duty, he returned to Britain, and was posted as an instructor to Turnberry.

On 15 March 1918, he was awarded the Croix de Guerre (Belgium), followed by the Military Cross, gazetted on 5 July 1918:

> Lt. (T./Capt.) James Martin Child, Manchester Regiment and RFC
> For conspicuous gallantry and devotion to duty. While leading a patrol he encountered four enemy scouts, one of which he destroyed. On another occasion he attacked one of two enemy two-seater machines which he encountered over the enemy's lines. He disabled the machine, and skilfully turned it towards our lines, where the enemy pilot was forced to land, and he and his observer were taken prisoner. On another occasion he attacked five enemy scouts, one of which he destroyed. He showed the greatest judgment and determination.

Captain James Martin Child MC was killed on 23 August 1918, when a DH9 flown by 2nd Lieutenant Archibald McFarlan, with Flight Cadet Andrew Anderson Hepburn acting as observer, collided in mid-air with another aircraft, causing both aircraft to crash and killing the pair outright. A report in the *Daily Record* (Glasgow) on 29 August 1918 stated: 'Their machine collided at a considerable height with another one, the occupant of which had a miraculous escape. The pilot of the second aircraft survived the impact and Child was on hand to help rescue him from the wreckage. A good deed done, he then set off on his motorcycle to return to Turnberry, but while riding along the Kirkoswald Road at a little after 6.20 p.m. he had an accident'. The Court of Enquiry that followed recorded that 'the cause of the accident was due to the Pilot being blinded by dust by a passing lorry, thus preventing him from seeing the tender.' A fractured skull killed him instantly.

There is reference to Child in the diary of 'Pop' Lewcock, which was subsequently donated to the Shuttleworth collection:

> Aug 23, 1918: Sopwith Dolphin collided with DH9 in the air. 2 Killed. Lt. ?, the Dolphin pilot climbed out on the tail and saved himself. Motor accident. Capt. Childs killed. Collided with Crossley tender when returning from the crash.

Captain James Martin Child MC is Buried at Chingford Mount Cemetery, Essex, along with his brother, Lieutenant Jack Escott Child, a Sopwith Camel pilot with 45 Squadron, who died of pneumonia during the influenza pandemic in 1918. James Child is remembered on the Turnberry War Memorial.

Major Cyril Edgar Foggin
Born in November 1891 at Newcastle-on-Tyne, son of William S. & Elizabeth J. Foggin. Cyril Foggin was one of Britain's pioneer aviators. In June 1911 he was apprenticed to Maison Borel-Morane, Oise, France, and gained his pilot's licence in 1912, flying a Bleriot Monoplane at the Eastbourne Aviation School, England. His flying training wasn't without incident though as the following *London Times* newspaper article of 25 September 1912 reports:

AVIATION ACCIDENTS - NARROW ESCAPE FROM DEATH

At Eastbourne early yesterday morning, Mr Cyril Foggin, of Newcastle, was flying in an Anzani Blériot monoplane from the local aerodrome, where he had been under instruction, when he apparently lost control of the machine, which fell from a height of 60 ft on to the grass at St Anthony's Hill. The monoplane was wrecked and Mr Foggin fell on his head but was so well protected by his safety helmet that he escaped unhurt.

He had his own Blackburn Type D Monoplane, built in 1912, and in 1913 he joined Armstrong Whitworth & Co. Ltd., Aviation Department.

Cyril enlisted with the 6th Battalion, Northumberland Fusiliers, in August 1914, stating his occupation as 'Aviator'. He transferred to the RFC as a non-commissioned officer in September, serving with the Expeditionary Force in Egypt until 27 September 1915 when he received a Commission as 2nd Lieutenant.

Moving to France with 1 Squadron, 2nd Lieutenant Foggin was wounded on 26 April 1916, while flying Nieuport 16 A118, when he drove down an enemy aircraft over Bailleul:

2nd Lieutenant Foggin on a Nieuport of 1 Sqn left the ground on seeing a hostile machine near Bailleul. He quickly overhauled it and fired at close range. Tracer bullets were clearly seen to strike the fuselage. Difficulty was experienced in getting the Nieuport slow enough to keep pace with the German machine without fouling it, and whilst 2nd Lieutenant Foggin was doing a half turn to take off speed, one shot hit his machine and one strut was shot through, the splinters wounding him in the left eye and he had to break off the fight. The German machine had six bombs, estimated at 40 pounds each, strung underneath, three on each side of the undercarriage wheels. The Albatros was observed by anti-aircraft artillery still going down and was thought to have been hit.

Returning to England, he served as an instructor at 12 and 25 Reserve Squadrons, 56 Training Squadron, and by October 1917, 1 School of Aerial Fighting at Ayr. While at Montrose (25 RS), Major Cyril Foggin saw a ghostly figure enter the officers' mess but did not initially report it, fearing he would be considered mentally unstable and he would be removed from his post, so kept it to himself until there were further sightings of the spectre by other officers and flight instructors. In all Foggin was to 'see' the apparition eight times, eventually filing an official report. (More can be read of this in *Scottish Ghosts*, by Dane Love.)

Posted to 41 Squadron RAF at Auxi-le-Château on 10 July 1918 as a Staff Officer, Cyril was killed in a car accident returning from a party at Dieppe on 29 July. Major Cyril "Billy" Marconi Crowe was the driver of the Crossley tender that skidded and crashed into a tree that night. Foggin and Captain Owen John Frederick Scholte died from their injuries the following day; three other passengers in the vehicle were also seriously injured. At the Field General Court Martial, Major Crowe was given a severe reprimand and lost seniority for driving a car against orders.

Major Cyril E. Foggin is buried in St. Riquier British Cemetery, Somme, France.

His Royal Air Force Service Records state that since joining the RFC he had flown 77 types of aircraft and since 1911 had logged over 2,500 flying hours. Interestingly, there is an airworthy Bleriot XI at the Shuttleworth Collection in Old Warden, Bedfordshire. This is the world's oldest airworthy aircraft. It was built in 1909 and is powered by a three-cylinder 'W form' Anzani engine.

The very Blackburn Type D Monoplane owned by Cyril E. Foggin is currently part of the Shuttleworth Collection and is Britain's oldest airworthy aircraft.

2nd Lieutenant Yvone Eustace Sutton Kirkpatrick

Yvonne E.S. Kirkpatrick

Born in 1899 at Shrewsbury in Shropshire, Yvone was the son of prominent civil engineer, Cyril Reginald Sutton Kirkpatrick and Pauline Elizabeth Courthope Last.

Joining the RFC in May 1917, he was awarded his Royal Aero Club Aviators' Certificate in September at the Military School, Ruislip, flying a Maurice Farman Biplane. Kirkpatrick was then posted to 1 School of Aerial Fighting at Ayr in April 1918, and 2 (Aux) School of Aerial Gunnery, Turnberry, prior to being posted to 203 Squadron, flying Sopwith Camels in France. He was credited with one kill and at least three shared.

Kirkpatrick was wounded on 31 July, suffering a gunshot wound to the arm, and was discharged to hospital and transferred to the unemployed list in July 1919.

After the war, Yvone Kirkpatrick was a schoolmaster at Wimborne Minster in Dorset, serving in the Territorial Force during WW2. He died in Wimborne, Dorset in 1975.

Gerald Joseph Constable Maxwell MC, DFC, AFC

Born in September 1895 at Beauly, near Inverness, the son of the Honorable Bernard Constable Maxwell and the Honorable Alice Fraser, Gerald was the nephew of Lord Lovat, founder of the Lovat Scouts. Educated at Downside School in Bath, at the outbreak of war he was commissioned in the Lovat Scouts on 4 August 1914 and served at Gallipoli, then Egypt, before returning to Britain, where he transferred to the Royal Flying Corps in September 1916.

Commencing his pilot training at the Central Flying School, Upavon, he received only 22 minutes of dual instruction before he soloed for

the first time in December. Obtaining his 'ticket', he was posted to 56 Squadron in March 1917, flying SE5a aircraft. In April 1917, the squadron was sent to France and he scored his first victory during his first patrol over the lines on 24 April 1917. A few days later his aircraft was shot down by anti-aircraft fire and he escaped unhurt.

Before returning to Britain in October 1917, he scored a total of twenty victories. On 22 November 1917, Captain Gerald Maxwell was appointed as a Fighting Instructor at 1 School of Aerial Fighting at Ayr. His first flight there was in an Avro 504, and during the next month he flew Avros and SE5s, passing on his experience of aerial fighting to his pupils.

On a typical wet and windy Ayrshire February day, Gerald Maxwell and Captain C. E. Foggin, known throughout the RFC as 'Foggy', flew from Ayr to Turnberry to fly the captured Albatros Scout which was there. They spent the night at Turnberry,

Gerald Maxwell

taking another flight in the German aircraft before returning to Ayr. The weather having cleared, a great deal of flying training was done at 1 SofAF. Gerald was practising aerial fighting with student, Lieutenant J. H. F. Baker, flying SE5a's when Baker, flying C5319, misjudged his distance and crashed. Gerald landed immediately and pulled the badly injured student from the wreckage.

In March, Captain Gerald Joseph Constable Maxwell was awarded the Military Cross:

For conspicuous gallantry and devotion to duty on many occasions. He has taken part in forty-three offensive patrols, in fourteen of which he acted as leader. He has destroyed at least three enemy aircraft and driven down nine others completely out of control. He has consistently shown great skill in aerial combats, and his fearlessness and fine offensive spirit have been a splendid example to others. (Supplement to the *London Gazette*, 7 March 1918)

Gerald Maxwell

On 11 May 1918, Captain Maxwell and 1 School of Aerial Fighting moved to Turnberry, amalgamating with 2 (Aux.) School of Aerial Gunnery, to become 1 School of Aerial Fighting and Gunnery. Gerald was in command of the SE5a group, the others being Camels, DH9s and Bristol Fighters. Training at Turnberry started three days later but was quickly halted when four aircraft crashed in one day, on 17 May – three SE5a's and an Avro. Gerald commented, 'Washed out flying in afternoon because very few machines.'

During June and July 1918, he rejoined 56 Squadron, scored 6 more victories and was awarded the Distinguished Flying Cross:

This officer has at all times shown exceptional skill and gallantry, and on numerous occasions has fought against greatly superior numbers. During the last six weeks he has brought down five enemy aeroplanes. Recently, he approached unobserved to within ten yards of three Fokker triplanes, one of which he shot down. He was chased for about nine miles by the remaining two until he met a formation of six Camels; these he led to attack some enemy aircraft, although he had only twenty-five minutes petrol left. (Supplement to the *London Gazette*, 3 August 1918)

L to R Leacroft, Atkinson, Gerald Maxwell, Taylor and Le Gallais

Gerald returned to Turnberry and became Chief Fighting Instructor, taking over from Major Gordon on 17 August 1918. On the first day in his new role, Monday 19 August, a student, 2nd Lieutenant Raymond Hinton Grove AFC, dived a Camel into the sea and despite an exhaustive search neither aeroplane nor body could be found and recovered.

Gerald Maxwell was granted a permanent commission as a Captain in the Royal Air Force on 1 August 1919, which he resigned on 14 February 1921, becoming a Director of Maxwell-Chrysler Motors Inc., and joining the Stock Exchange until he was called up in WWII, attaining the rank of Wing Commander. His younger brother Michael also served in the Royal Air Force during WWII and also became an ace.

Maxwell's son, William Michael Constable Maxwell, was killed in 1950 whilst on flying duty with the RAF in Hampshire. Gerald Joseph Constable Maxwell MC, DFC, AFC died in 1959 in Winchester, Hampshire, England.

Major James Byford McCudden VC, DSO & Bar, MC & Bar, MM

Perhaps the most well-known of the aces to spend time at Ayr and Turnberry was James Byford McCudden. Although he didn't train here, he was on strength for a time as an instructor.

Born in 1895 into a military family, James McCudden enlisted in the Royal Engineers as a bugler at the age of 15. After transferring to the RFC as an air mechanic, McCudden quickly rose through the ranks, and was one of the first who went to France with 3 Squadron in the first few days of the war. He served firstly as an observer and gunner, then he was selected for pilot training in January 1916, receiving Royal Aero Club Aviator's Certificate 2745 on a Maurice Farman biplane at Gosport on 16 April 1916.

Flight Sergeant James Byford McCudden was an exceptional pilot, with the reputation of being a perfectionist, and within days of receiving his aviator's certificate he became an instructor. By the time he was posted to France in June he had logged 121 flying hours, personally instructed forty student pilots and given 177 lessons.

Arriving at 20 Squadron as a front-line pilot in in July, McCudden was then reassigned to 29 Squadron scoring five victories before being posted as a Fighting Instructor on 'operational rest' on 23 February 1917. Returning to France in July, his time with 56 Squadron, flying the SE5a, McCudden brought down a further 52 enemy aircraft. He received his commission to 2nd Lieutenant, on 28 December, which came into effect on 1 January 1917, and on 6 February he was awarded the Military Cross for his fifth victory.

L to R, McCudden, Henderson and Maxwell

It was about this time that the War Office decided to use individual servicemen for propaganda and news purposes, and in conjunction with the *Daily Mail* ran a campaign to this end. On 3 January an article appeared on the front page about McCudden under the headline 'Our Unknown Air Heroes'. This was quickly taken up by other publications with titles such as 'Young Lionhearts of the Air' and 'Our Wonderful Airmen—Their names at Last', with accompanying pictures of McCudden and other aces, much to his embarrassment. He detested this 'lionising' and hero worship and felt that it would make him unpopular within the RFC and among his fellow airmen.

Returning to Britain in late February, he was once more posted as a Fighting Instructor and was awarded a Bar to his MC:

James McCudden and FB16D

2nd Lt. (T./Capt.) James Thomas Byford McCudden, MC, Gen. List and RFC
For conspicuous gallantry and devotion to duty. He took part in many offensive patrols, over thirty of which he led. He destroyed five enemy machines and drove down three others out of control. He showed the greatest gallantry, dash and skill.
Supplement to the *London Gazette*, 18 March 1918

The same month it was also confirmed that he had been awarded the Victoria Cross, Britain's highest award for gallantry:

2nd Lt. (T./Capt.) James Byford McCudden, DSO, MC, MM, Gen. List and RFC
For most conspicuous bravery, exceptional perseverance, keenness, and very high devotion to duty.
Captain McCudden has at the present time accounted for 54 enemy aeroplanes. Of these 42 have been definitely destroyed, 19 of them on our side of the lines. Only 12 out of the 54 have been driven out of control.

On two occasions, he has totally destroyed four two-seater enemy aeroplanes on the same day, and on the last occasion all four machines were destroyed in the space of 1 hour and 30 minutes.

While in his present squadron he has participated in 78 offensive patrols, and in nearly every case has been the leader. On at least 30 other occasions, whilst with the same squadron, he has crossed the lines alone, either in pursuit or in quest of enemy aeroplanes.

The following incidents are examples of the work he has done recently: —

On the 23 December 1917, when leading his patrol, eight enemy aeroplanes were attacked between 2.30 p.m. and 3.50 p.m. Of these two were shot down by Captain McCudden in our lines. On the morning of the same day he left the ground at 10.50 and encountered four enemy aeroplanes; of these he shot two down.

James McCudden at Turnberry

On the 30 January 1918, he, single-handed, attacked five enemy scouts, as a result of which two were destroyed. On this occasion he only returned home when the enemy scouts had been driven far east; his Lewis gun ammunition was all finished and the belt of his Vickers gun had broken.

As a patrol leader he has at all times shown the utmost gallantry and skill, not only in the manner in which he has attacked and destroyed the enemy, but in the way he has during several aerial fights protected the newer members of his flight, thus keeping down their casualties to a minimum.

This officer is considered, by the record, which he has made, by his fearlessness, and by the great service which he has rendered to his country, deserving of the very highest honour. Supplement to the *London Gazette*, 2 April 1918

Major James McCudden arrived at No. 1 School of Aerial Fighting at Ayr on 10 April 1918 having just been to Buckingham Palace on 6 April to receive his Victoria Cross from King George V and his promotion.

Shortly after arriving at Ayr, he learned of the death of his brother, John Anthony McCudden, also an RFC ace, who had been shot down and killed. His other brother also a fighter pilot, William Thomas James McCudden had been killed in 1915.

McCudden's perfectionism in flying also extended to all things technical. While at the front he was always 'tinkering' with his guns and working with his mechanics to improve the performance of his aircraft. He made a series of modifications, such as fitting high compression pistons to his engine and removing any excess weight. He also improvised by experimenting with equipment taken from captured enemy aircraft. While at Ayr he was able to indulge in unlimited flying and 'tinkering'. He enjoyed experimenting with new combat manoeuvres, usually with another instructor, and these mock 'dog-fights' were closely watched by the students on the ground.

It was at Ayr that McCudden first flew the Bristol M1c, flying C4955 on 15 April, he climbed to 10,000 feet in seven minutes, and then went on to 12,000 feet in nine. The next day he managed to reach 22,000 feet in thirty-five minutes, recording a speed of 99 mph at that height. These figures were above and beyond official performance figures for the aircraft, albeit the aircraft was lightly loaded and unarmed.

Instructors were encouraged to use the aircraft

outside duty hours, it was considered that if the students saw them being used as an everyday means of transport, it would promote an 'air-mindedness' and their confidence in flying would grow. It was at Ayr that Major McCudden and Captain James Douglas Latta 'borrowed' an Avro 504 with the intention of flying to Failford House, near Ayr, to visit the Latta family, landing in a small field in front of the house. Captain Latta's young sister Mary, was enthralled to meet this, 'Young Lionheart of the Air', who through the press propaganda had become the idol of feminine hero-worship, and coaxed the shy, reluctant McCudden to take her for a flight even though it was strictly against regulations for civilians to fly in military aircraft.

James McCudden

Heading out from Failford the Avro's engine abruptly misfired and stopped, McCudden realising that they would not make it back, immediately looked for a suitable place to make an emergency landing. His 'illegal' passenger, Miss Latta, was totally unaware of the events unfolding and thought that this was all part of the fun. McCudden selected a suitable place to land and glided in on his approach to touch down, noticing too late that there was a wide ditch across the field, right in front of him. With only time to shout a warning to Miss Latta, the aircraft undercarriage caught in the ditch, the propeller dug in and the Avro overturned, landing on its back. They both escaped unhurt, although this left McCudden with the problem of summoning help to recover his 'unofficial' flight, which meant that his CO would be informed. Returning by road to Failford House, McCudden telephoned Ayr to request transport for himself and a guard for the wrecked aeroplane. A Crossley tender was sent out from Ayr with driver Ruby Williamson, who drove back to the airfield with a rather subdued McCudden as passenger.

At the inevitable disciplinary meeting the next morning, James thought it best to confess all, and face his punishment. This was an awkward situation, for a distinguished flyer, a recipient of the VC to be hauled in front of his Commanding Officer for a severe reprimand, but he met his equal in Lionel Rees, also a VC, who had no anxiety about giving him a ferocious verbal dressing-down before declaring the matter closed. Rumour soon flew around the school that McCudden was guilty of some major misdemeanour and had received a severe admonishment, but very few ever found out why.

When No.1 School of Aerial Fighting moved to Turnberry in May, Major McCudden went with them. He commented on the clear weather that surrounded Turnberry and Girvan. Mr Bailkie the Kirkoswald schoolmaster lodged one of the instructors from Turnberry, with his family, and other visiting officers were always made welcome. McCudden and his colleagues were frequent visitors and 'played with the children by joining in their elaborate game of "Knights of the Round Table", armed with clothes props, and McCudden playing Sir Lancelot, they would cavort around the garden in childish glee. He also got a great deal of enjoyment tormenting the domestic servant, an old lady, whose broad Scottish accent he pretended that he could not understand, trying to persuade her to teach him the language.'

Ever conscious of his recent award of the VC, McCudden was anxious that he was not seen to be guilty of 'line shooting' and playing on his status, although he was naturally shy, some meeting him for the first time mistook this for aloofness. He was rarely to be seen in the mess, and in the evenings was busy with his writing, which perhaps his fellow officers, knowing nothing of his planned book, considered anti-social. He also kept a pet English bulldog called 'Bruiser', which had been a gift from a lady admirer. The ladies in the WRAF Clerk Section at Turnberry were not so enamoured though, and were quite fearful of this menacing looking animal, which had free range of the clerk's room. The door to McCudden's office was always open, and his dislike for paperwork could be seen as it was often filed straight from his in-tray into the wastepaper basket!

Parked outside his office was 'Pot Belly', a Vickers prototype FB16D. Given to him by its designer Harold Barnwell, it was for his personal use. The sight of such a unique aircraft at Turnberry obviously drew the admiration and attention of the students and McCudden would get very cross when pupils were seen to climb into the cockpit. Very possessive of his 'personal mount', he detailed Air Mechanic Tom Somerville, to take responsibility for looking after 'Pot Belly'. His job was to protect the aircraft from over-enthusiastic students and keep them out of it.

As at Ayr most of his flying around Turnberry was short local training flights, formation flying, cross-country sorties and aerial fighting demonstrations with other instructors closely watched by the pupils.

McCudden earned the utmost respect and admiration of the other instructors and pupils at Turnberry, as they observed his commitment his role, his attention to detail and the continual striving to improve the efficiency of aerial fighting tactics and enhancing the performance of his aircraft.

Major James McCudden took his last flight at Turnberry on 25 June 1918, when having received notice of his posting back to operational flying, he flew in an Avro to Failford to say goodbye to his friends the Lattas.

He was awarded the Distinguished Service Order as in Supplement to the *London Gazette*, 5 July 1918:

2nd Lt. (T./Capt.) James Byford McCudden, VC, MC, MM, Gen. List and RFC
For conspicuous gallantry and devotion to duty. He attacked and brought down an enemy two-seater machine inside our lines, both the occupants being taken prisoner. On another occasion he encountered an enemy two-seater machine at 2,000 feet. He continued the fight down to a height of 100 feet in very bad weather conditions and destroyed the enemy machine. He came down to within a few feet of the ground on the enemy's side of the lines, and finally crossed the lines at a very low altitude. He has recently destroyed seven enemy machines, two of which fell within our lines, and has set a splendid example of pluck and determination to his squadron.

This was followed by the Distinguished Service Order Bar. According to the Supplement to the *London Gazette*, 18 July 1918:

2nd Lt. (T./Capt.) James Byford McCudden, VC, DSO, MC, MM, Gen. List, and RFC
For conspicuous gallantry and devotion to duty. He attacked enemy formations, both when leading his patrol and singlehanded. By his fearlessness and clever manoeuvring, he has brought down thirty-one enemy machines, ten, of which have fallen in our lines. His pluck and determination have had a marked effect on the efficiency of the squadron.

On his journey to command 60 Squadron in France, he was killed in a tragic accident on 9 July 1918 at Auxi-le-Château. He was twenty-three years old.

Major James McCudden had flown in almost every aircraft type available to a British Fighter pilot in WWI, he studied and developed new tactics, and fought duels with most of the German Aces, scoring fifty-seven victories, twice scoring four kills in one day. He was never defeated in combat. While doing all this, he also found time to write his memoirs *Five Years in the Royal Flying Corps*. Major James Byford McCudden is buried in the Wavans British Cemetery, Pas de Calais, France.

2nd Lieutenant Thomas Charles Richmond Baker DFC, MM & Bar

Thomas Baker was born in Smithfield, South Australia, in 1897. Prior to joining the Australian Imperial Force in July 1915, he had been employed as a bank clerk.

Posted to the Western Front with an artillery unit in February 1917, Baker was awarded the Military Medal for his bravery in making repairs to a damaged communications line while under heavy enemy fire. He was awarded a bar to his decoration in June 1917, when he fought and extinguished a fire in one of the artillery gun pits, all the while with the risk of approximately 300 rounds of ammunition and high explosive igniting.

In September, Thomas Baker applied for transfer to the Australian Flying Corps as a mechanic, he was instead nominated for pilot training, which he undertook in the United Kingdom, and graduated 'Class A' at the end of March 1918 and was commissioned 2nd Lieutenant.

Arriving at Turnberry in May 1918, on 1 June he wrote home:

Dear all,
Just note the address. This is a School of Aerial Gunnery and Fighting, and a jolly good time I am having here too.

We, that is to say, three of us came up here from Chalford last Tuesday week. Until yesterday we have been doing gunnery, rearming machine guns on the ground, now we have commenced flying again and every time we go up, we have to do show flights, stunting, shooting etc. So, you can gather that the education in aerial warfare is very complete. This is a finishing school before going to France.

We are billeted in a hotel belonging to the Glasgow & Nth. British Railway Co. It is considered the most up to date 'pub' in Scotland and it really is glorious and luxurious. The Mess bill is rather stiff, but it is well worth it as I am sure a few weeks here will make anyone feel fit. It

is situated on the west coast not far from Ayr if you want to find it on the map and is a favourite pleasure resort. The hotel faces the sea and is about 300 yards from it. The view is magnificent, away to the left are cliffs and to the right, and in front a beautiful beach whilst out to sea is a huge mountain sticking out of the water. It is the well-known Aisla seu craig [*sic*] or some such name. Twelve miles across and to the right is a larger island and past that the mainland again.

The sunsets are magnificent although not so gaudy as in Aussie.

The weather is quite warm and although the temperature is not very high, it is moist and can't help feeling warm. Plenty of sea bathing though makes it A1 and as for flying the weather is perfect.

The Scotch mists are very funny and out to sea you can hear the foghorns. It is a common sight to see a convoy of merchant ships escorted by destroyers and 'Blimps' or in other words dirigible balloons, pass out to sea most probably from Glasgow.

There are here the best golf links in England [*sic*] and excellent lawn courts for tennis, croquet etc. I spend most of my time off playing tennis and it is jolly good after not having a game for such a long time.

We have from Friday midday to Saturday noon off. I went to Glasgow for once and had a very nice time. It is rather a decent place, Glasgow. This time I stayed right here and played tennis all the time.

We, two of us, have hired a boat for this evening and are going out to fish. Going to take some grub with us so it ought to be rather a jolly outing.

Received quite an amusing letter today from a VAD I met at Oxford. Three days before coming up here I flew to Oxford in my Camel Scout and did some vertical dives and climbing turns, cartwheels etc., just over where she lives, with the result that I am forwarded a cutting from the *Oxford Times*, which runs as follows: -

'The antics of a certain airman have caused some perturbation to the residents of the East Ward this week. On more than one occasion this pilot has flown his machine perilously near the roofs of the houses and has alarmed people of nervous temperament. It was obvious however, to those who knew anything about flying, that the aviator was not only a brilliant pilot, but that his machine was of wonderful construction and that there was therefore little likelihood of an accident. What struck the beholders as most clever was the manner in which, after the machine had dived to within a few feet of the housetops, it levelled and climbed again at terrific speed. Those who were not nervous marvelled at the pilot's skill and capabilities of the machine.'
No small compliment, I call that, what?

2 June
Had a jolly time yesterday evening. My pal and myself hired a boat and some fishing line from the neighbouring village and went a fishing, taking some sandwiches and etc.s with us. Had a delightful swim and a feed and then pulled out for some fishing and caught seven quite decent sized whiting.

A further letter to his grandmother on 12 June tells of 'the weather turning rough and the sea having "its back up," but that it is still jolly nice here. Had a delightful picnic the other day by a small loch not far up the coast. I have been aviating the elusive Camel quite a lot lately in spite of the wind which makes landing rather ticklish. Wind doesn't worry me one atom up top. I have nearly finished the course here. When I have, I suppose I shall be put in the pool for a little. I will go joy riding in a Bristol Monoplane then. I have not flown a Monoplane yet as you know they are washed out for action work. It

ought to be a good experience.'
On 17 June Baker writes from 4 Squadron AFC, France:

Now, what do you think of that! Last Thursday morning I received word to pack and clear out to the above address and here I am. Arrived day before yesterday evening. It is quite funny getting back, but not all bad. The actual fun has commenced now. I went up to the lines yesterday and had a look around. Looks pretty desolate from above though. I am going on my first show tonight which is looking after balloons or some such thing. You may bet your life the machines on this side are some bus! It is a jolly sight better life than when I was over here before. Live decently and are not bothered overmuch by shells etc. So much the better.
There is precious little to talk about being in the land of the censor again you see.
Well, so long, till next time.

During the following four months Baker was credited with twelve enemy aircraft and was promoted to Captain and Flight Commander.

In October 1918, 4 Squadron changed over from the Camel to the Sopwith Snipe as their primary aircraft.

On 4 November 1918, just before the Armistice, Captain Thomas Charles Richmond Baker age 21, was shot down and killed flying Snipe E8062. near Buisseral. by Staffelführer Carl Bolle.

In February 1919 he was posthumously awarded the Distinguished Flying Cross:

Supplement to the *London Gazette*, 8 February 1919
Lieut. Thomas Charles Richmond Baker, MM (Australian FC) (FRANCE)
This officer has carried out some forty low flying raids on hostile troops, aerodromes, etc., and has taken part in numerous offensive patrols; he has, in addition, destroyed eight hostile machines. In all

these operations he has shown exceptional initiative and dash, never hesitating to lead his formation against overwhelming odds, nor shrinking from incurring personal danger.

(MM gazetted 19th February 1917.)

Captain Thomas Charles Richmond Baker is buried in Escanaffles Communal Cemetery, Tournai, Hainaut, Belgium.

Flying Officer Henry Frederick Vulliamy Battle OBE, DFC

Born in London in 1899, the son of Harley Street Surgeon, William Battle, Henry joined the RFC in January 1918. Completing Advanced Pilot Training in March, and awarded his RAeC certificate in April, he was posted to Turnberry from 26 June until 20 July before proceeding as a pilot to 60 Squadron on the Western Front.

The following extracts are taken from his book, *Line!*:

Early in June six of us, including to my great pleasure the three Americans, were given four days graduation leave, on the termination of which we were posted to No 1 Fighting and Gunnery School at Turnberry in Ayrshire. I was particularly glad that 'Johnny', O. P. Johnson, was coming too because he was such fun. A tall, thin chap, with a tremendous sense of humour, he had the most infectious laugh. The other two in the party, not mentioned up to now, were Bains and Tommy Akin. Bains was British and had gone through his cadet training with me. Akin was a cheerful little fat Canadian.

I left behind at Beaulieu several good friends, including Jimmy (Fairy) Fairweather, whom I was to see again many times between the great wars.

During this leave I ordered one of the new RAF uniforms. The tunic was Khaki, similar to the infantry open-neck pattern, but a cloth belt with a brass buckle was substituted for the leather Sam Brown. No

shoulder straps were worn, ranks being shown in rings of khaki and blue braid on the sleeves. A brass bird and crown were positioned over the rank rings as a kind of substitute for the executive curl in the gold braid of naval officers. The hat was similar to the army peaked type only it had a black shiny peak and black band, and, on either side of the badge, which has not been altered since, the badges of rank of the wearer up to the rank of Captain. Majors and above wore more and more gold on the peaks of their hats in naval fashion. All officers could wear brown field boots if they wished, and all those of field rank usually did.

The first brass buttons had a bird and crown in the centre, surrounded by a rope edging, which was later to disappear. The bird and crown were taken from the Royal Naval Air Service under the impression, I believe, that the bird was an eagle. In point of fact, however, it started life as an Albatros. Even now, fifty years later, arguments still take place in the press as to the identity of this bird, retained in its original form in the cap badge.

A similar uniform in light blue, bearing gold braid for rank badges, was introduced at the same time. It could be worn by officers as mess kit during the war. Black boots were to be worn with it, and white shirts and collars replaced the khaki shirts of normal day wear. It was said at the time that the blue cloth approved for this uniform had been made for Russia but never sent there. It was much lighter in shade than the present RAF blue.

Whilst on the subject of clothing, this might be considered an appropriate moment for mentioning that, before leaving Beaulieu, I had my flying coat shortened up to knee level, as the voluminous skirt would get in the way and did nothing towards keeping me warm. Later, when I reached my first operational squadron, I was issued with one of the original pattern Sidcot suits, which enabled me to discard my leather coat.

The Sidcot suit then made it unnecessary to wear enormous thigh length boots, so I had mine cut down to just below the knees and used the material cut off as bedside mats, for which it was admirably suited.

But now let us return to the train. Upon the termination of our leave, the six of us duly arrived at Turnberry after a dismal rail journey in the pouring rain and were agreeably surprised by the magnificence of our new quarters. As the railway station formed part of the hotel, the whole of which had been requisitioned, as it stood, for an officers' mess, we stepped straight out of our train into the mess. It was a delight to live in a house for a change after our tents, and to use real beds and furniture in place of our camp kits.

The mess, which had been a golfers' paradise, was an enormous building on high ground which swept down to the sea, with the famous golf course spread out over the intervening space. The hotel was entirely staffed by service personnel, mostly WRAF. I started by sharing a room with one or two other officers, but as time went on, I succeeded in obtaining a room to myself, which was a luxury normally reserved for instructors.

Although the mess was very comfortable, and everyone did his best to make us feel at home upon our arrival, we sensed the general feeling of depression. It then transpired that the son of a very senior member of the RAF had been killed in a flying accident. He had gone up with another officer in a DH9 and it was said that they had been trying to change places in the air, a rather popular stunt at the time, when the machine got out of control, both occupants being killed. A parade ordered for the morning after our

arrival was attended by all ranks, when the bodies were taken away for internment. [Presumed to be the accident of 21 June 1918 in which Henderson and Redler were killed.]

We new boys soon discovered another cause for depression. There were no Dolphins at Turnberry. We were, therefore, all posted to the ground pool, in which we again worked on guns, both Vickers and Lewis, gun sights, the Constantinescu interrupter gear, and so on, mostly on the ranges.

Our working day was a full one:

08.25 - 10.00 hours On the Ranges
10.15 - 12.00 hours On the Ranges
12.20 - 13.00 hours Lecture
13.55 - 15.15 hours On the Ranges
15.25 - 17.00 hours On the Ranges
17.45 - 18.45 Lecture

The lectures were mostly on fighter tactics, Law and Administration, and we spent the evenings writing up our notes.

After some eight days of this work we were given a stiff examination which most of us passed fairly easily, except for the paper on law and administration. We had enough training on guns to last us a lifetime, but we had not done much to improve our knowledge of law.

One officer, not among the six from Beaulieu, was caught handing in someone else's notebook for his own. He was promptly put under close arrest, and for some time afterwards we other pupils had to take turns as Prisoners Escort. This entailed living in the same room as the prisoner for 24 hours on end.

Pupils who were Dolphin pilots and had finished the ground course now had nothing to do all day, as we were not attached to any of the training flights or groups as they were called. If we were not detailed for Orderly Officer or Prisoner's Escort, we could do what we liked except take leave, which was not permitted to pupils. As the wind was always blowing hard, we could not play tennis and so I tried my hand at golf. The links were well kept but I found the course very difficult, there being so many blind holes.

Four of us were lucky one weekend in being allowed forty-eight hours off to visit Glasgow. We had hoped to see a show that Saturday night, but the choice of entertainment was strictly limited. There was a morality play on at every one of the theatres. So that evening was a flop. Sunday was even worse, as no pubs were open, and we were forbidden by the management to play billiards in our own hotel. The time dragged painfully by until we were able to take the train back. Glasgow will always stick in my memory as the place where there were so many trams, one after another, that the only safe way to cross any main road was to board one and risk having to pay a half-penny before getting of the other side.

One day, as Orderly Officer, I had the ordeal of having to pay 400 WRAF. When I reported afterwards to the Accountant Officer and had to put my hand in my pocket to refund three shillings overpaid to some creature, he actually asked me if I did not think it had been worth it!

By now I was hopping mad at seeing practically everyone else flying except me. So, seeing a number of Bristol Monoplanes flying around looking so graceful in the air compared with the biplanes, I conceived a strong desire to fly one and, screwing up my courage, went in search of the Chief Flying Instructor who was the famous Major McCudden, VC, DSO, MC, MM, and found him on the tarmac on the aerodrome, which was situated to the north of the mess. A nice looking, clean-shaven young man, without 'side', he impressed me very favourably, particularly as he gave me the permission I wanted.

It did not take me long to go over to the HQ Group and borrow a Bristol, the first Monoplane I had flown. It had a 110hp Le Rhone engine and proved a delight to fly, my pleasure on this occasion being somewhat spoilt, however, by a loose throttle control which compelled me to keep my left hand permanently on the quadrant. In forty minutes, I gained the following impressions. The controls were light, effective and well-balanced, but the view downwards was poor, although forwards and overhead it was excellent. The makers had provided two slots at the main plane roots through which one could peep at the ground, but if one wanted to take a real look one could easily go over on to a wing tip. What really struck me was the apparently fragile construction.

The fuselage was an ordinary box structure incorporating four longerons, the whole being covered by stringers and fabric to give it a proper streamline shape, but it was the small section of the longerons, only about one-inch square, which astonished me.

I asked a friend in the Mess, who only recently had been serving in Egypt, what he knew about the Bristol Monoplane. He told me that it was not popular in the Middle East because it had a bad reputation for structural weakness. From experience gained later in Iraq, I can well imagine that the light wooden structure of this machine might have failed at times owing to a combination of wood-shrinkage, due to the intense summer heat, and insufficient maintenance facilities.

I suspect also that the Bristol Monoplane was not seen in France because of its short endurance, only one and three quarters of an hour, and its single machine gun, quite apart from any consideration of the rumours concerning its dependability.

It now became possible for me to borrow a number of these monoplanes, but they all seemed to suffer from engine troubles. They finally gave me one of my own to fly, painted with different coloured fishes' scales all over the fuselage. It was only given to me because no one could get its engine to run properly. This had a most annoying habit of running perfectly for five minutes then, suddenly, it would give a tremendous bang in the intake, an explosion that would almost knock the throttle out of one's hand. This blow back was probably due to a bent valve or weak valve springs, but I never had time to find out.

I was also able to get in a little flying on Avros and Pups, then, early in July our Dolphins arrived, and the Beaulieu contingent joined them in B Group. We now started our advance training in fighting, aerobatics and formation flying but, unfortunately, the machine I was allotted suffered from perpetual water leaks and overheating. However, I had one or two good scraps, one of which was with a Bristol Mono who took me up to 5,000 feet before breaking away. I could not understand why, the whole time I had him under observation, the pilot did all his turns in the same direction, (to the right I think). But this riddle was soon solved when I followed him in to land and saw the pilot get out.

It was the CO of the school himself, Lt.-Col. L. W. B. Rees VC, MC, complete with wooden leg! Very active in spite of his disability, Col. Rees was tremendously keen on fencing which he used to indulge when he could find an opponent. He showed us one day a pair of foils he had brought back from the USA curled up in a suitcase, from which they had emerged quite unharmed and as straight as on the day when they had been made. After leaving the service, this gallant officer devoted himself and his private means to philanthropic work.

Considerable interest was shown in a new Vickers single seat scout biplane of orthodox design flown up to Turnberry one day from Brooklands, in record time, by Major McCudden.

It was to have been fitted with a Hart engine, a water-cooled Vee type designed to allow a Vickers gun to fire through the propeller boss, but as this was not ready a Hispano Suiza engine had been fitted instead. McCudden did not seem particularly enthusiastic over this prototype, which never went into production.

A few days later McCudden left for France to take over command of 60 Squadron, but he was fated never to reach his destination. He was flying to his new unit and had engine failure whilst taking off from the depot where he had refuelled. His death was a great blow to the service.

Whilst my Dolphin was being cured of its water leaks, I was able to borrow an Albatros DV Vee Strutter. A typical Hun single-seat biplane, fitted with large six-cylinders in line water cooled engine of about 200 hp, it proved quite pleasant to fly though rather heavy on the controls near the ground. When I first went to get into the cockpit, I found a man inside, as I thought, having a bath! He was splashing away for dear life. It transpired that the doper had leaked badly when he had started up the engine, so he was baling out the petrol which had accumulated on the floor of the plywood fuselage. I liked the engine very much but regret not being able to remember whether it was a Mercedes or Austro-Daimler.

We Dolphin experts now got busy painting up our machines so that we could recognise each other in the air. I was busy giving the nose of mine a beautiful blue tinge with a white edging, when someone came into the hangar and seeing us all busy painting said, 'Now something is bound to happen. It always does when you paint a machine.' This cheerful charley was lucky to escape into the rain with only a little wet paint on his Macintosh. But he was right. We were all posted away that evening.

In September 1919, Henry Battle was granted a permanent Commission in the rank of Flying Officer. He served as Flight Commander with various squadrons, attending RAF Staff College in 1936, before joining 34 Squadron as Commanding Officer. He was awarded the DFC the same year. In 1938 he was on the Organization Staff of HQ Bomber Command, was awarded the OBE in the 1940 Birthday Honours, and in 1941 was Commanding Officer of RAF Swanton Morley. Henry Frederick Vulliamy Battle OBE, DFC, retired from the RAF with the rank of Air Commodore.

He died in 1981 in Suffolk.

Captain James Douglas Latta MC

James Douglas Latta

James Douglas Latta, usually known as Douglas, was born on 13 May 1897 in Hendon, Middlesex, the second son of James Gilmore Latta, from Troon, and Agnes (née Douglas) from Tarbolton in Ayrshire. James Gilmore Latta was an engineer who had worked for the companies of Andrew Barclay and G. & J. Weir, before becoming part-owner and managing director of the Scottish Stamping and Engineering Company at Ayr in 1920.

Educated at University College School, London, at the outbreak of war, James and his brother, John, enlisted in the London Scottish Regiment, transferring to the RFC in 1915. 2nd Lieutenant James Latta received his Royal Aero Club Aviator's Certificate on 16 November flying a Maurice Farman biplane at Military School, Norwich. Completing an Aerial Machine Gun Course at Hythe in January 1916, he was appointed a Flying Officer on 24 February 1916 and confirmed in his rank on 11 March. Posted to 1 Squadron, Expeditionary Force in France in April, flying Nieuport Scouts, Latta scored three victories in June, the first destroying an enemy aircraft over Wez Macquart on 1 June, followed by bringing down two enemy observation balloons on 25 and 26 June.

This action won him the Military Cross, which was gazetted a month later on 27 July. The citation read:

> 2nd Lieutenant James Douglas Latta, RFC, Special Reserve.
> For conspicuous gallantry and skill. On two occasions he attacked enemy kite balloons, and each time brought down the kite in flames. He has often driven off enemy aircraft, and his own machine has been badly hit.

In July 1916 he was transferred to 60 Squadron where he scored another two victories by driving down two enemy aircraft out of control on 31 August and 19 September.

In October Latta was appointed a Flight Commander with the temporary rank of Captain and was withdrawn from active service on rest leave in November which he spent at 28 and 43 RS. In May 1917 he joined 66 Squadron flying Sopwith Pups as a Flight Commander. He was shot down and wounded while on offensive patrol on 8 June, flying Sopwith Pup B1726. He force-landed at Voormezeele. Oblt. Bruno von Voigt of Jasta 8 claimed him as his 1st victory. Captain James Douglas Latta saw no further front line service after that and was struck of strength of the Expeditionary Force.

Between June and November 1918 Captain Latta served at Turnberry on temporary duty prior to being transferred to the unemployed list of the RAF on 2 February 1919.

During the war, Douglas's father, James Gilmore Latta, had rented the Ayrshire mansion house of Smithstone near Failford (at the end of the war he bought it and renamed it Auchenfail).

James Latta Snr., then a director of G. & J. Weir Ltd, was responsible for the aeroplane and engine department their Cathcart Works. He left the company at the end of the war and bought the Scottish Stamping and Engineering Company Ltd., Neptune Stamping Works, in Ayr which specialised in drop forgings. Both his sons John and James Douglas were also employed by the company.

James Douglas Latta joined the newly formed Auxiliary Air Force in 1924 and in October 1925 he became the Commanding Officer, with the rank of Squadron Leader of 602 (City of Glasgow) Squadron. He resigned his commission in May 1927.

In January 1927, after the death of his father, James Douglas subsequently became Chairman and Managing Director of Scottish Stampings and was president of the UK Association of Drop Forgers and Engineers.

During World War II, Scottish Stampings was a highly specialised producer of aircraft components and in 1939 Latta was one of the joint founders of Scottish Aircraft Components Ltd. in Glasgow. In the late 1940s and early 1950s he was Chairman and Managing Director of both Scottish Stampings and the Ailsa Shipbuilding and Engineering Company Ltd., Troon.

After WWII, Scottish Stampings returned to manufacturing vehicle components and drop forgings. The company was bought by GKN in 1953.

James Douglas Latta died in Lancashire in 1974.

Captain John Leacroft MC

The son of Dr John William Leacroft and Agnes Docker, John was born in 1889 in Derbyshire, England. He was educated at Aldenham School before going up to Pembroke College, Cambridge. He enlisted in the senior division of the University of Cambridge Officers' Training Corps on 3 February 1909, and was commissioned as a 2nd Lieutenant. John Leacroft joined the Army Service Corps at the outbreak of war in 1914 and served with them in France

and Egypt. He transferred to the RFC in 1915 and initially served as an Observer with 14 Squadron Egypt as part of the Sinai and Palestine Campaign. After completing pilot training, he was commissioned Flying Officer on 20 March 1917 and in May posted to 19 Squadron in France, flying Spad VIIs. During 1917, he scored fourteen victories, his first on 17 June 1917. Flying Spad B1535, he brought down an Aviatik east of La Bassée. On 6 July his second victory was an Albatros, brought down over Houthoulst Forest. He had been appointed the Flight Commander of 'C' flight on 22 July 1917 and by the end of September, he had brought down a further nine Albatros aircraft becoming a double ace, adding another four to his tally in October.

He was withdrawn from service at the front and returned to Britain, returning for another tour of duty with 19 Squadron in early 1918, which by this time had become the first unit to be equipped with the new Sopwith Dolphin. Leacroft scored another eight victories.

He was awarded the Military Cross (MC), as reported in the Supplement to the *London Gazette*, 18 March 1918:

> T./Capt. John Leacroft, Genl. List and RFC
> For conspicuous gallantry and devotion to duty. On one occasion he flew at a very low altitude in extremely bad weather and successfully engaged enemy troops with machine-gun fire, and on another occasion carried out a most valuable reconnaissance and engaged enemy troops from a height of 100 feet. He destroyed two hostile machines and has proved himself a courageous and determined pilot.

He was awarded a Bar to his MC on 22 June 1918. The citation read:

> T./Capt. John Leacroft, MC, Gen. List and RFC
> For conspicuous gallantry and devotion to duty. During a period of six months he has destroyed four and has brought

down completely out of control six hostile machines. In all he has destroyed eight enemy machines and driven down thirteen out of control. On one occasion, when leading an offensive patrol, his formation destroyed six enemy planes and drove another down out of control without suffering any damage itself. He has displayed exceptional qualities as a leader, and his patrols have always been characterised by vigour and dash.

He was presented with both his MC and Bar by King George V in an investiture at Buckingham Palace on 26 September 1918.

Although he was shot down twice, he survived the war and was granted a permanent commission in the Royal Air Force, retiring with the rank of Group Captain in 1937.

In June 1939, being in the Reserve of Air Force Officers, he was recalled to duty as a squadron leader and returned to the active list as a Wing Commander on 1 September 1939. He was then based at Cardington in Bedfordshire in 1942 as president of the Air Crew Selection Board. Retiring from the RAF in 1945, John Leacroft died in August 1971 at Bexhill-on-Sea, Sussex.

Captain Arthur Bedford Taylor

Born in Honiton, Devon, in 1878, Arthur Taylor joined the Royal Flying Corps from Garrison Duty abroad. He was promoted to a Commission from Warrant Officer, School of Musketry, having had years of experience on all types of machine guns. On joining the RFC, he qualified as an assistant instructor in gunnery at the RFC Armament School, Uxbridge, and was taken on staff of 2 (Aux) School of Aerial Gunnery, Turnberry in April 1918. He was transferred to the Unemployed List in January 1919.

Lieutenant Douglas Alfred Savage MC

Born in Oxford in 1892, Douglas Alfred Savage was the son of stationer, bookseller and publisher, Alfred Savage. He was commissioned as a 2nd Lieutenant in the Royal Warwickshire Regiment (Reserve Battalion) on 19 March 1915, and

saw active service in France, in a trench mortar battalion.

Transferring to the RFC in May 1917, Savage trained at Thetford and then 2 (Aux) School of Aerial Gunnery at Turnberry in November. He was posted to 62 Squadron in France in March 1918, flying the Bristol F2b.

On 26 March, flying with Observer 2nd Lt. Louis Thompson, they shot down an enemy aircraft, followed by two more on 12 April - an Albatros DV and a Pfalz DIII. A further two were brought down on 21 April, but Savage and Observer Thomson were hit by anti-aircraft fire near Armentieres while they were being attacked from behind by two enemy scouts. They opened fire at one of these, which fell vertically, and broke into pieces in the air. Both Savage and Thomson were injured in the forced landing of their aircraft at Blangy-sur-Bresle. His sixth victory occurred on 19 May, this time flying Bristol F2b B1336 with Observer Lieutenant E. W. Collis. They brought down an Albatros Scout over Bray, but again he was shot up and forced to land near Corbie. His seventh and final victory was on 2 June flying Bristol Fighter C953 with Observer Sgt. William Norman Holmes, when they shot down a Fokker DRI South of Pozières.

For his victories he was awarded the Military Cross:

T./Lt. Douglas Alfred Savage, Gen. List and RAF
For conspicuous gallantry and devotion to duty, especially on the following occasions. When on patrol attacked a formation of enemy aeroplanes, crashing one, while another fell to pieces in the air after a short combat. Attacked an Albatros, which he set on fire, and drove another down out of control. Attacked many ground targets from low altitudes.
(Supplement to the *London Gazette*, 26 July 1918)

Returning to Britain in late June 1918, Lieutenant Douglas Savage was posted to Turnberry, now 1 Fighting School, in August as a Bristol Fighter Pilot, and was appointed a Flight Commander with the acting rank of Captain on 2 October 1918. He left Turnberry in February 1919 and was transferred to the unemployed list in June.

His cousin Major Jack C. Savage was a famous aviator who pioneered skywriting after WWI.

Between the Wars, Douglas Savage was an aviary and aquatic dealer, but returned to military service in April 1940, joining the Royal Air Force Volunteer Reserve and was commissioned as a Pilot Officer in the Administrative and Special Duties Branch. He was transferred to the General Duties Branch in October and was promoted to Flying Officer. He retired from the RAFVR with the rank of Squadron Leader and was awarded the Air Force Cross in the 1945 New Year Honours.

Douglas Alfred Savage Died in 1967 in Oxfordshire, England.

Captain Joseph Dover Atkinson AFC, BSM

Joseph Dover Atkinson

Born in 1893 in Yorkshire, Joseph Dover Atkinson transferred to the RFC from the Royal Engineers in September 1916. He obtained his Royal Aero Club Aviators' Certificate on 26 February 1917 at Bradford. After training at Oxford and the School of Aerial Gunnery and Fighting, he was posted on 21 March 1917 to 29 Squadron flying Nieuports in France.

On 23 April 1917, 2nd Lieutenant Atkinson, flying Nieuport B1516, brought down an Albatros Scout out of control over Vitry and he was subsequently shot down by heavy artillery, but was uninjured. On 12 May 1917, he flamed a balloon whilst flying Nieuport B1584 and on 29 June sent down another Albatros Scout out of control.

Returning to Home Establishment on 31 August, he undertook a course of instruction in Aerial Fighting before being posted to 1 School of Aerial Fighting, Ayr, as a Scout Fighting Instructor.

On 26 December 1917 Atkinson was injured in accident at Ayr, flying Avro 504 C4427. He had side slipped down into a field to examine a wrecked aeroplane. The front stick came away in his hands and the machine nosedived from about 30 feet. From Ayr he moved to Turnberry in May 1918, where he remained as an instructor until January 1919. He was awarded the Air Force Cross the same month.

Captain Joseph Dover Atkinson was transferred to the unemployed list on 2 March 1919.

Joining the RAFVR during WWII, Atkinson was promoted to Wing Commander, and was awarded the American Bronze Star Medal in May 1946. He died in 1985 in Devon.

Captain Philip Edmund Mark Le Gallais AFC

Philip Le Gallais was born in 1893 in the Channel Isles, the son of Lt.-Col. Mark Le Gallais, (of The Royal Militia of the Island of Jersey) and Josephine De Schaefer. He attended Victoria College where he joined the Officers Training Corps, and then the Royal Militia of the Island of Jersey, as a cadet. In June 1915, 2nd Lieutenant Philip Le Gallais joined The Royal Sussex Regiment, before transferring to the RFC in April 1916, with the rank of Temporary Captain. He was injured on 17 January 1918, after which a Court of Inquiry was held. From 1 School of Aerial Fighting at Ayr he was posted to 1 Fighting School at Turnberry as a Flight Commander on 19 June 1918, and then transferred to 44 Squadron in January 1919, which was a Home Defence unit forming part of the London Air Defence Area. Within a few days he was again posted to the Airship Station at Luce Bay, back to 44 Squadron, then to 39 Squadron. Captain Le Gallais was granted a permanent commission in the RAF on 1 August 1919, which he relinquished on 15 October 1919. Le Gallais was awarded the Air Force Cross for 'an act or acts of valour, courage or devotion to duty whilst flying, though not in active operations against the enemy.'

Le Gallais's younger brother, Lieutenant Reginald Walter Le Gallais, also joined the Royal Militia of the Island of Jersey at the outbreak of war. In October 1915 he left the Militia to obtain his flying certificate, which he achieved within two months, before he was eighteen. Gaining his wings in the RFC, he was posted to France where he spent a year at the front on contact patrol work, before returning to England where he was flying small scouting aircraft on coastal patrol. He was killed shortly afterwards on 15 September 1917, in an accident caused by structural failure, near Faversham.

Lieutenant Colonel Phillip Edmond Mark Le Gallais died in 1958 in Jersey.

2nd Lieutenant Louis Marie Belloc

Louis Belloc was born in 1897 in Oxfordshire, son of the famous Anglo-French historian, author and poet, Hilaire Belloc, who was a one-time Liberal Member of Parliament for South Salford, and who had become a naturalized British subject in 1902.

Commissioned as a 2nd Lieutenant in the Royal Engineers, he was gassed on the Somme in August 1917, and was hospitalised in Manchester. He was attached to the Royal Flying Corps in February 1918. After initial training at Reading, he arrived at Turnberry on 3 August.

He wrote from Turnberry:

6 August 1918
I am now doing my last course before going to France (aerial gunnery). I got here on Friday evening, and will be here about three weeks; with luck, I will get out earlier because there is a great demand for Camel pilots. We are living in the lap of luxury here; our quarters are the G & SWR hotel (complete with all its furniture and fittings!!!) right on the beach and the food appears to be unlimited; we work very hard, but the work is very interesting indeed; the first week is spent on the ground learning everything about our gun and the remaining weeks is aerial fighting and firing (at targets in the sea).

In fact, his final course lasted one week, before he was posted overseas to 209 Squadron as a Camel pilot on 10 August.

On 21 August, 2nd Lieutenant Louis Belloc was flying Camel C59 when he overshot Bertangles aerodrome on landing and crashed into hangars, he escaped unhurt.

Five days later, 2nd Lieutenant Belloc was killed in action, aged 20, while flying Sopwith Camel F6028 on 26 August 1918. He was on a low-level flight attacking a German ammunition train near Cambrai, France when he was hit by anti-aircraft fire. The Camel was seen to crash into trees near Remy. He has no known grave and is listed on the Arras Memorial, France.

His father, Hilaire Belloc later wrote: 'I travelled here in 1918, in a voyage of sorrow, after the death of my eldest son, Louis. He had been a pilot in the Royal Flying Corps, and had died here, shot down while bombing a German transport column. In his honour, I placed a memorial tablet for him here, at Cambrai Cathedral.'

Captain Percy Jack 'Pip' Clayson MC, DFC

The son of John Henry and Georgiana (Darby) Clayson, born in 1896, Percy Jack Clayson attended the Royal Masonic School for boys in Bushey and enlisted in the Royal Navy in 1914. He served with the *Pembroke III* and *President II*, was rated an Air Mechanic 1st Class on 1 June 1915, and served in France from December. He was discharged from the Royal Naval Air Service, on 13 May 1917, to undergo training with the Royal Flying Corps in the Officers Cadet Wing. 2nd Lieutenant Clayson received Royal Aero Club Aviator's Certificate 5617 flying a Curtiss Biplane on 30 September 1917. While serving with 1 Squadron in France in 1918, he scored 29 victories flying the SE5a.

He was awarded the Military Cross, as recorded in the Supplement to the *London Gazette*, 22 June 1918:

T./2nd Lt. Percy Jack Clayson, Gen. List, and RFC
For conspicuous gallantry and devotion to

duty. When on low-flying offensive patrol, he engaged an enemy scout and shot it down, with the result that it crashed to earth. He has brought down several hostile machines, one of which he forced to land in our lines and has engaged massed enemy troops and transport from very low altitudes with machine-gun fire, inflicting heavy casualties. He has displayed the most marked determination, courage and skill.

He was later to be decorated with the Distinguished Flying Cross (Supplement to the *London Gazette*, 3 August 1918):

> Lt. (Temp. Capt.) Percy Jack Clayson, MC
> A patrol leader of great skill, and a skilful marksman, whose personal fighting successes have proved of much value to his squadron. Captain Clayson's patrol frequently encountered enemy formations in superior numbers, but invariably succeeded in inflicting serious losses.

He was invested with his DFC by King George V at Buckingham Palace. Clayson was then posted to Home Establishment. On 24 October 1919 he was granted a short service commission in the Royal Air Force with the rank of Flying Officer, serving during the 1920s with 6, 23 and 70 Squadrons. In September 1926 he was posted to the Aircraft Depot, Iraq, finally returning to the UK in 1928 to serve at the RAF Depot at Uxbridge. He was placed on the Retired List due to ill health on 16 April 1929.

Percy Clayson was the Chief Ground Instructor at the Civil Training Flying School in 1936, operated by the Bristol Aeroplane Company at Yatesbury.

Returning to active service in the RAFVR during World War II, Clayson was appointed Flying Officer in the Administrative and Special Duties Branch in December 1941. He was promoted to Flight Lieutenant in January 1944 and remained on the Air Force Reserve of Officers list until relinquishing his commission in 1954.

2nd Lieutenant William Arthur Rymal

Born in 1895 in York, Ontario, Canada, the son of John Warren Rymal and Elizabeth Griffin, William was employed as a clerk in the Canadian Bank of Commerce when he joined the RFC in September 1917 at Toronto and was appointed to a temporary Commission in January 1918. After training at 43 T.S. Chattis Hill in England, he graduated at the end of July and was posted to 1 School of Fighting and Gunnery on 31 August.

2nd Lieutenant William Rymal was killed on 5 September 1918, flying DH9 C1333 when the aircraft stalled and spun in next to the Public School at Maidens. He is buried in Toronto (Mount Pleasant) Cemetery, Ontario, Canada, and is remembered on the Turnberry War Memorial.

Flight Cadet Alexander McLean

Only son of William and Mary McLachlan McLean, of Cardonald, Glasgow, Alexander was born in 1899 and was educated at Bellahouston Academy. His father was manager of the Submarine Department of the Fairfield Shipyard, Govan. Prior to joining the RAF in November 1917, Alexander McLean was a marine engineer.

Posted for training to Turnberry, he was on the eve of obtaining his wings when he was accidentally killed whilst flying as an Observer in DH9 C1333.

The following was reported from the Court of Inquiry:

> The cause of the accident was due to the machine spinning into the ground through turning downwind and stalled spinning from 300 ft. The machine caught fire on impact with the ground.

Flight Cadet Alexander McLean is buried in Craigton Cemetery, Glasgow, and remembered on the Turnberry War Memorial.

Lieutenant Gerald Arthur Lamburn

Born in London in 1899, the son of journalist and editor of *Pearson's Weekly*, Francis John Lamburn and Aline Wood, Gerald was educated at St Cyprian's School, Eastbourne, and Haileybury College, where he served in the

School OTC from September 1913 to April 1916. He enlisted in the Artists Rifles, serving 85 days before being commissioned 2nd Lieutenant in the Royal Flying Corps in July 1917. After pilot training at Denham, Gerald was posted to 46 Squadron in November, flying Sopwith Camels.

On 28 December, Gerald and Lieutenant H. N. C. Robinson shot down a yellow Pfalz two-seater over Havrincourt Wood, firing about 200 rounds at 200 yards range. 'The E.A. disappeared into a cloud and immediately afterwards we observed him falling in flames. His right wing folded back, but then he disappeared from view about three miles west of Havrincourt, crashing to the ground near Gonzeaucourt in our lines.'

In April 1918, 46 Squadron were employed in air-to-air combat, and numerous strafing missions on German trenches, transport and artillery batteries, other targets included balloons and infantry.

On 10 April, Lieutenant Gerald Arthur Lamburn took off in Camel C1661 but soon had to return to the aerodrome when his aircraft was shot through during low level work on the 3rd Army front, he was uninjured. The next month Lamburn shared his second victory with Lieut. G. T. W. Burkett and Lieut. H. S. Preston, when they brought down a DFW over Merville. Ltn. Friedrich Roseler and Ltn. Diedrich Meyer were both killed in this action. On the same patrol, just south of Merville at 4,000 feet, Lamburn attacked a second two-seater, firing 80 rounds into it. The enemy aircraft did a stall turn and was seen to go down vertically, but owing to heavy artillery fire in the area, this victory was not confirmed. At the end of July, a further enemy aircraft was brought down over Merville, 'Diving on it from the front, he fired 100 rounds at 150 yards range. E. A. put his nose down and was last seen still gliding down at 1000 feet where he was lost in the mist.' Again unconfirmed. In mid-July he had to force land Camel D6497 in a shell hole near Bethune after suffering magneto trouble whilst on operational patrol. He escaped unhurt.

Posted back to Britain on Operational Rest, he arrived at 1 Fighting School, Turnberry on 29 August for duty as a Camel pilot. Lieutenant Gerald Lamburn was killed on 30 September 1918 flying Sopwith Camel F1410. He is buried in Girvan (Doune) Cemetery and is remembered on the Turnberry War Memorial.

Lieutenant James Henry Rattenbury Bryant
James Henry Rattenbury Bryant was born in 1898 in North Kona, Hawaii. He was the son of Mr and Mrs G. E. Bryant, of Kailua, North Kona, Hawaii.

Completing his course of training at Turnberry, Lieutenant Bryant was posted to 28 Squadron in Italy. While flying Sopwith Camel B5638 on a bombing raid on Campoformido in north-eastern Italy, he was hit by ground fire and crashed near houses. The bombs he was carrying exploded on impact. Lieutenant Bryant died age 20 on 4 October 1918 and is buried at the Tezze British Cemetery, Veneto, Italy.

Lieutenant James Stanley Brown
Born in 1889 in Nelson, Lancashire, son of John Brown and Ann Nelson, James was employed as a loom overlocker for a cotton manufacturer prior to the outbreak of war. He served in the 5th Battalion (Territorial) East Lancashire Regiment and was subsequently attached to the RFC in October 1916.

Lieutenant James Stanley Brown was killed on 20 October 1918 at Turnberry while flying DH9 C1372. He is buried in Nelson Cemetery, Nelson, Lancashire, and is remembered on the Turnberry War Memorial

2nd Lieutenant Charles Alexander Fletcher MM
Charles Alexander Fletcher was born in 1895 in Hadley, Shropshire, son of Charles Fletcher and Christina Forbes. In 1911 he was employed as a shop assistant for a tailor and outfitter and at the outbreak of war, joined the 6th Battalion Shropshire Light Infantry, serving at Gallipoli, where he was awarded the Military Medal. Transferred to the Worcestershire Regiment, he saw action in France and Belgium prior to joining the RAF as an Observer in July 1918. After initial training he was posted to 1 Fighting

School, Turnberry, on 5 October for a course of instruction, where he was killed while flying with Lieutenant James S. Brown, as an Observer in DH9 C1372.

2nd Lieutenant Charles Fletcher MM is buried in Hadley Cemetery, Shropshire, England and is remembered on the Turnberry War Memorial.

Lieutenant Alexander Crawford Dawson Anderson

Alexander Anderson was born in 1893 in Glasgow and served in the 2nd Dragoons (Royal Scots Greys) and Highland Light Infantry, before transferring to the Royal Air Force in 1918. In August he was posted as a DH9 pilot with the BEF in France until 1 October, when he was hospitalised with flying sickness. Posted to Turnberry on ground duties, he was Adjutant until May 1919, when he was transferred to the unemployed list.

Flight Cadet John Hughes 178259

Son of Hugh Hughes and Ellen Parry, of Tan-y-Bryn, Amlwch, Anglesey, Wales, John Hughes was born in 1900. A bank clerk in civilian life, he joined the RAF with the rank of Private, three days after his eighteenth birthday, in June 1918. After training, and now Flight Cadet Hughes, he was posted to Turnberry for further training on 9 November.

Flight Cadet John Hughes died on 25 November 1918 from injuries received after the force landing of DH9 C1374. He is buried in Amlwch Cemetery and is remembered on the Turnberry War Memorial.

Sergeant John Eric Lilley 86754

Born in 1899, in Forest Hall, Northumberland, to Henry John Lilley, mahogany merchant, and Lilian Sarah Evans, John was an apprentice engineer, having 2½ years' experience on marine engines when he joined the RFC as a private in July 1917. By August he had been graded as 'fit for Observer' but 'unfit as pilot', until January 1918, when he was re-classified as 'fit as pilot'.

Posted to Turnberry in early November, he was killed flying SE5a E3954 on 28 November, when the aircraft stalled on turning.

Sergeant John Eric Lilley is buried in Sunderland (Mere Knolls) Cemetery, Durham, and is remembered on the Turnberry War Memorial.

Corporal Thomas Allister King

Born in Lambeth in 1897, son of publican Albert Thomas King and Ellen Godwin, Thomas joined the Royal Naval Air Service as an Air Mechanic in May 1916 and transferred to the Royal Air Force in April 1918, with the rank of Corporal Mechanic. Posted to Turnberry in September, Corporal Thomas Allister King died whilst on leave, on 10 December 1918.

A report in the *Essex Newsman* of 21 December 1918 tells a little more about him:

SOLDIER'S SAD DEATH

The funeral took place on Monday of Corpl. Thomas Allister King, RAF, fighting section, only son of Mr and Mrs A. T. King, of The Yorkshire Grey Hotel, High Street, Brentwood. The death of this young soldier took place under peculiarly sad circumstances. He came home on leave to see his sister, who was dangerously ill, and who died, together with her child. While his sister lay dead, Corpl. King was himself seized with influenza, and pneumonia supervened, death taking place after only a week's illness. Deceased, who was well-known in the Brentwood district, had been in the Air Force for nearly three years, and was held in great esteem. The funeral was of a military character, as was attended by a large number of the townspeople.

Corporal Thomas Allister King is buried in Brentwood (London Road) Cemetery and is remembered on the Turnberry War Memorial.

2nd Lieutenant Charles Henry Albert Godfrey

Born in Croydon, Surrey, in 1896, Charles

Godfrey was the son of Albert Thomas Godfrey and Rosa Spraggs. He was employed as a telegraph messenger for the GPO prior to enlisting in the 16th London Regiment in February 1916, and was sent to France with the Expeditionary Force in July. Returning to Britain in August 1917, he transferred to 2 Cadet Wing of the RFC at Hastings in March 1918, graduating as Flight Cadet in July, and was appointed a Commission in the RAF on 12 November. After training at Reading and CFS, 2nd Lieutenant Godfrey was posted to 1 Fighting School, where he was killed on 11 December 1918 flying Sopwith Camel F1408, which stalled and spun in on Turnberry Airfield. 2nd Lieutenant Charles Godfrey is buried in Bandon Hill Cemetery, Wallington, Sutton, London, and is remembered on the Turnberry War Memorial.

2nd Lieutenant James Millikin (Milligan)
Born in 1892 in Ballyclare, Antrim, Ireland, the son of cattle dealer, Samuel Millikin and Hester Gilmore Smith, James was employed as a motor mechanic when he enlisted in the Royal Engineers (Inland Water Transport), transferring to the 1st Royal Irish Rifles, serving under the name Milligan. He embarked for France on 29 September 1916 and was appointed to a temporary commission with the 20th Royal Irish Rifles in October 1917.

In his medical records it states that James had been, 'blown up twice between March 21st and 31st 1918' and that 'he has been through the most heavy fighting.' He was admitted to hospital in France suffering from a gunshot wound to the eye.

James Millikin transferred to the RAF in August 1918 and after training at Reading, Eastchurch, Hythe and New Romney, was posted to 1 Fighting School on 28 December.

Second Lieutenant James Millikin age 26, was killed on 31 December 1918 in a flying accident at Turnberry, in DH9 E679. The Official Casualty Card states: 'Court of Inquiry. Cause of the accident was that the engine having failed, pilot endeavoured to turn back to the aerodrome to land, having lost flying speed, stalled into the ground.'

A report appeared in the *Ulster Press*:

POPULAR EAST ANTRIM SPORTSMAN
ULSTER FLYING MAN'S DEATH

The circumstances are painfully pathetic attending the death of Second-Lieut. James Millikin (12th Irish Rifles, attached RAF), which took place as the result of a flying accident at Turnberry, Scotland, on Tues. 31st ult. The deceased Officer – who was the eldest son of Mr and Mrs Samuel Millikin, Scoutbush, Carrickfergus – was transferred to the Royal Air Force in August last, and, after going through his course at Hythe and New Romney, he arrived at Turnberry on Saturday evening to take his last two hours flying in completion of his training. He made an assent on Tuesday 31st ult. As an Observer, the pilot being Lieut. Hillock of the Canadians, attached RAF, It appears, as the result of an inquiry which has been since held, that the engine cut out as it was leaving the aerodrome, and instead of keeping straight on, the pilot turned. The result was that the machine nose-dived to the ground, and immediately burst into flames.

The evidence went to show that the death of Lieut. Millikin was instantaneous, having taken place when the plane struck the ground, and before the fire broke out. The conflagration was so fierce that it was some time before the body could be removed. The pilot escaped with his life but is lying in a very critical condition in hospital at Ayr, his face, arms and legs being terribly burned, while he is suffering from severe shock.

Lieut. Blackman RAF has crossed to Belfast to represent the No.1 Flying School, Turnberry, at the interment of the remains of the deceased, and has conveyed the deep sympathy of his brother officers to the relatives in the great bereavement they have sustained.

He speaks in terms of high admiration of the military and social qualities of the late Lieut. Millikin, who was courteous, brave and enthusiastic. Second – Lieut. Millikin was 26 years of age, and joined the Army in June 1915, serving at first in the Inland Water Transport. He felt, however, that in this branch he was not taking a sufficiently active part in the war and he transferred to the Royal Irish Rifles, taking his commission and serving with them from October 1916 till August 1918. On 31 March of last year, he was wounded in the eye during the big German offensive. His friends felt that it was hard lines he should have been knocked out of action at this time, as he would doubtless have shared in the decorations awarded his brother officers in the subsequent successful operations.

He was a well-known and popular football player and athlete. He was for some time inside right for the East Antrim F. C., and an old Instonian. In sporting affairs in the Army, he took an active part. As a sprinter he ran second at Ulster last year, and although absolutely untrained, was only beaten by inches. When stationed at Poplar last year, he was approached by Millwall with a view to joining that club, but was unable to do so, owing to his approaching departure for France. He is a brother of Mr J. S. Milliken of Cliftonville F.C. and the East Antrim Harriers, a well-known personality in Ulster football and sprinting circles.

THE FUNERAL

With befitting military ceremonial, the remains of the late Second-Lt. James Millikin were removed from Victoria Military Barracks this morning for interment in Ballylinney Cemetery. The coffin, which was mounted on a gun carriage, and draped with the Union Jack, was followed by the chief mourners. They included Samuel Millikin (father), John, Samuel, Roy, Clair and Kenneth (brothers), Jas. Millikin, Jas. Alexander, Andrew Alexander, Samuel Smith, H. O'Hara and W. J. Ferguson (uncles), James Ferguson, Doagh; Lieut. B. Alexander, Pte. W. Millikin, RIR; and Pte. James Millikin, Australian Force (cousins).

Many lifelong friends of the gallant officer and of his father were also present, while the cortege was fully representative of the business community of Carrickfergus and Belfast.

A band with muffled drums and a firing party from the Northumberland Fusiliers escorted the remains.

As the sad procession wended its way along the Antrim Road to the mournful strains of funeral marches there were many pathetic incidents.

Impressive scenes took place at the grave side, where the service was conducted by the Rev. J. Y. Minford BA, Joymount Presbyterian Church, who also held a brief service in Victoria Barracks.

Many beautiful floral tributes were placed on the grave, including wreaths from the sorrowing family and the deceased officer's comrades at the Turnberry Camp.

Buglers sounded the 'Last Post' and three volleys were fired over the grave.

Second Lieutenant James Millikin is buried in Ballylinney Old Graveyard, Northern Ireland, and is remembered on the Turnberry War Memorial.

2nd Lieutenant Charles Alexander Hillock

Charles Alexander Hillock was born in Toronto, Canada, in 1890, the son of Alexander Hillock and Anna Maria Coleman. He married Marguerite Smith in June 1917. Before joining the RFC at Toronto, in February 1918, aged 27 years and 11 months, Charles worked for the Munitions Board.

Following training at Wycliffe College and Camp Mohawk, he arrived in Britain for further training and was posted to Turnberry on 21 November 1918.

2nd Lieutenant Charles Hillock died from injuries on 8 January 1919, at Ayr County Hospital, which he received in an accident on New Year's Eve 1918 when flying DH9 E679.

According to the Court of Inquiry: 'Cause of the accident was that the engine having failed, pilot endeavoured to turn back to the aerodrome to land, having lost flying speed, stalled into the ground.' He is buried in Girvan (Doune) Cemetery and is remembered on the Turnberry War Memorial.

Air Mechanic 1st Class Thomas Inglis

Thomas Inglis was born at Hyndford Bridge, Lanark, in 1889, son of Thomas Inglis, road surfaceman, and Agnes Robb. Air Mechanic 1st Class Thomas Inglis died in Ayr County Hospital of Spinal Meningitis, on 10 March 1919, while stationed at Turnberry. He is buried in Carmichael Parish Churchyard, Lanarkshire, and is remembered on the Turnberry War Memorial.

Major Frank Steel, OBE, made reference to reference to the flu pandemic.

The end of 1918 brought the devastating epidemic of Spanish flu which was sweeping throughout the world. The RAF station at Turnberry was, of course, affected and one of the early administration duties of the day was for the Medical Officer and Assistant Commandant to decide how many sleeping dormitories had to be converted to hospital wards. This having been done the remainder of the station on Armistice Day went about their duties, with varying degrees of excitement as 11 a.m. approached.

A group of the Flying Instructors who were all battle experienced veterans, wearing different numbers of decorations, decided to pay a call on the nearest RAF station at Ayr to which a formation flew and safely fired 'Verey flares' as visiting cards into the middle of the Armoury.

They then returned and, with the rest of the station, were getting ready for the signal to cease hostilities.

The drone of approaching aircraft heralded the arrival of the return ceremonial call from Ayr where pilots were not discriminating in their choice of targets. They fired 'Verey flares' among the ranks of parading airmen, and some of these flares flew through the open doors of the hangers, luckily without damage to any aircraft. One of the planes flew along the front of the hotel firing flares, one of which caused the immediate dispersal of two WAAF drivers from a tender which was standing at the front door. Another hit the stonework of the Orderly Room without breaking the windowpanes. After similar excitements the formation flew away, and the station dispersed to the various interests of those concerned, particularly alcoholic celebration.

The Commandant, who was no coward, as the VC/DC/MC ribbons on his chest bore witness, decided that discretion was the better part of valour and decided to visit his aunt for the day leaving the station in charge of the Chief Gunnery Instructor Major J. W. 'Stuffy' Gordon. He suggested a golf competition for the afternoon leaving 'Stuffy', 'Cupid' Knights, another gunnery instructor, and myself to make up the last party.

We had only played a few holes when, to our consternation, we heard the sound of an aeroplane engine 'revving up' on the aerodrome and saw an SE5 – the fastest fighter of the RAF take off. The pilot saw us on the fairway of the 5th hole – Whaup's nest – and came at us so low as to make us dive for cover in a bunker. In doing so he wrecked the telephone system which spread out from the clubhouse building to the various gunnery ranges. He then flew around the golf course trailing wire and was involved in various episodes. On the 11th green there were two balls, one of which his slipstream blew off the green and the other into the hole. A nice rest of the green for the player!

At another hole he caught the Station Medical Officer at the top of his swing and a piece of trailing wire wrapped itself round the head of his golf club and hurled it 100 yards away. He then flew out to sea in the direction of Ailsa Craig, a breeding ground for thousands of birds and very dangerous for any aircraft to approach.

'Stuffy' rushed off to the aerodrome, alerting the ambulances and fire services, and the motorboats which were moored in the neighbouring village of Maidens. We then waited for the inevitable tragedy.

In a few minutes the SE5 was seen to be returning unscathed and the pilot made a perfect landing far up the aerodrome. As 'Stuffy' went to see him to take the necessary disciplinary action, he was seen to clamber out of the machine and start going around the streets of Maidens for bits of trailing wire. He then hailed 'Stuffy' with 'Have some wire Major', presenting him with the wreckage of his own telephone system.

The pilot was a decorated Canadian instructor whom we had earlier passed outside the front of the hotel, sleeping off his celebrations under a bush. He said that when he had woken up, he had wanted to go up for a 'flip' to clear his head; found he could not see his instruments when he got up into the air and he wanted to be sick – as he was at 'Stuffy's' feet.

This excitement was not the last of the day as not long after another aircraft – an Avro 504 – was seen to take off in the hands of a young South African pilot who – though tea total – was intoxicated by excitement as the other pilot had been by alcohol.

He knew that if he made a perfect loop, he could actually gain a few feet in height, so he proceeded to demonstrate

this manoeuvre from ground level in between two hangars! He got away with it twice but on the third attempt he misjudged his distances and the protruding skid from the Avro touched the ground. The machine made several somersaults collapsing in a mass of wreckage in the middle of the aerodrome.

On clearing the debris, the pilot was found head downwards with his feet caught up in the control wires. When he was extricated, he was found to have sustained a slight cut above one eye. Otherwise unharmed, he was flying again within a couple of days.

Lt Ivan Beauclerk Hart-Davies
Born at Huntingdon in 1878, the son of Rev John Hart-Davies, Rector of Southam, and his wife Florence, Ivan was the grandson of the 8th Duke of St Albans. The first Duke was the son of King Charles II and Nell Gwynn. Educated at King's School, Canterbury, Ivan was an excellent sportsman – he played rugby and cricket, was a Scoutmaster, worked as a teacher and set up his own life and motor insurance business prior to the war.

An early aviator, Hart-Davies received his Aviator's Certificate in 1913 at the Grahame-White School, Hendon. He was also a keen motorcyclist, taking part in the 1908 Motor Cycling Club London to Edinburgh run on his 3.5 hp single-speed belt-driven Triumph, which had no clutch or suspension. In 1911, he broke the record for the Land's End to John o' Groats

Ivan Beauclerk Hart-Davies

endurance event for solo motorcycles, again riding his 3.5 hp single-speed Triumph. He covered the 886 miles (on the roads available then) in 29 hours 12 minutes. This was to be the last such event as his average speed was in excess of the then legal limit of 20 mph. The Auto-Cycle Union banned further official record attempts. Riding a 500 cc Triumph, he entered the 1912 Isle of Man TT Races; this event was boycotted by many of the famous manufacturers of the day due to the difficulty of the course and serious safety issues. Number 19, Ivan Beauclerk Hart-Davies was disqualified for 'taking petrol outside a depot'.

Lt Hart-Davies also won the Murren Cup for bobsleighing, despite his amateur team of motor cyclists never having ever bobsleighed before!

At the outbreak of war he tried to join up but was rejected due to his age, his flying experience enabled him to be commissioned in the Special Reserve RFC in 1916 and he became a Flying Instructor at 2 Auxiliary School of Aerial Gunnery at Turnberry. On appearing before a Medical Board on 2 July 1917 it was reported that he had been 'Flying daily at Turnberry and had been doing so for some months' and 'On account of age probably unfit for General Service as a Flying Officer but otherwise fit for General Service.'

Frederick Bristow and Ivan Hart-Davies

Lt Ivan Beauclerk Hart-Davies was killed on 27 July 1917, aged 39, flying Bristol F2B A7103. He took off from Northolt Aerodrome and when coming in to land the aircraft nosedived and crashed, injuring his Observer Lieutenant Miller.

A brother officer wrote, 'A gallant fellow whom we all liked immensely and are deeply grieved that he should have been fatally injured when he so much wished to go to France, where doubtless he would have won honours.'

Lt Ivan Beauclerk Hart-Davies was buried in Southam with full military honours.

4
WWI Aircraft

Within a decade of the Wright brothers' historic flight of a heavier-than-air flying machine, the aeroplane had been developed into a weapon of war. Initially, there was not much enthusiasm and the Chief of the Imperial General Staff had, in 1910, scoffed at the idea of using aircraft for military purposes as 'a useless and expensive fad', and some Cavalry Officers of the Army stated, 'that their only positive contribution would be to frighten the horses'. It was to this end that it was considered that aircraft could be useful, not as fighting machines, but psychological warfare. Zooming out of the sky towards the enemy, the loud roar of the engines caused panic amongst the infantry, flying low over their heads, harassing and intimidating the ground forces.

The Admiralty believed that the aircraft requirements of the Royal Navy would at most number 'two'. The arrogant attitude of the British War Office was astounding, as it was evident that aircraft could be used not only for reconnaissance and observation, but also for aerial photography and ranging artillery, as France had displayed at their Army manoeuvres of 1911, having at that time over 200 aircraft in military service. Still the British Army were reluctant to use the flimsy, unreliable, strut and wire biplanes, with a fabric 'skin' stretched over a wooden frame, but the Admiralty were quick to change their minds, seeing the benefit of using aircraft in an attack role, especially against enemy submarines or zeppelins.

As the advantage of aerial reconnaissance and observation was proved, the role of the aeroplane developed and a new range of aircraft were designed, manufactured and armed explicitly for use in aerial warfare, with enormous much needed improvements in performance and reliability being made as the war advanced.

In the early days of British military aviation, fierce competition had developed between the Army and Navy. The British air corps was reliant upon the French rotary engines for its aircraft, which were made under licence in Britain, and a more serious problem was the dependence on German manufactured magnetos. The Admiralty, considering itself the Senior Service, and therefore, superior to the Army Air Corps, started to develop its own air service, without official mandate. This led to the two wings of the early RFC being split, separately governed by the War Office, and the Admiralty, both competing for the scarce resources of aircraft design and manufacture that was then obtainable.

Fierce rivalry developed between the Army and the Navy battling for resources. This led the Admiralty to turn to private enterprise, which in consequence removed the naval wing from War Office control. Military contracts were given to firms such as Sopwith, Vickers, A. V. Roe and Handley-Page, who designed and developed new aircraft and engines. This fledgling British aeroplane industry had grown by the end of the war, to be the largest in the world, employing 2,700 workers.

On 4 August 1914 when Britain declared war on Germany, the RFC had only 84 aircraft on charge, and the RNAS had a mere 71 aircraft and seven airships.

The increase in aircraft technology was subsequently passed on to the higher training schools such as Turnberry, where the aircraft were superior and more powerful than those at elementary training schools and were intended to familiarise pilots on the types they would be flying in combat. Cadets were encouraged to fly every day and practice 'trick flying'. As the school developed, the number and differing types of aircraft seen flying over the west coast of Ayrshire increased. In the interests of brevity, only those aircraft are documented where there is evidence that they appeared at Turnberry. A short description of some of these follows.

The initial Establishment document gives a list of aircraft proposed for carrying out the training at Turnberry as below:

Machines (not yet finally approved)

Avros	24
Bristol Fighters	8
DH4 or 9	16
Sopwith Camels	20
SE5	12
Sopwith Dolphins	12
TOTAL	92

Royal Aircraft Factory BE2c
Designed by Geoffrey de Havilland and E. T. Busk the BE2 series of aircraft were built

Personnel and aircraft at Turnberry 1918

under contract by private companies and were first introduced into service in 1912. It was initially used as a reconnaissance aircraft and light bomber and was very successful as a night bomber. In May 1913, Capt. Longcroft flew Lt-Col Sykes in a BE2 from Farnborough to Montrose Aerodrome, a distance of 550 miles, in nearly eleven hours with only two stops. The return flight, aided by fitting an additional fuel tank, reduced his journey time to under eight hours, and requiring just one stop. This was done in a 24-hour period and in 1913 was the longest flight then recorded.

The BE2c was a complete redesign of the BE2, although using the same fuselage as the BE2b, with completely different wings and an aileron replacing the wing-warping used in earlier models. Production started in December 1914 using a 92hp engine developed at the Royal Aircraft Factory from the 70/80hp Renault powerplant. As the war progressed the BE2c carried a variety of weapons, most of the two seat versions were armed with a single mounted Lewis Gun for the observer/gunner. These mounts for these guns were made of metal pipes, an invention of Capt. Louis Arbon Strange, thereafter known as 'Strange Mounts'. Other armament carried was two 112lb bombs and ten Le Prieur Rockets for anti-zeppelin attacks.

In late 1915 with the appearance of the German Fokker Eindecker on the scene, the underperforming BE2c found itself outclassed. It quickly fell out of favour and was called 'Fokker Fodder'. The German aircrews disparagingly called it Kaltes Fleisch (Cold Meat). British Ace Albert Ball thought the BE2c 'a bloody awful aeroplane'. When reports of the BE2c losses reached the ears of Noel Pemberton-Billing, he made a

BE2c 4207 - one of 150 built

statement in the House of Commons attacking the Royal Aircraft Factory and their BE2c, stating that RFC pilots in France were being 'rather murdered than killed'.

The BE2e was a further development of the aircraft, with changes to the wings and tailplane to improve performance. It was intended to fit a new upgraded engine, but this engine didn't make it into production, so the improvement anticipated was sacrificed by using the existing engine used by previous versions. Over 3,500 of this type were made in total and it became one of the longest serving aircraft of the war.

Maurice Farman Shorthorn S11

First flown in 1913, the Shorthorn was a French aircraft built by the Farman Aviation Works.

seat armed with one .300 machine gun, and fully loaded the aircraft was able to carry eighteen 16lb bombs.

Avro 504A/J

Built by A. V. Roe & Co. Ltd., the Avro 504 was launched in late 1913. The 504 was superior to other aircraft of the time, mainly because of its stability and easy-to-fly characteristics. Although simple in its design, the 504 proved to be one of the most successful aircraft in the early days of the war. Whereas the 504 was not used much on operational service, it soon became the mainstay of the training schools, with both the RFC and the RNAS. Though its operational service was limited, the Avro 504 carried out the very first

Maurice Farman Shorthorn S11 with Captain Harry Butler

It was designed for reconnaissance and light bomber duties, and finally consigned to training stations. The S11 was a pusher type biplane fitted with a 70hp Renault Engine. Its maximum speed was 66mph at sea level. With a crew of two, the observer/gunner was placed in the front

tactical bombing raid on 21 November 1914, attacking the zeppelin sheds at Friedrichshafen. It was also the first aircraft to strafe ground troops and became the first British aircraft casualty to be brought down by enemy ground fire, on 22 August 1914. A development of the 504A, the

Avro 504A/J. A9789 at Tauchet Hill, Turnberry. One of 50 built by S.E. Saunders Ltd., Isle of Wight

Avro 504K, remained in service with the RAF until 1924.

Airco DH2
Designed by Captain Geoffrey de Havilland RFC, whilst employed as the chief designer at the Aircraft Manufacturing Co., Ltd., in Hendon, London, the DH2 was designed to meet the need for a single-seat, high speed, fighter of the pusher type. First flown in 1915 the DH2 was powered by a 100hp Gnome Monosoupape rotary engine, and armed with one .303 Lewis gun, which could be placed on one of three mountings in the cockpit. The pilot was able to move the gun in flight until it was realised that by aiming the aircraft at the enemy rather than the gun, it was permanently fixed on a forward-facing mount. A combination of poor pilot training at this time and the DH2 being quite difficult to fly, led it to be nicknamed the 'spinning incinerator' because of the high accident rate. Once pilots had learned to master its sensitive controls it was found to be very maneuverable in combat. The DH2

was introduced to the front lines in February 1916 and proved to be very effective until the appearance of superior German tractor types such as the Albatros D1. Retired from active service, the DH2 was used as an advance trainer until late 1918. The Commander of Turnberry, Lionel Rees, won the VC during the Somme Offensive when flying a DH2 of 32 Squadron he single-handedly attacked a group of ten German fighters, destroying two.

Vickers FB9
The FB9 was a pusher biplane designated a fighter scout, developed in 1915 from the FB5 or 'Gunbus', which was the first purpose-built fighter aircraft to see active service. Fitted with a 100hp Monosoupape Gnome engine it was capable of a maximum speed of 70mph and had a fuselage of high tensile steel with fabric covering the wings and tail. Manned by a crew of two, its role was that of an offensive fighter against enemy aircraft. It was armed with a .303 drum fed Lewis machine gun, fitted to a Scarff ring

in the front of the machine, giving the gunner a clear field of fire. By the end of 1915, the aircraft was already outdated by the arrival of the Fokker Eindecker and withdrawn from active service. Fifty were then delivered to Royal Flying Corps training units.

Vickers FB.9 A8620 - one of 25 built by Vickers Ltd. at their Weybridge factory.

Airship SS20

By late 1914 the heavy loss of British shipping due to increased German submarine activity had become critical. The Admiralty realised that by using Airships for observation and submarine spotting they could deter this threat and accordingly in February 1915, orders were given for the development and construction of a suitable Blimp. The specifications required that it should be capable of a speed of 40-50 mph, carry 160lb of bombs, wireless equipment and enough fuel for eight hours flying with a crew of two. The airship should be capable of reaching an altitude of 5,000ft and be of a simple design to accelerate production and speed up the training of the crews.

The new Airship prototype designated the 'Submarine Scout' class was ready for evaluation tests with five weeks of order. Each airship was quickly and cheaply produced by hanging the wingless fuselage of a BE2c aircraft beneath a simple balloon envelope. After a few modifications the SS class of airship entered service on 18 March 1915. With most manufacturing facilities being taken up with aircraft production at this time, airship

construction was transferred from Farnborough to Kingsnorth, Barrow and Wormwood Scrubs. Airship Stations were opened up, the first at RNAS Luce Bay near Stranraer in Wigtownshire, and soon a chain of bases were established around the British coast, from where these 'Submarine Scouts' could patrol in search of enemy submarines. The airship was not capable of carrying a large enough bomb load to attack enemy craft but had the advantage of a higher speed than most surface shipping and had a much farther range of vision. German submarines could sometimes be spotted just below the surface or by the wake left by their periscope, another tell-tale sign was a trail of oil on the surface of the water. Once the submarine had been spotted, the commander of the airship would radio back to base and request help. The airship section at RNAS Luce Bay patrolled the approaches to the Firth of Clyde and the North Channel, a section of the North Sea, the north of Ireland and escorted the Stranraer – Larne ferry and other shipping. A few days before the end of the war, one of Luce Bay's airships (SS12) was called to investigate gunfire heard to the south of Ailsa Craig lighthouse. Spotting oil on the surface of the sea, armed trawlers were quickly summoned to depth charge the area, although it was confirmed to be an enemy submarine, it managed to get away.

Airship SS20 at Turnberry

RE7 2256 Built by Austin

Royal Aircraft Factory RE7

Designed by the Royal Aircraft Factory, the RE7 entered service in 1915. It was a two-seat light bomber and reconnaissance aircraft developed from the RE5, able to carry heavier loads and was considered suitable for escort and observation duties. As an escort aircraft it proved unworthy of the task, due to the limited fire power available from the single Lewis gun. The RE7 was found to be more use as a light bomber although they proved vulnerable to attack because of their slow climb speed and low ceiling when loaded. The RE7 was withdrawn from active service and transferred to training establishments where they were the first aircraft to be used as target tugs, trailing a sleeve drogue for air to air gunnery practice.

Handley Page 0/400

This heavy bomber biplane was developed in December 1915 from the Handley Page 0/100, built to the Admiralty's specifications of December 1914, to design a 'bloody paralyzer' of an aeroplane capable of making overseas patrols with a heavy bomb load. The 0/400 was nearly 63 feet long and had a 100-foot wingspan (the wings could be folded back for storage) and was capable of carrying 8x250lb bomb load. Powered by two 360hp Rolls Royce Eagle V12 engines, it had a maximum speed of 98mph. It had a crew of three, a pilot and co-pilot in the cockpit and a gunner/observer in the open cockpit in the nose.

By mid-1918 the 0/400 was the main aircraft of the new Independent Bombing Force, and was the largest aircraft employed by the RAF during WWI.

Post war, the 0/400 was developed into various forms of passenger aircraft, at the dawn of commercial aviation at home and abroad. These commercial 0/400s operated a regular service under the flag of Handley Page Transport Ltd between Cricklewood and Paris during 1919.

Handley Page 0400 C3498 and Bristol M1.c C5014 with Shanter farm in background

Royal Aircraft Factory FE2b

Designed by Geoffrey de Havilland, the Royal Aircraft Factory (Farman Experimental) FE2b was a two-seater pusher biplane. (A pusher aircraft has the engine and propeller placed behind the pilot.) Before the invention of the interrupter gear which allowed the forward firing machine gun to fire between the propeller blades allied pusher types reigned in the sky; types like the FE2b, DH2 and the Vickers Gun Bus.

FE2d (a variation of the FE2b) with Harry Butler and Captain Anderson

Brought into production at the end of 1915, the FE2b entered service on the Western Front in January 1916 as a front-line fighter. The first production aircraft were fitted with 120hp Beardmore engines - later versions were upgraded to 160hp. Heavily armed and armored, the pilot occupied the rear cockpit and the gunner the front, giving his one or two Lewis machine guns an unobstructed field of fire of over 180 degrees. American ace Frederick Libby, who served as an FE2b observer in 1916 wrote: 'When you stood up to shoot, all of you from the knees up was exposed to the elements. There was no belt to hold you. Only your grip on the gun and the sides of the nacelle stood between you and eternity. Toward the front of the nacelle was

a hollow steel rod with a swivel mount to which the gun was anchored. This gun covered a huge field of fire forward. Between the observer and the pilot, a second gun was mounted, for firing over the FE2b's upper wing to protect the aircraft from rear attack.... Adjusting and shooting this gun required that you stand right up out of the nacelle with your feet on the nacelle coaming. You had nothing to worry about except being blown out of the aircraft by the blast of air or tossed out bodily if the pilot made a wrong move. There were no parachutes and no belts. No wonder they needed observers.'

The maximum bomb load for the FE2b was 350lb. By late 1916 The FE2b was outperformed by the arrival of the new German fighter types and was withdrawn from service as a day fighter, being used instead as a light tactical night bomber. Almost 2,000 FE2b's were built, mostly by private manufacturers, the type being retired in 1918. Today there are two reproductions of this aircraft, one in full flying condition, built by the Vintage Aviator Ltd., of New Zealand.

Armstrong Whitworth FK8

The FK8 was designed by the Dutch aircraft designer Frederick Koolhoven as a replacement for the Royal Aircraft Factory BE2c and the RE8. The first of its type, A411, flew in May 1916. A further fifty were ordered for the RFC with an eventual 1,650 being built. Being designed for use in reconnaissance, artillery spotting, and bombing, it quickly gained the favour of pilots, being stable, effective and dependable.

Armstrong Whitworth FK8 A2692.
The gunner is seen fitting his Lewis machine gun to the scarff ring mountingin the rear cockpit.

Nicknamed 'Big Ack', it was considered easier to fly than the RE8.

The aeroplane was armed with one fixed, forward firing .303 Vickers machine gun, one .303 Lewis gun, and could carry a 260 lb bomb load. The FK8 had dual controls, enabling the observer to take control if the pilot became incapacitated.

Sopwith Camel B9207 1F1 built by Boulton & Paul Ltd.

Sopwith Camel F1

The Sopwith Company had already designed several effective scout/fighters, such as the Sopwith Triplane and the Sopwith Pup. In 1916 there was an urgent need for the design and production of a much faster and more manoeuvrable single-seat fighter. Sopwith produced a prototype aircraft that was designated Sopwith F1 (F for Fighter). Powered by a 130 hp Clerget 9-cylinder rotary engine, and armed with twin synchronized machine guns, the Sopwith Camel is perhaps the most famous fighter aircraft of WWI.

The cockpit, engine, fuel tank and guns were fitted into the front two metres of the aircraft giving it a very close centre of gravity, The mass weight of the aeroplane, being centered in such a small area when combined with the torque of the rotary engine, made it difficult to handle for all but the most skilled pilots, but gave excellent handling in the air. To the novice however, the Camel often proved to be fatal, as shown at Turnberry, altogether hundreds of trainee airmen were killed when the aircraft went out of control at flying schools. It was joked amongst the men of the RFC that flying the Camel would get you 'a wooden cross, the Red Cross or the Victoria Cross'. There are only eight original Sopwith Camels left, one of these, B7280 F1 is on static display at the Polish Aviation Museum in Kraków, Poland. This aircraft was built in Lincoln by Clayton & Shuttleworth and had been shot down in 1918. It was then displayed in a museum in Berlin until the outbreak of WWII when it was sent to Poland and put into storage.

SPAD SVII

The SPAD SVII was a fighter aircraft, designed to replace the Nieuport 7, and built by Société Pour L'Aviation et ses Dérivés, which entered service in 1916. A small single seat aircraft, it became the principal French fighter able to take on the German Albatros. Powered by the 180 hp Hispano-Suiza engine, it could reach a top speed of 118 mph and was armed with a synchronized Vickers .303 machine gun. Although the SPAD was fast and difficult to shoot down, it had poor forward and downward visibility for the pilot. Allied services started using the new French fighter built under contract in England. The RFC was the first foreign service to employ them with 19 and 23 Squadrons and also at training schools.

SPAD SVII A9134 Built by Mann, Egerton & Co. Ltd., Norwich. Transferred from RNAS

RE8 A3432 Built by the Siddeley-Deasy Motor Car Company Ltd., Coventry.

Royal Aircraft Factory RE8

Designed by John Kenworthy for use in reconnaissance or as bomber, it was never a successful combat aircraft. The RE8 was introduced in June 1916 and was quickly withdrawn at the end of the war. It was considered very difficult to fly and gained the reputation in the RFC of being 'unsafe'. Despite this the RE8 remained a stalwart for reconnaissance and artillery spotting throughout the conflict. Nicknamed 'Harry Tate' by RFC aircrew, some of this type were on strength at training schools in the role of target tugs. Unlike other types, the RE8 was not popular with private fliers who purchased surplus aircraft after the war, so none appeared on the civil register. Out of over 4,000 built, there are only two original survivors from WWI, both on static display, one at the Imperial War Museum, Duxford, and one at the Brussels Air Museum. A replica was built in 2011 by The Vintage Aviator Ltd., New Zealand, and after being shipped to England undertook several flights at the Shuttleworth Collection. It is now on display at the Grahame-White Factory, part of the RAF Museum, Hendon.

Albatros D1

The Albatros fighter/scout was designed by Robert Thelen, and the prototype first flew in August 1916. The use of the Albatros over the front lines enabled the German pilots to win back air superiority from the Allied Fighters. Although not as agile as it's some of its opponents, the D1 had a significant advantage with its powerful engine, in its speed and armament. It was unusual, unlike other fighting aircraft at the time, as its fuselage was covered with sheets of plywood rather than fabric. This gave the aircraft great strength and rigidity. It was mounted with two forward firing synchronised machine guns, each with a magazine that held 500 rounds.

The Albatros D1 was for a time the best aircraft in its class, until the introduction of the Sopwith Camel, which being both fast and agile could outmanoeuvre the D1 and turn twice for a single turn by the Albatros. This aircraft became the favourite of such pilots as Boelke and Richthofen. Although Baron von Richthofen was most famous for his Red Fokker Triplane, most of his victories were scored in an Albatros, which he also had painted red. He was the highest scoring 'Ace' in service, with 81 'kills', during

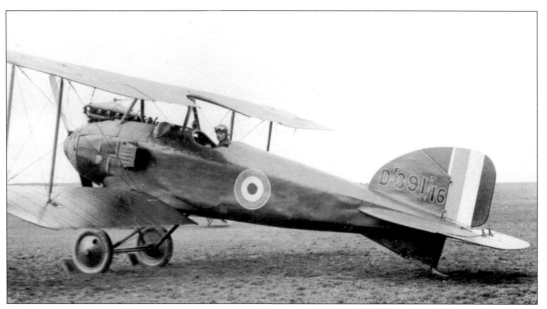

Captured German Albatros, (D1 391/16) XG6 had been flown by Lieutenant Karl Heinrich Otto Buttner of Jasta 2. Buttner had been forced down behind the British lines near Arras on 1 December 1916 by Captain G. A. Parker and 2nd Lieutenant H. E. Hervey, flying a BE2 of 8 Squadron RFC. When captured, the personal marking of Buttner, 'Bu', was visible on the fuselage of the Albatros, Lieutenant Buttner was taken prisoner at Arras. The captured aeroplane was used for evaluation and was nicknamed 'Albert Ross' by the RFC.
One of the pilots who flew it at Turnberry was J. B. McCudden.

Mock-up of 'in flight' SE5a B679 flown by Gerald Maxwell

1917. McCudden in his memoirs mentions his first encounter with an Albatros and was not impressed with its styling, 'It had a fuselage like the belly of a fish. Its wings were cut fairly square at the tips and had no dihedral angle. The tail-plane was the shape of a spade.'

Royal Aircraft Factory SE5a
The SE5 (Scout Experimental 5) was a single seat fighter designed in 1916 by Henry Folland, John Kenworthy and Major Frank Goodden of the Royal Aircraft Factory at Farnborough. An exceptionally stable, manoeuvrable plane which could be dived at a very high speed, the SE5 was developed into the SE5a, with the replacing of the 150 hp Hispano-Suiza engine with a 200 hp version. The SE5a was one of the fastest aeroplanes of the war with a speed of 138 mph, equal to any German aircraft at the time. Although not as agile in combat as the Sopwith Camel, it was considered much easier and safer to fly, especially for inexperienced pilots and had a superior performance at altitude. According to 'Dodge' Bailey, Chief Test Pilot of the Shuttleworth Collection, it had 'somewhat

similar handling characteristics to a de Havilland Tiger Moth, but with better excess power'. Over 5,000 of this type were built and it was considered the best British built fighter of WWI.

Sopwith Scout
Designed by Herbert Smith of the Sopwith Aviation Co. and given the official designation Sopwith Scout, the aircraft was quickly nicknamed the 'Pup', suggesting that it was the offspring of its larger predecessor, the Sopwith 1½ Strutter. The Pup was a single seat biplane fighter powered by an 80 hp Le Rhone rotary engine and armed with a single .303 Vickers machine gun mounted on top of the fuselage in front of the pilot. With excellent handling qualities, it was said that a good pilot could turn twice to once of a German Albatros. The Pup entered service with the RFC in December 1916, when after seeing one in combat Manfred von Richthofen stated, 'We saw at once that the enemy aeroplane was superior to ours.'

By spring 1917, the Pup was swiftly outclassed by the latest German fighters with their improved performance and increased

Bristol F2A A3303 Prototype fitted with 190 hp Rolls Royce Falcon engine with Shanter farm in the background

firepower. Sopwith was in the meantime re-arming, initially with Triplanes, then the Sopwith Camel. Although it was retired fairly quickly from front line service the Pup saw general use as a trainer, large numbers being transferred to training establishments. Trainee Pilots would progress from basic flight training in aircraft such as the Avro 504 and move onto the Pup as an intermediate trainer. The Pup was used in fighting school units such as Turnberry for instruction in combat techniques. Many officers and instructors earmarked Pups as their personal runabouts while a few were used as inter-station 'hacks'. Almost 1,800 Sopwith Pups were built, mostly by sub-contractors such as Standard Motor Co., Whitehead Aircraft and William Beardmore & Co.

Vickers FB16d

The FB16 was a very fast (136 mph plus) fighter prototype designed by Rex Pierson, originally intended to take the Hart air cooled radial engine. Having a very deep fuselage the aircraft was nicknamed 'Pot-Belly'. As the Hart engine was still only in the experimental stage and was proving unsuccessful, the aircraft was modified to take the 150 hp Hispano-Suiza engine. This new prototype was designated FB16A, but was destroyed in an accident, fatally injuring the test pilot Capt. Simpson RFC, in December 1916. A second prototype A8963, was built in January 1917 and the engine was changed to a 200 hp Wolseley-Hispano in June. Not finding favour with the authorities, no orders were placed for this type as it was 'not considered suitable in its present form for active service'. The main reason being perceived structural problems and maintenance difficulties, the streamlining made it very inaccessible for servicing with a front-line squadron. Meanwhile the RFC had placed orders for the easier to maintain SE5a. Harold Barnwell, Chief Engineer for Vickers Ltd., said that flying the FB16d cost him a new pair of trousers every time he flew it, as it always covered his legs in oil. Capt. J. B. McCudden VC, DSO, MC, MM, was much taken with this aircraft and was very impressed with its performance, so impressed that he managed to obtain sole use of the aircraft as his personal transport and took it with him when he was posted as a fighting instructor to Turnberry.

Bristol F2A

Designed as a two-seater reconnaissance aircraft

Bristol M1c C4940 Built by the British & Colonial Aeroplane Co.Ltd.

by Frank Barnwell, the F2A took its first flight on 9 December 1916. It was primarily intended to replace the BE2 series of aircraft. It was quickly found that, when fitted with the Rolls Royce Falcon engine, it had the performance of a fighter and was as agile and maneuverable as many single seat fighter aircraft. The F2A developed into the F2B. It was fitted with one fixed forward-firing Vickers machine gun, plus one or two .303 Lewis guns on a flexible mount in the rear cockpit. There were underwing racks allowing for a 12.9 kg bomb load.

Bristol M1c Monoplane

Another of Frank Barnwell's designs from 1916, the Bristol M1c monoplane was built to fulfil the need for a fast, single-seat fighter aircraft, with superior fire power, planned with the intention of being a replacement for the De Havilland DH5. The M1c was very advanced for its time and had outstanding acceleration and rate of climb and was very fast on straight and level flight. The aircraft handling and stability were good, but the main weakness was that the pilot's field of

vision below the aircraft was very limited, even though areas at the roots of the wings close to the fuselage were left uncovered. The pilot's access to the cockpit was made difficult by the cabane struts, though this same structure gave some crash protection. The Bristol M1c was much liked by experienced pilots, but the authorities had their misgivings. The main reservation was the mistrust of the monoplane configuration. Production was therefore limited, and the Bristol M1c was only ready for service in 1917. As few as 125 examples were built with only a handful in operational service in Mesopotamia and the Balkans. The remainder went to Flying Training Units, where due to its exceptional performance, and relatively high landing speed they were considered extremely unsuitable for students, and instead were chosen as the personal aircraft of Senior Officers and Instructors who had them 'customised' with elaborate and imaginative colour schemes. At Turnberry, some of these were painted in a variety of colourful designs with fish scales, intricate lattice work on the fuselage, sun-burst detail on wings and

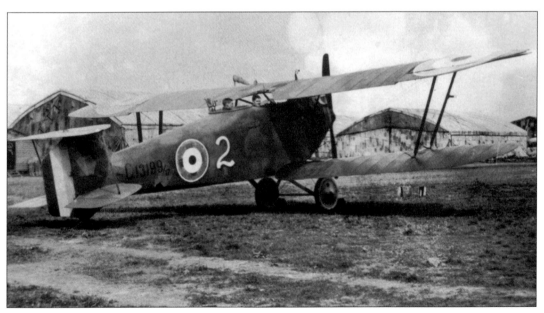

This aircraft, CL13199/17 was shot down by British anti-aircraft fire near Lestrem on 29 March 1918, flown to the United Kingdom on 26 April, and given the capture number G156. It belonged to Schlachtstaffel 38s and was crewed by Gefr. Bruno Karcher, pilot who died of his wounds the next day and Flg. Paul Schleuder, Gunner who was taken POW

tail surfaces (C5014), fuselage painted in a chequerboard of red, white and blue (C4995) a large skull and cross bones on the fuselage (C5019), a serpent winding around the fuselage, possibly in red or black (C5015), and a pattern which can only be described as psycho-patchwork (C5012). Some were silver doped instead of the standard olive-khaki colour, some even had coloured noses with faces painted on them. Only one genuine Turnberry M1c survives today, C5001, which was an amalgamation C5001 and C4964. This restored aircraft can be seen at the Butler Memorial Building at Minlaton, South Australia. *(See Captain Harry Butler)*. A full-size replica (C4918) can be seen flying on air show days at the Shuttleworth Collection, Biggleswade, Bedfordshire.

Hannoversche CLII

The 'HunBus' was designed in 1917 by Hermann Dorner and manufactured by Hannoversche Waggonfabrik to fulfil the role of an escort fighter, its primary duty being to protect reconnaissance aircraft over the front lines. The fuselage was similar to the Albatros and was also covered in thin plywood. Powered by a 180 hp Argus engine, it was capable of a speed of 103 mph. It was manned by a crew of two, pilot and gunner/observer and armed with one fixed, forward firing Spandau machine gun and one rearward firing Parabellum machine gun for the

Captured German Albatros (DV2359/17) G144 was brought down near Arras on 6 March 1918 by Lt. Bellway and AM1 Rose of 13 Squadron flying an RE8. Ltn. Otto Hohmuth of Jasta 23 was wounded and taken prisoner. G144 appeared at Marske when the station number '96' was added.

observer. Being a relatively small aircraft, Allied pilots often mistook it for a single seat fighter and discovered too late on attack that there was a second gun trained on them. Only 439 of this

Another Captured German Albatros (DV 1162/17) G56 was forced down by Lt. David Langlands flying a SPAD of 23 Squadron RFC at Poperinghe on 16 July 1917. Vzfw. (Sergeant Major) Ernst Clausnitzer of Jasta 4 was taken prisoner. It was the first intact Albatros captured and was flown to England and extensively tested before being sent on display to various RFC Aerodromes in Britain.
This aircraft was destroyed when it crashed into the Firth of Forth on 23 December 1917 killing the pilot Capt. C. F. Collett.

type were made.

Albatros DV

Introduced in to service in May 1917, with a 180 hp Mercedes engine and armed with two Spandau machine guns, the DV was plagued by a series of fatal crashes caused by structural failure of the wings. Attempts were made to correct this by the addition of small struts and wire bracing, but the accidents continued. Pilots were advised not to put the aircraft into a steep dive. The DV was the last of the Albatros fighters to enter operational service.

Fokker Dr1

Built by Fokker-Flugzeugwerke, the Fokker Triplane single seat fighter, designed by Reinhold Platz, became famous as the chosen aircraft of 'The Red Baron', Manfred von Richthofen. The Dr1 first flew in July 1917, and was found to offer exceptional maneuverability in combat, although found to be slower than contemporary

(DR144/17) G125 Flown by Leutnant Eberhard Stapenhorst was brought down by anti-aircraft fire badly damaged, inside British lines at Bailleul on 13 January 1918 where he was subsequently captured by British troops.

allied fighters. From the outset the Fokker Triplane was plagued with structural problems, namely the wings breaking up in flight. Powered by a 110 hp Oberursel engine, it was capable of reaching 115 mph. Armament consisted of two .312 Spandau machine guns. Unfortunately, the positioning of the gun butts so close to the cockpit, combined with insufficient crash padding, meant the pilot was exposed to the risk of serious head injury in the event of a crash landing. Because of the inherent structural problems as few as 320 were manufactured with

Sopwith Dolphin 5F1 E4466. One of 200 built by Sopwith Aviation.

production ending in May 1918.

Sopwith Dolphin 5F1

The Sopwith Dolphin was the world's first single seat multi-gun fighter, designed in 1917 by Sopwith's chief designer Herbert Smith to replace the Sopwith Camel. Although successful as a fighter, the pilot's view in the Camel was restricted by the upper wing directly above their seat. This was improved slightly by cutting out a small section of the centre wing. On the Dolphin, Herbert Smith decided to remove the upper wing centre section entirely and to fix the wing panels to a tubular steel frame, placed low over the cockpit. This gave the pilot a greater field of vision. The main problem with this new design was that the pilot's head and neck was left exposed above the top wing and unprotected, making him extremely vulnerable in nose-over crashes. This earned the Dolphin the nickname of 'The Blockbuster'. This issue was solved by fitting a cabane (crash pylon), on which were mounted two Lewis machine guns, firing at an upward angle, over the propeller arc, and also two fixed synchronized Vickers machine guns. The Lewis guns proved unpopular as they were difficult to aim and tended to swing into the pilot's face. Pilots also feared that the gun butts would inflict serious head injuries in the event of a crash.

Under the pilot's seat was the 20 gallon main fuel tank and a further gravity fed tank holding seven gallons was in the top wing section, just near his head. The Sopwith Dolphin was well liked by pilots, particularly as the cockpit, being slightly forward of the upper wing, meant the heat from the engine kept the pilot nice and warm. Powered by a 200 hp Hispano-Suiza engine, the Dolphin had a maximum speed of 131 mph. On the introduction of the Dolphin to front line service, much to the dismay of the pilot, several British and Belgian aircraft attacked the new type, mistaking it for a German aircraft. Understandably, Dolphin pilots were extremely vigilant and nervous near other Allied aircraft for the first few outings.

Airco DH9

The DH9 was a development of the DH4 tractor-type biplane, a two-seat day bomber, another of Geoffrey de Havilland's designs. The aircraft first flew in November 1917 and was powered by a 230 hp BHP engine (Beardmore-Halford-Pullinger, built by the Galloway Engineering Co. Ltd., Co. Ltd., of Tongland, near Kirkcudbright, known as the 'Galloway Adriatic'). After many problems with performance and production, the

Airco DH9 at Turnberry (Shanter farm in the background)

engine was further developed by the Siddeley-Deasy Motor Car Co. Ltd., and a lightweight version, named the Puma. was chosen as the main production engine. Armed with a forward firing .303 Vickers machine gun, and one or two .303 Lewis machine guns fitted to a Scarff ring in the rear cockpit, the DH9 was also capable of carrying a 230 lb bomb load. Camera and wireless/telegraph equipment were also fitted. The Puma-engined DH9 still suffered from lack of performance and repeated engine failures, which was improved by fitting a 375 hp Rolls Royce Eagle powerplant and the aircraft then became the DH9a. Over 3,000 production aircraft were built, the RAF finally retiring them in 1937.

Sopwith 7F1 Snipe

The Sopwith 7F1 was a single-seat fighter, designed by Herbert Smith in late 1917 for the Sopwith Aviation Company to replace the Sopwith Camel. Although smaller than the Camel, it was very manoeuvrable, and much easier to handle, with a greater visibility for the pilot. The Snipe also had a greater rate of climb and improved high-altitude performance, compared with its forerunner, putting it on more equal terms with Germany's newer fighters. The Snipe was powered by the 230 hp Bentley BR2 engine, the last rotary type to be used by the RAF. It had a maximum speed of 121 mph at 10,000 feet and an endurance of three hours. Its fixed armament consisted of two .303 Vickers machine guns on the cowling, and it was also able to carry up to four 25 lb bombs. The Snipe entered operational service in September 1918, a few weeks before the end of the war. Production ended in 1919, with just under 500 being built. The Sopwith Snipe flew in support the British Army of Occupation in Germany until 1919 and well as service during the Commonwealth intervention against the Bolsheviks in the Russian Civil War between 1919 and 1920, before being finally retired in 1926. A fully airworthy reproduction, Snipe E8102, complete with the original Bentley rotary engine, was built by The Vintage Aviator Ltd. in New Zealand.

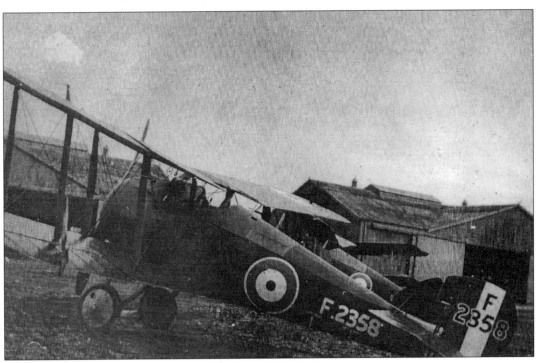

Sopwith built 7F1 Snipe

5
WWI Training

By 1916, air force casualties of the war had risen drastically. The life expectancy of an RFC pilot over the front lines was 23 days. In what was to become known as 'Bloody April', 361 aircrew and over 3,000 aircraft were lost.

Appeals went out to the infantry, artillery and cavalry for new applicants and civilian pilots/instructors were urged to volunteer. Despite being dubbed 'the suicide club' by officers of the regular army, the appeal for more men was very successful. There was no shortage of gallant, courageous young men to volunteer, grasping the exciting prospect of becoming one of the new 'Knights of the Air'. There were also many cases

of men being previously wounded in battle and considered 'unfit' for further duty, signing up for the RFC. The medical requirements seemed less stringent than in the army or navy. Some went on to become pilots, even though they had only one eye, or leg, and there were those with serious chest complaints and disorders of the stomach. The training programmes in place at the time struggled to keep up with the massive rate of expansion. This lack of adequate training became so critical that an ex-RNAS pilot, Noel Pemberton-Billing, pressed charges in the House of Commons of criminal negligence by RFC commanders. He accused the leaders of the

Gun Setting on Sopwith 1F1 Camel B9204

It was not until the introduction of the Smith-Barry Gosport training scheme in late 1917, that a basic course syllabus was established. The length of the course increased, and a higher training curriculum was introduced, although these courses were still liable to be shortened and adapted, depending on how desperate was the need for pilots at the front.

To improve the level of pilot training, the new system introduced additional qualification criteria, namely that the pilot must have flown at least fifteen hours solo, flown at least fifteen minutes at 6,000 feet, and made two night-time landings assisted by flares. The training scheme became increasingly specialised as it developed, with the establishment of 'Higher' training schools, such as Turnberry. Pupils were to undertake training in landing, bomb-dropping, aerial fighting and gunnery, night flying and formation flying. Before a pilot was posted on active service, it was anticipated that he would have spent at least 30 hours in the air, flying the type of combat aircraft he would use over the front.

Aerial Gunnery Training - early flight simulator

RFC of negligence, for the horrendous number of accidents that occurred at training squadrons, rather than in action, there being more fatalities in training than at the front. The lack of an organised training plan was proving to be very costly, in both pilots and aircraft. There were at least two dozen crashes per day at each training aerodrome, and on average one student pilot died every day in Britain and many more were seriously injured.

Not only was there no official training programme prior to the Gosport system, there were no purpose-built schools to accommodate the pupils. Golf courses such as Turnberry and racecourses such as Ayr, and many others throughout Britain, were requisitioned. These were ideal candidates, having a large expanse of relatively flat ground, and plentiful, nearby accommodation. Flying Instructors were also desperately needed. These posts were filled by men who were no longer capable of active service at the front, or on operational 'rest'. Who better to pass on first-hand experience of actual combat conditions? This was not a method welcomed by exhausted front-line pilots, who, having survived the horrors of aerial combat and living on the front line, were predictably averse to being killed by a clumsy student. Flying Instructors were known to nickname the student pilots the 'Hun', because it was reckoned they destroyed more aircraft than the enemy.

Captain Harry Butler gave an account to the *Pioneer* newspaper in March 1918 of the training procedures in place at Hythe and Turnberry:

After the student of aviation has done his theory course, providing he has been successful, he is sent to a training squadron to be taught to fly. This course extends over a period of three months. The pupil is taken up into the air, with an instructor, in a machine, which has dual control. The first few flights the pupil does nothing but watch the controlling movements of the instructor and admire the beauties of the world below. After a few flights the pupil is allowed to put his hands on the controls and feel the movements of the instructor. A few hours of this and he is allowed to take charge of one set of controls and is allowed to fly the machine and any faulty movements done by the pupil are checked by the instructor. The pupil is now allowed to do steep banked turns without sideslipping the machine, how to glide and how to land a machine, and how to take a machine off the ground. When a pupil has done a few hours of this, and considered safe, he is sent up to do his first solo – that is to fly an aeroplane by himself, without anyone else in the machine. This is the most exciting time of his aviation tuition; It very often happens that upon the pupil

Aerial Gunnery Training

landing the machine after his first flight, it is more or less smashed to matchwood, and unfortunately very often with fatal results to the pupil. The next stage for the successful pilot is to practice landings and flying in general, followed by a turn in a faster type of machine – one which has a little more vice, and not quite so tame. After he has done about 30 to 40 hours solo flying, he is sent to a finishing school and goes through a course of special flying, shooting, and fighting in the air, and aerial tactics. This course lasts a month and is taught at this school here.

The first thing he is taught is looping the loop. An instructor takes him up in a machine and loops the loop with him several times, allowing the pupil at the same time to have his hands on the controls and feel what movement takes place. After the pupil has looped a few times under the tuition of the instructor, he is next taught how to spin, roll and turn a machine, also formation flying, shooting with an aerial machine gun and bomb dropping. If the pupil is successful in passing the examination in all the courses mentioned, he is graduated and given his wings to wear and is gazetted as a Flight Lieutenant, and in most cases he very soon finds himself well on active service over the lines in France, with a fat old Hun manoeuvring to get a shot at him. Every airman trained costs the Government roughly from £6,000 to £7,000. That covers the salaries paid to the

Aerial Gunnery Training

Aerial Gunnery Training with Turnberry Hotel in the background.

corners of the Commonwealth, completed their training here before being posted over to France. With a staff numbering over 1,200 and 90-plus aircraft, the skies above this part of the Ayrshire coast had never seen the like before.

In a minute of the Air Council dated 27 December 1916 it is recorded that:

> Owing to the increase in the number of pupils that are now required to be trained in aerial gunnery, it has been found necessary to form two additional schools of Aerial Gunnery at Hythe and Turnberry, general sanction for which has been requested.
> The establishment proposed for these schools is the same as for the School of Aerial Gunnery, Loch Doon.
> It is essential that these schools should start working almost immediately, it is requested that early sanction may be given.

This was followed in June 1917 by a letter from Headquarters, Training Brigade, RFC, to the Air Board Office:

> With reference to the amended establishments for the Schools of Aerial Gunnery, which have been forwarded to you for sanction, I would point out that these have been prepared very carefully and it has been found impossible to produce one establishment to cover all three Schools, or even one establishment to cover No.1 Auxiliary School and No.2 Auxiliary School, as their functions vary very considerably.
> No.2 Auxiliary School of Aerial Gunnery at Turnberry carries out the actual aerial instruction for all fighting Pilots. This instruction is given on both the Lewis and Vickers Guns, and is given entirely to Flying Officers who have had ample experience in the air before going to the School.

instructors etc., but the most expensive part is the number of machines he crashes during his course of training.

Another difficulty was the shortage of skilled mechanics and engineers to service and repair the aircraft, guns and other equipment. It was estimated that 47 skilled technicians were required to keep a single aircraft flying. The RFC, by offering better rates of pay than the other services, recruited over 1,000 of the best mechanics in Britain.

Turnberry was the largest training school for aerial gunnery and fighting in Britain. Over three thousand young airmen, who came from all

Aerial Gunnery Training - firing at stationary target on Turnberry beach.

The training is progressive so far as the Gunnery is concerned, commencing from the Cadet Battalion and passing through the School of Military Aeronautics at Oxford or Reading, and the elementary and higher training Squadrons, on to No.2 Aux. School of Aerial Gunnery. All the ground firing on both the above guns is carried out before reaching No.2 Aux. School of Aerial Gunnery.

The establishment of this School is calculated on a basis of putting through 150 Pilots per fortnight, in air firing with both the Vickers and Lewis Guns.

Training at Turnberry was undertaken in two parts, first being an intensive ground gunnery course then combat exercises in the air. Aircraft silhouettes were marked out on the beach for air-to-ground gunnery targets, and some were placed on rafts moored out at sea. Trainee gunners were suspended above the target in a wooden cradle suspended from a steel rope and pulled along by a winch system. An early type of 'flight-simulator' was developed, which was a 'crate' fuselage section mounted on a circular railroad track, allowing the gunner to be made mobile whilst aiming at model aircraft targets.

Station dogfights were held, and camera guns were used to record mock air combats, with the instructors acting as the enemy, after which the pupil's camera-gun films were examined and assessed. Aerial gunnery practice at Turnberry consisted of students firing a Lewis or Vickers gun, with live ammunition, at a canvas target or drogue being pulled by a tow-plane (which could be easily identified by the bullet holes in its rudder), usually an ageing RE7. The bullets of the attacking student would be dipped in paint prior to flight and if he scored a hit, the bullet

Target Tug for Aerial Gunnery practice RE7 2256.

would leave a trace on the target. With each student having different coloured ammunition, it was easy to asses which cadet had the most 'hits.'

When 1 School of Aerial Fighting moved south from Ayr to Turnberry, courses were increased to train fighter-reconnaissance pilots and gunners, day-bomber crews and single-seat fighter pilots. The station was divided into four flights, one each of Sopwith Camels, SE5as, DH9s and Bristol Fighters.

One of the instructors who moved down from Ayr was James McCudden, who could often be seen in the skies above Turnberry practicing new combat manoeuvres with another instructor, and these duels were closely watched by the pupils.

There were also classroom lectures on all aspects of guns and gears. In fact, a book was written in 1918 by the Commandant of Turnberry and published at the *Observer* Office, Union Arcade, Ayr:

Commandants Address
The Importance of Gunnery Training

The Offensive Spirit

The importance of encouraging and infusing the offensive spirit into all Pilots cannot be overestimated. Pilots in every type of Squadron must be made to realise that, whatever their task, they will be required to fight – and the importance of taking the offensive at all times must be continually brought home to them. For Scout Squadrons a fighting spirit must be instilled into Pilots from the start. They must be taught that for fighting they exist, that their whole duty on Active Service begins and ends with fighting; that whereas in the case of other types – reconnaissance, bombing or artillery machines, fighting may be incidental to

their main task, in the case of Scouts it is their raison d'etre. Their first and all-important duty is to seek out and destroy the enemy wherever he may be found. To possess this essential 'Offensive Spirit' you must first have absolute confidence in: Your ability to Fly.
Your Machine.
Your Engine.
Your Gunnery.
That confidence cannot possibly be obtained unless you take a personal interest in, and thoroughly understand, your machine, engine, gears and guns. Before leaving the ground overseas you will satisfy yourself that your machine and engine are in order. You must realise that it is equally important to satisfy yourself that your gun is also in proper order.

Examples of Experts
It is interesting to note that all Pilots who have done well, such as captains Ball, Bishop, Rhys-Davids, Luchford, Fullard and many others, all took a personal interest in their guns, gears and sights, and although they did not in all cases actually clean their guns or fill their belts, they satisfied themselves that everything was done correctly, which undoubtedly gave them absolute confidence. The following examples will be of interest in this connection:

Aerial Gunnery Training - Lewis .303 - Turnberry Lighthouse in the background.

When in 60 Squadron Capt. Ball was at first laughed at, because he seemed to always have his machine in the butts in flying position. He was satisfying himself that his sights were correctly aligned, and it was soon admitted that there was method in his madness.

When 56 Squadron went overseas for the first time, Capt. Ball insisted on all the Pilots in his flight looking after their own guns and gears in every particular.

Capt. Bishop stated as follows, when asked for his opinion:

'A Pilot or Observer not knowing thoroughly his gun, gear and engine, had to sit on the ground and work day and night until he did. New Pilots should take this seriously to heart, because it at once exposes in them a lack of keenness that they have not paid any attention whilst being taught at home, and when they get overseas, they are soon very sorry for it. They must realise that they have got to know it thoroughly, and not be satisfied with just knowing something about it.'

Capt. Fullard, who is said to have over 40 Huns to his credit, told me that he always looked after his own gun, and boasted that he had fired over 5,000 rounds on his Lewis gun without a stoppage. To keep his hand in he often stripped the gun on his lap while in the air. He said he was quite confident that his gun would always fire when he wanted it.

Capt. Guynemeyer, when visiting one of the British Squadrons, pointed out that he attached so much importance to gunnery that the guns, gears and sights on his machine were his private property, and that he looked after them himself in every detail.

56 Squadron
Extract of Report from Capt. Maxwell and Lt. Barlow as to gunnery arrangements in 56 Squadron:
'Each flying officer filled and tested his

Gunnery Training Ranges with benches for re-filling ammunition drums.

own Vickers and Lewis ammunition. Flight Commanders had their ammunition filled and tested by the man in charge of that particular gun. Capt. Ball was very particular about his ammunition and filled it himself or got one of his flying officers to do it for him.

New pilots on arriving in the Squadron were sent up as often as possible to fire their guns in the air, to thoroughly acquaint the pilot with his guns and machine. A new pilot always did at least five hours flying before going on patrol.

When a patrol returned all stoppages were reported, and the guns in question tested on the range for cause of trouble. In this way it soon became a rarity for people to complain of stoppages. The pilot himself was always present when the gun was being tested and fired it himself.

Patrols on reaching the lines always tested their guns towards Hunland. Pilots were held absolutely responsible for stoppages on their guns, and it was very seldom that anybody returned from patrol with gun trouble.

Everybody in the Squadron took a personal interest in their guns, and if their guns were not going well, they experimented on the range with Spring Tensions and File, etc., until they did go well.

We were very particular about testing ammunition, and all discarded ammunition was fired on the range for stoppage practice.

After a little trouble had been taken with the guns to start with, it was extraordinary how few stoppages occurred, in fact people fired over 5,000 rounds without a single stoppage which could not be rectified at once, i.e. misfire, etc.'

Gunnery Training Ranges

Bristol M1c with Hythe Gun Camera. Shanter farm in the background

Gun Camera

Aerial Gunnery Training - Surprise Range No.2 - 200 Yds.

The personal interest was taken by the officers in 56 Squadron notwithstanding the fact that the Armament Staff, consisted of an Armament Officer, a Sergeant in charge of the guns, 1 man at Headquarters in charge of repairs to C .C. Gears, and 3 men in each flight.

56 Squadron, as you are aware, possess perhaps the best record in the Royal Air Force, and it is doubtless largely due to the fact that the importance of taking a personal interest in everything to do with their machines was fully realised by all.

Course at Turnberry

While at Turnberry it is your last opportunity of receiving instruction in Gunnery. From the moment you leave you will have to depend upon the knowledge you have acquired. It is, therefore, very important that you take every opportunity while here to collect all information possible.

You have been given complete notes on the principal subjects. Copy them into your notebook. The more complete and comprehensive you make your book now, the more you will appreciate it later, when you may wish to refer to it.

Some of you appear to think that you can keep all this information in your head, but you cannot without constant practice.

Unless you know your gunnery thoroughly you are not in a position to tell whether the armourers in your Squadron are doing what is right and necessary. In other words, you are not in a position to take the personal interest that is so essential to success and safety.

After Leaving Turnberry

Do not leave this School with the idea that having passed the Course, you are a qualified gunner, ready to tackle any Hun. You are far from it.

Constant practice is required, and it is your duty to take every opportunity to practice, both on the ground and in the air.

You must become absolutely mechanical with your stoppages, which can only be done by constant practice. Test your gears and the alignment of your sights at regular intervals.

Two-Seater Pilots

The closest co-operation between Pilot and Observer is essential. They must possess absolute confidence in each other.

When an Observer is allotted to you (you will probably keep the same one throughout), satisfy yourself that he knows his gunnery, as your safety very largely depends on him when you are attacked from behind. You must also be in a position to convince him that you know your gunnery, otherwise you cannot expect him to have confidence in you.

Overseas

Those of you who go to France as re-enforcement Pilots may find that after having passed certain Brigade tests, you are given little encouragement to keep up your gunnery.

Do not let that prevent you from following the example of the expert Pilots, i.e., taking the personal interest, which has been proved to be so essential to success.

Finally, you must all realise that your own safety and the safety of your Observer if you are a two-seat Pilot, depends entirely on your individual efforts.

Commandant Lt.-Col., R. Bell-Irving ended by stressing the importance of remembering everything they had learned whilst at the School of Aerial Gunnery, Turnberry, their life may one day depend on it!

Aerial Gunnery Training - Aircraft Silhouette on Turnberry beach.

Gunnery Training

1st Lt. Thomas Herbert on Deflection Range - Turnberry 1918.

Bell-Irving - Aerial Gunnery Training on rails.

6
WWI Accidents

It is an astonishing fact that of the 14,000 pilots who died during WWI, well over half were killed in training. Pupil pilots became almost matter of fact about sudden death in crashes - they even had a song about it:

The young aviator lay dying,
As 'neath the wreckage he lay,
To the Ak Emmas around him assembled,
These last parting words he did say.
'Take the cylinder out of my kidney,
The connecting rod out of my brain,
From the small of my back take the crankshaft,
And assemble the engine again!'

There is an old Arab saying, 'The grave of the horseman is always open'. The same could be said of these young pioneer airmen. Death was stalking, waiting for an engine misfire or wrong calculation, and with these young inexperienced pilots, the reapers harvest was bountiful.

Thirty-nine aviators were killed at Turnberry during WWI, or died later from their injuries, most are buried in Girvan (Doune) Cemetery. The townsfolk of Girvan bore witness to this great tragedy played out before them as one after another of these young 'strangers within the gates' were borne by motor transport down the six miles of road from Turnberry, the coffin covered with the Union Jack and wreaths. Other motors would follow bearing a party of Officers and other ranks of the fledgling Royal Flying Corps.

The conditions they were expected to fly in were grim, sitting in an open cockpit exposed to the elements of wind and bitter cold, temperatures could be as low as -50 degrees C (-58 degrees F), with slipstream speeds of

SE5a v Avro 504. Captains Latta and Le Gallais - non fatal.

SE5a v Avro 504 collision on Turnberry Airfield

100 mph and little protection from rain, pilots required a high degree of physical resilience. At the start of the war they were issued with the same uniform as Army motor transport drivers until in 1917 a special one-piece waterproof and insulated garment, the Sidcot Suit, was introduced.

Pilots would apply whale oil to their exposed faces, and lip balm in an attempt to stop their lips cracking, splitting open and bleeding when flying at high altitudes. Wrapped in layers of bulky clothing to prevent frostbite or hypothermia, often with steamed-up goggles, these young men were not only expected to fly in these conditions, but to fight as well. The trademark silk scarf of the aviator was not a thing of vanity, but to prevent the neck being chafed against the rough collar of his flying jacket as the pilot constantly turned his head, scanning the skies for danger.

Surprisingly, the pilots of the RFC were not supplied with parachutes at the beginning of the war, although they were issued to crews of airships and balloons. The first parachute jump from an aeroplane took place in St Louis, Missouri, in March 1912. R. E. Calthrop, a retired British engineer, had in fact developed the Guardian Angel, a parachute for aircraft pilots, before the war. Sir David Henderson, Commander of the RFC, was unwilling to give permission for them to be issued to his pilots. The Air Board decided against the measure. Officially the reason given was that the Guardian Angel was not 100% safe, it was too bulky to be stored by the pilot and its weight would affect the performance of the aeroplane. Unofficially the reason was given in a report that was not published at the time: 'It is the opinion of the board that the presence of such an apparatus might impair the fighting spirit of pilots and cause them to abandon machines which might otherwise be capable of returning to base for repair.' Pilots such as Major Mick Mannock became increasingly angry about the decision to deny British pilots the right to use parachutes. He pointed out that by 1917 they were being used by pilots in the German Air Force, French Army Air Service and the United States Air Service. In place of carrying parachutes, RFC pilots carried revolvers instead. As Mannock explained, unable to carry a parachute, he had a revolver 'to finish

Avro 504 D4421 - 'put it in the hangar when you get back!'

myself as soon as I see the first signs of flames.'

Another less mentioned aspect of early flying was the side effect of the lubrication system, which not only covered the pilot's goggles, but also punished them with constant severe diarrhoea.

We read in Duncan Bell-Irving's biography, *Gentleman Air Ace*:

> The only effective lubricant for the rotary engine was castor oil. As the huge mass of engine whirled around, spewing out a spray of castor oil, pilots and ground crew complained not only of the smell but of the fact that it affected their digestive systems.

The accident rate was horrendous, both in training and over the front. There were more losses due to pilot error or mechanical failure than there were in actual combat.

A letter which graphically describes the injuries sustained in an aeroplane crash was written to his parents by 19-year-old Lieutenant Bonham Hagood Bostick, of the First Oxford Detachment who trained at Turnberry in

March 1918. It was dated 24 May 1918 and was subsequently printed in a South Carolina newspaper in July 1918:

> Royal Air Force Central Hospital
> 5 Eaton Place
> London

'I am rather improved now and am writing to keep you buoyed up.' Says young Bostick to his parents. 'Don't worry about me, you know. I had a pretty close call of it, of course, but am getting perfectly all right again. I was at the front with a British Squadron of rotary motored scouts and accompanied by another machine of our squadron, was flying low behind the British lines.

'Left jaw broken, hole left cheek, left eyeball bruised, right cheek ripped open and hanging in ribbons, artery cut and spurting right cheek, hole in right foot, knee laid wide open, wound in right thigh, muscles of both calves torn and hanging, these were the principle injuries in combination with concussion of the brain.

Wreckage of Avro 504 (possibly near Kirkoswald) - no record of this accident can be found.

'I was so mixed with the engine that it took 25 with hacksaws and cold chisels to get me out, and then I had almost bled to death. But don't let us talk about wounds. They are so commonplace, but rather ghastly to you, I suppose.

'All this happened on April 11 and though I have been getting along rather slowly, I am much recovered and will get perfectly alright again. My jaw is set perfectly straight, and the awful wounds in my face have healed up. By massage scars will be practically taken out, and eventually there will be nothing left but a few whitish marks. Don't worry about me. My principle worry is being out of the swim for the time being. I was at the birth of our present air service, and I want to see it through to the end when it has outgrown the signal corps and even the army and becomes a new service on a footing with the army and navy. That is exactly what the English air service has done. First it has cut free from the Royal Engineers, and then the army, becoming in turn the late Royal Flying Corps and then the present Royal Air Force.

'I have given a good bit already and I intend to keep giving and to advance with the service, if possible. At any rate I am going to see the game through to the very end if God spares me that long. I am going to be one man with a full knowledge of duty done when this war is over if I pull through.'

Early in 1919 a committee of local people from Maidens, Kirkoswald and Turnberry was established to provide a monument at Turnberry, to be erected on the golf course, to 'perpetuate the memory of the Officers and others of the Royal Air Force who were killed while undergoing training at the Flying School at Turnberry Aerodrome.' To help raise the funds for this, Girvan Town Council held a flag day in Girvan and Turnberry Golf Club arranged an exhibition match between George Duncan, who was to win the Open Championship in 1920, and Tom Fernie, the Club Professional. The club met the cost of staging the event and a collection was taken from the spectators towards the cost of the war memorial. Lord and Lady Ailsa were very keen that it should be of Ailsa Granite. The chosen design was to take the form of a memorial cross, bearing the crown, RAF flying crest and a sword, the names of the dead to be cut in upright Roman lettering and the base and cross to be unpolished. Approaches were made to Ailsa Craig Granite Ltd. for the provision of the stone, but as the shaft and cross weighed about 7 tons, there were no appliances on the Craig to able lift that weight and the granite was sourced from elsewhere. The memorial was unveiled in April 1923.

A report in the local *Carrick Gazette*, 4 May 1923, gives details of the unveiling:

TURNBERRY WAR MEMORIAL
BRILLIANT AND IMPRESSIVE SPECTACLE

Stirring Address by the Marquis of Ailsa
The monument is erected on a rocky mound commanding a magnificent view of the isle-studded Firth of Clyde and is twenty-two feet in height on a square base with a tall pillar, surmounted by a double Celtic Cross, on one of which is sculptured the Air Craft Badge, and on the column, there is a device of the Crusader's Sword. It is absolutely original in conception and was designed by Colonel H. R. Wallace of Busby. It is composed entirely of grey granite, and the sculpturing and work of erection was carried out by Mr Robert Gray, St. Vincent Street, Glasgow. The unveiling ceremony took place at 3 o'clock on Saturday 28th April 1923 in the presence of a great crowd of people from the surrounding districts and from a distance.

Amongst those present were:
The Marquis and Marchioness of Ailsa, Culzean Castle; Lord and Lady Glenarthur of Fullarton; Sir James Bell, Montgrennan; Colonel H. R. and Mrs Wallace of Busby; Colonel and Lady Marjorie Dalrymple Hamilton of Bargany; Mr and Mrs Richmond of Blanefield; Mr and Mrs Strain, Cassillis House; Mrs Milligan, The Garth; Major Cockburn, Mr J. A. Bullevant, Hornsie; Lieut.-Colonel and Mrs T. W. R. Houldsworth

Avro 504 A9800 - South of Kirkoswald

Avro 504 D4404

of Kirkbride; Colonel J. C. Kennedy of Dunure; Mr and Mrs D. H. Kemp; Mr George McAlpine; Mr T. R. Fernie; Officers and Sergeants of the Ayrshire Yeomanry; Lieutenant-Colonel J. Milligan; Officers and Sergeants of the 79th Royal Field Artillery; Mr G. W. Browne and Officers of the Royal Scots Fusiliers, Ayr; President and Members of the Territorial Force Association, Ayr; Colonel Adair and Territorial Staff, Girvan Cadets (under Mr M. J. Finlayson) Maybole Boy Scouts (under local scoutmaster); Ayr Academy Cadets (under Captain Jamieson), Maybole Boys Brigade (under the Rev. Mr Williamson), Representatives from St. John's Masonic Lodge, Maybole; Representatives of Ayr Established and U. F. Presbytery, The Provost, Magistrates and Town Councillors of Ayr, Maybole and Girvan; The Parish Council of Kirkoswald; Wing Commander L. W. B.Rees, VC; The

Members of Turnberry War Memorial Committee; Miss Sinclair, Secretary of the WRAF.

Apologies were intimated from General Sir Charles Fergusson of Kilkerran; Vice-Admiral Sir James Fergusson and Lady Fergusson, Ladyburn; Brig. General and Mrs McCall; Lieutenant-Colonel and Mrs Kennedy, Dunure; The Countess of Cassillis; Rev. G. B. Allan, St. Edmund Hall, Oxford; Mr A. P. Henderson, Orchard Mains, Tonbridge, etc., etc.

The various bodies took up their positions beneath the monument, the brilliant uniforms worn by the Lieutenancy and the military officers contrasting vividly with the more sombre garb of the general public. The service was opened by the singing of the Hymn – 'O God our Help in Ages Past,' after which portions of the Scripture was read from Revelation VII 9-17, by the Rev.

Bristol F2.b C4841. 2nd Lt. F.G. Sutton and AM1 W.G. Hoyes both injured.

D. M. McLean, Kirkoswald, who also offered up prayer of intercession and remembrance.

In unveiling the memorial, The Marquis of Ailsa said: 'It is right and fitting that I should say a word to you, especially those of you who have come from a distance, regarding the circumstances (unique, I think of their kind) which were at the inception and origin of the monument which we have just unveiled. The Royal Air Force (School) of Aerial Gunnery and Fighting was formed here in January 1917, with Turnberry Hotel as its headquarters, and these links and surroundings for training purposes. From then upwards till February 1919, when the school was closed, some 5,000 officers, non-commissioned officers and men of other ranks of the Imperial, Colonial and American Air Services were trained in these

peaceful and historic surroundings for their part in the war. Many, after passing to the various theatres of war, were killed or died on active service, maintaining the glorious tradition for courage and efficiency which has made the Air Service one of the wonders of the world. Of these, no less than thirty-seven were killed here in sight of this spot in the course of their training at the school, which brought them into close and honoured association with the community of this place. Two more died of illness, due to exposure during their training. In a great many of our Ayrshire towns and villages monuments and memorials of various kinds have been erected in token of personal affection for those born and bred in our midst, who gave their lives in the war that we might live. In nearly all cases these memorials are to the communities in which they stand, the expression of a cenotaph or empty tomb, the token of bereavement of parents, of children, and of brethren, by ties of blood as well as of

Unidentified Sopwith Camel fatal accident with Lands of Turnberry Farm in the background.

Unidentified Sopwith Camel fatal accident on Turnberry Airfield.

life-long associations in the schools and churches of which they were members. To many they stand for an unknown grave. The memorial which we have just unveiled differs from those in origin and inception, in that it has been erected by the exertions of the people of Turnberry in admiration and affection it is true, not to their own kith and kin, and brought up here in this parish, or even in Ayrshire, but as a spontaneous tribute of admiration and honour to those who came here at the call of duty – "The stranger within their gates." To read the inscription on the base of this monument and the list of heroic dead whose names are recorded upon it, what they were and whence they came, tells us that they came here literally from the ends of the earth at the call of their Motherland. The whole world was their parish, the whole allied cause was their cause; they were representatives of our Empire across the seas, and of

our American kindred and allies on our defensive flank. They came here not on pleasure bent, nor for restful recreation on those "Elysian Fields" devoted once more to sport, but in the grim happiness and pride of their youth and high resolve to fit and quit themselves like men for the greater part to which they felt the call of fulfilling the destiny of their race - worthily of their blood. It requires no great stretch of imagination to arrive at the sense of romance which the presence of such men inspired in the people of this place - a small community unused to the realities and dangers inseparable from the daily life of the airman in his strenuous and nerve-testing training.

The people of Turnberry were daily witness of the daring courage and high morale, and of the cheerful service of these splendid youths. They were in constant and neighbourly association with them. They saw too often (and felt what they saw) the supreme sacrifice rendered on these very links. They realised the heroic effort going on from day to day in their midst, and they understood. Born of that understanding, and of their association with the members of the school, arose a community of sympathy and admiration which ripened as the days went on, and those who came as strangers became the honoured guests and pride of

DH9 C1270 - non fatal accident on Turnberry beach.

DH9 C1270

the place. It is not surprising that after the Armistice and the fighting school was closed, the dwellers here recalled the memories of their sojourn amongst them grateful for what they were, and for what they had achieved; proud of the service and the manner of it, and with a deep sense of the personal loss at the passing of so many gallant and promising lives in the tragic path to victory while here. And so it came about that at a meeting of the inhabitants of Turnberry and district held in June, 1919, it was resolved to erect a monument on the spot where their lives were given for their country, in order to perpetuate their memory and to remind future generations that these links must ever be associated with the sacrifice and achievement and example of one of the most gallant and sporting services which so greatly contributed to our victories on land and sea during the great war. This monument then, is a spontaneous tribute of the people of Turnberry to the gallantry and self-sacrifice of the Air Services of our King and Country. It is raised in the highest honour to those members of the Air Force who were killed or died near this spot, and is no less a tribute to all of those who passed through this fighting school of Turnberry, rendered the supreme sacrifice for King and Country on every battle front - "The whole earth is the tomb of heroic men, And their story is not graven only on stone over their clay, But abides everywhere with visible symbol. Woven into the stuff of other men's lives."

These words, written more than two thousand years ago, crystallise a truth which is at the foot of consolation. Heroic lives heroically rendered are lives not lost. The story of these heroic lives is graven on the hearts of the people here, and this stone of remembrance is but a symbol 'That their name liveth for evermore.'

Rev. James Muir, BD, Kirkoswald, then read the inscription and the names of the fallen, which is as follows:
"To the memory of the Officers, non-

DH9 C1270

commissioned Officers and Men of the Royal Flying Corps, Royal Air Force, and the Australian and United States Air Services, who gave their lives for their country while serving in the School of Aerial Gunnery and Fighting at Turnberry, 1917-1918". Their name liveth for evermore.

ROYAL FLYING CORPS
G. G. DOWNING, Lieut.
W. H. C. BUNTINE, 2nd Lieut., Sherwood Foresters.
C. A. COOPER, 2nd Lieut., Yorkshire Regiment.
E. C. HULL, 2nd Lieut.
R. S. McNAIR, 2nd Lieut.
J. STEVENSON, 2nd Lieut.
S. C. APPLETON, Sergeant.
C. W. H. BOWERS, Sergeant.
E. E. HALL, 2nd Air Mechanic.
H. TOWLSON, 2nd Air Mechanic.

ROYAL AIR FORCE
J. M. CHILD, Capt., Manchester Regiment.
I. H. D. HENDERSON, Capt., A. & S. Highlanders.
J. S. BROWN, Lieut., Lancashire Regiment.
C. A. FLETCHER, 2nd Lieut., Worcestershire Regiment.
J. MILLIGAN, 2nd Lieut., Royal Irish Rifles.

A. A. HEPBURN, Flight Cadet, Gordon Highlanders.
J. E. LILLEY, Sergeant.
J. S. TUCKETT, Sergeant.
T. A. KING, Corporal.
T. INGLIS, 2nd Air Mechanic.
H. W. ELLIOTT, Lieutenant.
G. A. LAMBURN, Lieutenant.
R. M. MAKEPEACE, Lieutenant.
H. B. REDLER, Lieutenant.
J. D. DUNBAR, 2nd Lieutenant.
C. H. A. GODFREY, 2nd Lieutenant.
C. A. HILLOCK, 2nd Lieutenant.
C. W. JANES, 2nd Lieutenant.
T. A. McCLURE, 2nd Lieutenant.
A. McFARLAN, 2nd Lieutenant.
W. A. RYMAL, 2nd Lieutenant.
J. HUGHES, Flight Cadet.
A. McLEAN, Flight Cadet.

AUSTRALIAN FLYING CORPS
H. R. H. BUTLER, 2nd Lieutenant.
R. H. GROVE, 2nd Lieutenant.

USA AVIATION
G. SQUIRES, Lieutenant.
R. B. REED, Lieutenant.
H. R. SMITH, 2nd Lieutenant.
G. A. BRADER, Cadet

The prayer of dedication, which was beautifully worded, was offered up by Mr Muir.

Captain Playfair, MC, said he was attending this ceremony on behalf of the Chief of the Air Staff and the Air Council, and he just wished to say how sorry Sir Hugh Trenchard was, that owing to stress of work, he was unable to attend personally. He had to express on his behalf, and that of the Air Council, how much they appreciate the motive which prompted the residents of Turnberry to set up the noble memorial which had just been erected to the members of the Air Services who gave up their lives here in the course of their training, by residents who were not related to them in any way, is the cause of extreme gratification to all members of the Royal Air Force. It is a tribute to their branch of the services of which they were justly very proud. It is also a permanent reminder to those of us who are still serving, that our efforts will be appreciated and not forgotten. The officers, non-commissioned officers and airmen whose names they saw inscribed on the memorial gave up their lives on active service just as much as those who died in the actual theatre of war.

It was their example and devotion to duty that spurred others on to perfect their training, so that they could the better take their part in the great conflict.

That their work was not in vain was obvious to those who commanded Royal Air Force Units in France and other theatres of war. He knew because he had British, Australian and American Squadrons under him in France, whose pilots had passed through this school and their efficiency was of a high order. In 1917 he spent four days at the school, and even in that short space of time he was struck by several small instances which showed how helpful the residents were to the new people that had come amongst them. It must indeed be a great consolation to the relatives and friends to see the memory of these men thus honoured, and to know in what esteem they were held. In conclusion, he most

DH9 C1270

heartily thanked all those who had worked for or subscribed to this memorial.

The Paraphrase – 'How Bright these glorious Spirits Shine' having been rendered by Maybole Prize Silver Band, Mr T. R. Fernie, on behalf of the Turnberry War Memorial Committee, formally handed over the monument to the care of Kirkoswald Parish Council, which was accepted by the Marquis of Ailsa as chairman of that body.

was a great test of morale in a man in his effort to fly, and he was glad to say that he never found a man in Turnberry School unwilling to go up, and he always found them willing and ready to do their duty. He had the privilege of seeing 1,000 men through the school, and he could assure them they all did their best. In conclusion he said a tribute to the American Airmen.

The Dead March in Saul was then played by the band.

DH9 C1270

Wing-Commander L. W. B. Rees, VC, in proposing a vote of thanks to the committee and subscribers, said he had the honour of being in Turnberry during the greater part of the time the fighting school was in existence, and had always received the greatest kindness and help from the people of the district. In listening to the names read out, he felt that there were very few of their relatives' present, but he knew they were very grateful to the people of Turnberry for the honour they had paid the Glorious Dead. He went on to say that fighting in the air was different in certain degrees from fighting on sea and land. It

'The Last Post' and 'Reveille' was sounded of by the buglers and pipers of the 1st Royal Scots Fusiliers, who also fired three volleys. The Royal salute was received by the Marquis of Ailsa, and the proceedings terminated with the singing of the National Anthem.

A number of beautiful wreaths were placed on the cenotaph, including one in the form of an aeroplane from the Air Council.

Footnote:
My grandfather, who as a young boy lived on North Threave farm on the hill above

RE8 D6840

SE5a on fire - with pilot posing beside smouldering wreckage!

Kirkoswald, spoke of a saboteur at work on the airfield who was responsible for most of the accidents. It was thought that the saboteur was sawing halfway through the propellers of aircraft, which would, in due course, fracture during flight and cause the engine to severely vibrate and break up. Upon discovery the man was summarily 'put up against a wall and shot'. I can find no evidence for this, except further rumour that he was actually a member of Sinn Féin, tried for treason and sentenced to death by hanging. Author.

SE5a fatal Crash - possibly Ayr not Turnberry. Note the unbroken window panes!

7
WWI Memorial

◆

IN MEMORY OF

Second Lieutenant James Stevenson
Royal Flying Corps
of Kirkwall, Orkney.
Age 20

and

Sergeant Christopher William Henry Bowers 932
Royal Flying Corps
of Birmingham, England.
Age 21

Sergeant Christopher William Henry Bowers 932

who died on Tuesday 1 May 1917

FE2b 6975

Court of Inquiry: 'Loss of flying speed produced by a very sharp turn causing the machine to stall at insufficient height for recovery from nosedive'. 'Nosedived from 30 ft. Burst into flames'. Occurred 300 yards north of Turnberry Lighthouse.

Grave of Sergeant Christopher William Henry Bowers 932

◆

IN MEMORY OF

Second Lieutenant George Squires
Aviation Corps, United States Army
of St. Paul, Minnesota, USA.
Age 22

who died on 18 May 1918

SOPWITH CAMEL 1F1 B9218

Stalled on turn at 100 ft, 1,000 yards south of Kirkoswald

IN MEMORY OF

Sergeant Stanley Chalmers Appleton
Pilot, Royal Flying Corps
of Warrington, Lancashire,
England.
Age 22

and

Second Lieutenant Walter Horace Carlyle Buntine MC
Acting Observer, Royal Flying Corps
4th Bn., Sherwood Foresters (Notts and Derby Regt.) of Melbourne, Australia.
Age 22

Seargent Stanley Chalmers Appleton

who died on
Tuesday 19 June
1917

FE2b A817
FE2b A817 stalled on turn at a height insufficient for recovery, west of Dalquhat farm, Kirkoswald.

Second Lieutenant Walter Horace Carlyle Buntine MC

Tread softly through the night of gloom,
Away with worldliness and mirth,
'Tis meet to 'weep with those that weep,'
A noble son has fall'n to earth,
A son of tender love and care,
Of spirit great and talents rare.
He needed no recruiter's voice,
His Empire's need he clearly saw,
And leaving all his heart held dear,
He bravely faced his country's foe.
For her he nobly fought, and won
A 'token' of his King's 'Well done!'

'Good-night,' beloved, sunny, smiling face,
The school-boys' hero for long years to come,
Long will the record of his fight in space
Live in their memory – 'Lyle, well done.'
Kindest of brothers, loving hearted son,
Rest from thy labours, 'Dear Lyle, well done!'
We bid thee not 'farewell' – 'Adieu!'
Thy loved ones did at God's behest
Offer their sacrifice 'to Him,'
Braved the fierce storm and stood the test.
Lived 'unto Him,' each loyal life
Dwells in God's heart above the strife
This poem, entitled 'A Humble Tribute to the Memory of Walter Horace Carlyle Buntine', was published in June 1917 in the Caulfield Grammar School magazine, just a few short days after Royal Flying Corps pilot Lt Lyle Buntine, MC, was killed.

♦

IN MEMORY OF

Second Lieutenant Cyril Ashley Cooper
Pilot, Royal Flying Corps
Of London, England.
Age 23

who died of injuries on Monday 29 June 1917

and

Air Mechanic 2nd Class Harry Towlson
Observer, Royal Flying Corps of Nottingham, England.
Age 30

who died on Friday 26 June 1917

FE2b 4926
Aircraft FE2b 4926 stalled during a turn and crashed east of Maidens.

◆

IN MEMORY OF

Air Mechanic 2nd Class Edward Ernest Hall MM 63054
Royal Flying Corps
of Cramlington, Northumberland, England.
Age 29

who died on Wednesday 20 July 1917

*Accidentally drowned. Body found on 25 July 1917 on rocks
50 yards south of Turnberry Lighthouse.*

Extract from death certificate of Air Mechanic 2nd Class Edward Ernest Hall MM 63054

◆

IN MEMORY OF

Lieutenant George Guy Barry Downing
Royal Flying Corps and 9th Bn.,
Welsh Regiment
of Beverly, Llanishea, Cardiff,
Wales.
Age 23

who died on Tuesday 4 September 1917

BE2c 5416
*Turnberry Airfield. Observer,
2/Lt. E. Allman was injured.*

IN MEMORY OF

**Second Lieutenant
Edwin Charles Hull**
Royal Flying Corps
of Bedford, England.
Age 23

who died on
Sunday 17 March 1918

*Casualty card for Second Lieutenant
Edwin Charles Hull*

SOPWITH CAMEL 1F1 B9222
*The court of inquiry is the of opinion that
the fatal accident was due to an error of
judgement on the part of the Pilot, and no
blame can be attached to anyone as the
machine was in perfect order before the flight
took place. The Pilot evidently stalled whilst
turning near the ground, resulting in a spin
from which he was unable to get out. Flat spin
from 70 feet. Turnberry Airfield.*

IN MEMORY OF

**Second Lieutenant Robert Schermerhorn
McNair**
Royal Flying Corps
of Kansas City, Missouri, USA.
Age 21

who died on Sunday 17 March 1918

SOPWITH CAMEL F1.1 B9204
On shore, 400 yards off Brest Rocks.

IN MEMORY OF

Cadet George Atherton Brader
Aviation Section, Reserve Signal Corps, US Army of Pennsylvania, USA.
Age 24

who died on 5 April 1918

SE5a C1762
Turnberry Aerodrome. Cadet George Atherton Brader was carrying out gunnery practice in SE5a C1762. He made an outward spin at 150 ft and the aircraft spun into the ground with engine running and caught fire on impact. The Court of Inquiry concluded that the crash was 'due to an error of judgement in that he turned downwind with insufficient bank too near the ground, probably mistaking ground speed for air speed, the wind being fairly strong'.

Front of Casualty card

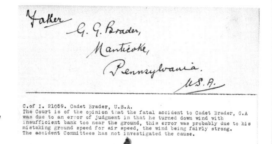

Back of Casualty card

IN MEMORY OF

Second Lieutenant Charles William Janes
Royal Air Force, County of London Yeomanry of Stoke Newington, London, England.
Age 23

who died on Thursday 11 April 1918

SOPWITH CAMEL 1F1 B9210
*Loss of control at 200 ft whilst firing at target raft, hit submerged rocks.
In sea near Turnberry Lighthouse.*

IN MEMORY OF

Sergeant
John Stabb Tuckett
Royal Air Force
of Birkenhead, Cheshire,
England.
Age 18

who died on Sunday 26
May 1918

SOPWITH CAMEL 1F1 B5560
Fell out in dive 100 yards south west of
Kirkoswald. The Court of Inquiry having
considered the evidence placed before them
are of the opinion the pilot met with his death
by pulling the machine out of the dive too
suddenly at the same time kicking on rudder
causing the machine to turn on its back, the
pilot falling out.

IN MEMORY OF

Lieutenant Reginald Milburn
Makepeace MC
Royal Air Force
of Liverpool, England.
Age 27

and

Second Lieutenant Thomas Albert McClure
Royal Air Force and Connaught Rangers
of Donegal, Republic of Ireland.
Age 20

who died on 28 May 1918

BRISTOL F2B B1178
Wings folded back in dive. Crashed 500 yards
west of Turnberry Airfield.

IN MEMORY OF

Second Lieutenant
Howard Rea Smith
USAS
of Newcastle, Indiana,
USA.
Age 30

who died on 27 May 1918

SOPWITH CAMEL 1F1 B7467
Spun in 1,000 yards north east of Kirkoswald.

IN MEMORY OF

Second Lieutenant
Howard Richmond
Henry Butler
Australian Flying
Corps
of Melbourne,
Victoria, Australia.
Age 21

who died on Sunday
2 June 1918

SOPWITH CAMEL 1F1 B9262
Stalled and spun in on turn,
Turnberry Airfield.

◆

IN MEMORY OF

Lieutenant Hugh William Elliot
Royal Air Force
of Biggleswade, Bedfordshire, England.
Age 20

and

Lieutenant Richard Brumback Reed
Aviation Section, US Service Signals Corps
of Van Wert, Ohio, USA.
Age 26

who died on 5 June 1918

AIRCO DH9 C1231
*Broke up in air,
dived into sea
which was 20 ft
deep at low tide.
Opposite Dowhill
farm, two miles
south of Turnberry
Lighthouse.*

*Lieutenant
Richard Brumback Reed*

*Left: Funeral Procession
of Captain Henderson MC,
Dalrymple Street, Girvan*

*Captain Ian Henry David
Henderson MC*

◆

IN MEMORY OF

Captain Ian Henry David Henderson MC
Royal Air Force and Argyll and
Sutherland Highlanders
of West Eaton Place, London, England.
Age 21

and

Lieutenant Herbert B. Redler MC
Royal Air Force
of Boschbeek, Newlands, Cape Town.
Age 21

who died on Friday 21 June 1918

AIRCO DH9 D1080
Glenhead Farm

IN MEMORY OF

Second Lieutenant John David Dunbar
Royal Air Force
of Saskatchewan, Canada.
Age 20

who died on 25 July 1918

SOPWITH DOLPHIN 5F1 E4437
Near Brest Rocks

IN MEMORY OF

**Second Lieutenant Raymond Hinton Grove
3131**
Australian Flying Corps
of Adelaide, Australia.
Age 26

who died on Monday 19 August 1918

SOPWITH CAMEL 1F1 B9212
*Engine failure in dive, spun into sea two miles
south of Brest Rocks.*

IN MEMORY OF

**Second Lieutenant
Archibald McFarlan 80139**
Pilot, Royal Air Force
of Ballasalla, Isle of Man.
Age 25

and

Flight Cadet Andrew Anderson Hepburn
Observer, Royal Air Force
of Dunfermline, Scotland
Age 18

who died on 23 August 1918

AIRCO DH9 C1334
*DH9 C1334 flown by Second Lieutenant
Archibald McFarlan, with Flight Cadet
Andrew Anderson Hepburn acting as
Observer, was involved in a mid-air collision
with a Sopwith Dolphin and crashed near
Minnybae farm, killing them both outright. A
news report in the Daily Record on 29 August
1918 noted that 'Their machine collided at
a considerable height with another one, the
occupant of which had a miraculous escape'.
The Pilot of the Sopwith Dolphin survived
the impact by climbing out of the cockpit and
on to the tail and saved himself. The body of
Flight Cadet Hepburn was found about one
mile away in Ballochneil Wood.*

IN MEMORY OF

Capt. James Martin Child MC
Chevalier of the Order of Leopold II. Croix de
Guerre (Belgium)
Royal Air Force and Manchester Regiment
of Leytonstone, Essex.
Age 25

who died on 23 August 1918

*Road Traffic Accident opposite High
Dalquhat farm while returning from the
crash of DH9 C1334. Captain Child attended
the accident of DH9 C1334 and the Sopwith
Dolphin and helped in the rescue of the
survivor from the wreckage. Returning to
Turnberry by motorcycle, whilst riding along
the Kirkoswald Road, he was in collision
with a Crossley Motor Tender
and was killed.*

IN MEMORY OF

Second Lieutenant William Arthur Rymal
Royal Flying Corps
of Toronto, Canada
Age 22

and

**Flight Cadet
Alexander McLean 110038**
Royal Air Force
of Cardonald, Glasgow.
Age 19

who died on
5 September 1918

*Flt. Cadet
Alexander
McLean 110038*

AIRCO DH9 C1333
*Aircraft took off crosswind, turned downwind,
stalled and spun in from 300 ft, adjacent to
Public School Maidens.*

IN MEMORY OF
Lieutenant Gerald Arthur Lamburn
Royal Air Force
of Chelsea, London, England.
Age 19

who died on Monday 30 September 1918

SOPWITH CAMEL 1F1 F1410
*Looped too near ground, Shalloch Park farm,
Girvan.*

IN MEMORY OF

Lieutenant James Stanley Brown
Royal Air Force and East Lancashire Regt.
Of Lancashire, England.
and

**Second Lieutenant
Charles Alexander Fletcher MM 35486**
Royal Air Force and Worcestershire Regt.
Of Hadley, Shropshire, England
Age 23

who died on 20 October 1918

AIRCO DH9 C1372
*Stalled on climbing turn and fell into sea near
mouth of Milton Burn.*

IN MEMORY OF

Flight Cadet John Hughes 178259
Royal Air Force
of Amlwch, Anglesey, Wales.
Age 18

who died on Monday 25 November 1918

AIRCO DH9 C1374

*Flight Cadet John Hughes
survived the forced landing
of DH9 C1374 after severe
engine vibration.
Tragically he was to die of
Injuries when his petrol-
soaked clothing caught
alight from a match in
Turnberry Hospital.*

IN MEMORY OF

Sergeant John Eric Lilley 86754
Royal Air Force
of Sunderland, Durham, England
Age 19

who died on 28 November 1918

SE5a E3954
Park farm, Kirkoswald

IN MEMORY OF

Corporal Thomas Allister King
Mechanical Engineer, RAF
of Brentwood, Middlesex, England.
Age 21

who died on 10 December 1918

Influenza

IN MEMORY OF

**Second Lieutenant Charles Henry Albert
Godfrey 347473**
Royal Air Force
of Kenley, Surrey, England.
Age 23

who died on Wednesday 11 December 1918

SOPWITH CAMEL 1F1 F1408
Stalled and spun in Turnberry Airfield

◆

IN MEMORY OF

Second Lieutenant James Millikin *(Milligan)*
Royal Air Force and Royal Irish Rifles
of Carrickfergus, Ireland.
Age 25

who died on 31 December 1918

*Second Lieutenant
James Millikin*

	No.	Name and Surname. Rank or Profession, and whether Single, Married, or Widowed.	When and Where Died.	Sex.	Age.	Name, Surname, & Rank or Profession of Father. Name, and Maiden Surname of Mother.	Cause of Death, Duration of Disease, and Medical Attendant by whom certified.	Signature & Qualification of Informant, and Residence, if out of the House in which the Death occurred.	When and where Registered, and Signature of Registrar.
Result of a Precognition. See Reg. of Cor. Ent. Vol I. fo. 143. January 31 of 1919.	1	James Millikin 2nd Lieut. Royal Irish Rifles attached to Royal Air Force (Single)	1918. December Thirty first 10h 50m A.M. On the Turnberry Aerodrome. (Usual Residence: No.1 Fighting School, Royal Air Force, Turnberry, Ayrshire.)	M	25 years	Samuel Millikin Farmer. Hessie Gilmore Millikin M.S. Smith.	Injuries to head and burns Aeroplane Accident As certified by Chas. S. Clegg, Capt. R.A.F. M.S., M.B., Ch.B. who saw the deceased post mortem	Hawke Stull Major R.A.F. No.1 Fighting School Royal Air Force Turnberry	1919, January 2nd At Kirkoswald L.S. Baikie Registrar.

Extract from Second Lieutenant James Millikin's death certificate

*Grave of Second Lieutenant
Charles Alexander Hillock 154732*

and

Second Lieutenant Charles Alexander Hillock 154732
of Toronto, Canada.
Age 28

who died of injuries sustained
Wednesday 8 January 1919

DH9 E679
Turnberry Airfield

◆

IN MEMORY OF

2nd Air Mechanic Thomas Inglis 148684
Royal Air Force

of Hyndford Bridge, by Lanark.
Age 30
who died on 10 March 1919

Spinal Meningitis

8
Inter War Period

The 'War to end all Wars' had brought drastic changes and upheaval to Maidens and Turnberry. With the closure of the aerodrome, work could start on reclaiming the agricultural land, and the disposal and clearance of the wartime buildings. Physically and emotionally scarred by the war like the rest of Britain, thoughts turned to reconstruction. Issues regarding compensation payments for the requisition of land dragged on as did material problems caused by the Military Camp, such as restoring the shortage of the water supply to farms in the area such as Ballochneil and Drumbeg, and replacing the fencing which was entirely due to the Military Camp being established.

Eager to resume agricultural production, the estate factor was hindered by the lack of official derequisition notices. Immediate possession of the land was desirable if a potato crop was to be harvested. The road to Turnberry Warren, which cuts across the golf course, was also still in the possession of the Military Authorities in January 1919, and it was noted that a very considerable quantity of material would have to be removed by this road, so it was unlikely to be vacated.

One property that was relinquished was the Dam House, near Kirkoswald. No longer required by the Belgian Refugees, the furniture was removed, and the house restored to the Curling Club. A request was made to the factor regarding the possible use of the building adjacent to the White House at Turnberry, which had been used as an auxiliary hospital. This building, put up as an emergency measure under the Defence of the Realm Act, did not comply with certain restrictions on the lands as to the character of buildings to be erected, and the land it was built on was let to Mr Bone, of Turnberry Lodge, for agricultural purposes only. Therefore, the application was turned down.

In February 1919, formal notice was received that the area west of the public road was derequisitioned. The enclosure fence along this entire area had been removed by the Military and, anticipating that the ground would be soon be put under crop, and the land on the opposite side of the road would probably be grazed by cattle during the ensuing season, the factor was anxious that provision was made by the Military Authorities for the prevention of trespass on the cultivated ground. Further land was relinquished at the end of the month, with the same problem of replacing the fencing. It was noted that Mr Marshall, tenant of Little Turnberry farm had been given permission by the Directorate of Lands to cultivate his land again, provided that the Government Buildings left on the property were not interfered with. The Lords Commissioners of the Admiralty surrendered the land and premises occupied by them on Ailsa Craig.

Turnberry

Members of Girvan Town Council met with the resident engineer, Lieut. Cleghorn RE, at Turnberry, in March 1919, to discuss the purchase of some of the huts. It was agreed to offer £35 for the sectional hut known as the Plane Store, and £15 for the sectional hut known as the Sergeant's Mess cookhouse. An agreement was also reached with the factor to buy fireclay pipes no longer needed on the aerodrome, namely, 2 - 30" socketed pipes, 12 – 6" bends, 10 – 4" bends, 12 - 6x4 junctions, 9 - 4x4 junctions and 15 – 4" Buchan traps, valued at £8. Huts were also purchased by the Education Authority for use at Maybole. An enquiry was made by The Marquess of Ailsa to the Air Ministry regarding the Motorboat Shed in Maidens Shipyard, whether he could acquire it after the Royal Air Force discontinued using it. He proposed to use the shed for the storage of the lifeboat, (which he had presented to the village of Maidens) and which was inadequately stored in the old shipyard sheds. It would also be used for repairing and overhauling the fishing boats belonging to the fishermen at the Maidens and laying up a small motorboat that belonged to his Lordship. He was of the view that as the primary purpose for which the shed would be used was one of national importance, and that the cost of dismantling the building would not be worthwhile, he would agree to make no claim for the site to be restored to its original condition if the Motorboat Shed was handed over to him without charge. This was duly approved.

An offer of £400 was submitted by the Estates for the following buildings left on their land, complete with their fittings, etc., Photographic Hut, Sail Makers' Shop and Dope Shop, Metal Workshop, No.2 Flight Shed, and the Power Station.

The factor wrote: 'The offer may be considered somewhat inadequate but, on the other hand, should the Proprietor not purchase these buildings they would have to be taken down and removed as the Proprietor could not be expected to allow outsiders to acquire them permanently, and the cost of removal would probably not be met by the sum realised for the materials.'

In July Rev James Muir, of Kirkoswald Parish Church, left to take up a Chaplaincy with the Army of Occupation on the Rhine, giving up the whole manse, etc., to the minister who was to take his place.

A rather bizarre newspaper article details the tale of the 'Wild Freak of the Semi-Nude Man in Girvan':

At a very late hour on Saturday night, the citizens of the Doune Park, at the south end of Girvan, were thrown into a state of alarm and excitement by the extraordinary conduct of a man - a stranger to the town – rushing about the beach in a semi-nude condition, and making several rushes into the sea as if he wanted to commit suicide. A crowd collected, and he was induced to come out of the water.

Two of the Girvan Police appeared on the scene, but as soon as the demented creature saw them, he made into a potato field and attempted to climb the wall into the cemetery, as he said he had a brother, an American Airman, whom he alleged had been killed at Turnberry School of Aerial Gunnery during the war, and who was buried in Doune Cemetery.

When placed under arrest he made a desperate resistance and gave the constables a good deal of rough handling and began tearing the little clothing he was wearing off his body. He was eventually overcome and placed in a motor and conveyed to Girvan Police Station, where he was medically examined and certified insane by Dr Scade and Dr MacDonald. He was afterwards removed to Ayr District Lunatic Asylum. He was not in a state to give any account of himself, but it has since transpired that his name is George Stewart, an American citizen, but his plea that he had a brother killed at Turnberry cannot be substantiated, as no one of the name of Stewart appears on the Turnberry Roll of Honour.

There was a very considerable shortage of houses at Maidens owing to the number of young men who returned from service and had been married. The factor wrote to Mr William Sloss regarding the urgent need for certain of the young couples being accommodated with at least one room, and while friends and relations of several had inconvenienced themselves by taking such in, there were still one or two not provided for, and although steps were being taken to provide additional houses it would be a considerable time before these could be built, or even temporary accommodation provided. The factor suggested in the meantime that the accommodation in various houses let to summer visitors might be made available for one or two of these couples. If the fishing industry was to be fully maintained the young men must be retained in the village, in view of the serious shortage of houses, the Carrick District Committee included in their Housing Scheme a proposal to erect four houses at the Maidens to provide for the wants of the community. Although the Marquis of Ailsa was unable to build houses himself, he was willing to grant every facility he could to forward the Housing Schemes. Under the Housing and Town Planning Act recently passed, local authorities had power to take land compulsorily for such.

Negotiations continued between the Estates and the relevant authorities throughout 1920, regarding the disposal of various buildings and the making good of fencing. Having purchased on behalf of The Marquess of Ailsa one of the hangars including the heating plant, comprising two large sectional boilers, 28 x 25 column radiators, 5 x 6 column radiators, and several hundred feet of piping etc., the factor offered this equipment for sale to the Ailsa Shipbuilding Co. Ltd. Apparently, it had never been used after it was tested, as the Armistice was signed very shortly on completion of the installation. This hangar was an all brick one, built alongside the road to Little Turnberry farm, and the one nearest the sea of three hangars. Several other

Maidens Bay, 1936, showing WW1 Motorboat Shed

brick buildings had been purchased and remained standing, on Little Turnberry farm, the tenant, Mr Marshall, intended to use these for the housing of the seasonal potato diggers.

Miss Mellroy of the Lady Artists Club, Glasgow, approached the factor with the view of using one of the buildings as a permanent studio. This was not possible, and it was suggested that she might be able to arrange some temporary accommodation with Mr Marshall.

The funeral of Lt. General Sir David Henderson KCB, KCVO, DSO, took place in Girvan (Doune) Cemetery on 1 September 1921. He had died in Geneva on 17 August, while attending a conference relating to the Russian famine in his role as Director General of the League of Red Cross Societies. The funeral took place with imposing military and civic honours. The casket was placed on a gun carriage, draped with the Union Jack, preceded by a firing party of the Argyll and Sutherland Highlanders. They were followed by a band of the Royal Air Force. The gun carriage was next in the procession, drawn by horses under a Royal Artillery driver. Immediately behind the gun carriage were the chief mourner and civic heads of the burgh. The route from the station was lined by a detachment of the Royal Scots Fusiliers under command of Major W. A. Farquhar, DSO. A large crowd lined the streets of the town as the cortege passed through to the cemetery. All businesses on the route from the station to the cemetery were closed, and blinds drawn at private houses as the remains passed through the town. The flags on the McMaster Hall and other public buildings floated at half mast, and the town bell was tolled. The solemn procession wended its way to the cemetery to the strains of the 'Dead March'. The provost, wearing his robes and chain of office, was accompanied by the magistrates, the members and officials of the Girvan Town Council, and Mr William Paterson JP, chairman of Girvan Parish Council. The Marquis of Ailsa, wearing his uniform of Lord Lieutenant of the County, and General Sir Charles Fergusson of Kilkerran, were among those present. The ashes of the General were laid to rest in the grave of his son, Captain Ian Henderson, who was killed in a flying accident at Turnberry in June 1918. After firing a volley at the graveside, the' Last Post' was sounded by buglers of the Royal Air Force.

David Henderson was born in Glasgow in 1862. Joining the military, he served as a member of the Nile Expedition of 1898 and fought in the Second Boer War where he was wounded at the siege of Ladysmith. Having already had a distinguished military career, he learned to fly in 1911, at the age of 49, becoming the world's oldest pilot at that time. He was appointed the first director of the new Department of Military Aeronautics, established in 1913, and at the outbreak of the First World War, he took command of the Royal Flying Corps in the field. This fledgling branch of the service made vast progress under his guidance; Sir David Henderson continually preached the belief that if ever war broke out aeroplanes would play an important part in it; he was the chief organiser and developer of our military aviation service, it could be said he was the true 'Father of the Royal Air Force'.

The matter of replacing the fencing which had been removed by the military was still unresolved, the tenant of the piece of ground on which the station buildings, repair shops, etc., were erected on the west side of the public highway, intimated that unless a suitable fence

was at once erected enclosing the buildings, he would be forced to give up tenancy of the land. A definite promise had been made by the authorities that a suitable fence would be erected to prevent stock straying, but this was never done, other than in a very temporary manner. The tenant had suffered considerable loss and damage to his stock in consequence, besides trouble and expense in endeavouring to keep up the fence. The fence along the public roadside had also not been replaced, and with the potato crop having been lifted, sheep or cattle would be put on the land, and in order to prevent the various stocks from becoming inmixed it was necessary for at least one fence to be erected.

On the night of 11 October 1926, a fire broke out at Turnberry Railway Station, as was reported in the *Evening Telegraph* the next day:

TURNBERRY LOSES RAILWAY STATION
FAMOUS GOLF RESORT'S
PLATFORM ABLAZE
Ayr Fire Brigade's Night Dash to Outbreak
Ayrshire's famous golf resort Turnberry has lost its railway station. An alarming fire gutted the hotel platform last night despite the efforts of the hotel fire brigade and a detachment of the Ayr Fire Brigade.

The alarm seems to have been given a few minutes before eight o'clock, and from all accounts, the fire originated in the stationmaster's room.

Fanned by the wind from the sea, the flames spread rapidly, and, although the hotel brigade was quickly turned out, they found a stiff task before them. Fortunately, the wind diverted the flames away from the hotel, so that no damage was done to the picturesque building, but it was found impossible to save the major part of the pretty station.

The covered passage from the station to the hotel was only slightly damaged at the platform end, but the adjoining erections, consisting of porters' rooms, booking office &c., were too firmly alight to save.

Fire Brigade's Dash

The Ayr Fire Brigade got a call to turn out about 8.15pm, and an engine dashed rapidly down the dark countryside inside half an hour. On arrival it was found that the hotel staff had four hoses playing on the station. It was evident that the station was doomed, and the expert firemen altered the line of action somewhat, and after a stiff fight, managed to extinguish the outbreak.

Turnberry Hotel is fairly well filled with guests at the moment, and the diversion, while it caused a little excitement, did not cause any real alarm to the residents. Indeed, a few of the guests lent a hand to fight the outbreak.

Had the wind been blowing the other way there is no doubt that the hotel would

In 1929, the London Midland and Scottish Railway Company proposed to close the line between Ayr and Turnberry to passenger traffic. A regular service would continue to be given from Turnberry via Girvan. The sub-committee felt that the establishment of one or two through carriages per day during the summer was essential to the proper success of the Turnberry Hotel, both as an hotel venture and as propaganda for the company as a whole for the Ayrshire coast. Reference was made to the unsatisfactory train service from the south to Turnberry and the Chairman undertook to follow the matter up and see whether improvements could be made in the future. The question of the train service to Turnberry was again referred to the following month, reporting that the number of passengers regularly travelling to Turnberry was so small that it would not be economical to run a train

Turnberry Railway Station

have suffered in common with the station. The damage is not so great as might have been expected. The trains are stopping as usual at the hotel today, and an emergency arrangement has been made for the entraining of passengers.

from the south direct to Turnberry, although it was stated that connection was given off the sleeping car train via Stranraer during week-days. It was further stated that in the last year the service was not taken advantage of to any considerable extent. It was pointed out though, that there was no direct connection, and that serious complaint was made by hotel guests who have to motor from Ayr. One of the committee members stated that a friend of his, who had been staying at Turnberry Hotel, was unable to get any train connection to London on a Sunday unless he motored to Kilmarnock, a distance of about 35 miles. The general performance of the hotel was causing some concern, the receipts at Turnberry had fallen to a very low level, and it was wondered whether an increase in prices had been one of the causes of the adverse results. The committee were seriously troubled that a mistake may have been made in converting Turnberry into an Hotel de Luxe. The arrangement in the main dining hall under which part of the dining room facing the sea was reserved for A la Carte guests had been reconsidered and it was proposed in future to bring that part of the dining room into the general dining room, thus getting over the difficulty of dividing the 'sheep from the goats'.

Turnberry Hotel, 1931

A programme of modernisation was undertaken, with the installation of running hot and cold water in 74 rooms, and in April 1930 the remaining rooms were similarly dealt with, thus completing the whole building. The cost of this work was £2,531 2s 10d. Central heating did not exist in every room, but it was considered that there was a sufficient number of heated rooms for the small amount of business done in the cold weather. In order to improve the light at the north end of the dining room it was recommended that

a fireplace should be removed and that a window should be inserted at an estimated cost of £300. The only service room at Turnberry Hotel was situated in the east wing, below the level of the main first-floor corridor, where most of the floor service took place. It was recommended that that certain alterations should be carried out at an estimated cost of £110. Due to an upturn in business the demand for floor service had increased and the food trolleys, which had to be lifted up nine steps, caused delay and bad service. Works were completed to remedy this. The refrigerator, it was decided, was also obsolete and it was proposed to replace it by a larger cabinet type at an estimated cost of £550.

The *Dundee Courier* of Saturday 16 August 1930, published an article extoling the charms of 'Maidens and Its Men':

A VISIT TO MAIDENS - a pretty name - is a quiet little holiday resort of fishermen's houses and picturesque bungalows on the Ayrshire seaboard. It is named after the Maidenhead Rocks that are prominent in the bay and form a natural bulwark against the inflooding waters. No lovelier spot could be found for a peaceful retreat from the roar and traffic of the city. Its situation makes it a favoured spot for holidaymaker and camper alike, lying as it does in sequestered beauty between the famous Culzean Castle - the home of the Kennedys - and the lighthouse at Turnberry. Maidens is rich in historical lore and romance, and full of 'characters'. The most familiar figure is the fisherman, who plies his ancient trade with industry and zest. Every Monday the fishing boats sail out of the harbour, encouraged by the children's cheers wishing good luck the hardy adventurers who fare forth to gather the harvest of the seas. They pursue their quest round the further waters of Arran through most of the week, returning usually by Friday night or Saturday morning. All the nets have to be gathered in before the Sunday comes round, for the fisherman believes that nothing but harm would come of his labours on the Lord's Day. Still, he is a great opportunist, holding that 'time and tide wait on no man,' and that he who ventures a full tide will not fail of his reward. Unfortunately, the reward is poor these days. Basketfuls of herring last week had to be emptied back into the sea; there was no market. One day the writer stood chatting to one - of keen, brave eye beside his boat before setting out; were watching a veil of mist slowly disappearing from the hills of Arran. 'The weather looks like settling,' was the remark made. 'Yes,' he said, 'it's going to settle; but I wish the herring market would settle.' There was a twinkle in his eye, but off he went, hoping for the best. They deserve well, those brave fishermen round our coasts who ply so faithfully in all weathers the ancient craft of the Apostles. Speaking of romance and legend of the place, we had an interesting talk with one of the oldest inhabitants, who sails his boat through the week and attends at the castle, and on Sunday is the precentor, and leads the Psalms in the hall where we worship, a quiet voice, full pride in the past glory of Maidens, said to me, 'Do you see yon hill?', pointing a little to south and up behind the village, 'yon's where the heather burned, and Robert Bruce, seeing it in the south of Arran, mistook it for the beacon and came across. He had to fight hard for footing at Turnberry Castle, but won through and got his object.' Then, taking a slow, meditative pull of his pipe, said, 'I've seen the heather burning on Arran on clear day myself, from this very spot. So, it's a true story, and no fairy tale. Man,' said he, 'they were great men in those days.' He assured me also of the truth of the story - that half a dozen years or ago - a ring was found at Turnberry, by a man digging near the famous golf course bearing the

initials R. B. and supposed to the ring of Robert Bruce, he hadn't seen it, but many of the villagers had, and it was sent up to His Majesty the King, who handsomely rewarded the finder. Another great Scot— Robert Burns - has a sacred place in the hearts of the good folks of Maidens, for at Kirkoswald School (the parish) the old residenter, with a gush of pride, told me that Burns finished his education under the old parish school dominie, Rodger. There are lots of Rodgers still in the village, he assured me. There's yin that's famous for his kippers. I said I had heard of him. Well, that's the same Rodger that educated Burns. On hearing such things of Maidens and great connections I lifted my hat and passed on, thankful in my heart that I was privileged for a short season to sojourn amongst them, for they are warm-hearted and most friendly people.

In 1934 the suggested provision of about eighteen lock-up garages on the asphaltic yard extending from the hotel coal bunker in the direction of the coppice was approved. It was also proposed that the golf clubhouse be improved, and that additional accommodation be provided for the golf club, and it was agreed that a scheme be prepared with detailed plans and estimates whereby:

(1) The whole of the existing Club House premises would be used for extending accommodation for Club members and Club purposes generally, including sleeping accommodation for the Club House attendant.
(2) The Caddie shelters replaced by a building which would also include new accommodation for the Professional's Office, Workshop, and storage room for Guest's golf clubs.
(3) New accommodation for male and female Caddies with suitable lavatories, be provided.

It was also agreed to recommend that the offer made by the Golf Club to contribute £300 towards the cost of the scheme be not accepted, but that the Club should undertake the responsibility of maintaining the internal decorations and furnishings of the clubhouse premises, and that the practice of the company in supplying fuel for heating the caddie's room be discontinued.

The installation of telephones in all the rooms at Turnberry Hotel was undertaken after protracted negotiations with the Post Office. The wiring commenced and a new switchboard was installed, the works hoping to be ready for the Easter guests. The total cost of the work was estimated at £3,080, with future annual expenditure on telephone services amounting to £348.

It had been agreed that an aerodrome be provided on land in the vicinity of the hotel, when the question of the free conveyance by rail of the personal luggage of those using the aerodromes to visit the company's hotels was raised. Having considered this point, it was decided that persons arriving at the hotel by aeroplane, other than those operated by the Railway Air Services were, from the railway point of view, in precisely the same position as those travelling by road - it was not considered that such luggage should be conveyed by rail at other than the full unaccompanied rates. The Railway Air Service was a domestic airline set up in 1934 by the four major railway companies, London Midland & Scottish, London & North Eastern, Great Western Railway and the Southern Railway, with Imperial Airways, the most important route being between London and Scotland.

Passengers travelling by Railway Air Services machines, in addition to being entitled to have 35 lbs of luggage with them, were able to take advantage of the Passenger's Luggage in Advance arrangements, in the same way as if they were making the journey by train, but such passengers were, of course, considered in a different category from those arriving by private aircraft, and it was thought that there was

no justification in making any extension of the concessions.

The planting of Broom at Turnberry was proposed so as to improve the outlook, although another outside problem was raised by committee member, Mr Taylor, who had recently returned from a visit to Canada, and mentioned that the smoke which emitted from the chimneys at Gleneagles and Turnberry was not only a nuisance but entirely out of keeping with the general surroundings of the hotels. He said that one would, of course, never see this sort of thing in New York, and if they had an automatic stoker it would put a stop to the nuisance. He suggested the smoke might be caused by using soft coals. It was agreed that this was a problem, but short of altering the boilers and the whole heating system, it would be too difficult and costly to improve this. Mr Taylor also added that the plumbing in the bathrooms at the company's hotels, Turnberry, Gleneagles and Birmingham in particular, was out of date. If somebody was having a bath in the room above, the occupant below was disturbed by the noise, and in this respect the hotels compared unfavourably with hotels in America. Plans were drawn up to provide sixteen additional private, modern bathrooms at Turnberry Hotel with loss of only one small bedroom and it was recommended that the work should be carried out at an estimated cost of £2,380.

With business picking up, there was a continued increase in the number of meals being served on the floors at Turnberry Hotel and that the counter accommodation in the American Bar was found inadequate. It was, therefore, recommended that, in order to maintain the standard of efficiency, another service room should be provided on the second floor over the service room on the first floor; that a hot plate should be transferred from St Pancras Hotel and that an electric service lift should be transferred from Gleneagles Hotel (both of these articles of equipment were out of use) and that a hole should be cut in the floor in connection with the installation of the lift and machinery, at a total estimated cost of about £75. The counter in the

American Bar should be turned round, a fireplace removed and that a service hatch should be provided in the side corridor at an estimated cost of £85.

The Chief Mechanical Engineer had pointed out that the supply of water from the spring at Turnberry would be entirely inadequate in the event of a serious outbreak of fire at the hotel, he, therefore, recommended that a pipe should be provided from the water main at the bottom of the drive to the hotel at a cost of £132, and it was estimated that water from this supply would have sufficient pressure to reach the roof of the hotel.

At the close of 1935, with all of the proposed upgrading works completed, Turnberry Hotel was able to offer guests:

80 Single Bedrooms
8 Single Bedrooms with bath
28 Double Bedrooms with large bed
1 Double Bedroom with large bed and bath
35 Double Bedrooms with twin beds
16 Double Bedrooms with twin beds and bath
34 Public bathrooms
7 Suites

Running hot and cold water and central heating.

The financial position as at 31 December 1934 was capital, £207,567.

At the beginning of 1937 the desirability of improving the 'Ailsa' Course was generally agreed, and it was suggested that the proposed alterations should be submitted to an outside expert for his observations, although there would be no obligation on the Company to give effect to any suggestions which he might make. It was proposed to approach Mr Robert Maxwell in this regard.

One of the committee members stated that he had been at Turnberry during Christmas week and had taken an opportunity of looking at, and playing over, the course, and in his opinion the alterations suggested would be a great improvement from the point of view of the hotel guests.

It was realised that all the improvements could not be carried out before the summer, and it was ultimately agreed that the first 5/6 holes should be altered as suggested, ready for the spring of 1937.

An early postcard of Turnberry Hotel, 1906

The question of accommodation for employees at Turnberry Hotel was raised. In the busy season the accommodation was grossly overcrowded, and that it was absolutely essential that steps should be taken to provide additional space. Rough sketches were prepared offering the following alternative schemes:

(1) The back wing of the hotel facing the railway line to be extended. (This would involve the back wing abutting about 20 or 30 feet to the north of the main wing in the front and would, it was thought, give an ugly appearance from certain parts of the golf courses.)
(2) The erection of a separate building on the car park, but entirely removed from the hotel. (This would leave a gap between the main hotel building and the proposed new building. The architect is of opinion that this would break the line and would not interfere to the same extent with the architectural appearance of the main building.)

It was further reported that the latter scheme would be the cheaper and that the cost of the work would amount to between £16,000 and £20,000. The erection of a separate building on the car park was approved.

Complaints were received regarding the lack of central heating in some of the rooms at Turnberry Hotel, and it was therefore recommended that central heating should be installed in 95 rooms at an estimated cost of £1,800. No expenditure had been included in the estimate in respect of new boilers, as it would not be required to increase the boiler capacity, until the proposed additional staff accommodation was completed. By January 1938, central heating had been installed in 35 out of the 100 rooms for which it had been ordered, and these rooms had been available at the previous Christmas.

The death of the Marquis of Ailsa was announced in *The Scotsman* in April 1938:

Noted Yachtsman-Peer and Landowner
HELD TITLE FOR SIXTY-EIGHT YEARS
The death occurred at his home, Culzean Castle, Maybole, Ayrshire, in the early hours of Saturday (9 April) of Archibald Kennedy, third Marquis of Ailsa. Lord Ailsa, who was 90 years of age, having been born at Culzean Castle in September 1847, succeeded to the title, which he held for 68 years, in 1870, and is in turn succeeded by his eldest son, Archibald Kennedy, Earl of Cassillis.

The family is an ancient one, dating back to Duncan de Carrick, who lived in the reign of Malcolm IV, whose reign began about 1150. The first Earl of Cassillis was created by James IV in 1509, and the twelfth Earl was created a Peer of the United Kingdom as Baron Ailsa in 1806, and a Marquess in 1831. In addition to the Marquessate, Lord Ailsa held the Barony of Kennedy (created 1457) and the Earldom of Cassillis (created 1509)

The eldest son of Archibald Kennedy, the second Earl, and of Julia, second daughter of the late Sir Richard Mounteney Jephson, Bt., Lord Ailsa was educated at Eton, which he entered in 1860. Joining the Coldstream Guards six years later as an ensign, he was promoted lieutenant in 1870, and retired upon succeeding to the Marquessate and estates. Exchanging an army career for a naval one, he was granted an honorary lieutenancy in the Royal Naval Reserve in 1874. In 1887 he became Lieutenant Commanding the Clyde Brigade, Royal Naval Artillery Volunteers, and in 1921 was made honorary captain of the Royal Naval Volunteer Reserve.

INTEREST IN THE SEA
Though he held many offices at different periods of his life, having been president of the Ayrshire Territorial Army Association, a J.P. for the county, and an extraordinary director of the National Bank of Scotland Ltd., the Marquess of

Ailsa's predilections were mainly for matters pertaining to the sea. It was as a yachtsman and shipbuilder that he was most widely known outside his own county. He was a yacht and boat builder and himself a designer of craft, and so keen was his bent in this direction that, as early as the '70's of the last century, he established a boat building yard near his seat of Culzean. Here were built to his own designs and under his superintendence some very successful small racing yachts and small launches.

Abandoning the enterprise at Culzean, he established a small shipbuilding yard at The Maidens two miles to the south. This business was started in 1884; and a number of yachts and steamers were built there. They included a 40-ton racing yacht, the *Clara*, and a beautiful steam yacht, the *Black Pearl*. The enterprise had to be abandoned owing to transport difficulties connected with the Ayr, Dunure and Girvan Railway, and, in 1886, the business was merged into the Ailsa Shipbuilding and Engineering Co. Ltd. at Troon which was turned into a limited liability company, with Lord Ailsa as chairman of directors.

As a competitor in yachting events few owners have been more successful than the Marquess of Ailsa, who took part in all the principal races along the west coast from Cowes to the Clyde. His first racing yacht was the *Foxhound* which he sailed from 1870 to 1873, during which period he won over twenty trophies. In 1874 he had the *Bloodhound* built to the designs of Fyfe & Sons, Fairlie. The Fyfe-designed 40-tonner became one of the most famous yachts in her class. Lord Ailsa sailed her until 1880, and with her won nearly twenty prizes, including the Prince of Wales's prize and a combination prize, valued at over £417. Fifteen of the prizes were first prizes.

YACHT SOLD AND REBOUGHT
In 1880 Lord Ailsa sold *Bloodhound* and acquired the *Sleuthhound*, which was also designed by Fyfe & Sons. The twenty or so prizes which he won with this yacht included two Queen's Cups and the King of the Netherlands Cup, in 1883. His Lordship's luck was, however, not sustained with *Sleuthhound*, for in 29 starts she won only twice, whilst her sister ship *Annasona*, won 14 first prizes. The *Sleuthhound* continued to race, but she never rose to the top of the hill as *Bloodhound* had done. In 1882 Lord Ailsa was again to the front with the slender little 3-tonner *Snarley-Yow*, and subsequently he built in his own yacht building yard at Culzean the 5-tonner *Cocker*, which did very well on the Clyde.

Rather than see his old boat *Bloodhound* broken up, Lord Ailsa acquired her again, had her reconstructed and afterwards sailed her every summer. The yacht ultimately perished by fire. Long after he had retired from the list of racing owners his Lordship was a faithful attender at Hunter's Quay whenever the 'Fortnight' came round.

During his career as a yachtsman, Lord Ailsa was a member of the Royal Yacht Squadron, Cowes, the most exclusive organisation connected with the sport. He was also a member of several Clyde clubs including the Royal Clyde and the Clyde Corinthian, which he joined in 1872. Moreover, he held several offices in the Shipwrights' Company, of which he was a Past Master. In 1904 Glasgow University conferred upon Lord Ailsa the honorary degree of LL.D.

PART IN LOCAL GOVERNMENT
Very naturally, Lord Ailsa played an active part in local government. He was, under the old regime, a Commissioner of Supply and a member of the Road Trust. Later he entered the Ayrshire County Council as member for Kirkoswald, and interested

himself in particular, in the Carrick District Committee and the Central Board. He was for nine years chairman of the County Road Board, and for a considerable number of years a member of the Maybole School Board. In 1919, he became Lord Lieutenant of Ayrshire, succeeding the Earl of Eglinton. He held the office until last year, when he resigned in favour of Sir Charles Fergusson of Kilkerran.

Owning about 76,000 acres, Lord Ailsa had the reputation of a model landlord, spending a large part of his income upon the upkeep of his estates and being deeply interested in the welfare of his tenants. He carried out a considerable scheme of afforestation at Culzean and was on one occasion awarded a gold medal by the Arboricultural Society. He interested himself in the introduction of Loch Leven trout into the numerous lochs on his estate, but the experiment of acclimatising them was not successful. He also bred trout imported from America, which grew to a great size in his policies.

A member of the Prestwick Golf Club from 1871, and at the time of his death the oldest club member, he had a course of nine holes laid out in the Deer Park at Culzean. Here he practised and became a good enough golfer to be elected captain of the Prestwick Club in 1889. It was on his initiative, and at his expense, that the first of the two eighteen-hole courses at Turnberry was laid out. The two Turnberry courses were afterwards taken over by the Glasgow and South-Western Railway Company, in connection with their hotel there. The Deer Park where Lord Ailsa's private course was situated was named from a herd of deer that his Lordship imported and kept there for some years. He also imported into his policies racoons and Indian native cattle.

PROMINENT AGRICULTURIST
As a landowner, Lord Ailsa was naturally prominent as an agriculturist. It was largely on his coastal farms that the now famous early potato industry was developed. For more than half a century he was connected with the Ayrshire Agricultural Association, was a generous contributor to its funds and served both as its vice-president and president.

The Marquess was also an enthusiastic gardener, and at Culzean Castle he succeeded in growing flowers which could be seen nowhere else in Scotland. The family seat is magnificently situated on a rock overlooking the Firth of Clyde. It was designed by Adam for the tenth Earl of Cassillis in 1777. When he was Prince of Wales, the Duke of Windsor once stayed there during a visit to Scotland.

The family to which Lord Ailsa belonged has played a great part in Scottish history. Among those who fell at Flodden was the first Earl of Cassillis, and another member of the house attended Margaret of Scotland when she married the Dauphin of France in 1436.

The Marquess's interest in the sea and seafaring had a traditional basis. One of his personal friends on one occasion asked him why the dolphin formed part of the insignia of the family. 'They tell me,' replied Lord Ailsa, with a twinkle of humour, 'that one of my ancestors married a mermaid.' Adding, 'Isn't that rather ridiculous?'

Lord Ailsa was remarkably conscientious in every duty which he undertook, or which naturally fell to him. At the same time, he was reticent and very sensitive, as well as modest – qualities in his fine, considerate character which were recognised by those who knew him.

PORTRAIT PAINTED
His portrait, painted a few years ago by Mr Fiddes Watt, was presented to him by the tenantry of the Cassillis and Culzean estates, on the occasion of his having completed over 60 years of ownership of

the estate. The presentation took place in the Town Hall, Maybole, in the presence of several members of the family and on this occasion the Marquess of Ailsa made reference to the connection of the Kennedy family with the sea. The Kennedys, he said had been at times shipowners for hundreds of years. One of the earliest of whom they had knowledge was Lord Kennedy, who, about 1490, built a large ship at Ayr. Nothing much was known of her except that she was called Lord Kennedy's 'pykhart.' What the rig of a pykhart was in those days no one knew, but it was on record that it took four days to launch her, a fact which was known by an old account for ale for four days for workmen 'who garred her flott.' 'It might be surmised, commented the Marquess, 'that this large provision of refreshment had something to do with the protracted launch.' Another shipowner in the family was Bishop Kennedy, the founder of St Salvador's College at St Andrews. He built the largest ship in Scotland at that period, and used her as a yacht for state purposes, and also as a merchantman as occasion demanded. She was eventually lost off the east coast of England.

FIGHT WITH THE FRENCH

A good seaman in the family was the 11th Earl of Cassillis, formerly a captain in the navy, who, in 1760, being in command of two small vessels, fought two large French frigates of double their size and power. After fighting all night, he drove them off and so saved a valuable convoy of food bound for Lisbon which was in a state bordering on starvation owing to the blockade. Fighters on land also figure in the Kennedy family. They include Sir Hugh Kennedy of Ardstinchar, who commanded the Scottish contingent under Joan of Arc at the siege of Orleans, and who was one of the Maid's favourite generals. In Covenanting days at least two Earls of Cassillis were on the side of

the people. One of them, because of his devotion to the Covenanting cause had the Highland Host quartered on him by Charles II. The other was a signatory of the National Covenant.

Lord Ailsa married first in 1871 the Hon. Evelyn Stuart, third daughter of the twelfth Baron Blantyre, but she died in 1888. There were four children of that marriage—the Earl of Cassillis who succeeds to the title, Lord Charles Kennedy, Lord Angus Kennedy and Lady Kilmaine. His second marriage in 1891 was to Miss Isabella McMaster, daughter of the late Sir Hugh McMaster, a missionary stationed in Kausani, North-West Provinces, and the two children of that union are Lord Hugh Kennedy and Lady Marjory Merriam.

A PRECEDENT

One of the most democratic and popular Peeresses in Scotland, Lord Ailsa's second wife created something of a precedent during the tenure of the Labour Government when, on the appointment of Mr James Brown MP, an ex-miner living in a two-roomed cottage, as High Commissioner to the General Assembly of the Church of Scotland, she proffered her services as Lady-in-Waiting to Mrs Brown. Lord Ailsa underwent an operation in April 1935 when he was the senior Scottish Lord Lieutenant so far as age was concerned. At the time of his death he was the oldest peer in Scotland - a distinction which had been his since the death of the Marquess of Huntly last year.

THE NEW MARQUESS

Well-Known Public Figure in Edinburgh The new Marquess, who is 65 years of age, resides in Edinburgh, where he is a well-known public figure. Educated at Eton, Trinity College (Cambridge), and Edinburgh University, he was admitted to the Faculty of Advocates in 1897. He practised in Edinburgh and also appeared before Parliamentary Committees at

Westminster, but on the outbreak of the South African War he left his duties in Parliament House for the front. He rose to the rank of Major in the Royal Scots Fusiliers (from which he retired in 1913) and in 1901 he acted as Intelligence Officer at Alkmaar on the Delagon Bay Railway. He was awarded two medals and five clasps for his services. In the Great War he was mentioned in dispatches while serving with the British Expeditionary Force. He is now honorary colonel of the 4/5th Battalion Royal Scots Fusiliers. He takes a keen interest in ecclesiastical affairs and has frequently been a member of the General Assembly of the Church of Scotland. Also, an enthusiastic Freemason, he recently established a record by completing a quarter of a century's service as First Grand Principal of the Supreme Grand Royal Arch Chapter of Scotland. Another interest is the study of Gaelic. He has a wide knowledge of Highland affairs, has been chief of the Gaelic Society of Inverness, and took an active part in the arrangements for the Gaelic Mod when it was held in Edinburgh. Last year he was elected president of the Royal Celtic Society in succession to the Duke of Argyll. He is a member of the King's Bodyguard for Scotland, Royal Company of Archers, Deputy-Lieutenant for Ayrshire, and a Fellow of the Royal Scottish Geographical Society. Last year he became chairman of the Edinburgh Good Government League. In 1903 he married Miss Frances Emily MacTaggart-Stewart, daughter of the late Sir Mark John MacTaggart-Stewart, who is a Dame of Justice of the Order of St John of Jerusalem.

FISHERMEN PALLBEARERS
Funeral of Lord Ailsa

The funeral of the Marquis of Ailsa, who died on Saturday at Culzean Castle, Maybole, at the age of 90, took place at the castle, and was attended by about 500 persons. A private service was held inside the building, and public service, conducted by the Rev David Swan, Maybole, in the courtyard. The Rev D. M. MacLean, Kirkoswald Parish Church, read the Scripture lesson and prayer was offered by the Very Rev Dr W. P. Paterson, Edinburgh. Among those present were the widow of the Marquis; the Earl of Cassillis (who succeeds to the title), provosts of Ayrshire burghs: clergymen in the county; members of the County Constabulary, and tenants. The coffin, which was draped in the White Ensign, was placed on a farm cart lined with branches of Scots fir and the funeral procession to the family burying ground was preceded by a piper. Fishermen from Maidens acted as pallbearers, the express desire of the Marchioness, and they had returned from the East Coast fishing for the purpose. Dr Paterson conducted the committal service. The Red Ensign (flag of the Mercantile Marine) flew half-mast on the castle battlements during the day, and the flag at Turnberry Golf Course, of which the late Marquis was president, was also half-mast. A memorial service was held in the Old Church of St John the Baptist at Ayr.

An accident occurred at Turnberry in April 1938, as reported in *The Scotsman* newspaper:

KILLED BY TRAIN
Blacksmith Thrown Thirty Feet
TURNBERRY TRAGEDY

John Duff, aged 58, Master Blacksmith, of Milton Cottage, Turnberry, Ayrshire, was killed last night near Turnberry Station. He was showing two gentlemen friends round the district, and the three men were on the railway, when a train from Girvan approached the station. Duff was thrown thirty feet in the air on to the railway embankment. His friends jumped clear as the train passed.

Stationmaster J. Wallace and porter W. Clanachan, at Turnberry, heard the train whistle and rushed out to find Duff lying at the bottom of the embankment. They rendered first aid and an ambulance, and a doctor were called, but Duff was dead before the ambulance arrived. Duff was well known in the district and leaves a widow.

It was decided to approve the erection of a separate building at Turnberry for the extra accommodation of staff. The cost of building a new staff house on the same lines as the out-houses at Turnberry was estimated at £3,200 and that the cost of additional furniture, wiring, heating, drainage, etc., was estimated at £1,400 – a total cost of about £4,600. This new staff house was intended to provide accommodation for 20 to 24 persons and would accommodate the upper female staff.

The emission of smoke from the furnace at Turnberry was still as bad as ever, and that while nothing had yet been done the whole question was being considered, but that difficulty was being experienced in overcoming the nuisance.

Turnberry Hotel and Golf Courses, with all of the improvements and modernisation were once again, a popular Scottish tourist resort, with wealthy guests arriving by air, train or private motor car.

With the onset of 1939, it was felt that the stabling accommodation at Turnberry Hotel was not now required, and it was advocated that it should be converted into garage accommodation at an estimated cost of £260, this additional accommodation would provide room for seven or eight cars. Little did they know, that before the year was out, the building would soon undergo another conversion, becoming a morgue.

9
WWII Airfield

Almost immediately on the outbreak of WWII, Turnberry Hotel was requisitioned by the Department of Health, for use as a civilian hospital. The hotel, closing on 30 October 1939, was taken over on 6 November. An agreement was reached that the Railway Company were to maintain the buildings, drainage, firing etc., the Department of Health were to maintain the gardens and grounds (taking over the Company's staff) paid for by the Government, and that the Company were to store the furniture in the hotel and to retain the linen. The hotel was requisitioned under the Emergency Medical Service scheme, whereby patients were discharged or evacuated from the many public and private hospitals and convalescent homes in central London, thereby freeing up beds in readiness for the anticipated air-raid casualties in the city. The estimated 250,000 casualties never materialised, and Turnberry Hospital was left caring for the displaced and locals alike. Most of the patients from London were elderly, suffering from senility and frailty. Some were convalescing from surgery or recovering from fractures. A number of the elderly patients died at Turnberry and were buried with the 'Civilian War Dead Scheme', in local cemeteries such as Dailly, Straiton, Kirkmichael, Kirkoswald and Girvan.

Turnberry EMS Hospital Nurses, 1939

It was decided to keep the Golf courses open, the maintenance of which was aided by the grazing of sheep which helped to keep the grass down - the opinion of the Railway Company was that petrol and labour should be used as little as possible on the maintenance of the golf courses during wartime and that the staff and land available should be used as far as practicable for the growing of food stuffs.

The Lands of Turnberry were requisitioned for a second time by the Air Ministry, after an evaluation of its suitability as a wartime airfield. AVM Sandy Johnstone quoted his diary entry for 14 February 1941 in his book, *Spitfire into War*:

> Asked to assess the prospects for constructing an airfield at Turnberry on the site of the golf course whence aeroplanes had operated during World War I. A strong onshore wind was blowing when I got there and I experienced a great deal of buffeting on all the approaches, particularly on the east-west run-in which entailed passing over the hotel. Certainly, I had doubts about its suitability, particularly as I'm told the place would be used to train 'ab initio' pilots. Before writing a report, however, I swapped my Spitfire for the Magister and took Mr Westacott with me to size up the constructional aspects, when my first impressions were confirmed after landing on one of the fairways only to be confronted by a stone monument on which were inscribed the names of those who had lost their lives during its previous spell as a training aerodrome. The resident engineer supports my belief that it is not a suitable site, and I am reporting accordingly.

But the report was ignored by the Ministry!

Again, the lands were requisitioned from Cassillis and Culzean estates:

FORM OF REQUISITION OF LAND
(AND BUILDINGS)
FROM THE OWNER AND OCCUPIER

To Cassillis and Culzean Estates, per
James T. Gray, Esq.,
Factor, Estates Office, Maybole.

The Owner and Occupier of the land (and buildings) described on the back hereof. TAKE NOTICE that in exercise of his powers under the Defence Regulations, 1939, the Air Officer Commanding, Coastal Command, being a person (or one of a class of persons) to whom the Secretary of State as a competent authority, by virtue of the said Defence Regulations has delegated his functions for the purpose, has given authority for taking possession of the land (and buildings) described in the said schedule, and that on behalf of the said Air Officer Commanding take possession of the said land (and buildings).

As early as March 1941, the Department of Agriculture and Fisheries were writing: '*The Air Ministry propose to construct a landing ground at Turnberry, for use with the aerodrome at Ayr. Part of the land on the links side of the road was to have been prepared under local arrangements as a relief landing ground for Prestwick, but the Air Ministry now want a full-size landing ground in this area. The proposals will involve closing of the road from Turnberry to Maidens, and the Air Ministry are in touch with the Ministry of Transport on this matter. The site is close to the coast in the vicinity of Turnberry Point where there is a salmon fishery. We would have no objection to the proposal if adequate treatment were given to domestic sewage or other effluent.*

And so, requisitioned once more, the Air Ministry sanctioned the construction of a full RAF Station. The Torpedo Training Unit at Abbotsinch, near Glasgow, had been using the landing strip at Turnberry on an ad hoc basis for practising torpedo dropping since 1940, and as their training area on the Clyde was now becoming congested with shipping, an alternative base was required.

Contractors moved in, with the purpose of having the airfield ready for use by February 1942, and extensive levelling operations completely flattened large areas of the golf course.

The contract for constructing Turnberry Aerodrome, including buildings, was awarded to the company of George Wimpey & Co., for the sum of £920,000 (equivalent to £55,184,046 today).

Turnberry was typical of a dispersed wartime airfield, having a large number of scattered sites around the locality. The term dispersed generally referred to locations outside and not immediately within the main airfield perimeter. These included communal and recreational sites, such as living quarters and sick quarters, for all ranks, male and female, technical sites and instructional sites. Other miscellaneous sites included administration, fuel dumps, radio facilities and gun posts. Buildings were usually temporary, brick built or prefabricated huts. There were four T2 type, fourteen Dorman Long Blister, and twenty-four Extra Over Blister hangars, and an aircraft shed left over from WWI was pressed into use.

The airfield site ran parallel to the hills behind Maidens and Turnberry. There were three concrete runways and surrounding perimeter track laid down, a typical RAF dispersed layout, the main runway below the ridge of high ground (North East to South West) was 04/22, 6,250ft. The additional runways were (North to South) 00/18, 4,500ft and (East to West) towards the lighthouse, 09/27, 3,900ft. Materials required for the laying down of the runways included 18,000 tons of dry cement and 90,000 tons of aggregate.

The Communal Sites were located around the village of Maidens to the north/east, with subsidiary sites on Blawearie Road.

To illustrate the size and scope of RAF Turnberry, a list of installations on the various sites as of February 1943, follows:

Airfield Site -

5 Aircraft Hangars	Type T2
Watch Office	
Fire Tender Shelter	
Floodlight Trailer and Tractor Shed	
Squadron Office	
Maintenance Block	Type E
Maintenance Block	Type D
Maintenance Block	Type F
Maintenance Servicing Squadron Offices	
Battle Headquarters	
Maintenance Wing Offices	
Armoury	
Main Stores	
Lubricant & Inflammable Stores	
Parachute Store	
Gas Clothing, Respirator Store	
Main Workshops	
Towed Target Store	
6 Technical Latrines	
Technical Latrines	WAAF
2 Office Latrines	
Gas Defence Centre	
Gas Instructional Chamber	
M.T. Sheds, Office and Yard	
M.T. Petrol Installation	
5,000 Gallons	
Bulk Petrol Installation	
Bulk Oil Installation	
Fuel Compound	
Station Offices	
Operations Block	
Photographic Block	
Guard & Fire Party House	
Fire Tender House	
Produce Compound	
Compass Platform	
Machine Gun Range	
Sleeve Streamer	
Works Services	2 Yards
Sub Station	
2 R.U. Pyro Store– 2 Compartments	
Speech Broadcasting Building	

AMWD Stores
Defence Unit – Officer's Quarters
Defence Unit – Officer's & Servant's Quarters
Defence Unit – Officer's & Servant's Latrine
5 Airmen's Barracks
Airmen's Dining Room
Airmen's Recreation Block
Airmen's Latrine Block
Sergeant's Quarters
Sergeant's Latrine
Bomb Stores
Fused and spare bomb area
Component Stores
2 Fusing Points
Flight Offices & Crew Rooms
Boiler
Pilot's Rest Room, Lockers & Drying Rooms
Flight Servicing Hut
2 Technical Latrines
AMWD Offices
Radio Stand-By Set
9 Air Raid Shelters
31 Hardstanding
3 Fighter Hardstanding
Workshops
Golf Club House (Requisitioned)
M.T. Offices & Store
Shoemaker's Shop
Cycle Workshop & Store
Fire Picket Qtrs. & Office
Post Office

Dispersed Sites

Site No. 1 -
Officer's Latrines for 1, 2 Huts
Officer's Latrine for 3 Huts
Officer's Quarters For 4 (Nissen) Huts.
Includes Servant Quarters
Sergeants Latrine for 2, 3 Huts
Sergeants Quarters (Nissen)
Airmen's Latrine for 2, 3 Huts
Airmen's Barrack Huts (Nissen)
Picket Post
2 Air Raid Shelters

Site No.2 -
Officer's Latrine for 4 Huts
Officer's Quarters (Nissen) Including

Servant's Quarters
Sergeant's Latrine for 4 Huts
Sergeant's Quarters (Nissen)
Airmen's Latrine for 5 Huts
Airmen's Barrack Huts (Nissen)
Picket Post
2 Air Raid Shelters
Sergeant's and Airmen's Ablutions
Site No. 3 –
Officer's Latrine for 4 Huts
Officer's Quarters (Nissen) Including
Servant's Quarters
Sergeant's Latrine for 4 Huts
Sergeant's Quarters (Nissen)
Airmen's Latrine for 3, 4 Huts
Airmen's Latrine for 8 Huts
Airmen's Barrack Huts (Nissen)
Picket Post
2 Air Raid Shelters
Site No. 4 –
Officers Latrine for 3 Huts
Officers Latrine for 4 Huts
Officer's Quarters (Nissen) Including
Servant's Quarters
Sergeant's Latrine for 3 Huts
Sergeant's Quarters (Nissen)
Airmen's Latrine for 10, 5 Huts
Airmen's Barrack Huts (Nissen)
Picket Post
2 Air Raid Shelters
Site No. 5 -
Officer's Latrine for 4 Huts
Officer's Quarters (Nissen) Including
Servant's Quarters
Sergeant's Latrine for 2, 3 Huts
Sergeant's Quarters (Nissen)
Airmen's Latrine for 5 Huts
Airmen's Barrack Huts (Nissen)
Picket Post
2 Air Raid Shelters
Site No. 6 -
Officer's Latrine for 5 Huts
Officer's Latrine for 4 Huts
Officer's Quarters (Nissen) Including
Servant's Quarters
Sergeant's Latrine for 4 Huts
Sergeant's Quarters (Nissen)

Airmen's Latrine for 5 Huts
Airmen's Barrack Huts (Nissen)
Picket Post
2 Air Raid Shelters
Fuel Compound
Site No. 7 -
Officer's Latrine for 4 Huts
Officer's Quarters (Nissen) Including
Servant's Quarters
Sergeant's Latrine for 4 Huts
Sergeant's Quarters (Nissen)
Airmen's Latrine for 5 Huts
Airmen's Barrack Huts (Nissen)
Picket Post
2 Air Raid Shelters
Site No. 8 -
Officers Latrine for 1, 2 Huts
Officers Latrine for 3 Huts
Officers Qtrs. (Nissen) Including Servants
Qtrs.
Sergeants Latrine for 4 Huts
Sergeants Qtrs. (Nissen)
Airmen's Latrine for 5 Huts
Airmen's Barrack Huts (Nissen)
Picket Post
2 Air Raid Shelters
W.T. Transmitting Station
H.F.D.F. Station
Instructional Site -
Instructional Block
Instructional Block Latrines RAF
Instructional Block Latrines WAAF
Torpedo Workshops
Torpedo Attack Trainer
2 Lecture Rooms
Fire Pool
Technical Latrines
Link Trainer Buildings - for 2 Trainers
Turret Instructional Building – 2
Compartments

*Post war - Turret Instructional Building and
Torpedo Attack Trainer.*

Loran Training (long range navigation)
Ground Equipment (ex-Bombing Teacher)
Tracer Trainer Hut
Communal Site -
Officer's Mess for 151
Officer's Bath House and Latrine
Squash Court
Sergeant's Mess for 286
Sergeant's Showers and Ablutions
Dining Room with Cinema Provision
Ration Store
Institute – Female NAAFI Staff
Grocery & Local Produce Store (Male NAAFI and Staff Quarters)
Gymnasium & Chancel
Airmen's Ablutions, Showers and Changing Block
Airmen's Latrine Block
Decontamination Centre Type 'M'
Medical Inspection Store & Barbers Shop
Stand by Set House

Fuel Compound
CO's Quarters
Destructor House
YMCA Building
Air Raid Shelter
Picket Post
Educational Hut (Re-sited from Aerodrome Site)
Post Office
Tailor's Shop
Store
Stores Huts
Latrine Bucket Emptying Platform
Sick Quarters Site -
Sick Quarters and Annexe
Ambulance Garage and Mortuary
Sergeants and Airmen's Ablutions and Latrines
Sergeants and Airmen's Barrack Hut (Nissen)
Picket Post
WAAF Communal Site -

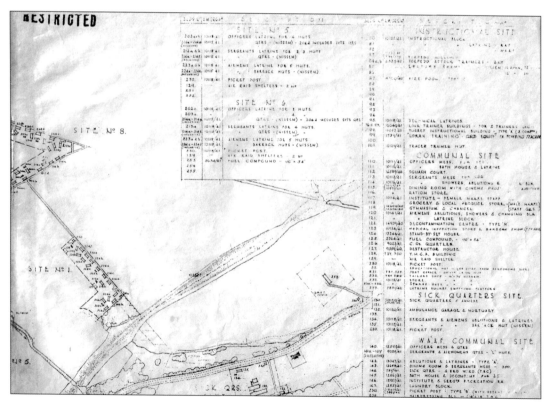

Sick Quarters and Communal Sites

Officer's Mess and Quarters
Sergeant's and Airwomen's Quarters 'L' Huts
Ablutions and Latrines 'Type A'
Dining Room and Sergeant's Mess
Sick Quarters – 4 Bed Ward
Bath House and Decontamination for 35
Institute and Sergeant's Recreation Room
Laundry Block
Picket Post - Type 'B' With Detention Room
Hairdressing Block
WAAF Dispersed Site -
Sergeant's and Airwomen's Quarters 'L' Huts
Ablution Block 'Type B'
Latrines and Drying Room
Picket Post
Site No. 9 -
Airmen's Quarters with Night Latrines and
Sergeant's Quarters 'L' Huts
Ablution and Latrine Block
Latrines and Drying Room
Picket Post
'Blawearie' (adjacent to Site No. 9)
Requisitioned Property

The construction did not run smoothly: correspondence winged its way to the Air Ministry and the Secretary of State for Scotland, from disgruntled locals: -

Sir,
I venture to write to you on a matter which for some time has been arousing much adverse comment in this neighbourhood. I refer to the wastage of public money on the aerodrome now in process of construction at Turnberry.

(1) As Minister of the Parish of Kirkoswald, of which the Turnberry district forms a part, I have occasion from time to time to pass the ground on which the aerodrome is being made, and from the out-set have been struck by the number of men who seem to have little or nothing to do. At any time, groups of men may be seen standing about idly, or watching some operation performed by others – in marked contrast to the steady

Aerial view of Turnberry, 1942

work of the farm servants in the adjoining fields at a much less wage. The matter has become almost a by-word, and, apart from the waste involved, is bound to have a prejudicial effect on the general morale.

(2) It is a matter of general observation and comment that there seems to be little or no effort to check waste of material. As illustrations I may mention:

(a) The case of petrol. Cars have been seen standing with no driver near them, their engines meantime running; I have first-hand information that, in the case of large tractors used for lifting turf etc., their engines are allowed to run on throughout the dinner hour period. Besides this it is believed that, in the use of both private cars and lorries, there is room for a greatly increased economy.

(b) The matter of cement. I am told on first-hand authority – amply confirmed – that a great wastage of this material is taking place. At Maidens Station where it is loaded from trucks on to lorries, the ground is covered with it – due partly to the bursting of bags, the contents of which there is no effort made to retrieve.

(3) There is the matter of wages. As to the general position I am not sufficiently informed to write definitely; but instances have been brought to my notice which would serve to show that the case of certain workers - both office staff and outside employees - payments are being made on an extravagant scale. The general opinion locally is that while the service rendered is poor quality the rate of wages is unduly high.

I regret, Sir, to trouble you with this letter and thus add to the many burdens I know you must have to carry; but it seems to me that there is here a matter which would repay enquiry. In these days when, from the highest quarters, the need for economy - even in the small things – is being urged upon all, the impression left by the way in which this work for the Government is being carried through is not a happy one.

The Air Ministry were quick to respond:

I have been asked by the Secretary of State for Air to tell you the result of a special inspection which has lately been carried out at the Turnberry Aerodrome site as a result of your letter stating that you thought that the work there was not being done in a satisfactory manner.

Your first point was that the methods employed resulted in a waste of materials and manpower. The work at present being done compromises the construction of concrete runways. This entails excavating which is being done with skimmers and graders, the removal of spoil by dumpers and the preparation and pouring of concrete for which mixers are employed. These are the methods normally employed on work of this type and it is not thought that there is any other method by which the work could be finished more quickly. An examination of the record sheets showing the output of the excavating plant and the concrete mixers showed that the figures compared most favourably with similar figures for other sites now under construction.

The only material which was found to be subject to some waste was drainage pipes. The process of loading, transporting and unloading this type of material inevitably results in a certain amount of breakage, but the quantity actually delivered undamaged on the site is recorded on the checkers' tally sheets and action taken to make sure that only the amount of undamaged material is charged against the cost of the work.

You stated that the men were being paid very high wages, but in fact the rates of pay to men working at Turnberry are in accordance with the appropriate uniformity agreement. On the occasion when the site was investigated there was no sign of slackness on the part of the men working there, but as the urgency of completing this work makes it necessary to employ all available labour, whether good or bad, it is possible that a portion of the men do not always show marked enthusiasm for this type of work. You will realise, however, that owing to the present shortage of manpower it is not possible to avoid making use of the labour available in the market at the present time.

You particularly mentioned that there seemed to be a large amount of cement wasted lying about in the yards at Maidens Station and this is due to the fact that there have been numerous cases of cement bags being burst in railway waggons and that it has been necessary to fill the loose cement into fresh bags which inevitably results in a certain amount being spilt. There is also a stock in the Station yards of approximately 70 bags of cement which had hardened due to exposure to weather during the journey from the manufacturers. The Cement Supply Company which provides the cement has a check-up at the Station and I understand that payment for the bags of damaged cement is being claimed from the Railway Company. The Air Ministry, however, is not directly concerned in this because payment is only made for bags of cement actually delivered to the site and a record of these is kept on the tally sheets.

The statement that lorries are left standing about with their engines running is probably attributable to the fact that the excavators and graders used for this work are driven by diesel engines and, owing to the difficulty of starting this type of engine, it saves time – and therefore labour – to keep them 'idling' when the machines are not actually at work. When the inspection was carried out no petrol driven engines were seen 'idling' except where there was full justification for

keeping them running.

The points dealt with above are those specifically mentioned in your letter to the Secretary of State, but I might add that the inspection of this site did not show that there was any waste of labour, material or time in the performance of the important and somewhat specialised work which is now being done.

Acres of concrete, tarmacadam, brick and concrete dispersal points, huts, hangars and transformer houses and technical buildings soon scarred the landscape. There were 55 dispersal points, 24 Blister type hangars, two full length and two half-length T2 type hangars were also erected along with control tower, bomb stores and admin offices.

The factor, like his predecessor Tom Smith during WWI, had to keep an eagle eye on developments, as land was being seized piecemeal. He notes that the contractors for the Air Ministry have entered into possession of part of the lands of Morriston farm without any Requisition Notice being served on the ground in question. This was very valuable agricultural ground and he was displeased that it was to be used for the erection of huts, etc., and that parts of Jameston lands were to be used for the same purpose: 'These areas in question are most valuable from the agricultural point of view, being flat, easily worked and in perfect heart, and it is the opinion of the tenants and myself that the huts could be built on less valuable ground such as the remaining part of the Golf Course or the ground at Balkenna a little south of Turnberry Hotel and equidistant from the Aerodrome to the proposed position of the huts.'

On Morriston farm, the contractors also drove a lorry through the locked gate, leading to the field which they were using as a dump, smashing it to atoms. which made it impossible for the farmer to have the use of his field.

The Air Ministry in reply to his complaint agreed that Messrs. Wimpey had no authority from the Air Ministry to use part of the field as a dump, or smash the gate, and that any claim

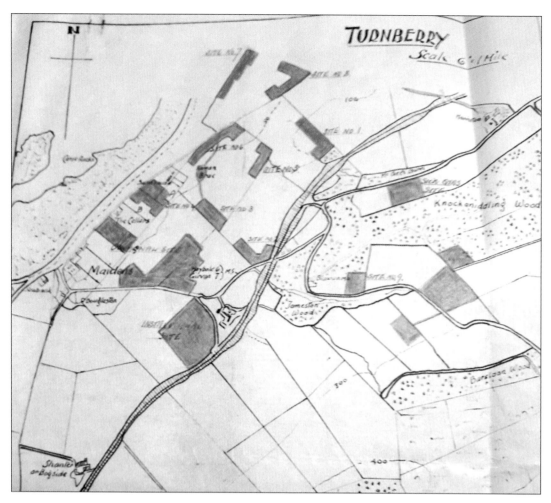

Land Requisition for Communal and Instructional Sites.

should be made against Messrs. Wimpey and not against the Air Ministry.

In July 1941, desperate to keep the golf courses open, the Railway Company ascertained that fourteen holes on the old course and two holes on the new course would still be available for play, the ground on which the golf courses were laid out was held by the company under a Deed of Servitude granted by the Marquess of Ailsa in favour of the Glasgow & South Western Railway Company; and stated that the land could only be used as a golf course and that in the event of the playing of golf thereon being discontinued the land shall revert to the Marquess. In light of this the desirability of keeping the golf course open was stressed and it was suggested that endeavour might be made to come to an

arrangement with the Turnberry Golf Club under which they could manage the golf course for the company during the period of the War.

The Hotels Controller and Estates Manager of the London Midland Railway Company agreed with the Lands Directorate, Ministry of Works and Buildings, a settlement based on the net annual profits derived from hotel working and receipts from the Turnberry Golf Courses. Under the provisions of the Compensation (Defence) Act 1939, an annual nett rental of £6,500 was payable by the Ministry.

In September the Air Ministry served a further Requisition Notice on the whole of the golf courses lying to the north of the path leading past the Club House. The land westwards to the beach, and between the Maidens-Girvan public

Aerial view of Turnberry Hotel

highway were to be taken into the aerodrome. As a consequence, only two or at most three holes were left. It was agreed that there was no alternative but to accept the position with all its consequences, and the playing of golf on the courses at Turnberry must be considered as at an end. The clubhouse was then closed, and the furniture, fixtures and fittings were sold off by public auction.

Maidens and Turnberry now found itself enveloped by the aerodrome to the south and accommodation sites and technical buildings to the north and east. Several local properties were requisitioned, Blawearie House was pressed into service as WAAF officers' accommodation, the golf clubhouse was appropriated for outside surveyors, construction and RAF personnel,

the airfield had its own sick bay and morgue, to the north by Morriston farm, separate from the main sites. The public road between Maidens and Turnberry was closed, and to travel between either village required a lengthy detour via Kirkoswald. As the airfield evolved, a contractor's hut on the airfield was taken over as torpedo photographic training building. Shanter farm became ground crew quarters as well as accommodation for the homing pigeons. RAF Turnberry was developed at a rapid pace as further land and buildings were requisitioned - in a very short time the area was unrecognisable, and was littered with the paraphernalia of war. The Technical Site was situated on the landward side of the main runway, and the bomb store was sited to the south, on Tauchet Hill overlooking the airfield. Although this land had been officially requisitioned, the factor took issue with the contractors cutting the timber thereon - he had been informed that the wood was to be left as it was for camouflage purposes. As the wood was now being felled, he felt it necessary to have a valuation made of the timber before much more of it was cut, and a claim submitted to the Air Ministry. Stone for the contractors was taken from Knoxhill Quarry, for which a royalty was paid to the estates.

Because of its excellent location and facilities, Turnberry was subsequently referred to as the 'best torpedo training station in the RAF.' Once again, the people of Maidens and Turnberry welcomed into their homes the young men of the Commonwealth and treated them as long-lost sons.

Substantial delays in the construction of Turnberry meant it didn't become fully functional until May 1942. The official 'daily diary', the Operations Record Book, which was kept on every RAF Station or Squadron/Unit chronicled events that took place. Depending on who was writing, some entries are full of description while others merely list the total number of flying hours and the weather. From these we can follow the happenings at Turnberry throughout the war.

FEBRUARY 1942
On 2 February 1942, RAF Station Turnberry was opened for the receipt of stores and equipment under the command of Squadron Leader J. H. Baird, with a small advance party of 1 Sergeant and 16 airmen. At this time the airfield was still under construction and the accommodation sites were not ready. There are no entries in the operations record book for a few days, until on 7 February a Spitfire from 58 OTU, Grangemouth, landed owing to a shortage of fuel. Turnberry was instructed by the CFI Grangemouth to send the pilot, F/O S. J. Godlewski, back to his unit by train. On the 9th, a draft of 69 NCOs and airmen arrived, and another stray aircraft, Swordfish W5903 of the Royal Navy, landed, the pilot having lost his way. He was only on the ground for 25 minutes, and having been given directions

he swiftly took off again. F/O Krul arrived by air on 13 February from Grangemouth to collect the Spitfire left by F/O Godlewski. The next day G/C Pearce DSO DFC and S/Ldr Dodd DSO MVO (both 17 Group) arrived by air to inspect progress of RAF Station Turnberry, followed a couple of days later by W/Cdr Lewin (17 Group) and F/Lt Morrison (Abbotsinch) to check on the installation of telephone lines, signals etc.

The first accident to occur was on 23 February at 09:45 when Blackburn Botha L6211 of 3RS Prestwick crashed on the foreshore and was destroyed. All of the crew were injured, the pilot, Sgt Shepherd, and Sgt Brown, died during the night. Sgts Mayho and Balmain were not seriously injured.

On the 25th, W/Cdr Kidd AFC, S/Ldr Clarke DFM, S/Ldr Howey and S/Ldr Dewey, from 5(C) OTU, arrived in Anson N9726, to inspect the progress of the site. W/Cdr Lewin (17 Group) arrived by road and S/Ldr James of the RAF Area Catering Division visited the station in connection with catering services.

MARCH 1942
The month of March 1942 found the station no nearer completion, although Turnberry continued to attract some stray aircraft. On 2 March, Hudson AM868, pilot P/O Ruddick landed looking for RAF Silloth, followed on 7 March by Hurricane V7338 with pilot Sgt Alexander on a delivery flight. Sgt Alexander had lost his bearings; he took off for Turnhouse fifteen minutes later. F/Lt. Dunn arrived from Prestwick on 15 March, in Oxford V4147, to investigate the Botha crash.

The rest of the month is busy with inspections by various officers of HQ No. 17 Group and Chivenor in order to prepare a progress report for the AOC, No. 17 Group.

APRIL 1942
On 1 April, S/Ldr Ross, Medical Officer, records in Station Sick Quarters O.R.B.:

That installed at RAF Turnberry is an opening up party consisting of 8 Officers and 126 other ranks RAF, a Local Defence Unit and 'A' Company of the 30th Bn. Royal Scots Fusiliers, with 4 Officers and 205 other Ranks. The RAF officers were living out or billeted in the villages of Turnberry and Maidens, the Army Officers Mess was in the Golf Clubhouse, the RAF other ranks accommodated in Nissen huts on No. 6 site and Army other ranks were lodged in Nissen huts on the Defence site. All lighting on the station was by oil lamps. On the Communal site the water carriage system was not in full operation, and the main sewer outfall went into Maidenhead Bay. Chemical toilets were being used, on No. 6 and Defence sites, the soil to be removed by the contractors. Ashes from the stoves were to be used for making footpaths.

The factor of Cassillis and Culzean Estates, James T. Gray, was faced with new challenges in the daily running of the estate caused by the occupation of the Air Force. In the first of many letters he was forced to write to the Air Ministry, Directorate of Works, he questions the actions of the contractors:

2 April 1942
Dear Sir,
Turnberry Aerodrome
I have today received a telephone message from Mr Marshall the tenant of Little Turnberry farm stating that workmen have taken possession of his garden and are digging a drain and laying a large pipe through the centre of the garden.

As no Requisition Notice has been served on this part, and as it would have been quite possible with a little adjustment to have laid the pipe clear of the garden, I shall be glad to know if you could let me have an explanation of this procedure and also if it is the intention of the Air Ministry to requisition the garden of the farmhouse in addition to the ground already requisitioned.

The Sergeants' Mess was on the Communal site used by all other ranks, rations were to be obtained thrice weekly from RASC Gailes and transported by road. The facility for ablutions on the Communal site was for a weekly bath at Turnberry Hospital (Hotel), by arrangement with the Medical Superintendent. For recreation there was a NAAFI canteen in operation on the Communal site and one football pitch at the north end of the aerodrome.

On No. 6 site there was one Nissen hut with a medical office, casualty room, dressing room, ward with two beds, and an office. All cases except those of very minor illness were transferred to Turnberry (EMS) Hospital. Water was obtained from a temporary (contractor's) supply on site and boiled before use. Food for the patients was carried from the Communal site. Station Sick Quarters had one Morris light ambulance and was staffed by Officer in Medical Charge, S/Ldr J. B. Ross, one Cpl Nursing Orderly, one AC2 N/Ord and one Cpl N/Ord on attachment from RAF Station Abbotsinch. The main water supply for the station came from Ayr County Council supply - and the first sample (tested at RAF Institute of Pathology and Tropical Medicine, Halton) was found to be satisfactory, free of biological and chemical contamination. W/Cdr Clifford, Air Ministry, arrived to inspect progress, and the next day a party consisting of S/Ldr Baird (Station Commander), S/Ldr Ross (Station Medical Officer), and F/Lt Sissmore (Station Equipment Officer) left to attend a conference at HQ No. 17 Group regarding Station requirements.

On 6 April Anson N6083 from West Freugh, pilot Sgt Handley force landed owing to fabric stripping from the port main plane, the aircraft was otherwise undamaged and once repaired left the next day. Two Trans-Atlantic Hudson's landed at Turnberry from Newfoundland; they had been bound for Prestwick. The first to land was Hudson FH278, pilot P/O Agate, navigator P/O Stephenson, W. Op Sgt McDonald, followed a few hours later by Hudson FH 274, flown by pilot F/Lt Cordell, navigator P/O St Clare-Millar and W.Op Sgt Metcalfe. Turnberry was instructed by the controller at Prestwick to keep the two aircraft grounded until the next day when both took off to complete their journey. S/Ldr Ross, Medical Officer noted that the crews were fatigued, but not unduly so.

Turnberry, still under construction, continued to be busy with inspections of the site and technical facilities almost every day. Air Commodore H. G. Smart CBE, DFC, AFC, AOC No. 17 Group, and W/Cdr Braithwaite, OBE of Abbotsinch arrived by road to check on progress and S/Ldr Devey with F/Sgt Coulson called in en route from Turnhouse to Chivenor, landing in Anson N5355, with pilot F/Lt Weir, to enable S/Ldr Devey to discuss technical details with the resident engineer. Tiger Moth T7914, with pilot F/O James and S/Ldr Dodd DSO, MVO, on board arrived from Turnhouse to inspect progress and further ascertain station requirements. The runway at Turnberry, as well as being used as an emergency landing ground, was now in constant use with flights arriving carrying RAF personnel daily. Anson N9853, piloted by P/O Hankins from West Freugh, arrived with a patient for Turnberry Hospital, the casualty was suffering from burns and a head injury, the duration of the flight was about twenty-five minutes and the patient was accompanied by an MO and Nursing Orderly. On arrival the casualty was lifted from the aircraft on a mattress and conveyed to the hospital by ambulance. On 18 April, Anson R3308 pilot S/Ldr Bland arrived with G/C Clifford who was posted to take over command of RAF Turnberry. D.H. Rapide X7372, flown by F/Lt Hankins flew in from Abbotsinch with a passenger, Mr Anger, sent to discuss the radio problems (Beam) with the station Signals Officer. On April 20, 1942, G/C G. R. M. Clifford took command of RAF Station Turnberry from S/Ldr Baird. The following day S/Ldr Harvey, P/O Walker and 73 other ranks as an advance party of 5(C)OTU arrived by train from Chivenor.

The first Hampden to land at Turnberry was AT234, pilot S/Ldr Gaine with W/Cdr Braithwaite OBE as passenger, the aircraft later took off for Abbotsinch. The first Beaufort was seen the next day when L9810, pilot F/Lt

Sharman, landed to collect W/Cdr Braithwaite OBE.

One of the more unusual aircraft to visit Turnberry was on 26 April, when Lockheed Vaught FN668 (known as the Kingfisher) flown by civilian test pilot Mr Wilkes landed, leaving again very soon after. Miles Falcon HM496, pilot G/C Kidd AFC, AFM with W/Cdr Lewin as passenger, landed on 29 April and Anson N5235 came in from Chivenor with W/Cdr Gilson AFC and P/O Penton, W/Cdr Gilson returning in N5235 to Chivenor the following day.

The last few days of April saw the arrival en masse of the aircraft and personnel of 5(C)OTU.

MAY 1942 5(C)OTU
The Officers' Mess was officially opened on 1 May and all officers not authorised to live out were accommodated in quarters. Training Beaufort L9838 crashed on arrival from Chivenor, the crew were uninjured, although the aircraft was a complete write-off. The station went about the business of organisation as aircraft from Chivenor continued to arrive over the next few days. On 5 May at 06.20 hrs a special train from Chivenor with S/Ldr Clarke DFM as Officer Commanding pulled into Turnberry Railway Station. There, five officers, five hundred other ranks, nineteen WAAF and eight civilians employed by the Bristol Aircraft Company disembarked. The number of RAF personnel was excessive for the number of Nissen huts available on the Dormitory sites, the surplus had to be accommodated on the Communal site. The WAAF personnel were temporarily lodged on the Anti-Aircraft Flight site, pending completion of the WAAF quarters. The increase of the number of personnel at Turnberry prompted the M.O. to occupy the unfinished Station Sick Quarters buildings, although there was no heating or lighting and numerous other constructional details outstanding. The Nissen hut on No. 6 site which had been previously used as S.S.Q. was to be kept on as a Medical Inspection Room, until construction of such a building was completed on the communal site. S/Ldr Ross M.O. transferred a patient with cerebral-spinal meningitis to

Heathfield Infectious Diseases Hospital, Ayr. This must have been a very worrying time for the M.O., with increased health risks associated with the sudden growth of the population of the camp, coupled with the inadequate water supply and sanitary conditions. Sanitary Squads were still digging pits for the Dormitory and Aerodrome sites. By mid-month, further medical orderlies were posted to Turnberry and Station Sick Quarters received an Albion Heavy Ambulance.

Within a week, the first training accident of 5(C)OTU at Turnberry took place. Beaufort AW357 crashed into the sea off Girvan. The pilot, Sgt Whitehouse and crewman Sgt Miles were both trapped in the machine which sank immediately (Miles was a student gardener at Kew from 1932-34 and returned to work in the Tropical Department in 1938, joining the RAF

in September 1940. He was listed as missing for four months until identified and buried. A wreath from the Kew Guild was among the floral tributes.) Sgt Marriot who suffered a broken leg and Sgt Hudson who was badly bruised were saved by the crew of a local fishing boat.

On May 14 an unofficial visit to Turnberry was made by Air Marshal A. G. R. Garrod CB, OBE, MC, DFC; Air Marshal P. Babbington CB, MC, AFC; Air Vice Marshal L. Hollinghurst CB, OBE, DFC; Sir John Abraham KBE, CB, (Deputy Undersecretary of State) and Mr Hamilton Kerr MP. The party inspected various sites and buildings and had lunch in the mess.

Other aircraft visiting Turnberry were an Airspeed Envoy flown by G/C Kidd AFC, AFM, also later that week Gloster Gladiator K7898, again G/C Kidd at the controls.

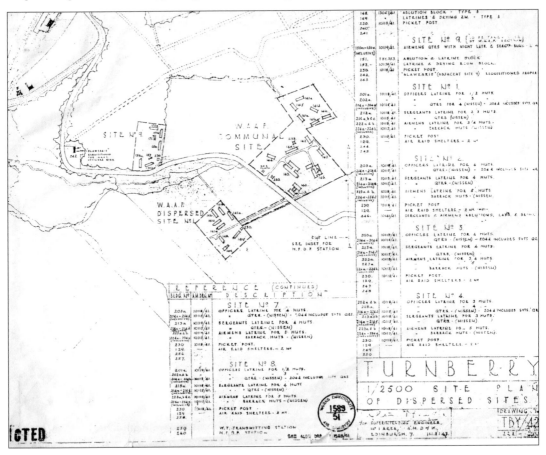

WAAF Site on Blawearie Road between Maidens and Kirkoswald

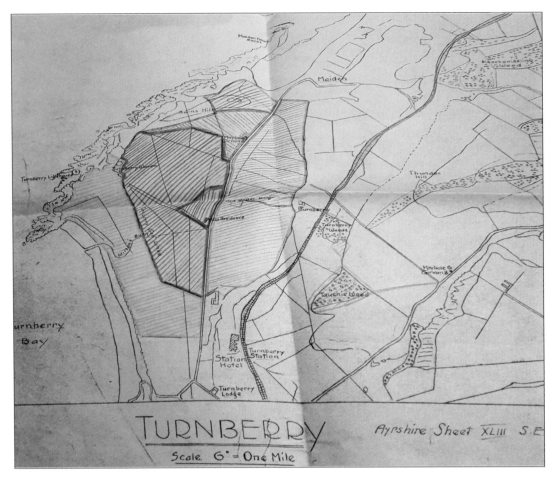

Land Requisition - Airfield Site.

Another fatal accident took place on 27 May when Beaufort N1023 crashed into the sea off Culzean Point. There were no passengers on board, pilot Sgt O'Donnell and aircraft were both lost.

JUNE 1942 5(C)OTU

On 2 June S/Ldr Eagleton and F/Lt Rome (HQ 17 Group) arrived by air in a Miles Mentor for consultations regarding training, returning to Turnhouse the same day. (HQ 17 Group was in the Mackenzie Hotel, Melville Street, Edinburgh)

Two more Hudsons landed after the long flight from Newfoundland, FH454, pilot P/O De La Paulle and FH461 pilot P/O Payne, after a short rest both aircraft took off for Prestwick.

Beaufort 'K' AW306 took off on June 6, with pilot Sgt Morris at the controls, the aircraft was later reported missing, and three days later, Beaufort L4493 crashed on the edge of the aerodrome, the pilot Sgt Montagnon was badly injured, he was given first aid at the scene and conveyed to Turnberry Hospital, there were no crew aboard.

The daily routine of an RAF Training Station began to take shape. Medical Lectures were given to each course, as well as inoculations to departing courses. Instruction in First Aid was given to the WAAF and air crash rescue methods were demonstrated regarding the entry and exit of damaged aircraft and the removal of casualties.

S.S.Q. dealt with a diagnosis of diphtheria, the patient being transferred to the Isolation Hospital at Heathfield, while all those he may have been in contact with were tested. On June 12, Liberator FK230 delivered a stretcher case to Turnberry, a patient with a cerebral hernia, from Iceland. The patient was accommodated in S.S.Q. overnight, leaving the next morning in D.H. aircraft 'Women of Britain' bound for the Head Injuries Hospital, Oxford. Liberator FK230 meanwhile had departed for Prestwick. Blenheim '2', pilot Sgt Batten attempted to take off on the wrong runway, the aircraft swerved and sustained a burst tyre, crashed and caught fire. The pilot escaped uninjured. The 15 June found a part of Beaufort N1023 washed up on the shore, the wreckage was examined by the M.O. but no body was found.

More land was requisitioned, this time on the hillside between Maidens and Kirkoswald, again the Factor writes to the Air Ministry, Directorate of Works on 18 June:

Dear Sir,
I now return the plan sent me with your letter of 16th instant in connection with the site on the lands of Kirklands farm.

The occupiers of the land are Messrs Gray and Martin, Kirklands farm, Kirkoswald, but Cassillis and Culzean Estates are the occupiers of the small plantation No. 1048. The UF Church Manse marked on the plan is feud from the Estates to Miss Susie Lee Rutherford, The Manse, Kirkoswald.

I trust this information meets your requirements and I return the plan herewith as requested.

On June 18, Beaufort L4515, pilot Sgt Fidgin, flying solo, crashed into the sea one mile north of Ailsa Craig. No trace of pilot or aircraft could be found. The next day a Botha 'alighted' in the sea, two survivors were admitted to Turnberry Hospital and a third was unhurt. There were still problems with the accommodation sites and the Ayr County Council Rats Officer was called to the WAAF communal site to deal with the vermin noticed there. A/Cmdre Aitken CBE, AFC, MC,

landed in a Vega Gull with W/Cdr Scarff and after refuelling took off for Islay (Port Ellen).

Lt-Col Villiers visited the station commander on defence matters, and G/C Clifford took off from Turnberry in Moth Minor X5122 to visit HQCC. There was another arrival from Newfoundland, Ventura AE892, pilot Capt. McNaughton, who was then directed to Prestwick.

On 22 June, problems arose regarding access to the requisitioned land, requiring the Commanding Officer, Group Captain G. R. Clifford to write to the factor:

Dear Sir,
The gamekeeper in Turnberry Cottage appears to be under the misapprehension that the shooting rights in Tauchet Woods are reserved to the Estate. As this is now requisitioned Air Ministry property, it is requested that the game-keeper be informed that only individuals properly authorised by the Air Ministry are allowed in the woods within the wire fence, and that as he does not come within that category, he is liable to arrest if found trespassing therein.

The body of Sgt O'Donnell was washed ashore at Dunure on June 23, 26 days after his Beaufort crashed. The next afternoon Beaufort M2, pilot Sgt Smit was seen to crash into the sea near Girvan, again no trace could be found.

In a reply to Group Captain Clifford regarding the trespass by the gamekeeper, James T. Gray, makes his position quite clear.

Dear Sir,
I have to acknowledge your letter of 22nd instant and note your remarks in connection with our gamekeeper in Turnberryhill Cottage.

I will instruct the gamekeeper that he is not to trespass on the Air Ministry property, but of course you appreciate that while the Air Ministry have requisitioned the ground, they have no right to the game and any persons found taking game from the coverts are liable to prosecution for poaching.

I have already had trouble with some people at Turnberry who were found shooting in the woods and I discussed this matter fully with Mr Fergusson, the Requisition Officer for the Air Ministry, when he explained to me the position and agreed that no one has any right to take game or rabbits.

I trust however, that there will be harmony between the personnel at Turnberry and our gamekeeper and that you will have no further cause for complaint.

Information was received on June 25 that the body of Sgt Morris (Beaufort AW306) had been washed up at Barra.

JULY 1942 5(C)OTU

The month of July began with adverse weather conditions, forcing the pilot of Hurricane MF36, P/O Monk, to make an emergency landing at Turnberry. He left three days later in a Miles Martin flown by F/O Southey, while F/Sgt Ahrnes flew out in the Hurricane. There was an outbreak of smallpox notified in the Glasgow area, but no cases had been reported nearer Turnberry than Ardrossan. S.S.Q. was short of staff, the M.O. reported to the SMO 17 Group, that the effective WAAF N/Ord. on strength was nil. An LACW, N/Ord. arrived on attachment from RAF Silloth the next day.

On 4 July, a search party set out from Turnberry after information was received regarding a wrecked aircraft on the hill of Shalloch-on-Minnoch, in the Carrick Forest. With about one hours walking still to go before reaching the crash site, the returning first search party, including the local policeman was met. The wreckage had been found, and the five occupants of the aircraft were all dead. At 23:45 hrs the search party from Turnberry turned back because of mist and darkness. Information was received that a corresponding aircraft had been missing from RAF Millom since 2 July. The next morning a further search party was sent out, returning later that day with the five bodies from the wreck of Anson N5297.

A request was made for an ambulance from RAF Turnberry to stand by at Girvan Harbour as a Swordfish had crashed into the sea off Ailsa Craig, ten minutes later S.S.Q. was notified that a Beaufort was likely to crash on landing, so Turnberry Hospital was asked to send an ambulance to Girvan, while the station ambulance waited for the Beaufort, which crashed as predicted. No one was hurt.

Training continued, as did the accidents. Hampden AE137 crashed and burned out near Kirkoswald, killing pilot Sgt McCormick. (This aircraft was wrongly numbered P2100 in ORB), and a burst tyre on landing caused Beaufort X to crash and catch fire - the crew were uninjured. Beaufort V crashed on take-off and was wrecked, the crew unhurt. Beaufort N1103 was reported missing, the crew F/Lt. Weir, F/O Bendrey, Sgt McFeeney and Sgt Evans. A 'phone message was received the next day from S.A.S.O. Northern Ireland that Beaufort N1103 had crashed and burned out and that all crew were safe. On 15 July, the crew of Beaufort X were involved in another accident when their aircraft overshot the runway. They escaped unhurt but shaken and were confined to ground duties pending further examination. Another Hampden, P2100, crashed into high ground near Maybole. The pilot, Sgt Rayner was killed, and the aircraft was a write off. Mustang AG504 from Ayr crashed at Meikle Letterpin, near Girvan. The body of P/O Clarke was recovered and taken to S.S.Q. by the M.O.

The last accident to occur in July was that of Beaufort W6480 which crashed into the sea four and half miles from Turnberry point, the pilot P/O Norton was killed and P/O Pike, P/O Dean and P/O Henry, were missing believed killed.

AUGUST 1942 5(C)OTU

The first entry for August records in the Station Sick Quarters diary that, at the ration store, seven hundred and forty-one pounds of mutton was found unfit for human consumption. At almost midnight on 3 August, Turnberry became host for ten Whitley aircraft and crews who arrived after a combined operations exercise. They left at intervals the next morning. During the night, a Hampden aircraft crashed on take-off, the occupants were all unhurt. Later that same night, Beaufort L4496 crashed in flames one and a half miles south of the aerodrome, killing pilot Sgt Highmoor and Observer Sgt Malcolm. Another Beaufort crashed the next day with no casualties.

During the next twelve days, no mention is made of the training taking place, but lots of aircraft movements are recorded. G/Cpt. Sheen delivered Hampden P1222 from Upper Hayford, and W/Cdr Willcox SMO No. 17 Group arrived to on inspection. G/Cpt. Kidd landed in Gladiator K7858, stayed for just over an hour before taking off for Turnhouse. Other arrivals were a Mohawk, a Miles Whitney Straight, an Anson and a Percival Vega Gull. The airfield accidents continued, on 15 August, a Beaufort crashed on landing, the crew all suffered minor injuries.

On 16 August, Beaufort 'U2' P1174 crashed into the sea one mile off Girvan, the pilot, Sgt Fardoe, who was flying solo, was killed. Three hours later Hampden H10 crashed on the Isle of Skye, injuring Sergeants Forbes and Harding, Sergeants Ross and Simrock escaped unharmed. More casualties were to follow, on 18 August, Sgt Burns crashed in a Hampden just south of the airfield. He was killed, and the other occupant concussed and admitted direct to Turnberry Hospital. Another Beaufort crashed on the aerodrome, this time with no injuries.

The next day Beaufort L4480 crashed due east of the aerodrome killing all crew, P/O Batterick, P/O Roye, Sgt Taylor and Sgt Vacheresse. Then, on 26 August, a Tomahawk out of Limavady crashed into the sea one mile off Culzean Castle, near to the north bombing area, the pilot F/O Oliver swam ashore uninjured and was allowed to proceed to RAF Maghabbery. That same day, the Marshal of the Royal Air Force, Lord Trenchard, GCB, GCVO, DSO, paid an unofficial visit to Turnberry and addressed the air crews and maintenance parties. Air Commodore H. G. Smart CBE, DFC, AFC, Air Officer Commanding No. 17 Group accompanied by G/C Kidd and staff officers paid an official visit of inspection on 27 August.

On 29 August, G/C Stone arrived in Anson N9904, taking off for Silloth later that day - he would return to assume command of RAF Turnberry on 31 August. In the meantime, another Beaufort was lost, crashing on Goat Fell, Isle of Arran. Sergeants Glay, Leyland and McLean were all killed. Two further accidents took place before the month was out - a Spitfire crashed on landing, the pilot was unhurt and fit to return to RAF Fraserburgh. An Anson force landed on the seashore, with no casualties and the aircraft was undamaged, although a note was made that the Morris ambulance sustained damage to a shock absorber whilst travelling over rough ground on the way to the scene of the landing.

SEPTEMBER 1942 5(C)OTU
G/Captain R. A. B. Stone assumed command and the first day of September records the horrific mid-air collision of a Beaufort and Hampden when on landing approach, from the north, on the NE-SW runway. Both aircraft crashed and burst in to flames. The crew of the Hampden, W/O Jones and Sgt Mewson were killed instantly. From the Beaufort, F/Sgt Goodson escaped with severe burns, W/O Drury was rescued by P. C. Gray (Ayrshire Constabulary) almost unrecognizable, and the body of L.A.C. Young was recovered when the fire subsided. W/O Drury died in hospital during the afternoon.

What must have been a harrowing time for the medical orderlies continued the next day when the remains of the three occupants of the Beaufort crash on Arran were received from the Royal Navy authorities. Another Hampden landing accident occurred, without injury, on the airfield.

The 3 September was the occasion of the anniversary of the declaration of war. There was a combined station parade and church service, which assembled at Station Headquarters, marched past the Station Commander and proceeded to divine service.

The following morning Liberator B24-123712 arrived from the USA, the pilot F/Lt. Roach, with a crew of nine, later took off for Prestwick. In the afternoon the funeral of Sgt Glay, Sgt McLean, Sgt Leyland, Sgt Mewson and W/O Jones took place at Dunure Cemetery with full honours. The 7 September added another eight names to the growing list of casualties. The Operation Record Book records that: 'Information was received that two aircraft from Abbotsinch operating from this station on night exercises crashed about the time indicated (22:00). Beaufort 'S' crashed near Machrihanish (Mull of Kintyre), all crew killed.

Pilot, Sgt Haydon, P/O Booker, Sgt Griffin and Sgt Grasswick. Beaufort 'Y' crashed on the island of Mull; all crew killed. Pilot Sgt Lutes, Sgt Hammond, Sgt Hargreaves and Sgt Francis. The Station Sick Quarters reports the receipt of a mutilated body with RAF slacks from the Girvan Police. The Medical Officer states that currently the strength of nursing orderlies is 50% of establishment.'

On 11 September Sgt Barlow was listed as missing, believed killed, after his Beaufort crashed into the sea off Maidens, the three other members of his crew were rescued by local fishing boats. The only other mishap recorded for September is that of an Oxford which swung

Aerial view of Turnberry Lighthouse, 1942

on take-off and damaged the undercarriage with no injury to the aircrew.

The Operations Record Book continues to log the arrival and departures of visiting officials. S/Ldr Beardslaw visited from HQ Coastal Command to discuss educational and welfare problems. W/Cdr McNeill and S/Ldr Hawley, HQ No. 17 Group, along with S/Ldr Barret 3S.G.R. arrived to set up a Court of Inquiry regarding the loss of Beauforts 'S' and 'Y'.

Sir Archibald Sinclair KJ, CMG, MP, Secretary of State for Air, landed on 15 September, in D.H. Flamingo R2766 'Lady of Glamis', en route to Hendon. This aircraft was utilised for the King's Flight, specifically for transportation of the Royal Family, and was also used to transport Air Ministry personnel of the Air Council.

The month of September ends with the rather curious entry on the 30th: 'Sir Ralph Glynn McAmmon, Viscountess Davidson, Mr Leach, Sir A. Pownell, Sir John Shute, Miss Ward, Sir John Wardlaw-Molne, Mr F. Wood (Liaison Officer, Air Ministry) and Mr St.G. Drennon (Clerk of Sub-Committee) visited this station to discuss expenditure problems with the Station Commander.'

OCTOBER 1942 5(C)OTU
SECRET
DESCRIPTION OF TURNBERRY AERODROME
(As at 1 October 1942)
Grid Ref.: (1-in map) G.S.G.S 3908. S. 695299. Sheet 82 *Ailsa Craig and Girvan*
Position: Lat. 55° 19' 45" N., Long. 04° 49' 45" W.
Locality: 7 Miles N. of Girvan.
County: Ayrshire
Nearest Railway Station: *Height:* A.S.L. 35 ft. Turnberry (LMS)
Magnetic Varn.: 13° 30' W.
Command: Coastal   *Function:* Parent OTU
Parent: N/A   *Satellites or RLG.:* Nil
Other Functions (if any): Also used by F.C. as Satellite for Ayr.
Permanent Landmarks in Vicinity of Aerodrome:

(i) *By Day:* Ailsa Craig, 1115 ft., 246° T., 12 miles. Lighthouse on W. side of Aerodrome.
(ii) *By Night:* Lighthouse on W. side of Aerodrome.
Landing Area:
(i) *Grass Surface: Dimensions:*
N. – S. 1,133 yds Fit for emergency use only.
NE – SW 1,900 yds Fit for emergency use only.
E – W 1,000 yds Fit for emergency use only.
SE – NW 1,400 yds Fit for emergency use only.
Surface Conditions after Precipitation: Ground becomes very soft, but drainage good, recovery quick.

(ii) *Runways:*
Direction: 029° T. 1,950 x 50 yds.
168° T. 1,450 x 50 yds.
082° T. 1,230 x 50 yds.
Perimeter Track: 50 ft.
Type of Surface: Tarmac texture coat.
Prevailing Wind: SW.
Obstructions: Hills up to 900 ft. within 5- mile radius. Higher ground in larger circuit.
Position of Wind Indicators: SW corner of Aerodrome (weathervane on roof of Watch Office).
Fog Prevalence: Occasional sea or industrial. Estimated maximum 15 days per year.
Type of Control: OTU. Type 2 Flying Control.
Facilities: *Petrol Storage:* 144,000 gallons aviation.
          Petrol Storage: 3,500 gallons MT.
          Means of Refuelling: Bowsers
          Oil: 5,000 gallons.
          Bomb Storage: 200 tons.
          S.A.A.: 300,000 rounds.
          Tractors: 6
          Jacks: All types.
          Cranes or Hoists: 1 Coles Crane.
Snow Ploughs and Rollers: 2 Johnson snow ploughs and 1 Gritter.
Night Landing Facilities: Goose neck flare path and Chance light.
Radio Facilities: 24 hours watch maintained both 2,410 kc/s. and 6,440 kc/s.
Accommodation: Scale: 2,173 personnel all ranks.

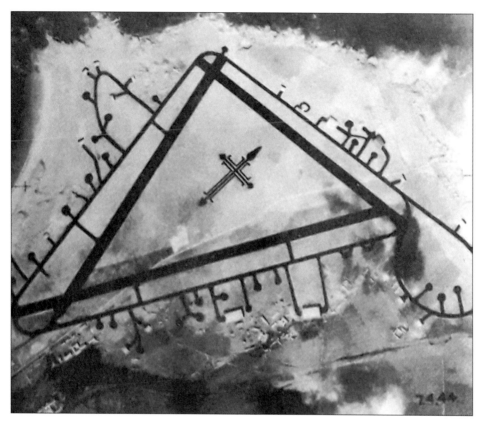

Turnberry Airfield

Essential Buildings: Completed.
All Buildings: Additional in course of erection.
Availability of Aerodrome:
(i) *For Emergency use only:* N/A
(ii)*For Full Operational use:* OTU.
Blind Approach: Being installed.   QDM: 220°.
Met. Facilities: Type 3 Station.
Communication other than W/T: R/T Teleprinter and Typex.
Dispersal: 24 Blister hangars, 34 hard standings.
Transport to Local Town or Village: Motor Bus.
Accommodation for Visitors: Hotels at Girvan, Messes.

The onset of October saw the trend of accidents continue; a Hampden landed in the sea off Girvan. Due to the prompt action of the Marine Craft Section the members of the crew, three in number, were saved. The next day a Botha belonging to RAF Prestwick crashed into the sea off Culzean Bay, again the Marine Craft Section at Maidens brought ashore three of the crew. One member of the crew was listed as missing.

On 3 October Defence Exercises were carried out at Turnberry. It is mentioned that this is a weekly occurrence and is greatly improving the standard of all ranks. Approximately 600 personnel marched past the Commanding Officer, G/Cpt. Stone.

A further Hampden accident on 7 October in the sea off Girvan, the crew of four brought ashore by the Marine Craft Section, one of whom died. On a lighter note a weekly concert was given by the station band in the NAAFI, which was greatly appreciated by all personnel. Defence exercises and battle practice continued, this comprised all types of drill, also armed and unarmed combat.

On 13 October another Hampden, this time

of the TTU Abbotsinch, crashed into the sea whilst on night flying exercise, two of the crew of four was brought ashore dead to Girvan by the MCS of RAF Turnberry. A 'slight' accident to a Hampden on landing, got a brief mention, all occupants uninjured. On 16 October a body was recovered from the accident in the sea on 11 May. The next day G/Cpt. Braithwaite, Commanding Officer of RAF Station Limavady visited Turnberry.

And battle training exercises continued.

Anson R9689, pilot F/Sgt Lloyd, landed his aircraft in the sea off the Island of Muck, Northern Ireland, on 19 October. The crew was rescued by the Marine Craft Section, Larne. On 21 October, pilot P/O Reneau in Beaufort AW360 ran off the runway and crashed into the sodium flare box. This was followed by other 'sundry' accidents. P/O Toombs in Hampden AE245 had the starboard undercarriage collapse on him due to failure of the hydraulic system. A Botha aircraft from Prestwick crash landed on the aerodrome, the crew of four was unhurt. Sgt Tapper whilst landing in Hampden AE128 collided with Beaufort X8937. On October 28 a Beaufort crashed into the sea one and a half miles south west of Girvan. The crew of four were brought ashore by the Marine Craft Section and given first aid in Davidson Memorial Hospital, then admitted to S.S.Q. Also, on the same day, Hampden P1284, with pilot Sgt D. Montgomerie, failed to return from a navigational exercise. It was later found that the aircraft had force landed in Northern Ireland. Not so lucky were the crew of Beaufort DD881. They too had left Turnberry on a night Navex, pilot Sgt Sherwood, navigator Sgt Ellis and W.Op/AGs Sgt Newman and Sgt Hancock. They never returned.

Remarks by G/Capt. Stone for the month of October were that the rainfall had been very heavy. Excellent work had been carried out by the Marine Craft Section, saving many lives by their efficiency, and preparation was in progress for the receipt of the TTU from RAF Station, Abbotsinch.

NOVEMBER 1942 5(C)OTU & TTU
The 3 November saw the passing out of fifteen Beaufort crews of 20 Course, who then proceeded to TTU at Abbotsinch for torpedo training. The following day 20 crews commenced training at Turnberry on 22 Course.

There was a visit from press reporters to write an article on the activities of WAAF personnel.

On 7 November battle flight activities were carried out in the morning, without incident, but the night brought the first casualties of the month. Beaufort L9865, with pilot Sgt Sutton, crashed into the sea soon after take-off on local night flying. The cause of the accident was unknown; although a search was made the two occupants of the aircraft could not be found. The next day, flying Beaufort DD876, P/O Ash stalled at about 30 feet when coming in to land and crashed badly. One of the crew was admitted to Station Sick Quarters. Administration duties went on with the assembly of a Monthly Selection Board in the Commanding Officer's office to interview applicants with a view to recommendation for Commission. Later that day the remains of one of the crew of Beaufort L9865 washed ashore west of the airfield, this was identified as Sgt Farnon.

Anson N9722 experienced engine trouble near Ailsa Craig and had to be ditched on 9 November. The crew of six took to the dinghy and was picked up by the motor vessel 'Frugality', owned by Mr J. Alexander, fishmonger of Ayr, and landed at Ayr harbour. The pilot Sgt Walters was admitted to Ayr County Hospital and the rest of the crew were brought back to Station Sick Quarters at Turnberry with minor injuries.

The accidents continued, on 10 November, Hampden AE379 crashed behind Turnberry Hospital during night flying. The crew of four, pilot Sgt Smart, Nav. P/O Greaves, W. Op/AG's Sgt Barnie and Sgt Shaw were all killed.

On 11 November a section of No. 1 TTU moved to Turnberry from Abbotsinch to provide an increase in torpedo training. They were housed in the LMS railway motor garage and

World War One hangars. The move was made very smoothly, although there was great difficulty in finding additional accommodation. RAF Turnberry, originally built for 1,400 personnel, found this number doubled.

A collision took place on the runway between Beaufort DD939 and Hampden P5331, The Hampden piloted by Sgt Moore was on No. III runway, stationary, preparatory to take-off, when the Beaufort piloted by Sgt Mann approached along No. I runway, stopped by the chance light and then turned the corner into No. III runway where it struck the Hampden.

Training accidents and the resultant injuries weren't the only hazard that the airmen had to face. On 13 November 52 personnel of No. 3 Squad and the TTU were sick with suspected poisoning from tea served by the WVS canteen. A sample of the tea was sent for examination to the EMS Laboratory at Ayr. Further fatalities occurred the next night when Beaufort AW384 crashed in a meadow to the north of Girvan Water by Killochan Castle. The crew consisting of pilot P/O Jolly, Nav. Sgt Fisher and W.Op/AGs Sgt Lunn and Sgt McGuire were all killed. The cause of the accident was unknown. A visit was made to RAF Turnberry by S/Ldr Birley of the Welfare Section of the Air Ministry to discuss welfare in general at the station. Another Beaufort, AW245, pilot Sgt Wood, made a forced landing on the airfield due to failure of the port engine, there was slight damage to the aircraft, and the crew escaped injury.

An inspection was conducted by Air Commodore Prawl and NAAFI staff when they visited on the 17 November to inspect NAAFI buildings and quarters. A discussion followed on the type of commodities received at Turnberry. Rear Admiral Phillips, G/Cpt. Beardsworth and W/Cdr's. Coleman and Craven also arrived, in a Lockheed Hudson from RAF Turnhouse, to examine the amenities of the station as a Torpedo Training Unit. They later departed to RAF Prestwick.

During the following days other visits took place, W/Cdr Greenhalgh landed from RAF Limavady in an Anson. G/Cpt. Kidd arrived in an Anson from Belfast; he later departed for RAF Turnhouse. More mishaps are recorded, Sgt Prosser swung off the runway in Beaufort AW380, the cause given as loss of revolution of the port engine. Sgt Brown made a heavy landing during night flying practice, causing the aircraft to swing badly resulting in the collapse of the starboard undercarriage. Anson R9808, piloted by P/O Challis, collided with a one-ton service lorry on the perimeter track, the aircraft swung heavily, and the airscrew struck the lorry causing it to overturn. Nobody was hurt. Another Beaufort aircraft was damaged owing to collapse of undercarriage whilst landing, the pilot Sgt Dick was unhurt. Beaufort AW354 piloted by Sgt Brown suffered damage when the port wheel of the undercarriage collapsed during landing.

In the midst of all this, rural business continued. A claim was forwarded by the factor of the estates to the Air Ministry in respect of Clachanton Cottage, which being in the middle of the airfield, had been demolished the previous year. The figure suggested was for £275, being £25 less than the value of the cottage adjoining. It was mentioned that this was 'a pretty fair figure, as the cottage had neither water nor sanitation in it and the cottage adjoining had both a w.c. and inside sink etc., so that the £25 of a difference makes a very fair price.'

The final fatal accident for November did not involve an aircraft at all, but was when a works flight lorry collided with an airman on a bicycle. The airman was partially deaf and did not hear the approach of the lorry when turning a corner.

The summary remarks by G/Cpt. Stone, for the month of November, were that the weather was mainly good; Eight Hampden crews on acclimatisation course passed out and were posted to 415, 455, 144, and 489 Squadrons, and the stage erected in No. 2 dining room was finished and ready for visits of ENSA concert parties.

DECEMBER 1942 5(C)OTU & TTU
With the arrival of December, 20 Hampden Course passed out from Operational Training to Torpedo Training; this consisted of eleven crews.

A Lysander aircraft from RAF Ayr crashed in the sea west of Turnberry Point; one member of the crew of two descended by parachute and was picked up within six minutes by a pinnace of the Marine Craft Section, Girvan. The crewman was given first aid at Davidson E.M.S. Hospital in Girvan, before being transferred to the E.M.S. Hospital at Turnberry. Beaufort DD871 took off on night flying exercises on 2 December, piloted by Canadian P/O Ash; the wireless operator requested a position, but could not be contacted to receive a reply. The aircraft did not return from the flight. All crew, P/O Ash, Nav. P/O Earls, W.Op. Sgt Newell and A/Gnr Sgt Goodman, were lost. In the station sick quarters record book, it is logged that on 3 December at about 23:00 hours a Beaufort aircraft with a crew of four went missing about two miles south of the aerodrome, a search was conducted for about three hours without result. A further search took place the next morning, again with no result. Could this have been a search for Beaufort DD871?

Beaufort AW360, piloted by P/O Gilmour, swung badly on landing and the undercarriage collapsed under the strain, damage was sustained to both propellers and the port mainplane.

The next few days saw a flurry of visits from some high-ranking officials; Air Commodore Smart CBE, DFC, AFC, arrived at Turnberry by air on a visit to the Station Commander, departing for Edinburgh by car. G/Cpt. Kidd OBE, AFC, AFM, HQ 17 Group and W/Cdr McNeill HQ Coastal Command, also arrived for discussions with the Station Commander. G/Cpt. Stone seemed to be in demand as the visitors continued to arrive. G/Cpt. McDonald Commanding Officer RAF Wigtown landed in Anson EG217, followed by S/Ldr Yellowley, Equipment Officer HQ 17 Group. Air Commodore Vernon Brown, Director of Accidents, arrived in Blenheim U5809 for discussions with the Commanding Officer of Turnberry, and he later left with G/Cpt. Kidd for Edinburgh by road, and S/Ldr Yellowley departed for RAF Stranraer by train. One can only speculate as to the content of these discussions, although it is interesting to note that having all arrived by air, they left by road and rail!

On 10 December, 20 Beaufort Course proceeded on leave. Eighteen crews were under instruction, sixteen passed out and two were retained for further training. S/Ldr Yellowley returned to Turnberry from RAF Stranraer, by train. Training continued, as did the miscellaneous accidents. Beaufort N1082, pilot P/O Sandgren, landed heavily causing damage to the fuselage and wing tip. Visitors came and went, S/Ldr Yellowley returned to HQ 17 Group by train and a conference was held by the Commanding Officer regarding a possible move of 5(C)OTU to Northern Ireland. Beaufort DD888 collided with a station truck, very minor damage was caused, and the aircraft was serviceable within 48 hours. The pilot was Australian Sgt Calmers. In the lead up to Christmas, even more visits and inspections are logged. Sqdn/Officer Pendlebury, WAAF, HQ 17 Group visited on inspection. Air Chief Marshal Sir E. Ludlow-Hewitt, Inspector General of the Royal Air Force, and W/Cdr Dennison arrived in a Percival aircraft from RAF Ayr for an inspection of the station. They both departed by air for Squires Gate. Another heavy landing by a Beaufort, L4456, pilot Sgt Cheley misjudging his height, caused damage to the stern frame and buckling of the fuselage. It is recorded in the Station Sick Quarters log that on 20 December the air ambulance 'Women of Britain' arrived at Turnberry to convey a patient to the London area, departing two days later. Next day, W/Cdr Edwards SOA HQ 17 Group and W/Cdr McNeill visited for discussion regarding recreation and amusements for personnel. In the evening a dance was held in the Sergeants' Mess. G/Cpt. Stone visited Edinburgh, and the following morning W/Cdr Edwards and W/Cdr McNeill returned to Edinburgh. Hampden P2091, piloted by Sgt Hoskins dived into the sea whilst on bombing exercise. Only one of the crew was rescued by a local fishing boat, slightly injured, and taken to Dunure harbour. An investigation was to be held. On the same day we have another Hampden, AT117, which crashed on the airfield when the undercarriage folded up on landing. The aircraft received slight damage and the engine was shock loaded. The pilot, Sgt Salmond, stated that the indicator light was green, and the horn was not sounding, thus leading him to believe that the undercarriage was down and locked into position for landing.

21 Beaufort Course consisting of 21 crews was posted from 5(C)OTU to TTU and Beaufort Course arrived. The Medical Officer stated that upon examination three pilots were of doubtful fitness as regards eyesight, but after further tests all were found fit for full flying duties. There was an ENSA performance in No. 2 Dining Hall entitled 'Ladies in Waiting. It was described as 'the first straight show at the station and greatly appreciated by all ranks.'

On Christmas Eve, all the messes and dining halls were decorated. A party was given by the NAAFI staff in the Institute. On Christmas Day there were the usual festivities - Sergeants visited the Officers' Mess and officers returned the compliment. All officers attended the airmen's dinner and later the WAAF's dinner. A party and dance were held for all airmen and WAAFs in the evening with presents being distributed.

On Boxing Day, it was back to duty as usual and work commenced on the dismantling of the training gear prior to the move of 5(C)OTU to Long Kesh, Northern Ireland. Another dance was enjoyed in the evening, this time hosted by the WAAF personnel. The decision to move 5(C) OTU to Long Kesh was the result of a desperate shortage of efficient torpedo bomber crews, and Turnberry was to become home to the newly formed 1 TTU using the staff and aircraft of the Torpedo Training Unit which had arrived from Abbotsinch in November.

On 28 December a farewell dance for 5(C) OTU was held in the Officers' Mess and an Anson aircraft from RAF Hooton Park crash landed on the airfield - the three occupants were unhurt.

The 29 December saw the departure of the Advance Party of 5(C)OTU on the move of the unit to Long Kesh, and 22 Hampden Acclimatisation Course qualified.

In the very early hours of the penultimate day of the year, information was received at Station Sick Quarters of a Catalina crash about four miles east of Ballantrae, on a hill of 1,200 feet, and of a crew of eleven, there were several trapped in the wreckage. Two Catalina aircraft FP184 and FP239 had taken off from Killadeas in Northern Ireland on a combined operational flying exercise and anti-submarine patrol over the Atlantic. Due to severe weather conditions FP184 was diverted to Stranraer, in a blinding snowstorm at night, after over ten hours flying. The Catalina crashed into Kilwhannel Hill, near Ballantrae. Fortunately, the aircraft did not catch fire, and the depth charges did not explode. Air Gunner Sgt Neville Mudd and another, Sgt Dean, although injured made their way down the hill to find help. A crash party from Castle Kennedy was despatched, changing their small ambulance for an Albion ambulance at RAF Stranraer, before reaching the farm on the outskirts of Ballantrae where Sgt Dean had managed to raise the alarm. A search party set out from the farm in gale force winds and reached the crash site about 01:30 hours. Four of the crewmen were found alive in the wreckage and two dead and one who was trapped in the wreckage and could not be reached for examination was apparently dead. Two of the wounded, Sgt Legget and Sgt Breeze, were freed from the wreckage and dispatched to Glenapp E.M.S. Hospital immediately. Sgt Jenkins and Sgt Cowie could not be released from the wreckage without the use of specialist equipment. The injured men were given Morphia and protected as best as possible from the elements. F/Lt Greenburgh and Cpl Lucas of SSQ Castle Kennedy stayed with the injured men until 07:30 hours when they were relieved, having been at the crash site since about 01:30 hours, by S/ Ldr Ross, Medical Officer from RAF Turnberry, along with local police officers. The extraction of the remaining two crewmen was completed at about 11:00 hours; they were carried by stretcher down the hill to the ambulance and taken to Glenapp Auxiliary Hospital. All the crew were now accounted for – eight in Glenapp E.M.S. Hospital and three dead. The rescue party returned to Turnberry about 14:00 hours. The other Catalina that took off at the same time as

FP184, FP239 also crashed, flying into a hill 17 miles north west of Omagh, County Tyrone, Northern Ireland - all crew were killed.

JANUARY 1943. 1TTU

On 1 January, 1 TTU was officially installed at Turnberry, using Hampden, Beaufort and Beaufighter torpedo bombers. The training consisted of dropping torpedoes on targets in the Firth of Clyde and in various lochs. The pupils sent to Turnberry were a mixture of those new to the art of torpedo dropping, but also crews on refresher courses, having failed to reach their full potential as torpedo bombers with disappointing results! A detachment of 1 TTU was also housed at Castle Kennedy. The transport for the move of 5(C)OTU to Long Kesh arrived; four three-ton trucks and four Queen Marys. S/Ldr Yellowley arrived from No. 17 Group to help in the movement of equipment.

The main party of 5(C)OTU left Turnberry from Maidens Railway Station at 06:40 hours on the 4 January. This comprised 2 Officers and 340 men. According to the Record Book the move went very smoothly. Instruction was given and parades held for the move of the second section of 5(C)OTU on 6 January, followed in the evening by an ENSA show in No. 2 Dining Hall, 'Palindrome', which was greatly enjoyed by all ranks. The next morning at 07:15 the train left on time, and everything was moved to schedule.

On 8 January the HQCC Military Band arrived at RAF Turnberry, under the leadership of W/O McDonald, compromising 1 W/O, 1 Sgt and 22 ORs. A concert was given in the evening. The band also entertained in the Officers' Mess, Sergeants' Mess and played at the church parade, leaving for Stranraer two days later. There was no flying mentioned over the following few days, but various meetings took place to discuss postal facilities, the moving of the Anson Flight to Carew Cheriton and a Mr A. Whyte, the Horticultural Advisor of the Air Ministry, visited the station to advise on favourable ground for the growing of vegetables. On 18 January Works Flight commenced the general cleaning of the Communal Site, which was left in a disgusting state by the contractors! Flying activities are mentioned with a Special Target attack which was made by four Beauforts and four Hampdens of 1 TTU. From assessments made it was considered that the ship would have been sunk after the first attack. The next day brought another fatal accident involving a Turnberry aircraft. Beaufort L9825 crashed into the sea during low flying practice 1½ miles west of Girvan. The cause was unknown.

On 21 January W/Cdr Coleman, Air Ministry and F/Lt Roylace, HQCC, visited Turnberry to discuss torpedo training. A conference was held by the Commanding Officer regarding the move of the Anson Flight and the rear party of 5(C)OTU to Long Kesh. On 25 January G/Cpt J. G. Elton DFC, AFC, arrived at Turnberry from HQCC to assume the duties of Station Commander vice G/Cpt Stone. This was followed by F/Lt V. W. Neate arriving from No. 17 Group to assume the duties of Station Adjutant, vice F/Lt Rickards. On 28 January G/Cpt. Elton and S/Ldr Empson visited HQ No. 17 Group to discuss matters relating to AMWD at RAF Turnberry. Six 3-ton vans and 2 Queen Marys arrived to convey further equipment of No. 5(C)OTU to Long Kesh and the Anson Aircraft departed for RAF Station Carew Cheriton.

The following day a report was received that an Anson aircraft piloted by S/Ldr Douglas W. S. Ireland had not arrived at Carew Cheriton. Search parties were detailed to look for the missing aircraft, but nothing was found, and was presumed lost. W/Cdr Tangye of the Air Ministry visited to discuss the possibilities of using RAF Turnberry as an intermediate aerodrome for ferrying aircraft. This meant an increase in strength by fifty maintenance personnel, and approximately twenty aircraft in addition to the present strength of the station. Also, in January 'B' Flight of 652 Squadron with Auster aircraft were detached to Turnberry for army co-operation duties throughout southern Scotland.

The summary for the month of January records that the Signal Section had completed the installation of Flying Control HF/DF station at Dunanhill, approximately 1½ miles south of Turnberry, except for electric power which had to be installed by the Ayrshire Electricity Board. Fifty per cent of signals personnel have been transferred to RAF Long Kesh to assist in re-opening of 5(C)OTU at their new station. A fitting party arrived on 18 January to install Standard Beam Approach; this equipment was fully operational on 31 January.

The weather for the month was mainly wet with heavy wind and low cloud, which made flying difficult owing to the near proximity of hills. Entertainment was progressing favourably; F/Lt Warden assumed the duties of Entertainments Officer and a Broadwood grand piano arrived from ENSA London. Class 'A' ENSA parties have been allocated to this station. General remarks were that the policy of the station has entirely altered due to the arrival of 1 TTU and departure of 5(C)OTU. The personnel on strength was much lower; allowing for more comfort due to non-crowding of sleeping sites. The conditions of the buildings at Turnberry were documented as bad. Although many efforts had been made, rain seeps into nearly all the buildings, which not only caused great discomfort, but raised the percentage of sickness amongst personnel.

FEBRUARY 1943 1 TTU
February 1943 begins by stating that flying consisted chiefly of individual and formation attacks on target ships until conversion of Beaufort and Hampden crews, which was due to begin on 5 February. The serviceability and flying time of the aircraft on strength are given as Beauforts 5 - 11.50 hours and Hampdens 3 - 7.40 hours, 3 Runners were dropped by Beauforts.

The 2 February brought Mr Furgerson and Mr Kemp, along with Mr Reade, Section Officer of Ayr to Turnberry to discuss building problems. Torpedo training continued with 2 Hampdens and 3 Beauforts attacking a 'C' class cruiser.

The Transport Section had to be moved to temporary quarters owing to the continual flooding of the M.T. offices and yard. The next day ten crews of 6 Hampden Course elected to convert to Beauforts, with pilot conversion

training taking place the following day. In Station Sick Quarters Records it is noted that a Wellington aircraft on ferry service crash landed at Kirkoswald (High Park farm). All the crew was unhurt, except one who had a cut finger. The next entry for 3 February was of a Sunderland which crashed off Ailsa Craig with fifteen on board, of these only S/Ldr Pierce (S.M.O. Wigbay) is mentioned as injured, having a compound fracture of the leg, and admitted to Turnberry Hospital. The fate of the other crewmen is unknown. Six crews arrived on 5 February (4 Hampden and 2 Beaufort) to form 8 Course. There was a taxiing accident involving Hampden P2075, pilot Sgt Morris.

Life under 1 TTU at Turnberry seemed to be relatively uneventful. There were the usual scrapes, Beaufort W6469 piloted by Sgt Brown crashed when the undercarriage collapsed after a one-wheel landing. On the social side a concert party from RAF Heathfield gave an excellent performance. On the thirteenth of the month, bad weather stopped flying. Mention is made of the difficulties encountered by AMWD in obtaining labour to carry out building requirements. The rest of the month records the progress of training, introducing night flying dual instruction, further visits and inspections and the enjoyment of a dance held at the WAAF site. Wellington aircraft from 7(C)OTU dropped three Runners.

G/Cpt. Stratton visited Turnberry for a conference with the Commanding Officer. He made a general inspection of the station and observed the difficulties experienced with contractors to obtain a completion of the work throughout the camp. The same day W/Cdr Lindeman DFC arrived from RAF Leuchars to assume the duties of Chief Instructor. Towards the end of the month there was another Beaufort which swung on landing, X8936, pilot Sgt Watt. February 1943 saw blessedly few accidents at Turnberry.

The Commanding Officer's remarks were that during February 1 TTU has become well established. Difficulty was experienced in fully equipping torpedo workshops owing to a lack of contractor's staff. He also states that the TAT is now in full operation. On a lighter note it is mentioned that the ENSA shows which have been given each fortnight, plus the additional shows from RAF Ayr and Stranraer, have given the personnel more entertainment, although the cinema projector which was authorized in November 1942 has not been started, owing to a lack of works personnel.

MARCH 1943 1 TTU

On 1 March, flying was busy with fifteen hours completed on Beauforts and six on Hampdens. The next day 11 Runners were dropped by Hampden aircraft, with one Runner lost. There was a taxiing accident to Hampden AW210, pilot P/O Ray, with no casualties. On 5 March Wellington aircraft from RAF Limavady on exercise dropped 5 Runners and the same over the following two days. S/Ldr P. W. Matthews and F/O A. A. Frazer, Canadian Liaison Officers, visited Turnberry to contact Canadian personnel. Another Beaufort L4461, pilot Sgt Colclough, swung on landing - there was slight damage to the aircraft but no casualties. This was on 8 March when the Beaufort course had a busy day completing seventeen hours flying. The following day S/Ldr Shaw arrived from RAF Limavady to assume duties as an instructor, and on 10 March four Wellington aircraft arrived in the move of 4 Squadron 7(C)OTU to form 2 Squadron at Turnberry. Training continued on 11 March, with one dummy torpedo jettisoned from a Hampden owing to engine failure. A Torpedo Training Conference was held at Turnberry attended by the following, Group Captains E. C. Kidd (SASO HQ No. 17 Group), H. Braithwaite (Limavady), H. Waring (Castle Kennedy), and J. G. Elton DFC, AFC, (Turnberry), Wing Commanders, R. E. Craven DFC (SASO HQ No. 17 Group), J. Greenhalgh, G.M. Lindeman DFC, R. G. Gaskell and S/Ldr's Harvey and Dinsdale; Lt. Commdre. Dimsdale RN and eight crew's ex Nos. 6 and 7 Courses departed to No. 2 OAPU. A further two crews departed to No. 2 OAPU on 13 March. Also, on this day a conference of station commanders was held at Machrihanish, which was attended by the Station Commander and

Chief Instructor from Turnberry. On 14 March A/C/M Sir E. Ludlow-Hewitt, Inspector General of the RAF, visited Turnberry. On 17 March there was a sudden influx of crews from RAF Long Kesh, 23 Beaufort crews to form No. 10 Course and 18 Wellington Crews to form No. 9 Course. The Station Commander and Chief Instructor visited the Royal Naval Torpedo factory at Greenock. Nos. 9 and 10 courses started training the next day. The station was visited by Lt Col Willows HQ 17 Group Defence Officer and Sgt Light, Sanitary Assistant HQ 17 Group. Airfield accidents continued on 20th, when F/O Armstrong made a belly-landing owing to an engine cutting on approach. For this he received a green endorsement.

Crews continued to arrive and complete training; six crews of No. 8 Course departed, two to 415 Squadron and four to 489 Squadron. The subsequent week was relatively quiet, exercises carried on with Beauforts and Wellingtons. There seems to be no Hampden flying recorded for the rest of the month. HQCC Military Band was posted to Turnberry from Abbotsinch for a few days before departing for RAF Leuchars on commencement of 'Wings for Victory' War Bond campaign.

The month ended with the terrible loss of Wellington LB223, on coming in to land the pilot, Sgt Hunter received a red flare. The aircraft had party completed the circuit with flaps and undercarriage down when it stalled and crashed. There were no survivors. The pilot and six others were killed.

On the last day of the month, seven crews arrived at Turnberry from 6(C)OTU to form No. 11 Wellington Course.

APRIL 1943 1 TTU

The 1 April was a General Holiday to celebrate the 25th anniversary of the Royal Air Force. The Commanding Officer held a parade and march past of all station personnel and a station dance was held in the evening. During the next week training went on in earnest, there is only three hours Hampden flying logged but the Wellingtons and Beauforts made up for it with approximately

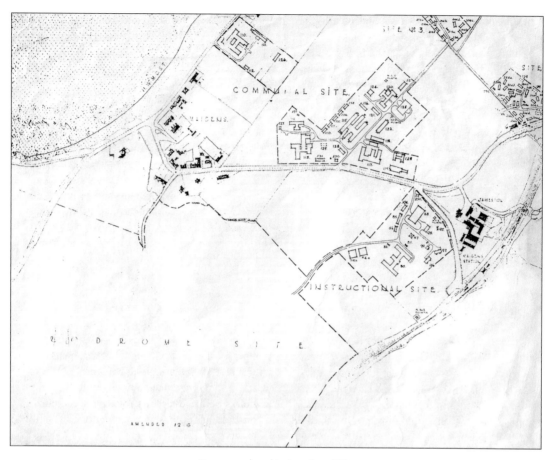

Communal and Instructional Site

98 hours and 117 hours respectively. S/Ldr Hughes (489 Squadron) arrived for a short torpedo course and two Polish crews arrived for torpedo training. On 7 March 12 Course (Hampdens) arrived. Beaufort DX114 was ditched by pilot F/O Taylor, Sgt Dearmer was killed.

The new Medical Inspection Room opened on the Communal Site and a letter was sent to S/Ldr Administration requesting that water be installed as soon as possible. S/Ldr Adams, Medical Officer (WAAF) visited the station regarding the condition of the living quarters of the WAAFs on the station and gave a series of lectures. Dr Powell and Mr Simms visited for a discussion with the Commanding Officer regarding an experiment to be held at the station.

S/Ldr Hughes had completed his training by the 12th and left Turnberry, there was a visit by S/Ldr McGrath, Intelligence Officer HQCC, and twelve crews of No. 10 Course departed to 303 FTU Talbenny. Fifteen crews of 13 Course arrived from 7(C)OTU Limavady. Eighteen crews of 14 Course arrived from 5(C)OTU Long Kesh. All three flights, Hampden, Wellington and Beaufort had a busy week flying.

On 16 March a Flying Fortress from the US landed at Turnberry en route to Prestwick.

There was an inspection of the station by A/M J. Slessor CB, DSO, MC, AOC i/c HQCC in company with AVM H. Smart CBE, DFC, AFC, G/Cpt. Grieve and W/Cdr Craven DFC. Hampden AT125, piloted by Sgt Cordingley, collided in the air with Wellington LB237 flown by P/O Ockalski, Polish Air Force, resulting in the Hampden diving into the sea. The pilot and crew were killed. The Wellington made a safe landing although fin and rudder were badly damaged. Wellington LB237 was later lost on 21 May 1943.

Other visits during April were S/Ldr Pendlebury WAAF Staff Officer HQ No. 17 Group to inspect WAAF quarters and Mr A. Whyte, Horticultural Advisor to the Air Ministry, to discuss and advise on agricultural problems. A proposed conference between the CO of the target ship, Commanding Officer RAF Station Machrihanish and COs, SIs of 1 and 2 TTU was postponed owing to adverse weather conditions. The last three days of April saw an accident every day: Wellington LB239 crashed into the sea, there were four injured and Sgt Treble W. Op/AG the only fatality listed as missing. Hampden P1198, pilot Sgt Batchelor, crashed into the sea, all crew unhurt and Beaufort DX997 swung on landing, pilot Sgt Bell, again no casualties.

May 1943 1TTU and 618 Squadron
The 1 TTU Operations Record Book for the month of May gives scant information for the first three weeks, a mention of a visit by W/Cdr Craven DFC, HQ No. 17 Group, S/Ldr Bunce and S/Ldr Worsdell, HQCC to discuss air tactics, followed on the 18th by RAF Gang Show given on the station. But, on the edge of the airfield, where Malin Court Hotel sits today, a detachment of 618 Squadron moved into a hangar there from RAF Skitten. The Air Ministry had authorised the formation of this new Squadron on 1 April 1943, equipped with Mosquito aircraft within Coastal Command. The squadron was to be worked up rapidly to a highly effective operational status within a few weeks, and all information regarding this was TOP SECRET! 'Only those personnel who had reason to see the aircraft should be permitted to do so, and the unit should be housed away from other units.' It was decided at an early stage to issue the squadron with 'practice rounds' of the weapon they would be using on operations and 54 of these were already being held under tight security at Turnberry having arrived at the end of April.

618 Squadron were conducting trials with a new weapon 'Highball' with which it was planned to attack and sink the German battleship *Tirpitz* at its anchorage in the fjords of Norway. Following the success of Barnes Wallis's 'Upkeep' bouncing bomb, which 617 'Dambuster' Squadron used to great effect against the dams in Germany's Ruhr valley, a smaller, spherical version 'Highball', was developed. Consisting of a depth charge, spun backwards at 1,000 rpm and released at sea level, the spinning caused the 'Highball', (or 'store' as it was known) to bounce on impact with the surface of the water and travel over a distance of 1,500 yards or so, therefore avoiding anti-mine booms and netting, finally reaching its target.

On impact, the 'store' rebounds, still spinning and falls underwater beneath the target ship. Filled with a charge of some 600 lbs of Torpex (an explosive 50% more powerful than TNT), the 'Highball' was set to explode with the use of a hydrostatic pistol which was pre-primed at designated depths. It was predicted that the explosion taking place beneath the hull of the target vessel, behind the outer armoured belt, 'Highball' would prove to be a most effective weapon against the enemy fleet.

The first ground crew of 618 arrived at Turnberry by rail on 7 May consisting of eleven NCOs and airmen, and the first five Mosquitoes landed on the 19th. This was the nucleus of the detachment at Turnberry; it was anticipated that a much larger establishment would be formed for the maintenance of aircraft undergoing trials, and that further detachments would arrive as the aircraft demands increased. Thirty-four ground crew were subsequently despatched from Skitten by rail to reinforce those already at Turnberry. A strict security cordon was put in place around the assigned dispersal. There were lots of aircraft movements as the Mosquitoes were modified for their difficult task. Most of these modifications were carried out at the de Havilland works at Hatfield or Vickers at Weybridge, with the crew staying in local accommodation before flying the aircraft back to either Skitten or Turnberry. Mosquito DZ534 flew in from Skitten and then on to Hatfield for modification, followed by

DZ539 and DZ543 which were delivered direct to Turnberry. One Mosquito on taking off, swung and spun through 90°, just missing the lighthouse. Leaving from Weybridge heading to Turnberry DZ493 swung violently and banked steeply, causing the wing to impact with a vehicle on the perimeter track, and then hitting a lorry, a van, and balloon cable, the aircraft crashed and burst into flames. The pilot, W/Cdr Hutchison was slightly injured. A further twenty Mosquito were sent to Turnberry along with a further detachment of personnel.

Meanwhile, the old French warship *Courbet* was anchored in Loch Striven to act as a target for practice runs. The first trials against the target ship were made by S/Ldr Rose during which ten 'stores' were released, five were lost due to mechanical failure of the release gear.

618 Squadron withdrew their detachment back to Skitten while the technical problems with 'Highball' were considered, difficulties with the release gear, the shattering of the wooden casing and the erratic behaviour of the bombs as they bounced across the water, especially in anything other than flat calm conditions. Another problem was that when two 'Highballs' were dropped together, they would sometimes collide with disastrous consequences.

The entries for 1 TTU change from the basic serviceability and flying times on 21 May. This logs a further dreadful accident to occur at Turnberry.

16 Course Wellington TB LB187 P/O Hollom 1st Pilot, crashed into the sea, 3 of the crew missing.

Wellington TB LB193 F/O Griffin 1st Pilot, crashed in sea, two Sgt W. Op/AGs missing.

Wellington TB LB 237 Sgt Forsyth-Johnson 1st Pilot missing, presumed crashed in sea, all crew missing.

S/Ldr. Gillen RC Chaplain visited in the afternoon.

The next day saw the start of the Ayr 'Wings for Victory' campaign, and aircraft of 1 TTU flew a display close to the seafront. On 24 May G/Cpt. Elton DFC, AFC, and a detachment of RAF and WAAF from Turnberry opened the Girvan 'Wings for Victory' week with an exhibition. A parade was held, and the HQCC band attended the ceremonies.

Flying displays were provided by Turnberry aircraft dropping concrete torpedoes into the sea from low altitude. On 26 May nine crews arrived to form 17 Course, No. 15 course proceeded on leave until 8 June owing to the unserviceability of aircraft (There had been no Hampden flying since the 17 May).

On the last day of the month, 16 Course – 8 crews passed out. The final entry of the Operations Record Book for May reads, 'Attack on Aircraft Carrier.'

JUNE 1943 1 TTU and 618 Squadron
The first of June records another 'Attack', this time on a major naval unit. Interestingly this states that four Beauforts took part, but there were only three listed as serviceable in the Operations Record Book. All Hampdens were grounded, reason unknown. The eight trained crews of 16 Course proceeded to Talbenny. On 9 June, 29 Beaufort crews and 13 Wellington crews arrived. On 14 June a single sentence records 'Sgt. Hunter and crew killed in Beaufort DW923.' Over the next three days, trained crews were posted out, seventeen Wellington crews to 303 FTU, ten Beaufort crews to 306 FTU and another eight Beaufort crews to 306 FTU. The courses were quickly filled with the arrival of ten Hampden crews, and Hampdens started flying again on 25 June. The only other entry for June is that one Beaufort crew was posted to 306 FTU. Mosquito DZ555, while flying on a fuel consumption test over the Northumberland coast, was hit by a shell from a Ground Defence Post. The aircraft managed to get safely back to Turnberry and landed on one wheel. 618 Squadron added to their aircraft compliment with de Havilland Dominie X7383, for use as general transport.

July 1943 1 TTU and 618 Squadron
July found Turnberry very active with all three flights, Wellington, Hampden and Beaufort, logging lots of flying hours.

As the problems with 'Highball' mounted, the 618 Squadron training at Turnberry was (temporarily) closed down. The Official Record Book records the departures of the Mosquitoes over the next few days, and the trials of 'Highball' were transferred to Reculver, in south-east England, where conditions were thought to be more favourable than Loch Striven. It was later discovered that one of the reasons for the disappointing results against the *Courbet* was that the buoy to mark the release point had been incorrectly placed at 800 yards from the target, instead of the recommended 1,200 yards. The effect of this was that the 'store', not having enough distance to lose its forward speed and height velocity, was damaged on impact with the target, which resulted in the malfunction of the detonating pistol.

Official visits continued with S/Ldr P. Matthews and F/Lt J. Andoff RCAF, liaison officers, arriving on the 5th. The same day, the Commanding Officer and Station Administration Officer visited the Regional Transport Officer in Edinburgh regarding new public transport services from the camp to Girvan and Ayr. The Commanding Officer then left for Gibraltar for a Court of Inquiry; W/Cdr Lindeman assumed command. Further courses came to an end and sixteen crews of 19 Wellington Course and fourteen of 20 Course went on embarkation leave prior to reporting to 303 FTU. Eleven crews of 15 Hampden Course also completed their training, five crews were posted to 415 Squadron, four crews to 489 Squadron and two to 455 Squadron. On July 12 Wellington HX744, being air tested by F/Sgt Ceha, force landed at Prestwick. The new bus service to Ayr commenced. On the 14th, the Group Torpedo Officer visited the Target Ship, and W/Cdr L. Shaw visited Turnberry regarding the resumption of the railway service.

Nothing else of much note is mentioned, on the 21st five crews of 26 Beaufort course arrived and were assigned ground day training, until the complete course arrived on 6 August. The next day, nine Wellington crews of 25 Course commenced training. By the 28th ten crews of 22 Hampden Course passed out – six crews to 415 Squadron, one crew to 489 Squadron and two crews to 455 Squadron. On the 31st, 24 Wellington Course completed their training early and passed out.

August 1943 1TTU and 618 Squadron
August finds the Operations Record Book divided into squadrons: No. 1 Squadron – Hampdens, 2 and 4 Squadrons – Wellingtons, and 3 Squadron – Beauforts. On 4 August more trainees arrive – thirteen crews for 28 Wellington Course. Nineteen crews of 23 Beaufort Course passed out and ten crews of 26 Beaufort Course took up their places. On the 7th, Wellington HX774 crashed in the sea and the crew of six lost their lives. On the same day Hampden X3026 crashed into the wood between Maybole and Turnberry, the crew of four perished. Training continued, without incident over the next week, until on the 16 August Hampden P5341 crashed into the sea off Dunure – three survivors and one killed.

G/Cpt. H. Waring arrived at Turnberry from RAF Castle Kennedy and the next day assumed the duties of Commanding Officer, G/Cpt. J. Elton DFC, AFC, being posted to HQ Coastal Command. Eight crews of 25 Wellington Course passed out and nine crews of 29 Wellington course arrived. Again, nothing of note to report until the next accident, that of Beaufort JM557 on 23 August, the crew of four lost their lives when it crashed into the sea. G/Cpt. H. Waring commented that Turnberry seemed to be a very organised station. Nine crews of 27 Hampden Course passed out, and thirteen crews of 30 Course arrived. On 26 August Beaufort EL134 ditched, the crew were picked up safely.

At the end of August, 618 Squadron reported on the trials from Reculver, 'As it had become apparent that the difficulties attending the production of the "store" as a perfect weapon were likely to be legion, and in order to ensure that the Squadron crews were kept together during these delays and thus avoid losing a certain amount of valuable experience, it was intended that the "stores" in their present state could be used against U-boats.' 618 Squadron had been grounded at Skitten, when news came through that the *Tirpitz* had been badly damaged by an attack of British mini submarines, so removing their proposed target. By turning their attention to U-boats, the Squadron would be able to continue its technical trials, remaining usefully occupied whilst efforts were made to perfect the performance of 'Highball'. To this end the 'stores' used for training were first modified by squadron armourers at Turnberry, and then transported to Skitten, where training was to commence. 'Valuable data was obtained from these Skitten practices, and in the evaluation of these trials, the project was eventually abandoned in the light of existing circumstances.'

There is nothing else logged for 1 TTU for the rest of August, only aircraft serviceability and hours flown.

September 1943 1 TTU and 618 Squadron
To start the month a further twelve crews of 31 Wellington Course arrived and commenced training. On the 10th it is noted that S/Ldr Custardson, Catering Advisor of HQCC, visited the station on inspection. (It has been previously noted that the spelling was Custerson, could this be a genuine spelling mistake or a measure of the sense of humour of the clerk?)

Information was received at Turnberry that exercise 'Nimbus' would be held in the area near the station and that accommodation would be required for aircraft and personnel. On the 15th nine crews of 29 Wellington Course passed out. The Signal Section and Maintenance Servicing Unit for exercise 'Nimbus' arrived on 18 September, prior to the arrival of twenty Boston and five Ventura aircraft on the 20th. The actual exercise took place early the next afternoon.

A small 'invasion' fleet made its way up the Firth of Clyde, past Arran and up into the Sound of Bute, laying smoke screens as it went. On the 25th four crews of 33 Hampden Course arrived.

In September 1943, Headquarters Coastal Command issued instructions for the grounding of 618 Squadron at Skitten and the disposal of aircraft and personnel. Detachments were to be sent to Turnberry, Shawbury and Wick,

with aircraft to be stored at these stations, and Squadron trials were to be continued at Turnberry, studying the impact, range and underwater trajectory of the special bouncing bombs. A party of maintenance personnel proceeded to RAF Turnberry, where they remained with the six aircraft flown there for detachment. It was proposed that 618 could perhaps be used in an anti-submarine role, this idea was short lived due to the operational difficulties presented by the short range of the Mosquito and the difficulty in finding the target submarines. Instead the unit was selected for carrier-borne operations in the Pacific attacking the Japanese Fleet. Trials at Turnberry continued to this end, S/Ldr 'Shorty' Longbottom, released ten 'stores' against the target ship *Courbet*, different types of 'stores' were used including some of the type trialled at Reculver. New types of detonating pistol were also tested. These trials were somewhat disappointing as valuable information was lost because not all of the 'stores' were able to be recovered for assessment.

The 26 of September was 'Battle of Britain Sunday' and a detachment from RAF Turnberry paraded in Girvan. On 29 September Beaufighter JL610, pilot Sgt Pritchard crashed en route from Castle Kennedy to Turnberry, the two crew and one passenger were killed. This tragic incident marked the move of 2 TTU to be absorbed into 1 TTU, comprising of 25 officers and 307 other ranks from RAF Castle Kennedy to RAF Turnberry. The last course to arrive that month were nine crews of 10 Beaufighter Course.

OCTOBER 1943 1 TTU
October found Hampden, Wellington, Beaufighter and Beaufort aircraft on charge at Turnberry. F/Lt Neep, Salvage Officer HQCC, visited the station to discuss salvage difficulties. The Operations Record Book for October lists few details of the activities for the month, a few visits of inspecting personnel, and thankfully no accidents.

NOVEMBER 1943 1 TTU
Nothing much to report for November, further

visits of inspection, no accidents. The Wellington contingent seems to have left, only Hampden, Beaufort and Beaufighters remain.

DECEMBER 1943 1 TTU
Again, a quiet month for 1 TTU, various courses completed their training and proceeded on leave, although four Wellington crews ex 31 Course were recalled from leave to receive plague inoculation. At Christmas a general holiday was granted to all personnel, Christmas dinner was served to airmen in the Airmen's Mess, which was attended by Officers and Senior NCOs. A special all-ranks dance was given in the Station Institute in the evening, which was greatly enjoyed by all personnel.

JANUARY 1944 1 TTU
Aircraft on charge in January were Beaufighter, Beaufort, Hampden, Oxford and Lysander. On 11 January W/Cdr E. Hutton arrived and assumed the post of CI vice W/Cdr Lindeman who was posted to 19 Group. 37 Course was due to complete their training on the 13th, but due to adverse weather conditions an extension was required. On 18 January, seven Beaufighter crews were posted from 132 OTU to form 39 Course and commenced ground training. Again, on the 27th, weather conditions hampered flying and 38 Beaufort Course required an extension. On the 28th Sir E. R. Ludlow-Hewitt, RAF Inspector General, visited the station from West Freugh and left the following day for Prestwick.

FEBRUARY 1944 1 TTU
Aircraft listed on charge in the Operations Record Book are Ventura, Hudson, Martinet, Oxford and Beaufighter. On 4 February, Capt. F. Hendler of the US Army visited the station and spoke on 'America and American Life'. The 8 February brought the first fatal accident of 1943, with the loss of F/Lt K. Noble, Staff Pilot, and F/O G. Whitehurst, of 39 Beaufighter Course.

By this stage of the war, the number of torpedo crews required was on the decline, the reduction of this training allowed the return of 5(C)OTU to Turnberry. On 15 February the main

party, consisting of sea and air parties arrived from RAF Long Kesh. The air parties consisted of:

Ventura FP629	F/Lt Jarvis and crew
Ventura FP380	F/O. Greenfield and crew
Hudson AM522	F/O. Pullan and crew
Hudson AM529	F/Sgt Simpson and crew
Oxford HN524	F/O. Westoby and crew
Oxford HN829	F/Sgt Reader and crew
Oxford HN843	F/Sgt Willshire and crew
Martinet HP164	F/Sgt Harper and crew

The same day six Beaufighter crew arrived from 132 OTU, East Fortune to commence training on 40 Beaufighter Course.

Mid-month, W/Cdr S. Leitch from HQ ATC visited Turnberry with reference to ATC summer camps. A further air party of 5(C)OTU arrived from Long Kesh:

Hudson 64	F/O. Harkness and crew
Hudson 77	W/O. James and crew
Ventura 87	F/Lt Park and crew
Martinet HP355	S/Ldr Naismith and crew

Four crew ex 38 Beaufort Course proceeded on attachment to 304 FTU Melton Mowbray pending posting on 17 February. Two more aircraft arrived from Long Kesh:

| Hudson D5499 | F/O. Rayner and crew |
| Ventura FP569 | W/Cdr Devitt and crew |

The rear party from RAF Long Kesh arrived on 27 February. Turnberry now trained general recce crews, with the addition of refresher courses for torpedo bombing, air sea rescue and aerial gunnery, but on a reduced scale. For this change in policy, diverse types of aircraft were used, such as the Ventura, Warwick, Wellington, Hudson, Beaufighter and sundry transport such as Anson and Oxfords.

MARCH 1944 1 TTU & 5(C)OTU
On 1 March, Commander Sears arrived at Turnberry from Sydenham en route to Ayr

(Aircraft Stinson 963). At 13:48 hours a crash was reported half a mile north-east of Chirmorie farm, near Barrhill -Thunderbolt No. 7638 owned by the Lockheed Aircraft Company, Warton, Lancs. A crash party was supplied by Turnberry.

Training at 1 TTU continued, six crews ex 132 OTU forming 41 Beaufighter Course, commenced torpedo training. Two crew ex 39 Beaufighter Course proceeded to 489 Squadron, two crews to 254 Squadron and two crews to 144 Squadron. On the 15th a further two crews arrived from 132 OTU and joined 41 Beaufighter Course. That night Ventura 87 crashed on landing, the undercarriage was not down, there were no casualties.

On 18 March, at the request of 13 Group, Turnberry supplied two Oxfords and a Hudson, (piloted by F/O Larkin, F/Lt Wilcox and F/Lt Bland), on an air sea rescue search for an aircraft believed to be in the sea between Lady Isle and Troon. The search was carried out in the area east of a line from Ardrossan to Ayr. The Oxfords were airborne at 09:40 hours and the Hudson at 10:37 hours returning at 10:54 hours and 11:34 hours respectively. The crews reported that they had searched the whole area without result. Another crash was reported at Loch Doon of a Hurricane belonging to Hurn, again a crash party was supplied by Turnberry.

Visiting aircraft in March were G/Capt. Cracroft AFC, SASO, 17 Group, who arrived in Anson LT186 from Squires Gate; G/Capt. Thomson arrived from Crail in Proctor LZ708, departing the next day for Wigtown. On the 27th Capt. G. Buxton arrived from Montreal en route for Prestwick in Liberator KL528. The same day Oxford NM348, piloted by F/O E. Russon swung on landing causing the undercarriage to collapse: there were no casualties. On 28 March eight Anson aircraft belonging to West Freugh and one Anson from Millom arrived at Turnberry, being diverted due to bad weather.

On 29 March Prestwick requested that Turnberry bring in a Dakota (292773) that was in the area of Ailsa Craig and short of fuel. The aircraft landed safely at 13:30 hours downwind. It had been in the air 16½ hours and had come from Marrakech in North Africa. The pilot, 1st Lieutenant Brown, expressed his pleasure for the assistance given to him by flying control at this station. The final aircraft to fly in during March was W/Cdr Devitt who arrived by air from Dyce in Mosquito DZ538.

APRIL 1944 1 TTU & 5(C)OTU
On 1 April 5(C)OTU at Turnberry had on charge Ventura, Hudson, Oxford and Martinet aircraft. Three Lockheed P-38 Lightning's (268/250, 134 and 106) arrived en route to Chalgrove flown by Lt Barnes, Lt Spencer and Lt Paterson. Another surprise visitor was Wellington 'Q' of 20 OTU, which made an emergency landing for minor repairs and refuelling.

On 5 April notice was received that Group Captain Waring, Commanding Officer of Turnberry and Squadron Leader Kerswell, Servicing Wing, had been Mentioned in Despatches. Air Vice Marshall H. G. Smart, CBE, DFC, AFC, Air Officer Commanding, arrived on a visit in Anson EG215 and then flew on to Turnhouse. 1 Hudson Refresher Course was completed. This comprised two crews of rest-expired instructors, who then proceeded to 7(C) OTU for further training.

Three Dakota aircraft (482, 607 and 063) flew in from Grove en route to Prestwick, piloted by Major Pengally, Lt Marr and Capt. Gloves, followed the next day by ten Ansons which had been diverted from Wigtown.

On 17 April Major Ongside in Dakota 113503 arrived from Renfrew heading for Prestwick, and two Venturas from Turnberry were diverted to Tiree due to the weather being unfit to land at base.

The next day saw the arrival of Air Vice Marshall H. G. Smart, CBE, DFC, AFC, Air Officer Commanding 17 Group, who flew in from Evanton in Anson EG125, Air Chief Marshall Sir Sholto Douglas KCB, MC, DFC, Commander in Chief, Coastal Command, arrived in Hudson FK745 from Tiree and Wing Commander Miller arrived in Domini X7383 from Machrihanish.

The following morning, Air Chief Marshall Sir Sholto Douglas and Air Vice Marshall H. G.

Smart departed by road for Longtown.

There were three further emergency landings in April, an Anson from West Freugh limped in with engine trouble, a Spitfire from Hawarden, and a Miles Master, all with mechanical problems.

MAY 1944 1 TTU, 5(C)OTU and 618 Squadron Aircraft on charge at Turnberry now included Beaufighter, Ventura, Hudson, Warwick, Oxford and Martinet. The crews posted in to the 4 Ventura Course, while awaiting the commencement of training, found themselves on the Hudson Refresher Training course, it was remarked that these crews had done no flying for a long time. Back at Loch Striven HMS *Malaya* arrived to act as a target ship for 618 Squadron, thinking to protect itself from the impact of the inactive 'stores', the *Malaya* deliberately created a list by altering its ballast, with the effect of exposing more of its lower armour-plated hull, in an attempt to protect its thinner-skinned upper hull. The trials began, despite severe bad weather, and in a swell of fifteen-foot waves, 618 Squadron managed to drop several 'stores'. It was noted in the Squadron records: 'There was considerable activity during the early part of the month to test the new sighting device and the air turbine gear which, it was hoped, would replace the engine driven gear. Release of stores were made against the target ship in Loch Striven, the trials with double release were fairly successful from the technical aspect, and certainly very impressive. The target ship was holed by a release by Squadron Leader 'Shorty' Longbottom, much to the consternation of the Senior Naval Officers witnessing the trials.' This was a demonstration watched by representatives of the Air Ministry, who had assembled on the *Malaya* with senior RAF, naval officers and civil servants. They were treated to an unexpected, close up view of the 'store', performing exactly as it ought to, losing height with each bounce and running straight on track, when in mid bounce it hit the side of the ship, and instead of the expected traverse down the hull and capture in the protective net, managed to tear a hole in the

bow. The elite gathering of VIPs ran for cover in all directions and red Very lights were fired to stop any further attacks.

At Turnberry, it was noted, F/O H. J. Shacklady and crew (3 Ventura Course) force landed at East Fortune. The engine cut out at 16,000 feet and the aircraft descended out of control to 6,000 feet manoeuvring violently. The engines then picked up and the pilot effected a successful forced landing.

Four crews arrived for Torpedo Training on 43 Beaufighter Course; 42 Course had completed their training and were posted to 144 and 254 Squadrons.

Wing Commander W. C. McNeil of 17 Group, Accidents Investigation, arrived at Turnberry as president of a court of inquiry regarding the accident of Ventura FP580

Mid-month, Ventura AJ181 with an instructor and pupil crew tipped onto its nose on landing due to the overzealous application of the brakes, there were no casualties. Training continued apace, a visit was made by Wing Commander Belena-Prozonoroski, who gave a lecture on Poland.

A conference was held at Weybridge, Surrey to discuss the recent 'Highball' trials at Turnberry. 'Of the "stores" released eight failed to hit the target and several fell apart. The failures were attributed to very bad weather conditions, and in one case mechanical defect. In all cases of "stores" running short, it could reasonably be expected that the second "store" to be released would register a hit, and that those failing by a short distance only, would travel forward underwater by virtue of their remaining velocity and eventually reach the target. The demand for an improved windscreen for the aircraft was accepted, and it was proposed that future aircraft should have armoured windscreens vertical and fitted with wipers and that a high boost engine such as the Merlin should be installed to ensure high speed at ground level. The film of the trials was shown to a highly impressed audience.'

On 22 May 1 TTU was absorbed into 5(C) OTU with Beaufighters replacing the Beauforts.

There was a crash landing on the aerodrome

of Ventura AE987 when the port undercarriage failed to lock, and the pupil pilot made the landing with the starboard undercarriage locked in the down position. There were no casualties.

Beaufighter Courses 43 and 44 completed their training and were posted to 254 and 489 Squadrons.

JUNE 1944 5(C)OTU and 618 Squadron
The Beaufighter, Ventura and Hudson Training Courses continued and was added to, with the move of the Air Sea Rescue Training Unit from Thornaby and 5(C)OTU took over Air Sea Rescue Training along with the Warwick Conversion Course. Ventura V FP360 swung on take-off, the pilot was unable to control the swing and the tail wheel collapsed.

618 Squadron continued training from Turnberry, six stores were released against the target ship by S/Ldr Melville-Jackson with W/Cdr Hutchison sighting, with successful results.

On 9 June, 22 Avro Lancasters of 5 Group landed at Turnberry, diverted by bad weather from reaching their base at Skellingthorpe and Bardney. Nearly all took-off the same day when the weather cleared. Hudson and Warwick Conversion Courses were completed, and the crews posted to 269, 279, 519 and 521 Squadrons.

Beaufighter EL255 dived into the sea, Pilot F/Sgt C. H. M. Foster and F/Sgt C. Cunningham suffered slight injuries. Sgt Lacey and P/O Alexander landed in two Halifaxes from Topcliffe.

The representative of the Archbishop of York, Rev. D. Railler, arrived for a visit, as well as Group Captain R. A. Garrard, Assistant Chaplain in Chief, followed by S/Ldr J. Tracy RAAF, RC Chaplain and the Chaplain from West Freugh F/Lt J. Doherty.

S/Ldr A.F. Binks, DFC, 618 Squadron, was air testing DZ577 when, during a tight roll, the starboard undercarriage leg was flung against the stops. The emergency system was used, but on landing the starboard wing dropped and the propeller and wing tips caught the ground. The aircraft was categorized AC.

The month of June ended with the loss of S/Ldr Bateson when Beaufighter JL835 crashed into the sea at Turnberry after suffering engine failure.

JULY 1944 5(C)OTU and 618 Squadron
A directive received from headquarters, Coastal Command, instructed 618 Squadron that it was to be re-equipped with modified Mosquitoes for aircraft carrier operations in the Pacific. All aircrew were sent for intensive training at Crail, practising dummy deck landing with Barracuda aircraft, before practising landings and take-offs from HMS *Rajah* just off the coast at Rothesay. The RAF pilots were the first to land on this type of carrier, and they were suitably feted by the Naval personnel. The training was completed satisfactorily, with only six of the Barracuda aircraft being damaged and one was ditched in the sea off Troon by F/Lt MacLean.

At Turnberry, the pilot of Warwick Mk I ASR BV333 undershot the runway and went round again, but unable to lower undercarriage or raise flaps, together with the starboard engine missing, belly-landed on the airfield.

Commanding Officer G/Cpt. H. Waring left to visit St Angelo in Moth Minor X5122.

Warwick Mk I BV355 swung violently to port after landing and the undercarriage collapsed, the aircraft caught fire and was totally burnt out. The pupils of 2 Hudson ASR Course had a 'live' drill when Hudson Mk III T9544 ditched at 02:40 hours when the port engine failed, and the pilot was unable to maintain height. The crew of Hudson 'O' were briefed to carry out a search in the area for the crew of Hudson 'R'. The search party was airborne at 05:36 hours and at 07:40 hours Flying Control informed the operations room that the crew of Hudson 'R' had been picked up by a merchant vessel off the coast near Ayr. Hudson 'O' was recalled to base. Warwick Mk I ASR BV406, force landed successfully on the aerodrome after the starboard oil cooler burst, causing a loss of oil and starboard engine seizure.

AUGUST 1944 5(C)OTU
More men arrived to start the Warwick Air

Sea Rescue Course and the Hudson Met. Course. Warwick Mk I ASR BV393 made a heavy landing causing the undercarriage to fracture, the emergency system failed to lower the undercarriage. The aircraft belly-landed, caught fire and burnt out, Category E. Hudson V AM843, while carrying out air firing exercise over the sea, the rear gunner shot up the tail unit of the aircraft, the elevator and tail plane was damaged by bullets.

Warwick & Lighthouse

SEPTEMBER 1944 5(C)OTU and 618 Squadron
Training by 618 Squadron continued at Turnberry throughout September; the aircraft were based at Turnberry with a detachment of Squadron ground crew for servicing.

Further attacks on HMS *Malaya* on Loch Striven were made by W/Cdr Hutchinson, S/Ldr Melville-Jackson and F/Lt Hopwood, each pilot releasing four, four and two 'stores' respectively. The target ship was listed slightly so that the stores would strike the armour plating, hopefully without a repeat of the previous mishap.

OCTOBER 1944 5(C)OTU and 618 Squadron
The training continued with the Torpedo Refresher Courses and the Meteorological Course. The Operations Record Book is mostly concerned with listing the visitors to Turnberry. On 8 October 618 Squadron was given notification of arrangements for embarkation. Flying to Turnberry for weapon release practice in Mosquito DZ648, W/O A. Milne and F/Sgt E. Stubbs were killed, and the aircraft completely destroyed when flying through cloud, crashed into the hillside at Bramsdale in Yorkshire. The 'Top Secret' weapons were later removed by special police.

The final trials against the *Malaya* took place on 12 October when two of the three Highballs hit the target and managed to pierce the hull, one found its way into the wardroom and the other landed in the admiral's pantry. The ship's captain reported that had the Highballs been charged with explosive, significant damage would have been done.

All of the 618 Squadron aircraft had been ferried to Renfrew by 26 October and within two days had been hoisted onto aircraft carriers HMS *Striker* and HMS *Fencer* at King George V Dock in Glasgow. On the last day of the month the ships sailed, arriving in Melbourne, Australia two days before Christmas.

A Hudson belonging to 519 Squadron, on an air transfer flight from Gosport to Skitten, swung on landing at Turnberry. The starboard undercarriage and nacelle structure were badly damaged. This was followed by Martinet HN944 which tipped on its nose, when a tyre burst when taxiing, causing propeller damage and shock loading the engine. The month ended with Beaufighter JM331 catching fire while taxying. Torpedo Refresher Courses were completed, and personnel returned to their Squadrons and the Hudson Met. Course, on completion, went on leave pending posting overseas.

NOVEMBER 1944 5(C)OTU
Training continued at 5(C)OTU with Hudson's, Warwicks, Beaufighters, Oxfords and Martinets on establishment. 6 Warwick/Hudson Course commenced with the arrival of nine Navigators and 27 W. Op/AGs. 2 Torpedo Refresher Course was completed, and their crews returned to their units. 49 Beaufort Course completed their training and were sent on leave pending posting. 3 Torpedo Refresher Course with ten crews started training.

Beaufighter VI JL832 carried out a successful forced landing on the aerodrome due to an engine defect, there were no casualties. Seven more Beaufighter crews arrived to commence 50 Course.

Martinet 1 HP335 went missing on an affiliation flight. The aircraft took off at 09:54 hours and was last seen at approximately 12:30

hours. The pilot, W/O Williams, and Drogue Operator L.A.C. Laurie, both missing. Wreckage was later discovered in the sea.

Pilots and navigators arrived throughout the month, completed their training and returned to their squadrons.

Hudson AM646 landed on one wheel, short of the runway, causing damage to the undercarriage, managed to take-off again and on the second landing the port undercarriage collapsed at the end of the runway. There was no injury to the pilots, both of 6 Hudson Course.

3 Torpedo Refresher Course finished training and returned to their Squadrons.

The Commanding Officer of RAF Station, Turnberry, wrote to the factor of the estate requesting permission to use the slipway at the old lifeboat shed at Maidens for the purpose of launching the RAF boats; this was granted on the proviso that the RAF would be responsible for any damage caused and that they would reinstate all fences, roads, etc. to their former condition when they left.

DESCRIPTION OF TURNBERRY AIRFIELD

As at 1 December 1944
Grid Reference: G.S.G.S 3908. (1"-in map) S. 695299. Sheet 82 (Scot.)
Graticule: Latitude. 55° 19' 30" N.
Height: A.S.L. 35ft.
    Longitude. 04° 49' 45" W.
COUNTY: Ayrshire
    Locality: 7 Miles N. of Girvan.
COMMAND: Coastal
Nearest Railway Station: Girvan LMS.

FUNCTION: Operational Training Unit (Parent).
AFFILIATED AIRFIELDS: Nil.
LANDING AREA:
Runways   Extensibility

QDM	Dimensions	Dimensions Remarks
222°	2,010 x 50 yds.	
181°	1,1400 x 50 yds	
275°	1,250 x 50 yds	

Type of Surface: 6-in. Concrete, ¾ -in. Tarmac.

PERMANENT LANDMARKS:
    (i) *By Day:* Ailsa Craig, 1,115ft., 246° T., 12 miles. Lighthouse on W. side of Airfield.
    (ii) *By Night:* Lighthouse.

PERMANENT OBSTRUCTIONS: Hills 600ft., within 2 miles. Hills 900 ft., within 5 miles.
FACILITIES:
Airfield Lighting   Mark II.
Beam Approach
Radio: QDM
Flying Control: Yes.
ACCOMMODATION: All buildings temporary.
Technical:

Hangars		Hardstandings	
Type	*No.*	*Type*	*No.*
T.2	2	100 -ft. diam.	31
½ T.2	2	Fighter H.S.	3
E.O. Blister	24		

Domestic:

	Officers	SNCOs	ORs	Total
RAF	195	296	1,854	2,345
W.A.A.F.	15	14	408	437

DECEMBER 1944 5(C)OTU

The month began with the arrival of crews for the Warwick and Hudson Courses. On 8 December Warwick Mk I BV295, when taxying and turning off the runway, caught fire due to a petrol leak from the starboard engine and the aircraft was totally destroyed, with all the crew injured. News was received that Warwick Mk I BV340 had crashed at Prestwick Aerodrome when it overran the runway after attempting an emergency landing on one engine. The instructor and pilot were killed, and six other crew were injured.

JANUARY 1945 5(C)OTU

At the start of the month, all Warwick aircraft were grounded pending notification. Group Captain Turner assumed Command of RAF Turnberry with vice Group Captain H. Waring. A Hudson aircraft overshot on landing in poor visibility causing damage to the undercarriage, the pilot was unhurt. 51 Beaufighter Course was unable to complete training on time due to the poor weather conditions. A visit was made

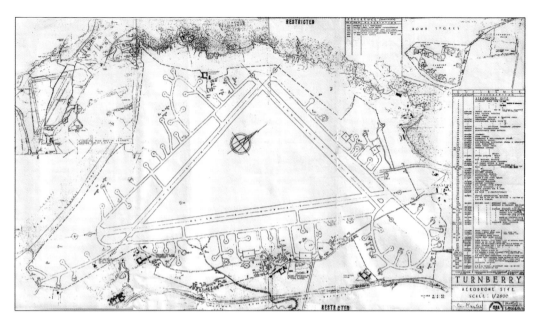

Aerodrome Plan

by F/O H. Stringer, the Personnel Counsellor
for the Canadians, from RCAF District HQ in
Edinburgh.

Beaufighter JL618, piloted by the CI, W/Cdr
Sandeman, force landed near High Drumdow
farm, near Turnberry. The starboard engine failed
to respond to the throttle control. Extensive
damage was sustained and W/Cdr Sandeman was
slightly injured. It was noted that the accident
was due to careless servicing.

FEBRUARY 1945 5(C)OTU
Beaufighter VI JL649, pilot F/L Johnstone,
overshot on landing – tail unit damaged. Court
of enquiry revealed that pilot placed too much
reliance on the use of his brakes, which, in this
aircraft were below average.

On 11 February, Hudson aircraft FK502 dived
into the sea two miles off Turnberry point. Pilot
F/O Beagley; Navigator P/O Pewsey; W. Op/AGs
Sgt Mariner and Sgt Foster; all killed.

Plans to initiate night attacks on target ship
with flares for Beaufighters, (Torpedo Refresher
Course) were deferred, owing to enemy
submarine activity. The Naval Authorities banned
the use of the Torpedo dropping area until further
notice.

618 Squadron, now in Australia, held
discussions throughout January and February
with the British Pacific Fleet, but no decision
could be reached as to how and when the
squadron be put into action. This state of affairs
dragged on until June, when a directive was
received from the Air Ministry that 618 Squadron
was to be disbanded forthwith. This notification
came as a shock to all ranks and, after all their
training, 618 Squadron was denied the chance to
make its intended contribution to the war effort,
and never received a formal squadron badge. The
squadron was disbanded with all speed, the unit
aircraft and equipment were offered to the RAAF
and the 125 Highball weapons they had in their
possession were to be statically exploded. Top
Secret, special equipment was destroyed by being
dismantled and taken from Sydney in a Naval
launch to somewhere in the South Pacific Ocean,
where it was dumped overboard. Squadron
Leader Rose witnessed the inglorious end of 618
Squadron.

MARCH 1945 5(C)OTU
Training continued on Beaufighter and
Warwick aircraft, including night exercises
with Beaufighters. Instructions were received to

postpone 11 Warwick Course for fourteen days in order to convert five crews from 179 Squadron and two crews from 6 OTU from Wellington to Warwick aircraft.

The Right Honourable Viscount Stansgate (William Wedgwood Benn, Secretary of State for Air, August 1945 – October 1946) visited the station and stayed two days taking discussion groups and debates for all ranks. Politician William Wedgwood Benn was a decorated WWI airman, who although aged 37 at the outbreak of war, was awarded the DSO, DFC, and the Bronze Medal of Military Valour (Italy) for his exploits. In 1918, he was the pilot of a machine which had been adapted for a special mission, this was the first military parachute drop, delivering spy Alessandro Tandura behind enemy lines. He joined the Royal Air Force Volunteer Reserve in May 1940 and served as aircrew on several operational flights as an air gunner, aged 67. He resigned his Commission in August 1945, as Air Commodore.

By the end of March, 53 Beaufighter Course was completed. With the termination of this course 102 torpedoes were dropped in succession without loss and a special letter of commendation was received from the Air Officer Commanding, 17 Group.

The month closed with a record number of flying hours and no flying accidents of any description.

1,241 hours 40 minutes – total flying hours for the month.

APRIL 1945 5(C)OTU
Warwick, Hudson and Beaufighter courses continued.

All was quiet until the middle of the month when an accident occurred to a Beaufighter aircraft piloted by F/O Mills. The undercarriage was selected instead of flaps on landing, the pilot was unhurt, but a summary of evidence taken against him.

Two Beaufighters, whilst on a formation attack, collided. F/O Nagley and W/O Smith, Navigator, of one aircraft were missing believed killed when the aircraft plunged into the sea. W/O Hopkins, pilot of the other aircraft made a successful landing at Turnberry on one engine. A green logbook endorsement was recommended for this NCO.

Owing to the ban on shipping in the torpedo area, a target ship had not been available for the crews of 54 Beaufighter course to complete their training, and operational requirements necessitated their immediate posting to squadrons without dropping runners.

April closed with information received from higher authority that no further flare dropping exercises for the Beaufighter aircraft will be made and that disposal instructions are to be issued for the Wellington flare dropping aircraft.

Total Pupil Flying Hours for the month:

Warwick	413
Hudson	274
Beaufighter	330

MAY 1945 5(C)OTU
With the end of the war in Europe in sight, operational training began to wind down. Lt-Col Brown, Capt. McKay and F/Lt Guy of the Intelligence Corps, RAF station Prestwick, visited Turnberry to discuss customs and security measures in regard to visiting aircraft from overseas. A concert party was given by RCAF, 'All Clear'. F/Off. Fahrenholtz was in charge of the show. On 6 May, instructions received from HQ Coastal Command that 13 Warwick and Hudson Courses were to be cancelled and the personnel sent on indefinite leave, 101 Navigators Holding Course was cancelled and further input restricted and further directives were received that 12 Hudson Course was to be cancelled and 11 Course brought up to Fortress standard before being cancelled. Aircraft disposal instructions would be issued.

Beaufighter RD426 crash landed on the beach whilst practising single engine flying, only minor injuries were received by the pilot.

On 25 May, 10 Gliding School was formed at Turnberry under 66 Group, Scottish Reserve, with a Slingsby Cadet TX.1 basic training glider.

Total Pupils Flying times for month:

Warwick	347
Beaufighter	33
Hudson	171

JUNE 1945 5(C)OTU and CCFIS
More discussions were held at Turnberry when W/Cdr Leitch ATC visited regarding the proposed summer camps.

Coastal Command Flying Instructors' School was to transfer from satellite St Angelo to RAF Turnberry under command of S/Ldr Stubbs. This unit was to function as a lodger unit at Turnberry and were busy getting ready to leave Northern Ireland, with a 'great deal of work carried out. The packing is almost finished, and all the aircraft are ready to be flown away.' A few days later, 'The weather was very bad all day making it impossible to commence the unit move to Turnberry. The heavy equipment was loaded and despatched to Enniskillen Station. A terrific party was held in the Officer's Mess following a guest night.' On 6 June the weather was fair with a few scattered showers. Thirteen aircraft were flown across to Turnberry after a very fine farewell 'beat-up' and formation. Some pilots had to make two trips as there were only nine pilots for the aircraft. 'The Bishop', the unit's tame jackdaw, and the collie-dog flew across. CCFIS settled in and the Buckmaster carried out some flying at Turnberry, F/L. Balshaw being converted by the Bristol Test Pilot. S/Ldr Stubbs and F/Lt Wootton were busy at St Angelo handing over buildings and clearing the place up. The main rail and boat party, under F/Lt A. Laney, left Enniskillen for Turnberry on 8 June, S/Ldr Stubbs and F/Lt Wootton visited Killadeas to pay their respects to the Station Commander, the Chief Instructor, and others who had done so much to help and make their year at St. Angelo so very pleasant. They then left by air in a Beaufort after a 'beat-up' of Killadeas Mess.

The rail and boat party arrived at Turnberry the next day; the complete unit was now there apart from the heavy equipment. All the personnel were given the weekend off so as to be able to settle down to the new airfield and some more hard work afterwards. Mr Birks and F/Lt Wootton flew in a Beaufort to Filton. The changeable weather dictated the subsequent daily activities, entries read: 'The weather was very fine all day. There was no work at all, except for a duty clerk. Most of the personnel remaining behind spent the day on the beach.' Total flying time – Nil. Followed by: 'The weather was very rough all day, with a strong wind from the sea.' Work continued getting the aircraft ready for the arrival of 13 Course. F/Lt Wootton was at Bristol getting 'genned-up' on the Buckmaster aircraft and the Centaurus engine. Group Captain Turner, the Station Commander, visited the school. F/O E. Ridgway, an old instructor, returned to pay a short visit before returning to Australia. Total flying time – Nil. Again: 'The weather is cold and stormy.' No flying took place, but the aircraft were serviced. 'The Unit was now settled in, but many reforms will be required before we shall feel really comfortable.' Bicycles were issued to the officers and airmen. Total flying time – Nil.

11 Warwick Course was finally completed after being postponed from March.

Information was received from HQ 17 Group that all surplus Hudson crews comprising pupils of 9, 10, 11 and 12 courses were to be prepared for aircrew re-allocation to ground duties and to be posted in due course, to Haverfordwest for disposal.

Warwick BV351 force landed, Pilot F/S Hayward uninjured, and Beaufighter aircraft piloted by P/O Loosemore struck runway with port propeller on take-off and landed without further damage. Disciplinary action is to be taken against the pilot.

All Canadian aircrew, a total of 43 airmen, proceeded on posting to RAF Station Bircham Newton, for disposal. Instructions were received from 17 Group that flying instructors surplus to the establishment plus 10% were to proceed to RAF Haverfordwest for disposal.

Pupil Flying Times for the month:

Warwick	274 hours
Beaufighter	241 hours

Anticipating that RAF might now be ready to vacate the airfield, a meeting of the Hotels

Sub-Committee, of the Scottish Committee of the London Midland and Scottish Railway Company, held in the middle of June, reported, 'The Chairman raised the question of the desirability of facilities again being provided at Turnberry for golf and suggested that one of the prominent golf architects should be asked to advise on the possibility of laying out an eighteen-hole course on the land at present available including the land on which the company has an option.' It was agreed that before anything was done the vice-president should ascertain whether there were any prospects of the Air Ministry relinquishing in the near future the golf course land, they at present occupy.

The staff of Turnberry airfield carried on, CCFIS records 'The weather was again very rough all day with a very strong wind blowing from the sea'. No training was done, inclement conditions lasted over a week. F/Lt Balshaw, F/Lt Foster, F/Lt Sorby, F/Lt Crowe and W/O Gunter flew in a Wellington to Bristol, landing at Filton. They took with them Mr Burrows, the Bristol Representative, and are going to tour the factory and workshops of the Bristol Aircraft Company. F/O MacLeod and F/O dos Santos flew in a Beaufort to St Angelo to collect a few belongings left behind in the move. Notification was received that S/Ldr F.H. Stubbs had been awarded the AFC in the King's Victory Birthday Honours. F/Lt Balshaw and the other instructors returned from Filton. F/Lt Crowe flew the Beaufort back. F/O MacLeod returned from St Angelo. The weather was very bad in the morning with rain and strong wind, it cleared in the afternoon. Three instructors played golf for the station team.

The first part of the heavy equipment arrived from St Angelo and unloading, and unpacking commenced. 'B' Flight at St Angelo are preparing to move to Alness. Preliminary arrangements were made for the start of 13 Course.

Still the weather was very rough with a high wind. There was no flying carried out. The Unit was now organised so as each weekend from lunchtime on Saturday was free for all until Monday morning. Officers were spending free time playing golf; the airmen seemed to enjoy the beach and cycling.

Every day the weather was mentioned, 'again rough with a strong wind from the SW.' 25% of the airmen, under F/Lt Balshaw attended church parade.

More heavy equipment arrived and was unloaded. Preparations were made for the incoming 13 Course – unfortunately a small one – two Beaufort and two Mosquito students.

Finally, on 19 June, 'A beautiful day, clear sky and very warm.' Air Vice Marshal H. G. Smart visited the school in the afternoon and toured the camp with the Group Captain and S/Ldr Stubbs. The officers played a golf match in the evening.

Now the weather had cleared all serviceable aircraft were air tested – three Wellingtons, three Beauforts and one Master. The afternoon was devoted to organised games – football, swimming, cricket, hockey and golf. Total flying time was 3 hours and 40 minutes.

Members of 13 Course arrived, F/Lt Bazalgette and W/O Newman for Beaufort training and P/O Mansfield and W/O Parfitt for Mosquito training. During the last week of the month the weather was good. On the Saturday, lectures and discussions were held in the morning. The unit closed down at lunchtime, resuming on Monday, when flying took place throughout the day with a total flying time of 7 hours and 5 minutes. F/O Bastable arrived for the Master/Martinet Course. Unserviceability did not permit Mosquito flying. On 27 June, the weather was good, and flying was carried out until lunchtime. The afternoon was devoted to organised games. The majority of personnel carried out physical training on the beach, then bathing in the sea. Unserviceability again prevented Mosquito flying, but flying was carried out by the Beauforts.

On the penultimate day of June, the weather, which was fit in the morning, closed down in the afternoon. Flying was carried out in the morning and early afternoon. F/Lt Balshaw flew a Wellington to Filton, returning with Mr Birks the Bristol Test Pilot.

The next day a visit was arranged to Hullavington for certain members of the unit to visit the Handling Flight to pick up information relating to the Buckmaster. A very interesting discussion on many points was held with W/Cdr Fryer presiding. F/Lt Balshaw, F/Lt Foster, F/Lt Wootton, and Mr Birks were in the Wellington which flew to Hullavington. Flying was carried out in the morning before the Unit closed down at mid-day until Monday morning.

Total flying time for the month – 96 hours, 15 minutes.

July 1945 5(C)OTU and CCFIS
A Syllabus for Beaufighter training was devised to cater for a maximum of eight crews on a six-week course.

Instructions were received from Headquarters 17 Group that their modified Warwick Course consisting of crews drawn from Wellington personnel was to last six weeks instead of twelve weeks, will be started.

The Accident Investigation Branch visited the station, and an inquiry was made into an unsuccessful lifeboat drop. The recommendations of this investigation, namely that the drop should be at 700 feet instead of 500 feet, were forwarded to HQ 17 Group. CCFIS report that F/Lt Balshaw and other members of the party who visited Hullavington returned. The Buckmaster was air tested but proved to be unserviceable. F/Lt Smith arrived for 13a Course on Wellingtons. F/O Bastable completed 11 Master/Martinet Course. CCFIS flying was carried out in the morning of 4 July; the afternoon being devoted to organised games. The Buckmaster was again air tested but proved to be still unserviceable. F/Lt Balshaw took an aircraft to St Angelo to see members of 'B' Flight prior to their departure for Alness. The next day F/Lt Balshaw returned from St Angelo. The weather was good throughout the morning but deteriorated thus preventing night flying from taking place. Unserviceability of the Mosquito aircraft again interfered with flying. W/O Folley took a Beaufort to Filton for the purpose of picking up spares for the Buckmaster. F/Lt Johnson, F/Lt Wootton, Mr Birks and Mr

Stone also proceeded to Filton. The weather, on 9 July, which had permitted flying for CCFIS in the morning, deteriorated after lunch. The afternoon was spent in discussions with students. The Mosquito developed trouble whilst flying and is again unserviceable awaiting spares. That same day, Hudson aircraft FK406 piloted by F/L Pullham was damaged on landing at High Ercall. The pilot was unhurt.

Mid-month, a very successful station sports day was held, with Station Headquarters scoring the most points.

Fifteen aircraft and crews from 254 Squadron arrived for special torpedo trials at Turnberry -these were carried out successfully. A formation of twelve aircraft of 254 Squadron attacked three destroyers. HMS *Clare* was holed by a circulating torpedo. A CCFIS Beaufort returned from Filton with aircraft spares and S/Ldr Stubbs returned from leave. The weather was good and night flying was carried out by 13 Course. Total flying time - Day 10 hours and 5 minutes. Night – 6 hours and 55 minutes. Making the most of the nice weather, a sports afternoon was held by CCFIS which was well attended, mostly by swimmers. The next day course flying was carried out. The two Beaufort students were tested and cleared, and both departed for their unit. The CCFIS instructors were finding things a little dull with no flying to do. On 14 July, although the weather was favourable, there was no flying on account of the unserviceability of the aircraft. The repairs to the Buckmaster were progressing and it was hoped soon to be airworthy. Three days later the Buckmaster was air tested but after less than one hour flying, again became unserviceable. An aircraft flown by F/O MacLeod was sent to Filton for spares. More aircraft were despatched on a maintenance mission - one was flown to Charterhall to pick up spares for the Mosquito, another to Bristol for parts for the Buckmaster. In the meantime, the afternoon was devoted to the station sports. The next day flying resumed, devoted to Wellington, Master and Buckmaster flying, and in the afternoon solely Master flying. The Buckmaster was u/s (unserviceable) after the first landing with tyre trouble. In the morning of the 20th, the weather closed in and was unfit for flying although a limited amount was carried out in the Master. The Mosquitoes were still u/s, and the Buckmaster was still awaiting tyres. CCFIS was plagued by a mixture of poor weather and u/s aircraft for the rest of the month. Daylight flying had been in progress in the Mosquito, Buckmaster and Master and night flying was laid on for the Mosquito but had to be cancelled because of bad weather. The Buckmaster again became u/s due to tyre trouble. When the weather improved there was Mosquito and Master flying. The Buckmaster was still awaiting spare tyres. Mosquito night flying was commenced but not completed; the aircraft went u/s with brake problems. This was the pattern for CCFIS for the remainder of July until the last day when *'The weather was excellent. A Mosquito flying category test was carried out, as was Buckmaster flying.'*

The month ended with RAF Station Turnberry competing against RAF Station East Fortune in the Inter Station Athletics meeting held at Rosyth. Turnberry were beaten by a narrow margin in the Inter Station Swimming competition at Rosyth by RAF Station Wig Bay.

On 30 July, when Beaufighter aircraft RD42 piloted by F/S Seabrook was making a dummy attack on HMS *Glasgow*, one engine failed, and the aircraft was ditched unsuccessfully. Both Pilot and Navigator F/S Ley were missing believed killed.

5(C)OTU total pupil flying times:
Warwick   143 hours
Beaufighter   104 hours
CCFIS total flying times: -
Day   184 hours.
Night   11 hours and 5 minutes.

AUGUST 1945 1 TTU and CCFIS
5(C)OTU was disbanded at Turnberry on 1 August 1945, but reformed back into 1 TTU and began training all over again. With the reduced need for training aircrew, attention was turned to recreation; RAF Turnberry beat Headquarters No. 17 Group in the area final of the inter-station golf

competition at Prestwick. At the beginning of the month, CCFIS were awaiting the arrival of a new course, which duly arrived, the weather was fit and 14A Course commenced, compromising four Wellington, two Mosquito, two Beaufort and one Master. The day was spent by the pupils signing in, collecting parachutes, reading local orders etc. The next day was also good and flying commenced, the Wellington, Beaufort and Mosquito were detailed. A Beaufort was sent to Bristol for Buckmaster parts and the unit had a long weekend off. Back to work on the Monday, the weather was fit so Wellington and Beaufort flying was laid on. The course was progressing satisfactorily, although the Mosquito was unserviceable in the morning. Training continued when the conditions allowed, when flying was cancelled a programme of lectures took place or the students participated in organised games.

Instructions were received from 17 Group that the Warwick ASR commitment was to move to RAF Station Kinloss from Turnberry on 16 August. CCFIS Mosquito, Wellington and Beaufort flying was detailed. The lecture programme was proceeding satisfactorily. A Mosquito burst the port tyre on take-off necessitating a wheels-up landing, there was no injury to the pilot. The aircraft was categorised AC. The Master Course was completed, the pupils cleared and returned to their unit. Back to sport and RAF Station Turnberry defeated RAF Charterhall in area inter-station cricket competition.

On 14 August an airborne lifeboat was dropped by Warwick HG116, the drop was unsuccessful and an investigation into the circumstances was made. A new establishment for RAF Station Turnberry dated 18 July was received, the policy being Torpedo Training Unit.

The next day, VJ Day was declared, RAF Turnberry closed down for two days. A travel ban was declared for a radius of twenty miles. A dance was held in the evening in the Station Institute.

The war was now over, but training carried on, Beaufort, Wellington and Mosquito flying was detailed. The Mosquito became u/s late

RAF Turnberry Marchpast

morning with tail wheel problems. A full afternoon flying was completed.

A thanksgiving service was held in No. 3 Hangar on 19 August at 10:00 hours, followed by a march past. The salute was taken by the Station Commander G/Capt. C. H. Turner.

Hampered by poor weather conditions for the remainder of the month, flying was difficult, so link training and lectures were carried out by students. 254 Squadron arrived to undertake refresher training at Turnberry. Notice was received that RAF Turnberry was to liaise with 18 Group and investigate suitability of RAF Tain for torpedo training. Time was allocated to organised games and Turnberry beat RAF Bishop's Court, Northern Ireland, by 6 matches to nil in the divisional semi-final of the inter station golf competition.

On the last day of the month, RAF Turnberry was defeated by RAF Tiree in the inter station cricket competition, but Turnberry defeated 57 MU in the inter station golf competition.

Total flying time for month:
Warwick   57 hours
Beaufighter   30 hours

The last entry for August reads: 'The Buckmaster was flying.'

SEPTEMBER 1945 1 TTU and CCFIS
On 3 September the Operations Room at Turnberry was closed down. All briefing now took place in Flying Control. S/Ldr Harrower, Senior Met. Officer, 18 Group, and F/Lt

Routledge, 18 Group, paid an initial visit to Turnberry on matters of general policy and supervision upon Turnberry being transferred from 17 Group to 18 group.

Course 15A (CCFIS) assembled consisting of two Mosquito, two Wellington and two Beaufort pupils. The first day was allotted to warning in, reading orders and the issue of equipment. The usual problems arose when, 'Weather doubtful, full programme detailed, p.m. allocated to organised games. Flying cancelled early morning, low cloud.'

The RAF & WAAF sports teams visited RNAS Heathfield, results: WAAF relay race first place, RAF relay race second place.

CCFIS reported weather dull, thick haze, flying detailed. Both Mosquito aircraft u/s. One awaiting spare flame trap, one undergoing acceptance check. The remainder of September chronicles the frustration of the bad weather, totally unfit for flying on most days. The aircraft serviceability was also a problem, so little was accomplished. Time was filled with lectures and training students on the Link Trainer, and more organised games. The Mosquito and the Buckmaster were continually unserviceable, so even when the weather was fine, flying was cancelled. On 20 September, in showery weather, flying was detailed in the now serviceable Mosquito 'W', again this was curtailed when F/Lt Johnson carried out a SE landing due to engine failure. Whilst on approach, starboard engine had vibrated severely and emitted clouds of steam. The propeller was feathered and a successful landing made. On examination, it was found that a connecting rod had fractured, and the bearing cap was forced through the crankcase. Night flying was planned but bad again weather necessitated cancellation.

The next day a gale warning was issued, with heavy rain and wind gusting up to 60 mph, no flying was done, and the unit closed down for a long weekend. Four crews from 254 Squadron arrived for 4 Beaufighter Torpedo Refresher Course and, the storm having passed, night flying was carried out and successfully completed.

G/Cpt. C. H. Turner represented Headquarters; Coastal Command versus Technical Training Command in a Golf match held at Lytham St Annes on 25 September. The CCFIS in preparation for the move to Tain sent F/Lt Crowe in a Beaufort to inspect the accommodation there. While at Turnberry flying training was completed on 15A Course, and 1 TTU were scaling down with the posting out of airmen to other duties. 254 Squadron left CCFIS and returned to its own station on completion of the course.

October 1945 1 TTU and CCFIS
A conference was held at RAF Station, Turnberry, to discuss the move of 1TTU and CCFIS to Tain, heralding the closing down of the station. The training programme was reduced, and again the weather and u/s aircraft hampered flying. More lectures and organised games! The weather brightened mid-month and flying was done nearly every day, at least while the aircraft remained serviceable, the Mosquito having problems on a daily basis. It was proposed to send the Buckmaster to Bristol to collect spares, but this trip was later cancelled due to rain and low cloud, proceeding a couple of days later. On return the Buckmaster was flown to Prestwick for a display, grounded there due to bad weather, returning after three days.

Farewell Menu 1945

November 1945 CCFIS and 1 TTU
CCFIS received a signal from HQCC that the move to Tain, was to commence on 12 November. 'Authority given for the disbandment of CCFIS w.e.f. 29/10/1945' and the formation at Tain of CCFIS to an aircraft establishment of 2 VE Sunderland Mk V, 2 Liberator Mk VIII, 2 VE Warwick Mk II, 2 VE Buckmaster, 2 VE Mosquito Mk III and 2 VE Master Mk II. A further signal instructed that 'Detachment of Marine Craft from RAF Turnberry to 22 MU Silloth, being discontinued and establishment deleted. Two marine tenders and one Dinghy at 22 MU to be prepared fully serviceable for movement by road at a later date.'

S/Ldr E. Withy was posted to Turnberry with the duties of Senior Administrative Officer, presumably to take charge of the 'stand down' of the airfield.

On 9 November, the Advance Air party of CCFIS proceeded to Tain and arrived successfully. Two days later a church parade was held in the morning, daily inspections were completed, and all aircraft loaded.

On 12 November 1945, 1 TTU, and the main air party of CCFIS departed for Tain, consisting of ten aircraft. The rest of the day was spent loading freight trucks, and the main party proceeded to Tain by rail in the afternoon. Two days later the rear air party flew out, ending all military flying at Turnberry.

10
WWII Airmen

As during WWI, men and women came to Turnberry from all corners of the British Commonwealth and Europe, from all backgrounds and trades. Some arrived to complete their final training before being posted on active service, others as instructors, having already been 'bloodied' in battle. Once again, the local people welcomed these strangers and took them into their homes and hearts. The memories entrusted to me have been written as told, in their own words. This is their story, of how they came to Turnberry, the friends they made and lost, and where their time in service took them when they moved on. Here I have recorded their stories, many being published for the first time.

Harry McDowell
Early in the war, I was employed as a civil engineer with Wimpey Construction and went to Prestwick Aerodrome, where it was decided that the runway needed reinforcing, to be able to cope with the Atlantic Ferry traffic that was arriving in large numbers. After a short spell I helped with the start of construction at RAF Ayr, Heathfield, until 1941, when I was sent to Turnberry to 'lay out' the airfield there. Basically, we were given the plans and told to get on with it. It was a standard type RAF airfield, so, me and two other chaps arrived on the golf course and began hammering in wooden pegs

Harry McDowell

and marking out the runways and buildings according to the drawings.

The first task was to level the golf course and demolish a detached house that was situated beside the road, right in the middle of what was to be the aerodrome. This property had only been built in the early thirties and was in use as a guest house. A temporary barrier was put across the road at the edge of the village of Maidens and again at Turnberry. In effect, the road was closed to the public for the duration.

We carried out our instructions from a wooden office erected near the road by Turnberry Hotel, and along with personnel from R. & A. K. Smith, Consulting Engineers, transformed Turnberry Golf Course into RAF Turnberry. We laid out the main site and installed services such as water and sewage disposal. The latter was carried out into Maidens Bay by means of a pipeline that is in use to this day. We had problems with this sewerage pipe because there is not enough of a gradient from the airfield site to carry the waste away. The Station Sick Quarters and mortuary were discretely built a distance away from the aerodrome, as were the WAAF quarters.

I well recall the first aircraft to land at Turnberry, a Blackburn Botha, which tragically hit a road-roller belonging to the contractors.

I met my future wife, Doris McQuiston, while I was at Turnberry. She was secretary to the manager. A local girl, whose family had been in the village for generations, in fact her grandfather had been a yachtsman for the Marquis of Ailsa. The family home was Rocklea, in Ardlochan Road, Maidens, and was next door but one to the Officers' Mess.

The arrival of the airfield brought the village to life, there were great dances in the officers' mess and the ne'erday parties were legendary.

There was a great social life with so many people suddenly appearing there also came a cinema and concerts performed by E.N.S.A.

Work continued on the airfield and buildings began to take shape, all types and sizes. Buildings to house the photographic section, workshops, classrooms and of course the accommodation, including a separate 'house' for the CO and even a squash court. One building designated as a torpedo workshop had the fuselage of a Beaufort placed inside. This was used for familiarisation training.

At this point I thought that I would like to join the RAF myself, and so applied and was accepted. I was then sent to St David's in Wales to work on another new airfield while I waited to be called up. Eventually, after a brief period, I was sent to South Africa to do my flying training.

After training I was posted Transport Command flying Dakotas. I mainly did para-dropping and glider-towing.

When the war ended, Doris and I married, and I gave some thought to resuming my old job with Wimpey. However, I realised that I had a liking for flying and wondered about the possibility of carrying on commercially. So, although still in the RAF I decided to try to obtain a civil pilot's licence and went to London to sit the necessary exams and passed. I realised that a lot of other chaps had the same idea and thinking that it would improve my chances of employment if I had a Navigator's Licence as well. Once again, I studied, returned to London and obtained the navigators qualification, and one month later, left the RAF.

I then enjoyed a couple of weeks of leisure living at Rocklea before I contacted Wimpey again. I was asked to visit their head office in London, where I was made most welcome and told that they had just the job for me. When I asked what and where, I was told, 'We are building a new Civil Airport at a place called Heathrow' (I had never heard of it). I thanked them for the offer and said that I would need to speak to my wife and think it over for a couple of weeks. I headed for Euston Station and home, and whilst waiting for my train I went to

the bookstall and on impulse bought a copy of *Flight Magazine*. On the journey north I read the articles and finally came to the back pages on which there were situations vacant. There was the job I fancied, so the very next day I posted my application, and was lucky enough to get the position, so started my career as a Commercial Pilot at Croydon, never giving Heathrow another thought!

By 1948 I was based in the Basler Hof Hotel in Hamburg. I spent my time flying ex-RAF Handley Page Halifax bombers on the Berlin Airlift, carrying food, not bombs. Early in 1949 I accepted a job with British European Airways, and there I happily remained, flying many thousands of passengers including royalty, notabilities, film stars and the folk next door. I retired as Senior Captain, British Airways, having lived a schoolboy's dream.

J. D. Thomson

I was employed as an electrician with James Scott & Co. of Dunfermline, who at that time were one of the largest, if not the largest electrical contractors in Scotland. I started my apprenticeship with them in October 1941 and I think it was in December that I was sent down to Turnberry.

There must have been about twelve or fifteen electricians and apprentices on the site and Scott & Co. had another contract at a seaweed factory between Turnberry and Girvan, then run by a company known as Cefoil Ltd. The site was well developed when I arrived. The main contractor was Messrs Wimpey who were busy laying the runways. The main runway was roughly the line of the main road between Turnberry and Maidens, and the secondary runways formed a triangle with the main runway with the right angle being somewhere about the lighthouse. I worked in various areas of the aerodrome such as the large T2 type hangar near the control tower, the Control Tower itself and the WAAF site on the Kirkoswald Road and the Link Trainer building.

Most of us were in 'digs' in Girvan and travelled by train to Turnberry. There was a train which ran from Glasgow direct to Turnberry in the morning and returned in the evening carrying a lot of the workers. I cannot remember much about the layout of the site or the technical details of the airfield on that first visit. I do recall that I was in three different 'digs' in Girvan, always moving around to try and find decent grub at a time when the rations were minimal, and our 'piece' consisted of four slices of bread, two with hard cheese and two with fish paste photographed on! However, I was transferred away to another job after a couple of months, I think it was up to a seaweed factory at Barcaldine near Oban, which was being developed at that time.

Round about June 1942 I got word to report back to Turnberry to assist a cable-jointer on the airfield lighting project. This time I was in 'digs' in Maybole and travelled to the site by bus. This airfield lighting job was designated 'contact lighting'. It involved installing a high-tension cable along both sides of the main runway and tapping off with transformers at regular intervals to light fittings flushed into the tarmac. This was a sophisticated and expensive form of lighting for a temporary airfield as there was another system called 'Drem' lighting (after Drem Airfield) with which I was involved elsewhere. The 'Drem' system used less expensive rubber cables and much simpler light fittings.

The summer of 1942, or was it 1943? was wonderful as regards weather. The sun seemed to shine every day and we were out in the middle of Turnberry Airfield enjoying it all.

The main type of aircraft stationed there seemed to be Beauforts and Beaufighters which were used for torpedo training operations. While we were engaged on the airfield a squadron of the new Mosquitos arrived and they were dispersed away from the main airfield, for security purposes I presume. While we were on the airfield lighting project, we were supposed to fill in the cable track and joint holes whilst there was flying on the main runway. We were supposed to evacuate the area with all our equipment whenever an aircraft was taking-off or landing, but this hardly ever happened. We would be busy down a joint hole, or more likely lying

sunbathing, when a sudden roar would herald the arrival of a flight of Beauforts, which would land and bounce their way with tyres smoking towards the end of the runway. We were never bothered by the Mosquitos as they took-off and landed with great speed and efficiency.

I do remember many accidents on the actual airfield, but there was one which happened so dramatically that I shall never forget it. One evening when we were heading towards our bus to take us back to Maybole we saw a Beaufighter making his approach over the sea towards Maidens, heading for the main runway. We remarked that he seemed to be taking it very slowly when suddenly the aircraft nosedived into the sea and disappeared. I don't know whether the aircraft or crew were ever recovered.

After I left Turnberry, I was sent to quite a few more RAF and RN stations such as Prestwick, Stornoway, Crail, Balado Bridge and finally Evanton where I left to join the RN in 1944.

Andrew McMillan

I was born and bred in Girvan and had six brothers and two sisters. When I left school, I became a baker, but in the early 1940s, I left to become a timekeeper for Wimpey Contractors at Heathfield where they were building the airfield. It was mostly Irish labourers, and we were told to be on the lookout for 'Ghost Workers', a scheme whereby men who didn't exist were drawing wages! Later that same year I was transferred to Turnberry where they had started work laying runways etc. I recall walking along the fifth fairway when a torpedo flew right over our heads and landed in the rough. This shot had apparently been fired from a ship who was having problems with range finding! The first time I saw a Spitfire was on the road at Turnberry - it had made an emergency landing with engine problems. There was a Canadian Sergeant called Redmund who ditched his aircraft just off Ailsa Craig and was picked out of the water by an air sea rescue launch. He had a girlfriend who worked in the draper's shop in Girvan.

I joined the RAF in February 1941 and was

attached to a Spitfire squadron in the Western Desert, who were providing cover for our front-line troops below the Marath Hills, Tunisia. The Germans and Italians held the hills. I was wounded and sent to Tripoli to be patched up and was given one week's convalescence leave at the rest camp. One morning I decided to visit the canteen for 'Tiffin' but found that it was mobbed and there were no seats available. I eventually saw a table occupied by four aircrew, but one of the chairs was tipped up, leaning against the table. I asked if it was alright for me to sit down, and on hearing my accent, the crew enquired where I was from. I told them I was from Girvan and that they had probably never heard of the place, to which they all smiled and said they knew it well and that it was a dump! Every one of them had trained at Turnberry. They allowed me to sit down, and then proceeded to tell me that the empty seat had belonged to their navigator who had been killed that very morning, apparently they had been walking out to their aircraft when he saw a vacuum flask just lying on the ground and had stooped to pick it up, it was booby trapped and exploded in his hand, killing him instantly. I was stunned when they told me he was a Girvan man called Campbell Ingram. He had been one of my best friends at school, and we had many happy times together. They asked that I pass on their condolences to Campbell's mother next time I was home, but when I saw her, I couldn't bring myself to speak to her. She looked so drawn and tired. I have felt guilty about that all my days. Perhaps, though, some things are better left unsaid. Campbell's brother Lawrence Ingram was also in the RAF, he was a tail gunner and was awarded the DFM.

When I had recovered from my wounds, I was attached to 69 Squadron at Malta, and again discovered that most of the aircrews had passed through their training at Turnberry. We then landed in Italy after Celerno, and I was put in charge of extra messing at Pompei.

From there I went to Claggenfort in Austria where I was part of the advance party, and we requisitioned an avenue of terraced houses. They were lovely buildings, full of pine furniture. One

of our duties was to search for weapons, etc. At one house I was watched all the time by the Haus Frau, who stood with her arms folded and followed me from room to room. In the bedroom, hidden at the back of the wardrobe I found a full Nazi dress uniform, complete with sword. The woman told me her husband was an Ober Lieutenant and was now in the laager. I continued my search and found a pistol underneath some blankets on top of the wardrobe. The Haus Frau explained that her husband had told her to keep it, in case the Russians came. I took the gun out of its holster. I had never seen the likes of it before and while examining it I pulled back and released the bolt. I then saw a button on the side and on pressing it the magazine fell of, not thinking I then pulled the trigger and to my horror the gun went off. Luckily the weapon was pointing toward the floor, the bullet travelled between the Haus Frau's legs, ricocheted of the hard-wooden floor into the wall. The Haus Frau ran out screaming and was never seen again. When I handed the gun in, I was told it was a Mauser 45.

At Claggenfort I was responsible for the messing of about 40 personnel. Here we were busy dismantling surrendered Messerschmitt aircraft.

My staff consisted mainly of POWs, displaced persons and deserters. They did general chores in the kitchen and cut firewood, etc. In return they were fed and given cigarettes. I even formed them into a male voice choir that was quite successful.

Only one of my brothers, James, joined the RAF, he lied about his age to get in and survived 32 operations on Lancasters. He told me that he was terrified when they were on raids over Germany, mind you, he was only seventeen years old then. My eldest brother Hugh spent the war manning anti-aircraft guns in Kent. Another sibling, John was a miner, but ran away to join the Army in 1935, so was a regular soldier with the RSF at the outbreak of war. He was on patrol in France, walking down a country lane at night when he saw a puddle glinting in the moonlight, he walked around it to save getting his boots wet and stepped on a mine. He lost a leg. William joined the Black Watch and while on patrol in Germany one of his party was wounded. They called for a stretcher bearer, but before they went very far the stretcher bearer was also hit. William took his place, but before he'd gone three paces he too stepped on a mine and lost both his legs.

Robert Logan

I was born in 1920 at Turnberry and on leaving school went to work at Shanter farm with my father. Shortly after the outbreak of war, we heard rumours that there was to be an airfield built at Turnberry. I recalled the stories of the First World War airfield told by the old worthies of the village and wearied to see it coming. There was nothing in the Maidens at that time, so an airfield would be quite exciting.

Robert Logan and fellow Observer Corps personnel

Soon afterward Wimpey's contractors arrived with huge earth scraping machinery to level the ground, there were other sub-contractors too. I was glad to see life coming to the place. Early on I volunteered for service with the Observer Corps, to start with we had no uniform or special clothing, just an armband. Our post was situated on the hill behind Turnberry Hospital (Hotel). We had four full time members, who usually did the day shifts and we part-timers who did the night shift after our day's work on the farm - shifts were 8 pm until 4 am. We had to report everything we saw to Prestwick control; we had no contact with Turnberry control at all. We had to record everything and were told nothing. At first our post was pretty basic. We had a paraffin

Brothers Robert and David Logan

stove to keep us warm, but it was not much use; then we had a coal fire, it was freezing at night. We were then issued with wellies and a balaclava. There were other Observer posts at Barr, Barrhill and Dunure and we kept in contact with them by radio headset, tracking aircraft or shipping from sector to sector so that the whole area was covered in this way. Nothing moved in the skies without it being watched. I still remember the weather reports we had to give; Roger Charlie code meant blue sky and cloud. We had training in rifle drill, aircraft recognition, which was ongoing, we had to identify silhouettes of different types of aircraft. Japanese aircraft were given women's names, probably because their real names were too difficult for us to pronounce. We also had some crash rescue training and were shown how to get aircrew out of a crashed aeroplane, where the hatches were, etc. On one night shift a Wellington flew straight towards our post, it was on fire, and coming straight for us, I didn't know what to do. It just missed us and managed to land okay, I saw it parked up next morning, the tail section was burnt and blackened, it seems a flare had caught in tailplane. The only other near miss I had was when I was riding my motorbike down Shanter Road and I collided with an airman who stepped out in front of me, I was thrown off and skidded down the tarmac, losing the skin off my knuckles on the way, the airman was hurt as well.

The airfield grew, and the road was closed

between Turnberry and Maidens. If you wanted to go to Girvan from Maidens, you had to go via Kirkoswald. There wasn't much evidence of strict security, farmers went about their daily business, although I had clearance to enter certain areas of the aerodrome. Italian and German POWs worked at Shanter and helped with the farm work. One, called Helmut, had been shot on the Russian front - he enjoyed showing the scars of his wound. Some spoke good English. They were kept in the German camp at Greenan. The Italian POWs were housed at Coral Glen in Maybole - this area was nicknamed Baloney.

I got to know quite a few of the aircrew, and they used to show me around the aircraft, sometimes I even got to go up for a flight, although I didn't like the Beauforts much as you had to sit on a cross member of the main frame, which was quite uncomfortable. The Beauforts always seemed to be nose heavy and came in to land at a great speed. As the site got bigger, the facilities improved. We got to go to ENSA concerts in the Sergeants' Mess and get our haircut at the camp barbers. The base even had its own church, where Redgates caravan park is today.

The social life of the Maidens took a turn for the better, and the local pubs were always full of airmen. We made friends with a fair few of them, mostly Canadians. On a Friday night there were usually dances on in Richmond Hall in Kirkoswald, dancing to McKay's Band. There was no drink allowed at the dances. Public transport wasn't what it is now, and we used to pushbike everywhere, sometimes three to a bike. One night I had a contest with a Canadian pilot to see who would escort a young lady home from a dance, we tossed a coin for it, and the Canadian won. I used to go to the pictures. There were lots to choose from then - there were two cinemas in Maybole, and two in Girvan. I usually went to the cinema to get some sleep after dancing the night before!

The mess was always a lively place to be. The airmen used to serenade local girl Nessie McDowall, singing 'Turnberry Bay' to the tune of the Irish ballad 'Galway Bay'. The NAAFI

beer was rotten. Lots of local women worked in the NAAFI, where we played games such as housey-housey. Women didn't drink in pubs then, not nice ones anyway. I remember one Christmas Eve there was an Irish pilot, who had a wonderful singing voice, had the whole mess entranced by his version of Bing Crosbie's 'White Christmas'. Later that evening, he took off in his aircraft, only to crash into the sea. He was killed. I remember him to this day, a large, cheerful, red head of a man.

I always wanted to join the RAF and to fly aeroplanes, but what I saw at Turnberry put me right off. There were many, many crashes at Turnberry. Some said it was the air turbulence coming off the sea that caused them, I don't know. Some said that the aircraft were not airworthy. I know of one pilot who refused to fly because his aircraft had problems and he was promptly put in the guardhouse. Another pilot was detailed to take it up, and immediately crashed. The accidents at Turnberry were horrific, sometimes it seemed that there was one a day. Friends we had made were suddenly not there anymore.

It was a very busy airfield, all types of aircraft coming and going, crews changing all the time. One-night loads of Lancasters arrived and in the morning Shanter Road was choked solid with all these aircrew going to the mess for breakfast. I recall the nights we heard the German bombers going over on their way to blitz Clydebank, the noise was deafening, the windows rattled, even the stove in the lookout post vibrated. We used to call enemy aircraft 'Away Doos' (pigeons) and our own aircraft 'Home Doos'. We had real homing pigeons which were billeted at Shanter farm with the ground crews. These pigeons were carried on aircraft as messengers in an emergency, a primitive method of communication, but one that worked.

With all the wartime shipping going up and down the Clyde, there was always something to be found on the beach. I found a cap with the badge of the Saskatchewan Light Infantry, on another occasion a canvas bag containing whistles, some half made, the name Hans Klein was stamped on them. These possibly came from a U-boat.

As I've said, the crashes were awful. I was standing on the railway bridge at Shanter farm one Sunday morning, watching a Ventura take off, when it suddenly swerved, and overturned, right in front of the control tower. I heard there were eleven people killed in that aeroplane, but we were never told any details. Many others just crashed into the sea. The ones that crashed on land were either on landing approach or on take-off. Usually engine failure.

I also witnessed a collision between a Beaufort and a Hampden that were coming into land, just opposite where Malin Court Hotel stands today. The Beaufort seemed to come in steeply and smashed into the top of the Hampden, which was also on final approach. The aeroplanes immediately caught fire. The local policeman managed to drag some of the crew from the wreckage and roll them on the ground to extinguish the flames. He burnt his hands very badly. I remember seeing a glow in the sky over toward Ayr. I later learned that two Beaufighters had collided, killing all on board. A Beaufighter crashed at Morriston farm. There were three killed in that accident. There was an air mechanic on board. Three Beaufighters had already managed to land, but the fog came down and the fourth overshot the runway and crashed in the field beyond Morriston - you can still see the gap in the hedge today.

Hardly a night passes without me dreaming of the airfield and the men that were there, I close my eyes and the memories come rushing back.

There are lots of the lads from RAF Turnberry buried at Dunure (Fisherton) Cemetery, lads from Canada, Australia and New Zealand. The older locals still tend the graves and remember.

Jan Sloan

I was born into a well-known Carrick fishing family in a house on what is now Shanter Road in Maidens. In those far-off days, houses had names, not numbers. Being at the top of the brae ours was, predictably, 'Hillcrest'. When the Wimpey men arrived and began to build the

Jan Sloan

airfield, I had just finished primary school and was starting high school in Maybole. There was a large building left over from the first war, called the Sailmakers' Hall, along the Turnberry road where we used to have Sunday School parties when I was wee. Along with other buildings on Warren and Clachanton farms nearby, it was demolished to make way for the runways.

The airbase had a huge impact on village life, for up until then Maidens had been quite isolated. The war was more than half over before the social side of it impinged on my consciousness. I remember knitting mittens and squares for blankets and balaclavas and my mother helping in a canteen in the local church hall. The buses were always full – indeed, sometimes you didn't get on at all, and once I rode to Ayr with a WAAF perched on my knee.

For a teenager it was all very exciting, rather frightening, and sad too. We went to concerts and church services and dances. The best dances were in the sergeants' mess and you had to be given a printed invitation - I still have one of mine. In those days we were naïve and, in many ways, innocent and the future was uncertain. I remember going to dances in the Richmond Hall in Kirkoswald and being walked home or riding home on the cross bar of a bike - two miles in utter darkness. Those young airmen were often total strangers, yet nothing untoward ever happened. It was as if there was an unwritten code of honour amongst them. No doubt we gave our mothers a few anxious nights though. At the time I don't think we gave much thought to the risks they ran flying the planes and it was always a shock if someone you knew got killed. Life was very much a day-to-day affair. When I went out fishing in my rowing boat, I'd quite often end up with two or three extra crew. The Canadians used to bring chocolates called 'Smiles 'n Chuckles' or little discs of sugary maple syrup.

My Dad used to bring home lots of fresh fish each weekend and this was much in demand. He sent me to deliver some to a Squadron Leader who lived in a house in Turnberry and I thought I would take a shortcut straight up through the base. All went well at first, but at the second picket post the guard got really nasty because I had been allowed through. He let me go only after many phone calls and a big fuss. When my Dad came home at the weekend, he found that the terribly loud noise of the planes roaring in to land just behind our house, when they were doing "circuits and bumps", kept him from sleeping. So, he used to go out and stand on the coal bunker and watch them. Sometimes if they forgot to reel in their trailing aerial, they would rip a few slates off the roof. I'm sure he never thought of claiming compensation. If a plane went into the sea near the village and the fishermen were at home, they would row out to search for survivors - but I don't think they ever found any. They did, however, manage to pick up boys from aircraft that deliberately headed for the boats if they had to ditch.

One day, walking down the road from Blawearie, I saw a cloud of smoke which obscured our house. When I got home my mother

was out with the stirrup pump hosing the place down. The house was covered in glycol and in the field behind us was a huge pile of burning wreckage. The military policeman had ordered my mother to leave the house, but she was afraid it would catch fire. In fact, she had been ironing in front of the window and saw the Hampden and Beaufort collide. As far as I know all the aircrew were killed.

On another occasion I was having a late breakfast, having been out to a dance the night before, and was looking out the window across the Firth of Clyde. A Hudson doing torpedo training flew too low and suddenly plunged into the sea. One of the crew, Jimmy Jenkins, was the lad who had brought me home from the dance just a few hours before. None of the crew was saved.

My mother had a soft spot for Australian boys because both her brothers had emigrated there. She adopted one crew, starting with the navigator Bill Wong Hoy, a Chinese-Australian who charmed us all. He, in turn, brought the two other Australians of the crew – Ken Abbott and Stu' Wright … and the rest is history. I married Ken's cousin and am still in touch with Bill.

Turnberry airfield changed our lives dramatically. It scattered the local girls to Canada, New Zealand and Australia … even to England … and the village had changed forever.

Douglas Andrew
My father was a fisherman, and I was born in 1930. When the airfield came to Turnberry, it was a cause of great excitement to us schoolboys. With shouts of 'There's a plane coming', we would run across the fields to watch it come in to land. I remember the first aeroplane I saw on the airfield running up its engines to full power while a couple of the aircrew sat on the tail to hold it down.

The first I remember of the aerodrome was the buildings of the First World War, which were used by farmers for storage, being demolished to make way for the new hangars. There was a navvy camp at Kingshill, and they used to have concerts on a Friday night. There wasn't much in the village at this time, not even a pub, the estate wouldn't allow it.

The airfield surrounded the village of Maidens, and the whole area was taken over for the training of aircrew. There were targets in the sea just off the shore at Balkenna and the aircraft used to drop smoke bombs on them. The 'planes would come in low over the village to land.

During the war, the place became alive with people from all over the world. We had airmen from other countries and evacuees from Glasgow, who even brought their teachers with them. There seemed to be so much going on - I remember the convoys sailing down the Clyde, streams of ships coming and going. Local families made friends with the men stationed at RAF Turnberry, and there were a few romances going on! Local women ran a teashop in the church hall and served home-baking to the servicemen. Our family didn't do too bad as far as food was concerned, fishermen got extra rations. My father spent Monday to Friday at sea.

My elder sister, Sheila, had become friendly with a pilot called Sergeant Taylor. He used to visit and blether with my father about fishing every Friday night without fail, and he would come and have dinner with us. My father was a very devout man and attended the church every Sunday of his life. One day when he was out fishing, just offshore, he watched an aircraft come in to land, when it crashed and exploded on the hill behind the airfield. He told us that he turned and said to his mate, 'There goes some mother's poor son'. Sergeant Taylor didn't come to visit as usual all that week. Finally, for Sheila's sake, my father telephoned the Sergeants' Mess and was told that Sergeant Taylor had been killed. It had been the accident that my father had witnessed from his fishing boat, but no one from the RAF had come to tell us.

There were lots of crashes. I recall one sunny summer Saturday afternoon when a Spitfire crashed into the sea behind the Maidens rocks. The tide was out, so the fishing boats couldn't get a rescue party to it. After a short while a small motor launch arrived, and I watched as they searched the area, eventually they gave up and

I remember the launch coming into the harbour wall, all they had found was a boot.

Many aeroplanes went into the sea. The pilots didn't like the sea when it was calm, it was hard for them to judge the horizon. Sometimes the wingtip of the aircraft hit the water and the plane would crash. There was an Irish pilot I remember who tried to ditch his aircraft near to the fishing boats because he had engine problems. His wing-tip caught the waves and the 'plane spun in.

A local policeman was awarded a medal for trying to rescue men from a burning aeroplane. This was on the airfield. Two 'planes had collided when trying to land. The policeman had very severe burns to his hands. He was very brave, but sadly the man he had rescued died later that day. There was a famous singer who was killed while training at Turnberry, his name was Bob Ashley. Everyone was talking about it; we were all very sad when a 'plane crashed. The villagers had made friends with the lads, and had taken them into their homes. Most of the local men, being fishermen, had joined the Navy and if it hadn't been for the airfield, the place would have been desolated.

The whole area was turned over to the war effort. There was a munitions factory at Girvan and a huge military base at Cairnryan. There were Italian POWs billeted at the Turnberry end of the airfield and Germans were kept in a camp at Doonfoot. These prisoners of war worked as general labourers and helped with the harvest.

After the war, some of the accommodation blocks were used to house displaced persons and POWs awaiting repatriation. I remember seeing the men in the village, wearing tunics with D.P. printed on them.

When I left school, I went to work at Shanter farm on the airfield. By this time the war was almost over. I left Shanter to work at Auchincruive for a couple of months and when I returned, everything had gone. The accommodation site had been demolished, and most of the airfield buildings too. The village became an isolated fishing harbour once more.

I got a job with Baird Bros. and demolished the munitions factory at Girvan prior to building what is now Grant's Distillery.

I shall always remember Sergeant Taylor, and many of the other airmen, some of whom came back and stayed after the war. Sergeant Thorburn was one - he taught at Maybole High School, and married a local girl, as did many of them. Quite a few village girls married aircrew and moved away, so now Maidens has connections all over the world.

Sam Wright
The following is taken from Sam Wright's book, *Up, up and away:*

Turnberry, to me was rather like a 'Finishing School', to put the final touches on both pilot and navigator, although we really had no specific duties to learn, or perform and went along mostly for the ride. And it was quite an exciting ride, to sweep along around the Firth of Clyde, with our pilots manoeuvring to get that desired optimum position from which to drop the deadly torpedo. My pilot was Jeff McCarrison from Auckland, New Zealand, and we both greatly enjoyed the training and the hospitality of the local people.

The 'torps' dropped in these exercises were not, of course deadly, but known as 'Runners'. They were 18 inches in diameter- as against the 21-inch diameter naval torpedoes- and the head was filled with water, to get the correct balance. They were fitted with a small plywood tail to ensure the correct trajectory in flight and set to run under the target ship. After a suitable time interval, the air was automatically expelled from the nose and the torpedo rose to the surface, to be recovered by a team using a small motorboat.

The pilot had to fly to very strict parameters to ensure a successful drop, and this, together with the rather obvious need to have the correct position in relation to the target, demanded a flying skill rather more than most types of operations. As the whole process was carried out at extremely low level, the initial approach to the target being made at some fifty feet above the sea, each drop demanded a full supply of adrenaline, and sweaty palms were not entirely unknown, particularly as the view from the navigators seat in the back revealed the 'deadly dangers' of the rapidly moving sea seemingly just beneath ones bottom!

The best approach to the target was from sixty degrees off the bow of the ship, in order to get the best 'comb', - if I remember the correct term - in that the closing angle of ship and torpedo would provide the best possibility of a hit. The aircraft had to be flown at 195 knots, 150 feet above the sea, straight and level, and the torpedo released at 1,000 yards from the target. Put all this together as a 'workload' for the pilot and it is safe to say that he had his hands full! Oh! I nearly forgot to mention that when, on actual operations, as all of this was carefully being established, the gunners on the target ship had not been idle, and had by this time drawn an ominous bead on the rapidly approaching foe and seemed to derive considerable pleasure for the opportunity to check on their marksmanship.

And another thing became apparent. With the aircraft flying at 195 knots, it took about ten seconds to cover the 1,000 yards to the ship, so there was no possibility, after dropping the torpedo, of turning tail to retrace the approach path as the gunners on the ship would have had a sitting target. The plan was, therefore, to keep going over the target and hope that the gunners had by this time 'chickened out' and got their heads down.

The runway was situated right on the coast, with take-offs immediately over the sea, and into the broad expanse of the Clyde. It was pleasant flying, and I see from my logbook that Mac and I did thirty-five 'sorties', in just under one month, and on one day we did five very quickly in a row. The Island of Arran was a good backdrop to our flying, with the 'sugar lump' ... the small island of Ailsa Craig always very close to our 'dummy runs' ... and to determine the accuracy of our pilots in this intriguing technique, our aircraft were fitted with several cameras synchronised with the torpedo dropping mechanism. The target ships were not large, but in true naval tradition they had the imaginative names for ships which reflects the perhaps more intellectual or sensitive standards of the naval officers. In this regard, there are many legion stories of the semaphore exchanges between ships, often in the most disastrous or hazardous situations, which capture so pertinently, the humour and inimitable style of the naval officers. Perhaps this is a legacy of the naval heroes of the past, like Cook and Nelson, who exercised with little apparent effort, a management and leadership style which could only be the envy of lesser mortals.

To return to the theme of target ships. I see that we attacked in quick succession, His Majesties ships, *Malahne, Heliopolis, Cardiff, Ramillies* and *Dido* ... what naval encounters of the past such names evoke. On return to base, the cameras were quickly relieved of the exposed film, which was processed and then analysed by several young, and I must say, very attractive, well-educated and intelligent Women's Royal Air Force Officers. (I think there was not one man of our course who did not harbour intense feelings of desire for these young ladies, who, though professing not to notice, would have been dumb in the extreme not to be aware of our flushes and 'tongue-tied-ness' when they debriefed us as to the results of the drops.)

Each exercise had an abbreviation, for simplicity, and we did many 'ALTs' ... Attack Light Torpedo, which was innocuous enough. We then progressed to 'ARTs' ... Attack Running Torpedo. We then tackled the more difficult flying for the pilot in the exercise, 'FALTs' ... Formation Attack Light Torpedo, but when this was done with running torpedoes, the abbreviation was, for its day somewhat rude, particularly when uttered in the presence of these sweet and desirable, but innocence unproved ... at least by any of our course members, as none admitted to any liaison with them, innocent or otherwise, ... conversation was even more circumspect.

Our activities, on joining an operational squadron in Coastal Command, would entail close co-operation with naval forces, and to further our knowledge of the navy, it was arranged that we would visit a navy submarine at the docks in Glasgow. This was looked forward to with pleasure, as a break from the flying and a close look at a submarine greatly interested all of us. We were a mixed group, Mac and I were both NCOs at the time, but the hospitality of the navy, usually fairly rigid in the distinction between officers and 'other ranks' held no sway in the welcome we were given and the tour of the submarine.

There was not one of us who would have willing ventured to the depths of the ocean in the vessel that we saw. Even to be on board, tied to the bollards made some of us squeamish. The narrow access ways, the cramped working

17B Course

conditions and the area in the bows where the torpedoes were in their tubes was unbelievably cramped. To crawl through the water-tight doors, aware that one could be trapped, in emergency situations, made the blood run cold. The Navy men responded to our perhaps obvious concern by gallantly suggesting that our role in the hostilities was far more life- threatening, but I 'had mour doots'.

Having been thoroughly scared by all this, then followed the good news. We were all, irrespective of rank, invited to the wardroom for drinks. The crew obviously had our measure, no doubt by previous visits from Turnberry trainees, and the drinking that followed was a most unequal contest. Perhaps they had the edge, being on their own ground, so to speak, but what followed is still a total mystery to me, beyond any hope of recall and perhaps better left in that alcoholic limbo which protects us all from too much embarrassment!!

Another 'contact' we had with the Navy was during our low-level flights, playfully dropping torpedoes at our target ships. This area was the scene of much activity, being the entrance to the major port of Glasgow, and shipping movements were heavy and constant and on occasion we would come across a Royal Navy Aircraft Carrier. Our aerodromes had runways mostly at a minimum of 1,000 yards in length, and this, to the navigator, who took part in the delicate process of getting the 'bird' ... as the Americans would say ... on to the ground, seemed not overly generous on some approaches in poor weather or cross-winds.

To see the aircraft carrier decks of some 200 yards as a place to land an aircraft seemed to be a bit irresponsible, and more so as that surface was performing gyrations in all planes. The skill of the Fleet Air Arm pilots, to make a precise approach, at the right speed and height and with the nicety of making connection with the arrester wires, made us respect our counterparts. I made no point on harping on the practice in the Navy, of the navigator ... properly 'Observer' as we Beaufighter navigators were so designated ... being regarded as the captain of the aircraft. I

was certainly of no maturity to adopt this role, which was based on the aircraft performing reconnaissance and tactical functions, rather than the expectation that we in the Beaufighter Wings would get very aggressive and display brutal and ruthless destruction of the enemy with never a qualm.

One Australian pilot and his navigator had a most unpleasant experience. Each of our flights was of a short duration, being to take-off over the sea, manoeuvre to position to carry out the drop and return to the aerodrome. This crew taxied round to the end of the runway, the pilot opened the throttles to get airborne and the aircraft gathered speed down the runway. As subsequently recounted by the somewhat shaken (now read 'traumatised') pilot, the aircraft failed to respond to him pulling back on the stick. With the end of the runway fast approaching, he resorted to fairly hefty trim on the elevators to raise the nose.

It did raise the nose, but so violently that an immediate application of downward trim was demanded. Realising that the elevators were not performing normally, this unfortunate pilot had to maintain control solely with the small amount of trim available to him. He flew an erratic circuit and managed to get the aircraft back on the deck, but he was extremely fortunate and displayed great skill in emerging from a situation which has undoubtedly been the downfall of many pilots.

To escape from the Beaufighter by parachute, a not very attractive option in the light of day but perhaps the final recourse to survive, both pilot and navigator were each provided with a hatch to the rear of their seats. This was hinged, to be forced open by the action of the slipstream when a wire was pulled. They were somewhat prone to open, untouched in flight and with a mighty bang, and of course, to the immediate concern of the crew. We very seldom, in this part of our training, ventured up to a height of more than several hundred feet and to save time and effort, we tended not to bother putting on our parachute harness, I usually threw mine onto the hatch at the back of my seat.

But on one take-off, just as we crossed the

coast and over the sea, my hatch flew open, in a cloud of dust and debris, and my harness was sucked out and disappeared into the Irish Sea. After this incident, it seemed rather prudent not to stand on the hatch unless duty necessitated, such as a quick tune up of the radio equipment, the Marconi 1154/1155. However, the formalities of service discipline and the strict application of rules and regulations demanded that, despite what could have been my early and tragic demise, the ponderous application of investigation, findings etc., should be carried out. The outcome was that I should be charged with 'Loss of Air Force Property', and I was found 'Guilty as Charged' and made to pay for the lost parachute harness ... I think it was worth about six pounds.

The verdict rankled somewhat and for a long time I felt hard done by, but in later maturity, I accepted that such an action was proper to the maintenance of order, but I could not help but compare this extremely modest loss of equipment, due to my misdemeanour, with the loss of a complete Beaufighter due to 'finger-trouble' on the part of a casual type pilot!!

Our accommodation was a bit spartan; I recall we were housed in Nissen huts, with ablution facilities consisting of outdoor troughs, but the Ayrshire coast in this vicinity is delightful. During the war years, in the service, one of the first priorities after moving to a new camp was to reconnoitre to find a good source of food and, of course, alcoholic refreshment. The former we found close by our quarters, in a quaint little village called Maidens ... no, I make no comment on the high improbability of any such find there, food being the object of the exercise ... where we were able to get supplementary food and relaxation when not rostered to fly. We had by this time 'bonded' not only with our pilots but also with other crews, and this phase, of relaxing together in this small village is recalled with particular pleasure.

A photograph in my logbook refreshes my memory of that time and captures the smiles of pilot Merv Thorburn, RNZAF, and his navigator, Charlie Brown, pilot Harry Bennet, RNZAF, and his navigator, Nick, myself with Mac my

pilot and another RNZAF pilot whose name I cannot recall. Fate is strange in the game it plays; it tosses us hither and thither and no rhyme nor reason is ever discernible. Looking at this photograph, taken over fifty years ago brings back the instant of its capture and all the warmth of feeling and comradeship that is reflected in this fading print.

Merv and Charlie survived the War - Merv to marry a young lady from Maidens, raise a family and become engaged in trawler fishing on the coast nearby, and after many happy years, including obtaining a degree from Glasgow University, succumbed to sickness while still comparatively young. Charlie married when with 254 Squadron, raised two children, built up a successful chicken farming business, and at an early age was struck by Multiple Sclerosis, forced to spend nearly twenty years in a wheelchair and died at the age of sixty-four. But double tragedy fell on Charlie's family, with his son Mervyn also contracting MS, condemned to a wheelchair in his mid-twenties and also died very young.

Harry Bennet and his navigator Nick were posted to the Far East and were lost in the jungle there, I think without trace. The remaining pilot of that group was also posted overseas, and I have no information as to his fate.

Having been polished to the ninth degree, by so many training courses, we were now adjudged ready to go into action, and we awaited the details of our posting to a squadron with much expectation. For me, it had been a particularly long haul, from October 1941 to March 1944, and although I had seen and experienced a small sample of being on the receiving end of the 'horrors of war' and though in my role of navigator would not personally demand that I 'pull the trigger' in retaliation, I would certainly be getting a good view from my rear seat of giving the 'Hun' a taste of his own medicine.

I remember not whether, at this stage, I had built up any great animosity toward the foe, our type of air warfare was, in any event somewhat impersonal, a far cry from the trenches of the First World War bayonet charges when the louder the shriek when going into action the more likely the prospect of success.

I saw no need to practice screaming.

Ron Banyard

I have been in Australia for almost forty-five years, but during the 1939/46 War, I was for five and a half years in the RAF, serving part of my time at Turnberry. In fact, after undergoing training to become a wireless mechanic, my first posting was to Turnberry about January 1942. At that time the airfield was being converted from the former golf course and many of the buildings were not yet constructed and the only Air Force personnel there were those who accompanied my group of about fifty or so and a small administration section. There must have been cookhouse and cooks etc., I suppose.

Although the golf course was largely destroyed and covered by the runways and perimeter roads, when we arrived, there were still about three or four golf holes and fairways remaining. The officer commanding the station was not at that time an RAF Officer, but an

Ron Banyard

Army Colonel, there was little Air Force activity apart from those supervising the building of the basic structures. The Colonel was a golfer, who couldn't stand to see the destruction of the magnificent course and he fought to retain what was remaining of the few holes and fairways. Of course, it was a vain fight and one he was destined to lose in a very short time.

When I arrived at Maidens, the little fishing village adjoining the airfield, in 1942 and stood on the beach there, it was mid-winter, and whilst the Glasgow airport from where we had driven had been under a thick blanket of snow, at the Maidens, due to the influence of the warm Gulf Stream, it was like a warm spring day. We were allocated Nissen huts which were only a few yards from the beach and a little north of the village. A caravan park now occupies that site. The transmitting station where I helped maintain the radio transmitters still stands on the hill overlooking the village and airfield. The tall aerials have been removed, something the local farmer wanted to do many years ago when his cows burnt their noses on the live aerial leads.

Deryck McCusker

I joined the RAFVR local unit at Derby in June 1939, to commence training as a W Op/AG. After initial training I was sent to Prestwick in the middle of January 1940. I have not forgotten being marched along the snowy, icy lanes around Monkton and Rosemount, trying to keep in step whilst slipping and sliding on the difficult footing. Our attempts at some other tricky footwork was our version of the Scottish dancing with the local girls in the town hall. They pulled no punches, and we were whizzed around until we were dizzy, but it was all good fun and leaves some pleasant memories.

I joined my first Squadron, 224 (Hudsons), at Leuchars in Fife, principally engaged in anti-shipping operations in Norwegian waters and alternating with Convoy Escort Protection to our own Merchant shipping until December 1941. I was fortunate to be part of a very professional operational crew, with Pilot/Navigators Tony Timms and 'Bunny' Brown. They were both always most accurate and positive. It is my opinion that this very satisfactory situation existed because of the Pilot/Navigator classification of the early RAFVR aircrew members, and also the decision Tony and Bunny made to fly and navigate alternately.

I found myself about to set course for a further stint of Instructional work at Turnberry, where once again I was to meet up with Peter Leigh.

Meeting up with Peter was a piece of good fortune. He had joined the RAF as a regular airman sometime during 1937, as 637407 AC 2 P. R. Leigh, and then by some circuitous route had evolved into a W Op/AG. Somewhere along the way he too had been bitten by the flying bug. He was a great self-improver and discovered that various subjects interested him, and he could be a source of quite useful knowledge. He was quite a studious type of person.

I do not know where he did his wireless operator training, but he must have completed his course about a year earlier than myself. By then it would have been obvious that a war was brewing up, and Pete would have volunteered to remuster as air crew, with a spell at gunnery school to follow.

Maybe Canada figured in his training programme. I don't know, but somehow, he found himself in the USA quite early in the war. I am speculating wildly about what happened next. And this is pure speculation.

It was about 1935/6 that the Government looked in the direction of the USA as a possible source of aircraft with which to expand the strength of the RAF. The Chief of Air Staff at the time was Sir Cyril Newall, and he persuaded Arthur Harris - who later became the head of Bomber Command, - to undertake a review of suitable aircraft available in the USA, which led to the purchase of Hudsons and Harvard Trainers.

The Harvard was very useful in the early part of a pupil pilot's training programme, because it was quite likely to be the first aeroplane which required

him to remember to lower the undercarriage before landing.

It may seem an essential survival precaution to ensure that the wheels had been lowered before attempting a landing, but to a pupil pilot with so many other things to remember all at one time when in a highly nervous state, it was the Harvard which introduced this additional aviation hazard into his life - the retractable undercarriage.

The Hudson required to be adapted for a military role. The Lockheed planners and engineers worked a minor miracle by producing a plywood mock-up of the modifications they intended to incorporate for the front end of the military conversion in only one day. Arthur Harris and his study team were both impressed and convinced that the Lockheed designers and engineers could very adequately achieve the modifications and design changes necessary to convert their aeroplane into the military reconnaissance aircraft that the RAF required.

However, this is a long way away from where we left Peter Leigh, but it may be relevant to his story, because when these Hudsons were ready for delivery, someone was required to fly them across the Atlantic.

As yet, America was not involved in the war. One of the methods of providing assistance, whilst still claiming to be not involved was to have equipment delivered to us from Canada.

In the case of the aircraft, they were towed across the border into Canada, and then flown

618 Squadron Ground Crew

over by a variety of skilled pilots from all sources. The 'crew', or Wireless Operator could be supplied from the wireless training establishments in Canada. They were due to return to the UK anyway.

I admit that there is a fair amount of speculative reasoning in all of this, but how else could I have met Peter early one morning on the airfield at Limavady in Northern Ireland, as he stepped out of a Hudson on a delivery transit flight, having just made a trans-Atlantic crossing from Canada.

I became interested in this aircraft circling the airfield with wheels and flaps down, flying in and out of low cloud, obviously intending to land. I had noticed that it had no gun turret. I continued to watch until the plane landed and taxied up to the control tower where I was standing. And who should dismount but Peter Leigh.

The turrets were fitted on this side of the Atlantic. Pete was gone the next day to complete the delivery to Speke, near Liverpool. I didn't see him again until we were both at Turnberry, where we were engaged on instructional flying with trainee crews, principally trainee wireless operators.

Trainee navigators were accompanied by navigational instructors, and the trainee pilots were under guidance from a staff pilot.

Pete and I were both aware of the dangers of being lost in the air, and how the anxiety level increases rapidly with the less experienced aircrews. The pupils forget everything they've been told, and a state of panic sets in. It is necessary to reduce the panic state to get them thinking logically and methodically once again.

One example of this can occur when the navigator gets his calculations wrong, and he realises that the aircraft is not where he has calculated it should be. A rapid back-check on his plotting fails to reveal the error, and the skipper is asking for his next course. By now, in a panic, he admits that he is lost.

The pilot must fly the navigator's courses accurately without deviations to course or speed; whilst the navigator must glean every available scrap of information about changes of wind-speed and direction and make compensating corrections to accommodate these fickle unseen forces, especially when flying for long distances over a vast featureless ocean to achieve a successful rendezvous with a solitary, very vulnerable merchant vessel in mid ocean. They needed all the protection that we could provide, thus it was imperative that we made a successful rendezvous.

If they were where they said they would be - we found them.

In our instructional capacities, Pete and myself felt strongly that we could influence the course of events for crews that found themselves above cloud, out of sight of the ground, and in doubt about their position. In other words – 'lost'.

Back in our two-man billet we had discussed the danger of the disastrously misguided procedure, apparently adopted by some crews, blindly descending through cloud to try to locate their position, and the almost inevitable tragic consequences of flying into high ground in the vicinity.

Pete and I worked out a pattern for our lecture-room advice, and an explanation of our simulated lecture-room exercise. We preached that they must resist the temptation to reduce height, hoping to catch a glimpse of the ground, and possibly recognise a familiar landmark. More difficult and dangerous at night, of course.

Stay at a safe height, 5,000 feet minimum for the whole of the UK. This does two things:
It removes the danger of flying into high ground.
It restores a calm atmosphere within the aircraft.
By reducing the tension, cool minds will be able to resolve the situation.
To resort to prayer at such moments can be extremely negative and ineffectual, but training puts you in a positive state of mind and rarely lets you down. Whether it paid any dividends for the crews we instructed, we never knew, but it was the one way that we could pass on the experience we had accumulated.

Our message to them was:

If you are lost, 'GO UP: NOT DOWN'.

The simulated part of the exercise was a standard RAF signals procedure. We did not invent it. It already existed, QGH.

We were only concerned to demonstrate that it worked and could be a life saver.

The lecture-room simulation was to ensure that the procedure was used properly.

Having practised it on the ground, the next phase was to get them into the air and get in some confidence building airborne practice.

The airborne part of the exercise would be carried out in daylight, in conditions of normal visibility, so that the pilot could see the ground and the airfield, and yet follow the procedure without 'cheating'.

By following the directions from the Ground Station, given to him by the W Op, he can see that the procedure is bringing him onto the correct heading and height to make a safe landing, which will require only slight adjustment as (in an actual event), he gets his first sight of the runway.

In the simulated exercise the 'cloud' must be imaginary.

To carry out the exercise more than once, required the pilot to perform an overshoot manoeuvre in place of an actual landing. Thus, in the interest of safety the entire QGH exercise was enacted with the altimeter set 1, 000 feet higher than the correct setting for the airfield.

The whole object of carrying out the airborne simulations in good weather, was to give the pilot enough confidence in the procedure so that he will readily trust the method on those occasions when it has to be used for real.

It worked just as well at any RAF airfield with a Direction-Finding W/T Station. All that was required was for the aircraft W Op to tune to that station's D/F frequency, and request QGH.

It may all appear rather primitive and difficult to imagine, when compared with modern miniaturised equipment, with duplicate, and triplicate, back-up systems.

For about the first three years of the war, it was all that we had. The QGH system was well

used, but when H2S (radar with rotating scanner) was introduced, the procedure became a visual one, by displaying the terrain below on a Cathode Ray Tube.

The accommodation for instructional staff at Turnberry was on No. 2 site by the main gate. It was nothing palatial, just Nissen huts divided into four separate sections. Each quarter section had its own door, an iron stove and two bunk beds. Pete and myself bunked in together. Can you visualise it happening today? There would be plenty of suggestive remarks and comments about two chaps bunking in together!

A dead-end railway line terminated in a coal compound on our site. Despite the severe Scottish weather in the winter months, we were never short of coal for the stove. And throughout the coldest of days, we kept that stove at a temperature which was just below the melting point of iron. Above our hut the seagulls circled on the rising thermals of hot air. We assumed that the coal stocks were the property of the RAF, but in fact they belonged to the local coal merchant a Mr Naismith.

This last bit of information only became known to me in the summer of 1994, when I returned to Maidens on holiday. We stayed in a guest house on the shore, which was next door to a Mr George Naismith, who was the son of the wartime coal merchant. I was smitten with remorse during our stay there, and determined to make my confession, about pilfering his father's coal stocks at an opportune moment, but sadly the ideal opportunity slipped by and was lost.

The only other item of interest I remember from Turnberry was the large number of night landing accidents when the short runway was being used. In most of cases they were Hampden aircraft which ran off the end of the runway, coming to rest on the rocks below the lighthouse. I felt that the high ground inland exerted some psychological pressure on the mind of the pilot, causing him to feel that he was somewhat lower on his approach than was prudent, which had the effect of making him land further along the runway. To make effective use of this runway it was necessary to make use of the full length, to allow time to bring the aircraft to a safe stop. Wasting the first few yards was a recipe for disaster, and almost certain overshoot beyond the end of the runway to land on the rocks in a heap. I remember coming down to the airfield one morning to see that two aircraft had gone off the end during the night, and the undercarriage of the second one was astride the rear end of the first one to go off. Running out of runway was a frequent occurrence at Turnberry.

In the Sergeant's Mess we had 'waitress service'. The food was brought to us at the tables at mealtimes, and there was one very busty WAAF, who served in the mess. She was the prototype 'busty blonde'. She must have observed that the fighter pilots wore their jackets with the top button left undone, to proclaim themselves a race apart from the rest of us. 'Blondie' could obviously see other benefits by adopting the same practice. She was a big girl and getting bigger! Of necessity that top button remained unfastenable, despite many generous offers of 'assistance' from the boys. Whoever assigned her for duty in the aircrew sergeant's mess must have been in a very 'morale-boosting' frame of mind and should have been given a medal.

At mealtimes we sat on long bench seats, four persons each side of the table, normally, but sometimes an extra body would increase that number to five. If we were sitting quite tightly packed together, with little space between each person, she was quite adept at manoeuvring a plate of food as well as that very ample bosom through a quite inadequate gap. If one felt a pressure on one's shoulder, and turned to see the reason for this, it was this bosom carving an access like a rampaging barrage balloon. It could be quite exciting!

Hey Ho! We had to make do with simple pleasures in those days.

Whether this had anything to do with one airman having a wild wayward moment, I don't know. But he described to me one evening how he had been down amongst the sand dunes at Maidens, lying in the sun, when he was approached by a very wet 'pin-up' of a girl. She

had gone down for a swim and emerged from the water with her wet swimsuit under some considerable strain (did I not say that she was a big girl!) She made her way over to where said airman was sunning himself and suggested that he should lend a hand with drying her hair. Well, that was his story! Apparently, things developed, not unexpectedly along predictable lines, and the episode appeared to have made quite a considerable impression on the chap in question!

A couple of days later, although I didn't immediately connect the two events, the airman was exhibiting quite a degree of unrest and agitation. He fidgeted around before finally blurting out that he was feeling distinctly uncomfortable around the groin area. He was rapidly beginning to fall apart mentally; which was quite unlike his normally calm and placid nature.

Clearly his sexual encounter had had a devastating effect, and he was now starting to panic. My suggestion that he should report sick and get a proper medical opinion only increased his state of panic. But, I argued, if there was a real problem it was most unwise just to hope that it would go away. I dug in. I said I could not stand by and let him drift into a potentially much more serious condition, which might permanently damage his health. I tried to impress on him that a serious medical disability could develop from something that was easily treatable if attended to at an early stage. I went as far as telling him that I was not prepared to stand aside and see him blight his life, when by reporting sick now, he could prevent all the anguish. I reminded him that is was an offence to conceal something of this nature. Finally, I said that if he did not report sick, I would report him as sick! That did it, he reported sick the very next morning, and although no immediate solution was forthcoming from the staff in the sick bay, he was to report to a hospital unit at Peebles for further examination and tests.

I calmed him down about information reaching his mother. I told him to write to her saying that he was going on detachment for a short period (which in a way, he was), but would be returning to Turnberry in just a few days, and in the meantime, she should continue to address his mail to Turnberry, and that it would be sent on to him for the period of his detachment. I was to attend to this little matter, he was happy with this arrangement, and it calmed his mind on that score. Within three days he was back on the unit at Turnberry, very relieved, and able to report that he was free of infection. The discomfort he had experienced was some sort of strain, due, it seemed, to over-exertion! It was mind-boggling stuff! What had he been up to! Well no damage was done, and it was soon all forgotten.

By February 1943, my time in the instructional role had come to an end. My next move was to return to 1 Radio School, Cranwell, to commence a six-month course to remuster as a WO/M (Wireless Operator/Mechanic) prior to conversion onto Halifaxes.

I was then posted to 58 Squadron, Halifaxes until VE Day. Transferring to Transport Command, I completed the Conversion Course to Dakotas, but my release group came up quite early and I left the service.

Despite an offer of employment with the National Airline, BOAC, I returned to being a Rotogravure packaging printer back in Derby. I no doubt turned down a far better salary, but I thoroughly enjoyed my years in printing, ending my working life as the Technical Director and feeling fully satisfied and compensated; and the job satisfaction had been immeasurable. I feel privileged to be able to say that.

David Phillips
I joined the Royal Army Medical Corps in 1936 when I was eighteen years old, and was stationed in the Middle East, Egypt and Palestine until 1942. Having spent a long tour over-seas, we were allowed the privilege of requesting a posting near our home town, so having been raised in Brighton, I asked to be sent to the huge military hospital at Nettley near Southampton or Millbank in London. Eventually a list of postings went up on the board, and there next to my name was 'Turnberry'. I had never heard of the place! I didn't think it was near Brighton!

David Phillips

So, after a typical army train journey, a lot of 'Hurry up and stand still', I arrived at Turnberry station on a cold, bleak November day. The sight that greeted me was awesome, the grandeur of the large hotel almost took my breath away. The railway station was beautifully kept, leading straight to the courtyard of the building.

The hotel had become a busy, bustling military hospital, caring for troops who were convalescing. All the oil paintings and silver had been removed and were replaced by standard army issue equipment, and one or two of the rooms were kept locked, but apart from that, the hotel still exuded luxury. We looked after patients from all over the area. There was a detachment of Royal Engineers at Cairnryan working on the 'Mulberry Project'. They kept us occupied, along with casualties that arrived by train from Ayr. There were all sorts of patients, general surgical cases, chaps with appendicitis etc. and even a psychiatric ward where we kept the 'nutcases', genuine or otherwise. Some of them were just trying to 'work their ticket'. We didn't often deal with casualties from the airfield, they were either taken to Station Sick Quarters or the mortuary.

The officers had private rooms and other ranks were communal, but comfy. Having the use of the hotel kitchens, we had the best of food. I still vividly recall sitting by the window in the dining room of the hotel and watching the 'planes coming and going. I always used to sit at the same table in the corner, looking out at the runway on one side and over the Clyde on the other. Many a lunch was eaten watching the Swordfish aircraft dropping torpedoes, and ferry aircraft arriving from across the Atlantic.

Every morning at 10.00 hours prompt, all the Medical Orderlies would have a tea-break, and there would be a crowd of us running down the steep flight of steps of the hotel to the Caddie Shop which had been turned into a small canteen by the ladies of the Church of Scotland. Here we would enjoy homemade cakes and buns before tackling the long climb up the steps and back to work.

Although I was only posted to Turnberry for eight weeks, the place has had a life-long effect on me. I fell in love with the area and it was here I met the local lass who was to become my wife, Jean Devoy. She was standing by the harbour at Girvan, waiting for a bus to Ballantrae. I made small talk and then plucked up courage to invite her to a dance at the Sergeants' Mess the following Saturday night. She agreed, on the condition that it would be a foursome, with her friends Betty and Bob Sloan.

The dances at Turnberry were very popular, with live bands and everyone had a good time. We used to get a local taxi to take us to Girvan. This was operated by a man called Sanny Campbell. He was quite a character; his taxi was in a sad state of repair, held together by string in places. The fare from Turnberry to Girvan was half-a-crown, usually about a dozen of us would pile in and share the cost!

My girlfriend's father gave me a pushbike, and when visiting, I would park it behind the police office. More often than not, when it was time for me to pedal back to Turnberry, my bike had been 'stolen'. So, I would walk back and find my bike propped up against the wall of my billet. The lads got to know where I left my bike in

Girvan and would just 'borrow' it!

During the war, the local people made us very welcome. There was a lady in Maidens who often gave us a cup of tea and a bun. She even did our laundry for us. I became friends with the Lane brothers who were resident electricians on the airfield. Bob Lane's wife was the village post-mistress. The small harbour village had quite a cosmopolitan atmosphere about it, there were lads from all over the world wandering about.

I left Turnberry to be sent to France and then forward into Belgium, Holland and finally Germany. All the time I worked in field dressing stations or on field ambulance duties. There was no such thing as post-traumatic stress counselling in those days, you just had to get on with it.

When I left the Army, I returned to Girvan and 'collected' my Ayrshire lass, and we moved to London. My wife couldn't settle, and missing the peaceful life of the west coast, we came home. It was a very different place to war time, I found it impossible to get housing. I was told when applying for a council house that because I was English that I didn't have a chance, and to go back to where I came from. It was the same story with employment. At first, I rented a holiday home in Bourtreehall, whilst cleaning windows and doing odd jobs. I was then given the opportunity to purchase an electric chimney-sweeping machine for the princely sum of seventy pounds. There was no way I could afford that amount, until a kindly lady, Sarah Kidd, offered to give me an interest free loan. I took possession of the machine, and with the work I obtained cleaning chimneys in and around Girvan, was able to repay her in a very short time. After a while I was able to purchase a house, at 34 Bourtreehall, which cost me £750. My wife and I settled down to a very happy married life and raised our family.

John (Jack) Rainham

I first came to Girvan as a very young boy in about 1939. I originally came from Glasgow - I was sent to Girvan during the war to live with relatives. My mother came from Girvan and my father was a Glasgow man. My mother and father stayed in Glasgow; he was a male nurse. I remember hearing the German bombers heading up the Clyde at night towards Glasgow and then seeing the sky turn crimson with the fires. I was terrified for my parents, so much so that I didn't talk until I was ten years old.

I recall visiting a house with my parents between Ballantrae and Girvan, I think owned by a cousin of my mother. It had stone walls painted white, a stone floor and a tin roof. We were waiting for a car coming to pick us up and take us back to Girvan, when suddenly there was a lot of noise and soldiers and police turned up with the surviving crew of a cargo ship which had been torpedoed just out in the Clyde. The ship's captain was Dutch and the crew German. They were brought into the house while waiting for transport. The captain was a large, well-built man with a beard and moustache, dressed in black jacket and bleached trousers, with a black cap and black wellington boots. He spoke perfect English. The rest of the crew only spoke German. One of the crewmen had a badly burned hand, so my father tended it for him, he was in a lot of pain. The lady of the house gave them all tattie soup and home-made rolls, all washed down with tea and whisky. I remember them sitting around the big range in the kitchen which was well stocked with driftwood from the shore and gave off such a great heat that their wet clothes were soon steaming. The next morning a lorry arrived from Cairnryan to collect them. The British soldiers treated them cruelly, until the lady of the house turned on them and told them off, saying that they were now Prisoners of War and should be treated as we would want our lads to be treated. The police also told the soldiers off.

I recall walking along Coalpots Road one day (which was a dirt track surrounded by fields then) between my gran and my mother, both holding my hands, when a German aircraft came speeding over the Byne Hill at low level followed by a Spitfire, who chased it out to sea. The noise of the guns was petrifying, and I thought we were all going to be killed. The Spitfire shot him down just off Ailsa Craig and the German aircraft crashed into the sea in flames. I can still see it

now, red, yellow and black smoke and to this day I can hear the sound of the guns.

My aunt worked at Turnberry during the Second World War, when the hotel was a hospital. She told me a story when one night after a shift she was walking down the road in the blackout when she fell over something. Terrified she ran back to the hospital! In the morning when the milkman arrived, he told everyone that they had found a dead cow lying on the road, my aunt said, 'Well when I found it, it was still warm! It wasn't me that killed it!'

The morgue for the hospital was down at the garages, where they used to lay the bodies out. There were ten marble slabs and it was very cold in there. I used to wait there for my aunt, to get the bus back to Girvan with her sometimes. When she finished work, she would meet up and socialise with the young pilots, until one night they were not there. No explanation.

One night, two orderlies were tasked with taking a body on a trolley down the hill to the morgue, they had the bright idea of jumping on the trolley and riding down the hill with the deceased. But the hill being steeper than they thought, the trolley soon went out of control, overturned and they were all thrown off in different directions. It took over an hour in the blackout to find where the body had landed and was eventually found propped up against a tree looking out to sea.

I worked at Turnberry Golf Course in 1967. I remember I found a live bomb at the fourth tee of the Ailsa Course - this was on a Saturday at 8:00 am. The Bomb Disposal Squad blew it up at 14:00 p.m. The bomb was German and had the date 1943 stamped on it. The following week I discovered a live mine which was connected to the seabed by a chain. This was a job for the Navy Disposal guys who towed it out to sea and detonated it. Soon after this we were working on the beach near the Ailsa Course when the tractor dragged up ten bags of TNT - we had been scraping sand to fill the bunkers when we uncovered them. More digging and we ended up with 22 bags! We had to fill pails of water to pour on the bags to keep them cool as it was a hot summers day, until the army came and disposed of them.

I wish I could remember more, but then, maybe not.

Ron Brighty

At the age of eighteen I was employed as assistant timekeeper at the largest factory of Smedley's fruit and vegetable canneries, in my hometown of Wisbech, Cambridgeshire. One day I was ambling back to my office from the fruit factory when I was joined by an apprentice, also headed in that direction. I asked him where he was going and upon receiving his reply that he was going to volunteer for the Royal Air Force I said that that sounded like a good idea and I joined him in his walk down to the local recruiting office! That was in 1940, and the RAF took a deep prayerful breath and sucked me in a year later.

I was trained in Blackpool, square-bashing resort of approximately 30,000 raw sprog airmen, and completed my W OP training at Compton Basset where I was recommended for my AC1

Ron Brighty

status … fell flat on my face by getting just one symbol wrong in Morse plain language: my personal language at the time wasn't so plain.

After a whole two weeks posting to Felixstowe, within easy reach of my girlfriend (now my wife) in London, I, and a London guy who became my pal for the rest of our RAF career, were posted to that unknown, foreign, ungodly place, Turnberry!

When we arrived, before taking 'K' broadcasts, when the station became operational, we were put on fatigues clearing the mess on the ground around SHQ. While busily engaged in this delightful task, we were treated to a rapid series of minor explosions all around us which scattered gravel and dirt over everything, including some of us. The Royal Navy, practicing light gunfire out on the Clyde, either had a warped sense of humour or a lousy set of gun crews.

The only other time I came near to combat at Turnberry was on a one-week battle course with the Royal Scots Fusiliers, in their domain near the woody hills south-east of the runways. Any broken branches you may find in the trees on those hills you may attribute to my clumsy climbing when installing practice field telephone cable. I also achieved AC1 and LAC status in short order whilst at Turnberry. All this was, of course, prior to the transfer of we blessed eight pioneers to Kirkoswald, my memories of those delightful days now follow....

This is not a wartime story of combat and violence: it isn't even about aircrew heroics. It is about ordinary ground crew, in non-exciting jobs, who become separated from Air Force routine and the tragi-glamorous flying fraternity, and who travelled afar in strange places. It's about the everyday things they did and the things they saw.

What I write is what happened, from memory, from letters and from scribbled notes. Some of the memories are vivid, some faint: but all of them add to the colour and form, of a nostalgia born in a youth half a century past.

I stood on the high brae in the pasture, leaning on the cabin fence, my forage cap tucked under the shoulder strap of my RAF tunic. Behind me in the DF signals cabin I could hear Neil tapping away at the Morse key as he wound up his spell of duty.

As Neil ended his key tapping, the roar of engines overhead confirmed that the exercise was finished: the Hampden swung low in its glide approach to the unseen distant runway. A Beaufort followed, higher and less noisily. I watched the two of them drop below the trees on the distant hill.

A black smudge appeared over the distant trees and, fifteen to twenty seconds later, there came a muffled 'CRRUMP' and a faint, concussive thump. Then again silence but for the Mavis song and the sigh of the breeze. Slowly, a smudge rose on the still air, a straight black finger, spreading and drifting into a flat, dark haze.

We heard later in the day: the Hampden, in its low shallow glide, had neared the cottages of the village in its runway approach. Above and behind, the Beaufort in its steeper glide approach, had failed to see the Hampden below, and had not followed radio procedures. It dropped suddenly in its final approach – directly into the wing tanks of the Hampden.

The fiery, tangled, screeching tomb of airmen hurtled over the cottages, missing their roofs by a scant three hundred feet, and made a spectacular, horrifying landing on the runway approach.

The men in the aircraft died doing their job because a plain human being made a tragic mistake. In different circumstances it could happen to anyone of us – we would only have to be in the wrong place at the wrong time.

The Corporal kicked open the door of the Nissen hut and switched on the lights. Four naked bulbs hung from the arched metal above and shone starkly upon the bare necessities of comfort. A lifeless coal burning stove sat in the centre of the hut and bleakly accentuated the chilled reality of our new home. A row of metal wire sprung beds, each holding a forlorn little stack of folded blankets, was meticulously spaced along each cold, sloping 'wall'.

We filed in, tired, hungry, and miserable, and threw our kit bags and small packs onto beds at random. Derek slung his onto the bed next to mine.

The Corporal said, 'Breakfast at 07:00 hours sharp; pleasant dreams.' He smirked and was gone. We already knew it was almost a mile and a half back to the mess! I looked glumly at Derek, and he looked sourly back.

'What a dump!' he said.

It was our second joint posting after completing training as RAF W Ops/DF. We felt more like pawns in a chess game between two lousy players. Our first posting after completing training hadn't been so bad. Felixstowe was a well-established camp with a hopping NAAFI (Navy, Army, and Air Force) service canteens, affectionately referred to as 'No Aim, Ambition, or F------ Initiative'!) nearby, pleasure spots, and only about seventy miles from London where our girlfriends lived. Travel wasn't so difficult in those war years that we couldn't manage a weekend jaunt with the help of a sturdy thumb and a sympathetic driver.

We had settled into station routine and, free of the restrictions of training camp, were beginning to think that the Air Force wasn't so bad; we were looking forward to our first weekend pass.

The dream lasted all of five days!

The notice on DROs in the mess carried a posting list, and we were on it. We repacked our kit bags, boarded a truck and rumbled off to the railway station. We went to London all right, but only to board the 23:00 train for the long, slow journey to Glasgow! The train rattled slowly out of Euston Station; the dim lights of its corridors barely bright enough for us to find a place on the floor to sleep. As usual, the corridor was crowded and uncomfortable, but we found a spot near the toilet door and arranged our kitbags in strategic positions to reserve our bedding down spots.

At Ayr an RAF truck had picked us up and brought us to this God-forsaken spot on the Clyde. The day had been spent dolefully hanging around a spanking new SHQ, still smelling of wet concrete, lounging in boredom in an incomplete, plaster streaked signals room, gazing out over empty, rain-swept runways to the lighthouse on the headland. There wasn't even a 'plane to relieve the monotony. Of the Clyde or the surrounding countryside, we could see nothing.

Our only relief had been a Signals Sergeant, regular Air Force type, who sauntered in to demonstrate his superior knowledge. The granddaddy of all disreputable forage caps sagged from the centre of his head and around his ears, and a non-regulation black moustache supported the bulbous protuberance which passed as a nose.

He lounged against the bench in the battery room, poured a measure of sulphuric acid from a carboy, topped up with water and downed the concoction with apparent relish.

'Tastes a bit like lemonade,' he said.

We had hopes he might die on the spot to add a touch of zest to the entertainment, but he gave a snaggle-toothed grin, a long look of condescending superiority, and lounged away to some other fruitless pursuit. That Sergeant must have been in the Air Force a very long time indeed.

It was dark when the Corporal collected us after supper and trudged with us to the Nissen hut we now found ourselves in. That night, as we fell into an exhausted sleep, we were not at all happy to be on Maidens Beach, near Turnberry on the Ayrshire coast.

Next morning, I dressed, grabbed a towel and my small pack, and shambled along the hut to look for the toilet and wash house. The Corporal's snores vibrated from the end cubicle as I passed and opened the outer door. I stepped from the end of the Nissen hut and squeezed my eyes shut in the sudden light: My mouth remained as open as my staring eyes which were trying to take in the scene before me. The rain had gone, and a cool, clean, salty taste lay on the soft breeze. Twenty yards beyond a stretch of tall, springy grass was a clump of dark, weathered pines on a softly rising sand dune. Beyond them, fringing a narrow beach on the left, the ground undulated its way through clumps of gorse and broom, then lifted to a thickly wooded crest. There, it broke and scattered down into the water in a jagged headland pointing a rocky finger out across the gently rolling waves.

The air was clear and the sky a new-washed blue, and my eyes followed the pointing rock finger across the blue-green waters of the Firth which rippled placidly into the distance. There, stretched the misty, ethereal Mull of Kintyre, with Goat Fell on Arran etched against it. A little further to the south the domed hump of Ailsa Craig floated on a thin sea mist.

I knew none of these names on that first morning, but in those few first brief, wondering moments, the despondency of the past few days vaporised like the mists. I didn't realise until later that it was love at first sight.

Within a few weeks the new runways of Air Force Station, RAF Turnberry, were in operation and the roar of Beauforts and Hampdens, swinging low overhead and out across the firth, became familiar background music for our reason for being. We quickly settled into the routine of Coastal Command, and the rhythmic chirping of Morse signals became our second language, linking us with men we never knowingly met but whose Morse key signature styles were like familiar voices.

Other voices were not quite so familiar. In six months of air force life I had been thrown into contact with half the dialects of the British Isles and now understood almost half of what was being said. But here, on the quiet Ayrshire coast, outside the confines of the RAF station, I was lost. The Ayrshire dialect has a smooth, soft brusqueness about it: it sings and lilts, and strings almost intelligible English words together into completely alien syllables. It was delightful, and soothing, and utterly without meaning to me: I had to find an interpreter soon or spend my entire off duty life in the NAAFI!

That was still the state of affairs when I dropped into the NAAFI a month or so later for a beer and cigarettes. I lounged against the counter gazing abstractedly through the smoke haze when I spotted Derek, sitting at a table and talking to a strange airman. I noticed he had 'sparks', but he was from another billet and on a different watch from ourselves, so nothing else was familiar about him.

I strolled over and dropped into a spare chair. The stranger was almost bald and had a long face with a pleasant look about it, and something in his eyes that said his partial baldness was vastly premature. It turned out later that he was three months older than my advanced age of nineteen years, eleven months!

'Hi y' old ------!' said Derek. 'This is Neil ... he got here just before we did.'

I was about to make some comment when Neil spoke, and I knew why he had got there first: he'd had a head start. It was the thickest Scottish accent I had heard inside the station and spoken in a soft husky voice. I had no idea of what he had said. But during the garbled interchange which occupied the next hour or so Neil had the advantage. I passed several comments about the Scots' strange impediment of speech, not realising that he understood English perfectly but was not prepared to use it. I am sure he insulted me in many ways, but as I didn't know what he had said, I couldn't retaliate. The last thing he said as I rose to leave was, 'Y'rrrrr no bu' a ***** Sassenach!'

Somehow the grin on his face and the softness in his voice belied the insult that I knew was in there somewhere. I felt sure that the guy liked me and hoped I'd see more of him - if only for use as an interpreter. I could not foresee that the lives of Derek, Neil and myself would be bound together for the next four years.

I walked slowly down the road from the NAAFI, headed for the beach and my Nissen hut. I shuffled my feet through the small pebbles and shells and breathed deeply again of the salty air. Halfway back to the hut, I raised my head to watch a Beaufort roar low overhead, sweep across the beach and out over the water before swinging over the Turnberry lighthouse for a landing. I stopped to watch its graceful banked turn, about three hundred feet above the waves and two hundred yards or so out. Then it all happened, just as though I was watching a silent movie.

A sudden burst of fragments from the port engine, a rapid upward flick of the tail, and the aircraft disappeared in a surge of foam. I stood, frozen to the spot in horror. A gentle soughing of breeze, a sort of swish of the waves, were the only sounds, strangely out of place in the sudden silencing of the engines. I saw something bobbing in the water, drifting slowly northward, but it was indistinct and unrecognisable.

I started to run along the beach, hoping that the object might be washed in against the rocky headland past the hut and that it might just be a survivor. I passed the hut, breathless, and dashed the last two hundred yards to the end of the beach. I clambered up the rocks and stumbled across them to where they disappeared in the surging foam; and I waited in the spray, watching for the floating object to appear.

It seemed that an eternity passed before I saw it. Bobbing in the water, less than fifty yards from the rocks, the lifeless undercarriage of the aircraft floated slowly northward and out of sight. I just stood there, helpless and stunned.

I shivered as I climbed back down to the beach and made my way to the nearby Nissen hut. I had known the sound of bombs dropping close to me, had felt their concussion, and had seen the devastation they caused. I knew that people were dying violent deaths daily, in homes and in combat. Yet, the lonely futile death of one airman in surroundings of great beauty on a bright spring evening seemed, somehow, more poignant, more tragic.

Swiftly and subtly, the Ayrshire coast, Burns country, and its people crept into my heart and took over. The dialect began to make sense and I found myself using previously unknown words and an irresistible urge to sing my sentences. To Neil and other Scots, I knew, I was still a ***** Sassenach and spoke like one: but I was a changed person.

It was just after breakfast as I was leaving the Mess that Derek approached, wearing his best look of doom. 'Have you seen the DROs?' he said: and without waiting for an answer added, 'We're posted again ...... *****'. As Derek hardly ever swore, it was clear, that he too, had begun to like the place.

A trip to SHQ revealed that we, Neil and five others had been posted to some place called

Kirkoswald to set up a DF station. As we were the only DF operators on the station, and fully familiar with goniometers we were just being told to do our job; but still it was a ***** bind.

The others had gone on ahead when Neil, Derek and I piled our kit onto a small truck and climbed aboard. Derek was using more *****'s than was good for him. The truck pulled out from SHQ, through the little village of Maidens, with its one general store-cum-post office, where the post mistress was always 'out of cigarettes' unless you happened to be aircrew. We rumbled past the NAAFI and headed north for a mile or so, then swung into a steep hairpin climb, up and away from the coast road.

The panorama of the Clyde with its misty island backdrop dropped from sight as we climbed through gentle hilly folds of farmland, headed only SHQ knew where. Derek had stopped swearing only because, like me, he had sunken into a morose silence. It was the journey from Euston all over again except that, now, there were three of us.

We had travelled hardly three miles before the truck entered a tiny village, neatly ranged on either side of a quiet highway. We looked with disinterest as we passed old stone buildings, a farm on the left, a butcher's shop and a pub called the Shanter on the right. The truck pulled off the road and stopped at the side door of another pub, the Kirkton Arms. Our interest rose slightly. The driver came to the back of the truck and lowered the tailgate.

'O.K.- this is it!' he said.

We clambered down and hauled our kitbags to the ground. Beside the door stood a woman, slightly greying, but still youthful in looks and manner; when she spoke, it was with a quiet, well-educated English accent. 'Bring your things in here and be careful of the door,' she said.

The first impression wasn't good. Neil raised an eyebrow at me as we entered the room, and Derek started to look glum again. Inside, four beds were neatly arranged in four corners of the room and a fireplace occupied the centre of one wall. Two windows, and another door, leading to the pub, were the only other items of interest:

except for a very small airman with a pixie face and a shock of brilliant red hair who had already thrown half his kit onto and around a bed near the pub door.

I dropped my kitbag onto a bed under one of the windows, and Neil and Derek took possession of the other two. The woman had disappeared into the pub. 'Looks a bit of a dragon,' said Derek.

'A wee bit crabbit,' said Neil. 'Dinna think she wants us.'

'I'm Dougie!' said the red-haired airman.

We acknowledged him and decided he'd do and, introductions completed, we unpacked our necessities and sat on our beds waiting to see what the next move would be. Five minutes passed, and we were beginning to feel utterly neglected when the woman reappeared. She studied us for a few seconds, then smiled. 'Would you all like a cup of tea?' she asked.

It was like the sun rising on a spring morning. That smile was warm and welcoming, intent on our welfare and comfort and we knew, suddenly and instinctively, that her remark about the door as we entered was out of concern for us and our heavy kitbags, not for the varnish on the door as we had thought. We had never enjoyed a cup of tea more.

Mrs G. and her husband, a tall, balding, quiet spoken Englishman, proved to be a second mother and father to us during the following weeks and months. We had our own little dining room, with real tablecloths, near the bar, a private shower house and toilet behind the pub and our own door to the outside world that we could use at any time, day or night, no questions asked.

'Hoo aboot this?' exclaimed Neil. 'Nae ***** guard hoose tae go through!!'

We felt superior to, and a little sorry for, the other four who were billeted in the Shanter: they were comfortable and well cared for, but they weren't 'family', as we were.

Few service experiences could equal the one we now found ourselves in. Not only were we still in beautiful, dreamy Burns country, but we were off the station with little or no need to visit it, we'd been provided with a couple of bicycles, and we were billeted in a pub! It was the start of an idyll.

The DF cabin with its thirty-foot antennae was up on Archie's cow pasture near the Klondike Brae. Archie, or 'Airrrrchie' as I soon learnt to pronounce it, was 'Maister' of the village, the community's biggest landowner and roughly the equivalent of an English village squire. We didn't find this out immediately, because the entire village went into a cloak and dagger routine for the first week or so. The sudden uncalled-for influx of eight airmen, total strangers and no less than seven of them blatant Sassenachs, was a profoundly momentous event, and the villagers had no intention of accepting the intrusion without close scrutiny and deliberation.

So, as if by unspoken agreement, they all adopted the guise of 'Dour Scots' and built a psychological wall around themselves. If we said, 'good morning,' they treated the salutation with quiet suspicion as though questioning the fact or, perhaps, our right to comment on it. Their verbal response was either non-committal, non-intelligible, or non-existent.

As we passed the old stone houses on our way to the local store/post office we would see a window curtain hurriedly drop into place to conceal a vigilant housewife, teacup in hand and eye at the ready. Or maybe a door, open just a crack, would swiftly and silently close as we approached, or the hum of conversation in the pub's bar would pause and quieten as we entered.

At first, this negative response to our presence, to our very existence, was not only disconcerting: it was downright discomforting. But after a week and a half of it our offbeat sense of humour took over, we could barely suppress a grin whenever we spotted a subversive cover-up, and wee, shock-haired Dougie, who had no control over his emotions, would break into insane giggling fits. We had to drag him into our room at the pub with indecent haste to avoid offending the villagers.

The stand-off ended suddenly and unexpectedly.

By the roadside, near Airrrrchie's midden,

was a gate leading to the byres. One evening, a calm warm evening with a low breeze and the contented lowing of nearby cattle behind me, I sat on the gate, filing away at a piece of aircraft Perspex. It was a popular airman's pastime to whittle shapes like hearts out of Perspex, and to heat sink into them crowns or eagles cut from tunic buttons, to give to their wives or girlfriends.

Filing finished, I dropped the stuff into my tunic pocket and looked around me. There across the road was the Shanter, with warm comfortable sounds bumbling out through the partly open bar window. Next door, the closed premises of Sammy Thomson, the butcher, gave no hint that here once stood the schoolhouse which seventeen-year-old 'Rabbie' Burns attended.

Behind me to the left, set in a steeply sloping graveyard, were the ruins of the 'Auld Kirk', and beneath the quiet peace of its surrounding grasses lay the mortal remains of Tam o' Shanter and Souter Johnnie, and the maternal grandparents of Rabbie himself. Nearby, directly opposite the Kirkton Arms, the ruins of an old smithy echoed the long-gone laughter of Burns, Tam and Souter Johnnie: for on this spot stood the howff of Kirkton Jean, favourite carousing spot of the trio.

I didn't discover all this until later, but I felt in the peace of that village, and the calm of that summer evening, a timelessness that I didn't understand but was beginning to love: and again, I wondered at the twist of fate which had brought me to this peace during a time of world strife.

Dusk was deepening as I climbed down from the gate to stroll across the road and back to the Kirkton, our affectionate name for our pub home. Next to the Kirkton's yard and opposite the door of our room was an old white house with a worn stone doorstep. Standing on the step was a white-haired old lady, probably in her late seventies. As I drew level with her, the impossible happened: she spoke to me!

'Guid Evenin',' she said. 'It's a braw nicht, is it no?' For a moment I was too stunned to reply: a villager, and one of the oldest respected inhabitants at that, had made the first salutation. There was twinkle of youth in her eye, and a song of friendship in her voice, and we lapsed into comfortable conversation about nothing in particular. At first, I found her softly lilting, smooth brogue difficult to understand, but slowly the music of it began to make sense.

Suddenly she said, 'Wil ye no' come ben the hoose for a wee bite and sup?'

I followed her into a small neat room, with a huge fireplace almost filling one wall. An ancient wooden table with tasselled cloth, a few wooden chairs and an old, well preserved couch made up the furniture. Over the fire, suspended from a chain from a swinging cast iron bracket was the griddle, a circular metal plate, a full thirty inches in diameter: a warm sweet smell of baking spread a soft air of comfort in the dim light of the room.

The old lady bade me, 'Sit doon an' mak yer'sel comfortable.' Then, 'Will ye tak a girdle scon' or a tattie pancake?'

Being completely ignorant as to the true identity of either I accepted a plate of both out of politeness, and a steaming mug of tea. The scones and pancakes were delicious, with the earthy, homely flavour of the food of days long past; and the tea was even better than Mrs G.'s!

I don't know how long we talked, or recall what we talked about, but her soft voice, the glow of the fire, and the wholesome food had a soothing, soporific effect on me. I rose to the surface as she picked up my plate and mug and walked over to a small sink in the corner.

'Weel, it's time I ganged tae ma bed,' she said. We said, 'Guid Nicht' on her doorstep and I walked slowly over to the Kirkton. It had been a remarkable evening, a wonderful evening of total rapport between an old Scottish lady and a young Sassenach airman, people from different worlds and different centuries: and although I wasn't aware of it as I fell into a happy sleep that night, it was the evening the village opened its heart to us.

From the evening that Granny Sloane invited me into her cottage for a 'bite and a sup' things changed around the village. Whether granny had set the pace and the village fallen in step, or whether the time was just ripe, we never found out. But curtains stopped moving and doors stopped closing; and the villagers started to greet us instead of avoiding our greetings. We were welcomed into gossip sessions around the bar and would pass a leisurely few minutes for a crack outside the post office.

We came to know the locals as friends: the Templemans who ran the store, Wullie (Willie) Porter the village 'Pollisman', wee Wullie Sloane who lived with his granny and worked in the gardens up at the Marquis of Ailsa's place, Culzean Castle, and Georgie Birrell, the overseer for 'Airrrrchie' Gray's farm teams. All became part of our daily lives, and the wonder of it was that we had become part of village life.

The only exception was Archie himself. He remained a shadowy, unapproachable, dour Scot, seldom seen, never entering the village gossip or social activities, and as remote from the eight airmen as Ailsa Craig itself.

We lived in two worlds. On the one hand we felt like villagers and began to almost talk like them: on the other, as we left the road and trudged up through Archie's yard, past the byre and up the steep cow path to the top pasture, the village door closed behind us. In the tiny confines of the squat DF cabin where we sat at our watches, the whole Firth of Clyde, the Mull, the Irish Sea and the Ayrshire coast from Greenock to Stranraer were compressed into the rhythmic squeaking and cheeping of Morse signals. Three thousand, five hundred square miles of roving aircraft crammed into a hundred square feet of cabin space.

Our job was to provide directional bearings to aircrew in training, to help them develop techniques which could prove crucial in returning to base from later bombing or torpedo raids. I learned in that cabin that failure of those techniques could produce tragic results.

I was relieving Neil and as he shrugged into his tunic and hitched his small pack onto his shoulder he said, 'Och, there's nae activity the nicht. Ye can probably get yer' heid doon for a bit.' He stuck his head out of the door and paused as a damp blast hit him. 'It'll dae ye nae guid though: a doot ye'll sleep in the ***** racket!' and he disappeared into the darkness.

Unknown Course

The wind howled and sighed round the cabin and I made myself comfortable at the console, log pad in front of me and pencil in hand. I pushed the headset forward, half off my ears and made a couple of experimental twists of the frequency dial. Hearing nothing but static - 'hash' we called it - I resigned myself to a dull watch.

I'd written a couple of letters home and was beginning to doze. The dim lighting, the crackling and hissing of the headset and the moaning of the wind outside the cabin fed my drowsiness, and my head would nod and sink toward the log on the console desk.

Then, from a semi-sleep, I heard it.

'Dah-dit-dit, Dit-dah-dah-dah-dah, Dit-dah' ....'D1A'.

However sleepy a wireless operator might be, whatever condition the mind, the slightest hint of his own station's call sign would bring him down to earth. I grabbed the tuning dial with my left hand and the pencil in my right.

I strained my ears and gently coaxed the dial minute fractions left and right. There was nothing but static. Then again, so faint and unstable that

I almost thought I was imagining it, came my call sign. There were other dots and dashes, but so indistinct and drowned in crackling that I could make no sense of them. Now it became a challenge, and all of my senses were tuned to deciphering the radio signal message.

Slowly, by combining the scattered pieces of signal I had written in the log, I deciphered the call signal of the transmitting station and recognised a request for a QDM.

I requested the long dash signal I needed to take a bearing, but a minute or two passed before the aircraft received my repeated message and started to transmit the required dashes. Their faint whistle rose and faded in the confusion of static, and I could feel the tension growing in my face and shoulders as I swung the knob of the goniometer through a thirty to forty-degree arc, trying to find a null point where the signal became weakest.

A first-class bearing was a recognisable null point within a swing of six degrees, a third class within a swing of twenty degrees. I couldn't tie the signal down to within a swing of forty degrees, but I transmitted the bearing I obtained, such as it was. Twice, I managed to obtain a similar bearing, and after several attempts, the aircraft twice transmitted the 'Received' signal. Meanwhile, radio reception had grown steadily worse until only the odd Morse symbol in a whole message was readable: I could recognise the tone of the aircraft transmitter intermittently, feebly sounding through the hash, but useful communication was lost.

For an hour I listened, and fine-tuned the dial, hunched in concentration over the set but I heard nothing more. The last faint 'R', telling me that the aircraft radio operator had received the bearing, was the last radio signal I heard until the storm had abated and the RAF main ground station opened up for general transmission and exercises at around 07:00 hours.

At a Court of Enquiry, two weeks later, I learned what had happened to the aircraft. There had been two of them, not attached to our station, flying over the Irish Sea and bound for a station north of us. The aircraft, Beauforts, each carried a crew of four men, and only one of the aircraft had a functioning WT transmitter, the one being used to 'talk' to me.

Had they taken my bearing immediately upon receiving it they would have swung north-east, past the southern tip of the Mull of Kintyre, and arrived safely at our runways. Tragically, the aircraft had continued trying to get a confirming bearing and had flown further north before following the bearing I had given them. On that black, blustery night eight young airmen died in a fiery crash, high up on the slopes of Goat Fell, on Arran.

I was the only operator to hear their signals, and the last earth-bound person with whom they 'spoke'.

The presence of the Royal Air Force in and around the village gave a new dimension to village life. The villagers learnt a lot of new words, and foreign language from as far away as London and the Midlands; and the pace of village life picked up to meet the Air Force's demands by at least two percent.

We found some new dimensions ourselves.

Old Mrs Campbell who worked in the Kirkton's kitchen for instance. Mrs Campbell was a placid, happy soul with a brogue as soft as velvet and a droll sense of humour. We'd regularly give a hand with the dishes - the only time and place in my life that I've actually enjoyed doing dishes! The atmosphere of the dark, warm little kitchen, the bustling sounds from the bar, the sense of belonging - everything about the place made the Kirkton a home away from home and taught us what kind hearts and friendship really mean.

Of course, we weren't always really on speaking terms. I once asked Mrs Campbell where I could find a dry teacloth for the dishes. Her hands stopped, pink and glistening in the soapy water, as she tilted her head at me, a little quizzical frown on her face as though she were considering what language I was speaking on that particular day.

'We'el ye ken, they're in the press in the lobby.' She pronounced 'lobby' as 'lawby', with a delicious singing accent on the first syllable. The glint in her eye and the quirk at the corner of her mouth said, 'Figure that one oot if ye can!', and I was left to my own devices to find the teacloth in the cupboard in the hall.

On another occasion I asked her if she knew where wee Wullie Sloane was as I wanted to see him about a promised guided tour around the grounds of Culzean Castle where he worked in the gardens. She half grinned and jumped at the opportunity.

'Och, he's awa tae the fairm t' fetch the breckan.' she said and, seeing my stunned look, added, 'Ye no ken whut a breckan is?'

I shook my head and said, 'No; how do you spell it?'

Her face showed no expression as she answered, 'Och, it's easy. It's B-R-ECH-ECH-A-N!!'. It took me two days to find out that a 'Breckan' is a horse collar.

I got my own back a couple of days later. Happily rubbing away at a teacup, I was singing quietly to myself one of the songs I had learned in school. We had learned it in dialect, as it was written, and that's the only way I knew how to sing it.

'Bawnny Chairrley's noo awa', Safly o'er the friendly main......' I yodelled. 'Monney's the hairrt will break in twa, shuid ye no' come bacch again.......'

I became aware of a sudden suspenseful silence and a complete absence of motion at the sink. I turned to see a face registering paralysed astonishment. I smiled blandly and asked if there was anything wrong. Mrs Campbell's glazed stare remained fixed for a few more seconds, then she heaved a deep, breath and said, 'Ah dinna ken the meanin': ye haver like a Sassenach and sing like a Scot!'

It was my one triumph, and I never enlightened her.

Mrs Campbell was one of those jewels one finds in tucked away villages, uneducated, unpretentious, and filled with a wisdom no University can teach. 'Airrrrchie' was another one, although for many weeks we had only the word of the villagers that he actually existed. We walked through his farm every day, past the white stone house, up the cobbled cow path to the byre, and across the hillside pasture to the cabin: but we never saw him.

I probably spent even more time around the farm than the others, for I regularly helped with the chores in the wheat fields, like bundling and stooking wheat, a pastime made particularly pleasant by the company of a tiny, dark haired Land Army girl I had taken to strolling the braes with or escorting by bus to the movie house in nearby Maybole. It was an innocent and delightful affair, spent mainly in joint admiration of the hills, the birds, and the gentle, soothing essence of the environment into which the war had thrust us both.

One late afternoon I was ambling by the auld stane dyke, off duty and enjoying the pleasant warmth of the sun and the earthy smells around me. As I passed the yard of the house, a voice spoke softly. I started for the voice was unexpected and its owner not immediately visible: it was well modulated with a refinement of accent that was not of the village.

'It's a braw evening,' it said. It had a familiar ring, but this was no Granny Sloane.

By the side door in the cobbled yard stood a small, slender man, probably in his late sixties or early seventies. His thin hair was grey, his tweed suit was smoky grey and, somehow, his very presence exuded an air of reserved greyness.

Even as I acknowledged that it was indeed a 'braw' evening, the shocking realisation came that here, at last, I was face to face with the 'Maister of the Village'......'Airrrrchie Gray'! I mumbled a few pleasantries and he responded with a quiet and pleasant courtesy before he delivered the second shock.

'Would ye tak' a bite o' tea wi' me?' he asked.

I must have responded in the affirmative - I don't remember: I only recall that a few minutes later I was sitting in the cool farmhouse, my back to the window in its thick stone wall, discussing topics, now forgotten, as though I had known Archie all my life.

A neat and equally grey little housekeeper materialised, spoke quietly with Archie, and softly disappeared back to the kitchen. Just as swiftly, she reappeared with fresh bread, scones, farm butter and homemade strawberry jam, and a huge steaming pot of tea. In the rationed austerity of wartime Britain, I had found the meaning of ambrosia.

We talked of many things during tea, of the war, of my home, of the Air Force, and of the local Burns' heritage. Too soon, I realised that the room was growing dim and that a blue dusk was settling outside. I knew instinctively that it was time to leave, and I rose to bid Archie goodnight and my deep-felt thanks for a very special evening.

Archie accompanied me to the door and shook hands. He looked at me with a calm steady gaze, and with just the shadow of a smile said, 'Ye'll no' be botherin' the girrls too much an' keepin' them frae their worrk?'

I suddenly realised that the canny Scot had saved the punch line for the end. I smiled back at him and promised. I never saw Archie again.

In our off-duty periods we would often migrate from the kitchen, where we had been helping Mrs Campbell, to Mr G. where we would help in the bar - on both sides of it. It was a tiny bar, comfortably holding a total of about fifteen people, and there were always a few airmen gossiping with the regulars. At least, that was true during the week.

Sunday was a different matter. On Sundays the throng overflowed the bar, spread out into the corridor and into the cobbled yard, and came perilously close to taking over our bedroom. The swelling of the pub population seemed to bear a direct relationship with the unusually heavy flow of traffic through the quiet village: most of the traffic seemed to have urgent business at either the Kirkton or the Shanter.

I was too polite to mention it for a time, but finally my curiosity won out. Mulling over a pint of bitter one quiet Monday evening, when business was remarkably slow compared with the previous day, I asked Mr G. about the Sunday phenomenon.

'It's the law!' he said.

I gulped and choked over my bitter. 'To drink heavily on a Sunday?' I spluttered.

'Not at all,' said Mr G. 'It's just that the pubs in Scotland are closed for normal business on Sundays.'

'You could have fooled me!' I said, a dazed expression on my face.

'That is, except for bona fide travellers,' added Mr G.

The light dawned. If the law said that only bona fide travellers could legally buy a drink on Sundays, that would provide an irresistible incentive to travel. Therefore, half the population of Girvan, around seven miles to the south, spent their Sunday afternoons in Maybole, a few miles to the north, and Maybole residents returned the compliment by becoming temporary residents of Girvan for the day. Driving being a thirsty occupation, Kirkoswald, about midway between the towns, provided a perfect refreshment stop for 'bona fide' travellers.

My respect for the canny Scots leaped by several points.

Our double lives went on placidly through that delightful summer, and only our work and the occasional leave in London and my beloved Cambridgeshire Fens reminded me that a war was in progress. I regularly cycled the thirteen or fourteen miles to Ayr, past ancient, ruined Crossraguel Abbey where the caretaker, a resident of Kirkoswald, once allowed me to mow the lawn in the cloisters to help restore my circulation on a cold Autumn day. I had been standing at the top of the tower, sketching the ruins, and mowing the lawn did us both a favour.

Many a time during that idyllic summer I pedalled through the little village of Alloway with its low thatched cottage, birthplace of Robert Burns. Forty-six years would pass before I saw inside that cottage, closed for the duration of the war. Before reaching the cottage, I would pass the Auld Kirk, the site of Tam o' Shanter's supposed encounter with the witches, and near the kirk I would pause in my journey and turn off the road to a narrow track that ran by the river Doon. I would lean my bike against the timeworn stonework of a hump-backed bridge over the river: and from a parapet of the Auld Brig o' Doon, in the soft summer gloamin, I would watch the swift flight of swallows skimming the water between the steep wooded riverbanks. I didn't know until years later that Burns himself had done the same thing in the same spot, or that the place on which I stood was the fabled spot where the witch had grabbed the tail of Tam's mare, Maggie. As Burns wrote it:

Ae spring brought off her master hale,
But left behind her ain grey tail.

It might be thought that none of these dreamy reminiscences have anything to do with the war or the Air Force. They are, in fact, an integral part of the whole experience, for without the war, and the enforced displacement of men from their familiar environments, their lives would never have been enriched by places and people they would never, themselves, have chosen to visit.

So, it was with us. The separation from our homes and families, unpleasant as it had been at first, had resulted in new friends, new scenery, and a totally new and exciting environment. In a few short months we had become a part of that environment and it would never be quite the same again, for us or for those whose lives we had entered. We had become accepted into the homes of a warm-hearted people, we became accustomed to the mock-serious admonishments of 'Wullie', the village 'pollisman', for riding our bikes without lights at night, and we had become participants in, rather that the subject of, local gossip. At a local birthday party, I even survived the crazy, whirling, noisy, joysome ordeal of several eightsome reels after eating a crippling chunk of Scots dumpling.

Christmas came and went, and Hogmanay riotously ushered in the New Year, 1943. The sleets and rains of January and February, howling and biting their way in from the Clyde, couldn't dampen our love for the place, or weaken our sense of belonging: it took Derek to do that!

I was sitting on my bed on a dull, cold afternoon in late February, writing a letter home. The door opened and in walked Derek. He paused dramatically before closing the door behind him and gazed at Neil and me briefly, a mixture of despair, bitterness, frustration and a dash of suicide on his face. I would never have guessed even Derek could have pulled off an expression like that. He paused for effect then, like the crack of doom, came four words.

'The axe has fallen!!'

I thought it sounded a little theatrical, but obviously something mildly serious had happened. You never quite knew with Derek.

Then he told us, and Neil and I assumed expressions almost as gloomy as Derek's. We had been posted, out of Kirkoswald, out of Turnberry, out of Scotland, apparently to some nameless, God-forsaken place in the English midlands, closer to home - and far away from what we now recognised as being a wartime paradise.

With doleful faces, we packed our kits, made a few, last, pathetic farewells, and sat down to wait for the station truck to collect us. After an hour or so it arrived and disgorged three hateful, brand new faces, usurpers of what was rightfully ours - or so we felt: mere Air Force wireless operators without a vestige of rapport with what we now felt to be our village, or our friends. We climbed aboard the truck and watched in misery

as Kirkoswald faded into misty rain behind us.

We remained in Turnberry for a week or so, performing routine wireless operator duties, and reluctantly and resentfully sliding back into Air Force routine with SHQ, the signals office, discipline, saluting, noisy NAAFI and all the other horrors thrown in. The day came that the three of us, still forlorn, climbed aboard a truck for Ayr: and the Isle of Arran, Ailsa Craig and the Mull o' Kintyre were hidden, veiled in grey cloud as they had been on that first day. I felt that they were crying at our departure.

We found ourselves in Newbold Revel, near Rugby, being indoctrinated in the mysteries of Japanese Kata Kana Morse code and procedures, as members of a mobile intercept unit in 'Y' Service, the WT Intercept Branch of MI5: and there we were, locked into the 'Secret Service' for the rest of the war. To wind up, I was recommended for my 'Tapes' at the end of the course because I had achieved high marks in all subjects and, unlike my first AC1 test, the Japanese plain language Morse was pure gibberish to me on paper, so I couldn't anticipate and make that one stupid mistake. As a matter of fact, I could take Kata Kana Morse at over thirty words per minute while I was restricted to twenty-two in English plain language. My biggest obstacle was in the fact that I am a fast reader and couldn't help scanning ahead the whole time: in Morse, that was fatal!

Upon arrival in North West Frontier Province in India, 140 miles from the Khyber Pass, I was told that 'Tapes' were only available to those with 'Seniority' in India and six others were promoted instead of me: my language at this news was very plain, but they kept me happy by making me a Corporal, Acting, unpaid as one of eighteen chosen ones to remain there for a time as instructors to around three hundred and fifty Royal Indian Air Force sprogs. We trained them in virtually everything, Japanese Morse and procedure, electricity and magnetism, radio theory, foot drill, and rifle drill, and how to behave like airmen instead of a motley assortment of Sikhs, Hindus (all Castes), Anglo Indians, Muslims and tribesmen! I received my paid tapes just before my return to England, three years later!

The two friends from Turnberry who remained attached to me for the rest of our Air Force career were Derek Briscoe, a Londoner, and Neil Semple from Stirling who enjoyed insulting me as much as I did him from the moment we first met! I have not seen, or heard of them, since our demob in 1946, much to my regret.

My association with aircraft didn't end at Turnberry. When I moved to Canada in 1954, I was employed by Avro Aircraft where I became Experimental Flight Test Instrumentation Laboratory Supervisor. I was associated with the ill-fated 'Arrow' aircraft for the whole of its history, and I and my staff of forty engineers and technicians were responsible for the operation and maintenance of all the experimental flight instrumentation. The continuous multi-channel tape recorder system, telemetry system, continuous trace oscillographs etc., were the most advanced flight test systems of the time, and occupied the entire missile pack, 17 feet long, 9 feet wide and 3 feet deep. Not widely known is the speed that the aircraft achieved the day before the programme was cancelled. 'Spud' Potocki, Chief Test Pilot who made that flight, told me as we walked back despondently to Experimental Flight Test on the day of cancellation: 1,480 mph in level flight at 35,000 feet! Jan Zurakowski, former Chief Test Pilot and WWII fighter 'Ace' told me three years later that the next aircraft off the line, No. 6, with the new 'Iroquois' engines, would have almost certainly done 2,000 mph. This was in 1959!

Kirkoswald probably saw many changes of faces in the airmen that became part of its life over the next few years. But it could never go back to that day in May when the first eight strange airmen arrived to become, first, an alien breed, then, in so short a time, accepted members of the village: 'Our boys' as I overheard one villager say. Over sixty years later, I still miss them.

Joseph Cosby

I was born in Bristol at the end of 1920, and both my parents died when I was quite young, so I was taken into the care of an older stepbrother. When I left school, I trained as an electrician and got a job as a cinema projectionist at the Park Picture House in Whiteladies, near Filton, where the Bristol aircraft factory was. I remember that when the war started in earnest, I was able to go up onto the roof of the Cabbot Cinema and watch the German planes as they approached to bomb Filton.

I was called up in 1941 and sent to RAF Hitchin to do my trade training, before being posted to 5(C)OTU at Chivenor in Devon. When the unit was transferred to Turnberry, I went too.

I remember the train approaching Turnberry station, we had to wait until some lady took her washing off the clothesline which was stretched across the track. Eventually we pulled in beside the platform at the hotel.

We were given accommodation in Nissen huts situated by the firing range, close to the hotel, conditions weren't too bad, it was a case of having to like it!

A taste of things to come was provided when one of the first aircraft to attempt to land at Turnberry, a Blackburn Botha, crashed into a road roller, we all thought this was a good start, little did we know it was to be the first of many!

Once we had settled in, I became part of the ground crew of 'F' Flight. Each flight had its own team of engineers, electricians, riggers and instrument technicians. I had quite an easy job, just routine things like checking and replacing blown bulbs. More technical things like magnetos were the responsibility of the engineers. On our section, which was on the west side of the airfield, we looked after aircraft like the Fairey Battle, Lysander, Miles Master and Blenheim. Basically, we would perform routine checks and then fill in form 700.

Occasionally, I would work in the workshops which were on the technical site at the other end of the airfield. It was here one day that I witnessed the accident between the Beaufort and the Hampden. I think that the Beaufort was on a test flight, because there was a member of ground crew aboard. I recall hearing the crash and dashed outside, I looked toward the sky and saw, what I thought were birds flying about, then I heard the sound as the oxygen bottles exploded. It had been bits of shrapnel filling the air, not birds. There was nothing we could do to help.

There were lots of other crashes around the airfield, at Balkenna near the shore, and on Mochrum Hill, between Maidens and Maybole. I heard that there were over ninety-eight crashes in just nine months. I also recall seeing a Blenheim crash at the end of the short runway, it seemed the pilot suffered loss of power on take-off, apparently caused by some sort of air pocket just by the lighthouse.

We were never told very much, on a need to know basis only. One night we were called back on duty and told to get every aircraft serviceable as soon as possible, every little detail had to be checked, we weren't told why. Speculation was rife, we all thought that there must be something big on. We never found out what it was! In 1942 I went to a fun-fair in Girvan, and there met a local girl called Agnes Brown. She had gone to the fair with a Corporal, but she came home with me! We were married in October of the same year. My best man was Tom Hylands from West Ham in London - he was later posted to S/toon. I never saw or heard from him after that. Once I was married, I moved out of the base and into Maybole, to 34 School Vennel. I was never much of a dancer, most nights out were spent at the pictures, but usually I just had a quiet night enjoying the comforts of my new home.

After two years at Turnberry, I was posted to Lough Earne in Northern Ireland with 201 Squadron. Here I worked on Sunderland Flying Boats, they were the best aircraft I ever had to work on, of all the types, they were my favourite, although, as usual with Coastal Command postings, it meant that we had to wear a raincoat all the time! I was able to experience a flight over Lough Neagh, in a flying boat, a memorable trip, apart from another 'joy ride' in an Anson, that was the only time I took to the air. From Lough Earne I transferred to Ballymena and Liberators

of 59 Squadron.

The RAF, having no further use for me, gave me over to the army. We had no choice but to go, the only concession we had was that we could choose which regiment to join, with the right to request to serve with a relative. I volunteered for the Somerset Light Infantry, with which my stepbrother was a regular soldier, having joined up before the war. At this time, he was Colour Sergeant E. T. Bailey, serving in Italy. Somehow, I ended up in Germany, the war was by this time over, and I can honestly say that I never saw an angry German, I never got to see my brother either, not until I was demobbed in August. For a time, I was stationed near Belsen, though I never went to see it, I spent the last months of the war showing films for the troops.

When I returned to Maybole, my wife and I settled down to married life, our home was in one of the pre-fabs at Whitefaulds. In all, we lived there for 33 years, the best home we ever had. I worked on a coal lorry, bagging coal, and then for Callaghan's the plumbers before being employed as a driver with the local council on the refuse lorry. I became a member of the Auxiliary Fire Service, I had over eighteen years' service as a volunteer fireman and reached the rank of Sub-Officer. I would have stayed on and received my long service medal after twenty years, but the council bosses told me that I just couldn't dump the 'ash cart' whenever the siren went! I must say though, that I saw worse things as a fireman than I ever experienced in the RAF or army.

Terry G. Parker
After spending three months at Chivenor, North Devon with 5(C)OTU, the unit was instructed to move to Turnberry in Ayrshire in April 1942. It took two days to load everything aboard a train. We finally pulled out of Wrafton Railway Station, Chivenor at about 7 am, after breakfast on a Wednesday morning. Our route was via Barnstable to Exeter, but we were diverted because there was an air raid on Exeter at the time, so after three hours we finally got to Crewe. All we had to eat was a packed lunch consisting of two cheese rolls. We stopped at Crewe for

Terry G. Parker

an hour and we all managed to get a cup of lukewarm NAAFI tea. By this time, it was dark, and all the lights were out on the train, there was no heating in the carriages at all and we had to pull the blinds down at the windows to ensure a proper blackout. We went on for about another three hours and came to a halt just outside Carlisle, there was another air raid warning, so we sat it out. We finally arrived at Carlisle Station at 6 am, we were treated to a cup of tea and a plain roll, setting off again, to eventually arrive at Turnberry about 11 am.

We had to unload our kit then line up for roll call, by the time this was over we had missed lunch, which had finished at 2 pm, we were told we would have to wait until teatime at 4 pm before we could get fed.

Meanwhile we busied ourselves collecting our blankets and finding our allocated Nissen hut. By the time we had a wash and a shave teatime had finally arrived.

The next morning, we got up at 8 am and went down to breakfast, we were stopped by an airman at the mess door who said, 'Sorry, breakfast starts at 6 am and finishes at 7 am.' I said to this chap on the door, 'How many were in for breakfast this morning?' 'None,' he replied, 'You are the only ones here.' 'Well,' I said, 'We want our rations, there are fifty of us here, plus five Sergeants and two officers. We have been on this camp for two days and we want our proper rations, I bet you have been flogging them to the civvies on the camp, I've noticed a lot of

them hanging around here, what are they waiting for?' 'I don't know.' he said. I replied, 'Okay, I'm going to get our CO he'll sort this out.' The airman looked worried and told me to wait there while he went and got the cook. A scruffy looking chap appeared at the door, smoking a pipe and rather huffily snapped, 'What do you lot want?' I asked if he was the cook in charge of messing and what was his rank. He retorted 'That he was not in the Air Force, but that he worked for Wimpey's Contractors and did the cooking for them.' 'I sleep here as well,' he added. Flabbergasted, I informed the pair of them that I was getting the CO.

Unfortunately, at that time we didn't have a CO, so one of our chaps went to the Sergeants' hut and came back with two Sergeants ready to do battle for our breakfast. By the time we arrived back at the mess both men had vanished. We didn't see the scruffy Wimpey cook or the airman on the door again. I think he was a deserter hiding out. There were some suspicious types hanging around.

On the Friday morning at about 11 am, the aircraft began to arrive, the first one landed ok and the lads were on each dispersal point ready to receive their charges. As the second aircraft touched down, a motor roller belonging to the gypsy people who were doing the turfing, started

Terry G. Parker

Terry G. Parker

to cross the runway, the wing tip of the plane hit the roller and the aircraft slewed round. The undercarriage collapsed, and the aircraft came to a stop just off the runway. The five crew members were a bit shaken up, and they had a few bruises. It was lucky that the plane didn't catch fire as there was no fire crew on the airfield at that time, as it was not considered operational. The chap on the roller jumped clear and escaped with a few cuts. We never saw him again after that day. Six more aircraft arrived without incident and were formed into A, B, C and D flights.

I remember that Turnberry Hotel was built on a hillside overlooking the large golf-course which had recently been carved up and the runways laid down. The Nissen huts, which were our living quarters lay further up the hillside on the road towards Ayr. The hangars lay at the foot of the hill, with the approach road leading down to the perimeter road and on to the main camp buildings and workshops. We had some very cold weather, and very little coal for our stoves in the freezing Nissen huts. The gypsies on the airfield asked if we wanted to buy some coal from them, so we had a collection in our hut, put in a shilling each and bought five one hundred weight bags of coal. That kept us warm for two weeks. On Saturday evenings we used to walk to Maybole to the dance, this was until the army chaps came with their studded boots which ruined the dance floor, so the dance was stopped. If we were lucky enough to get a day off, we walked to the top gate, which was called Wimpey's corner after the contractors who built the airfield, and got a bus to Ayr for the day. The last bus back was at 7 pm.

Early one morning two of our chaps were walking along the seashore when they came across boxes and boxes of fish which had been washed up on the beach. The boxes were taken to the cookhouse and we had fish for every meal of the day for weeks! Those chaps were not very popular as you can guess!

After a while the camp began to fill up with lots of Air Force personnel, three WAAF flight mechanics arrived on the scene, we named them Faith, Hope and Charity. They went to work in the Hangar. Ha-ha! that was a laugh from the start. The fitters, riggers, sparks, wireless, radar, and instrument chaps all worked together, I don't think they took kindly to this invasion of 'women' into their domain.

A couple of the locals were fishing one night at high tide, when feeling their lines go taut the all thought they had a bite. When they reeled them in, they found no fish but sailors hats! The local Police, ARP and Home Guard were summoned but there was nothing they could do in the blackout. It was morning before the full picture could be seen. All along the beach there were the bodies of sailors caught up in the barbed wire defences. It was said that RMS *Queen Mary* going at full speed up the Clyde had rammed a British Destroyer.

[This was probably the incident on 2 October 1942 when the Queen Mary was returning to Glasgow from New York carrying 20,000 American troops. On the last leg of her journey she was joined by the light cruiser HMS Curacoa providing anti-aircraft escort up the Clyde into Greenock. As was usual the Queen Mary was on a standard zig-zag course to confuse U-Boats, when there was some confusion and the two ships found themselves on a collision course. Both captains thinking the other would give way stayed on course with tragic consequences. The Queen Mary sliced the cruiser Curacoa in two causing her to sink rapidly with the loss of 338 men from a crew of 439. Being under strict orders not to stop for anything due to the U-Boat threat, the Queen Mary carried on to Greenock, despite taking in water. Author]

As I mentioned before, the airfield had four flights, A, B, C, and D. I was detailed to D flight along with a squad of other chaps. Some of them were a lot older than me, they used to jibe that they were in the RAF when Pontius was a Pilot!

A small section was formed to deal with minor problems out on the flights, to save the aircraft having to be towed in the hangar for repair. We named this section the 'snag gang', which consisted of two Sergeants and eight fitters. We worked two shifts, early until 3 pm and 3 pm until finish of daily flying. So, on each shift we had one Sergeant, two engine fitters and two airframe fitters. We had a large Albion, canvas-covered lorry with a complete workshop on board. It was a job to get it started in the mornings, we had to swing on a starting handle, there were no electric starters then. We had to move down to the airfield, into a large Nissen hut, separate from the main camp. This suited us fine! We would drive down to the Mess for our meals in the lorry. We had a lot of bad weather days in which no flying was done, although there was always something to amuse us. In the evenings we would cycle down to Girvan to the WRVS canteen, which was situated in a large house. The downstairs had been converted into a hall with tables and chairs and a counter for tea and coffee urns. We became quite friendly with some of the young ladies there and I was invited to a large farmhouse one evening by a young lady to meet her grandparents who she was staying on holiday with. Her grandfather was the biggest man I have ever seen, he must have been six feet eight inches tall and he had a sixty-inch chest! I thought I was tall at six feet three inches.

He wore a large kilt, that could have been used as a bell tent and when we shook hands his hand felt like a leg of mutton. We got on very well, and after enjoying a large roast dinner, we sat and talked, he asked me if I knew anything about engines and when I replied that I did he said, 'Come along wi' me laddie.' I followed him outside to a large barn, the young lady, who was Anne Cameron, came too. Grandad led us to the corner of the barn where stood a large tractor, turning to me he said, 'It willnae start', I turned

the engine over but there was no luck, it wouldn't start. I looked under the fuel tank and found that the fuel pipe had corroded, I managed to get the pipe off and told grandad that I would take it away with me and get a new one made up in our workshops.

We went back into the house and settled down to supper and a cup of coffee. Afterwards, I took my leave and promised to return the next evening. I skipped my tea at the camp that night!

The following night I cycled down to the Girvan farm and fitted the new pipe and flushed it through, the engine started first time and grandad drove the tractor out of the barn, around a field and back into the barn! I had made his day, he asked what he owed me, and I replied, 'Nothing, as long as I can come and see Anne.' He replied that I 'would be most welcome any time.'

The weather began to change and subsequently there was a lot of flying, so I was unable to get to the farm, but one of the lads took a message to the WRVS canteen. It was over a week before I managed to get away and visit Anne. Her holiday with her grandparents was over and she had to return to Glasgow on the Saturday. She told me that her grandad had telephoned her parents and that I had been invited to stay for a weekend at their home in Queen's Park, Glasgow. She said she would meet me at the bus station.

My next free weekend I travelled up to Glasgow and there was Anne waiting for me as she promised. I had a lovely time and met Anne's parents, had dinner with them and then we went to a cinema near Mount Florida. I stayed until the Sunday evening and then caught the bus back to camp.

I telephoned Anne's parents and thanked them for the nice weekend, and they told me that I could visit whenever I was free.

The following weekend, again the weather closed in and all flying was cancelled, so I telephoned Anne's parents and said I would visit on Friday and arrive about midday. I then went to see one of the cooks who owed me a favour and he gave me four pounds of sugar, one pound of golden syrup a large tin of corned beef and a dozen rashers of bacon. I managed to get the lot into my large backpack. Again, I was met at the station by Anne. We got to the family home in Queen's Park just in time for tea and then we went to the cinema. When we got home, we were given two tickets for a boat trip on Loch Lomond the following day. It was a lovely trip, but very cold, we were glad to get home that evening, but, then Anne's parents took me aside and told me that as Anne was going away to university soon it would not be possible for us to see each other again.

On arriving back at camp, I telephoned Anne's parents to let them know I had got back safe and sound, they thanked me for the food, told me to look after myself and wished me good luck. I popped down to the farm to see grandad, but he told me he knew nothing about what had happened, and I never saw Anne again.

Quentin Wilson

As a boy, I had walked with some friends from Maybole to Maybole Shore and we were on our way home in the late afternoon, near Pennyglen farm, when we saw an aircraft flying low and more or less level, along the valley from the sea. The aircraft was an Avro Anson, painted partly or completely yellow. The engine noise was not normal – a sort of high-pitched wailing sound that told us immediately that something was wrong. The Anson went on up the valley and disappeared over the hill towards Maybole. We hurried home as fast as we could, to find that an aircraft had crashed on the Memorial Park.

Looking down from the top of the hill near the monument we saw that the aircraft had crashed on the golf course, near where the second tee is now. There were men standing round the wreck. The crash site was compact and there was no fire. [Avro Anson DG861]

Annie Townsley nee Hyslop

I was born in Girvan in 1926 and had one older sister and two younger sisters. I started work in the Town Chamberlain's office as a clerk when I was fourteen, having left school that June. The Town Chamberlain then was Mr Blake. In

Annie Townsley

1939 the McMaster Hall in Girvan caught fire and burned down, at that time the town only had an auxiliary fire service and a wheelbarrow for carrying water to extinguish fires. The fire master was John McEwan; he also worked for the council and used to come into the Town Chamberlain's office to collect his men's wages. One day he mentioned that he was tasked with finding two girls to join the fire brigade and asked if I would be interested. I thought it would give me something to do as my elder sister was already in the ARP, so said yes. My friend Connie Stewart joined with me, although I was one stripe up, being a leading Firewoman. There was no formal training as such, but we did have a uniform. Most of the time we girls looked after the office and paperwork, answered the phone, etc. We had a siren that we used to call out the firemen during the day and later they had bells installed in their homes. More men joined the brigade, although Connie and I still did night shifts on our own, we felt quite safe in those days and used to walk to and from the fire station during the black-out. The fire station was at Laganwhilly.

We all enjoyed the fortnightly dances at the hotel, especially the live bands, up to eight of us used to squeeze into one taxi on the way home. I remember one dance, called the 'Palais Glide', to the tune of the 'Lambeth Walk'. On one occasion on the dance floor, I was kicked under the chin by a chap who was most embarrassed!

I remember a crowd of us had gone to the Byne Hill for a walk and we were watching a plane flying south, it was no higher than Ailsa Craig, suddenly it dived into the sea, that would be about 1942.

The firemen from Girvan were called out to aircraft crashes quite often. The worst crash was a Wellington that exploded on take-off. It seems the plane's undercarriage caught on the steep bank of a burn, overturned and caught fire. The firemen had to remove the airmen from the scene after putting out the fire. I remember Fireman McCready was severely ill when he returned to the station and had to take some time off work to recover. It must have been awful.

Once, we got the fire engine stuck in the sea. An aircraft had crashed on the beach at Girvan, and caught fire, but the firemen couldn't get near it for exploding ammunition. We couldn't rescue anyone and had to let the fire burn itself out, then douse it with sea water. While waiting for the fire to die down, the tide came in and the fire engine got stuck, we had to recover it the next morning using a farmer's tractor. There were two RAF launches kept at Girvan for rescuing aircrew, I seem to remember there was a small one kept for inshore rescues and a larger one for the open sea.

My father was in the army serving overseas, and one of my sisters was attached to a Signals Regiment. I joined the army in 1944, having finally persuaded my mother I would be alright, and was sent to Newbattle Abbey, Dalkeith, near Edinburgh to do my training.

I had wanted to be a driver. I had a 'shot' at driving the fire-engine and thoroughly enjoyed it, but I had to settle for the Army Pay Corps and was posted to London. Our office was in the Lilywhite's building opposite Harrods and I used to enjoy sightseeing walks around the city. I was sent to work in the POW section and was responsible for keeping the wages of our POW troops up to date. Some of them had a sizeable lump sum to come home to. There were rows of us girls and we each had a ledger with the men we were responsible for. Sometimes we received pay books that were in a terrible state, covered with blood, and packets of letters from deceased POWs, these were sent to the 'Dead Section' in the basement.

I hadn't been in London long when I was challenged in the canteen about the three upside down stripes on my sleeve and my length of service, I stated that I had been in the army for three months. I was informed rather curtly that I got a stripe for every year's service, not one a month. I promptly told them that I had earned them by serving in the Fire Service. I recall speaking to a girl who asked where I came from and when I said Girvan, asked if it was anywhere near Turnberry, as her brother had been killed there – he was a pilot.

One morning as I was sightseeing, watching the Changing of the Guard, a doodle bug went over, it hit the Chapel in Wellington Barracks during morning service. There were lots of air raids, and another morning when I was cycling up Constitution Hill, the siren went, and I sheltered against the wall of Buckingham Palace until the raid was over. A few days later, I passed that very same spot and was shocked to see that the wall had been demolished by a subsequent bomb! I stayed in the Church Army Hostel, in Praed Street. There the windows were smashed by the bombing, so I moved to Lennox Gardens, just off Sloane Street. I still remember the strange noise of the V2 rockets coming over London, it made the blood run cold. I enjoyed going to chapel at Chelsea Barracks, it was a lovely service accompanied by a Military Band.

Ronald B. Wood

The RAF was not my first service, before the war I was on the purser's staff of the P&O Shipping Co., but then, that is a story in itself!

However, due to a bout of diphtheria I was returned home to recuperate, and it was during this period that I was informed by the Air Ministry that they were urgently looking for volunteers for aircrew duties. I replied straight away, and instead of the navy it was RAF Coastal Command for me!

Our Beaufort crew were on Course 15 which commenced OTU training at RAF Chivenor, North Devon, on 16 March 1942. We were well into our training when it was decided to transfer 5(C)OTU to Turnberry. We flew up on 4 May 1942, along with other crews whilst the ground personnel took care of the equipment in general.

My pilot was a Scot, Lt Stewart, who lived in South Africa, so was a member of the SAAF. The navigator was Sgt Robin Wescott and the other W Op/AG was Sgt John Kendall. Lt Stewart was returned to South Africa early in our operational tour on health grounds, that was the last we heard of him.

Turnberry was a lovely setting, and I have no doubt that golfing enthusiasts were mortified to see such a grand course marred by the addition of a long runway through the centre. The site was rather widespread, so we were issued with bicycles to help us get around.

On 26 May 1942, just at the end of our training, two crews of trainees and a crew comprising of instructors were told to get fitted out with operational gear. This was a surprise. We were despatched to RAF North Coates. On take-off from Turnberry we lost power in our starboard engine and instead of climbing steadily past the hotel we found ourselves heading for the lighthouse! Even to this day, whenever I watch golf on TV from Turnberry, I recall those anxious moments. We reached North Coates safely and the engine was duly checked whilst we had a meal. Afterwards we were told that the operation had been cancelled and we could return to Turnberry. Rumour had it that we were being sent to make up numbers for the 'Thousand Bomber Raid' on Cologne. Thankfully, a Beaufort Torpedo Bomber was not considered a suitable aircraft for that occasion.

My recollections of Turnberry are happy ones, I enjoyed being beside the sea, at one stage I even did some canoeing. We used to practise our Radar readings using Ailsa Craig, the reflection of the granite rock formation was very similar to that of a ship.

After leaving Turnberry we were posted to RAF Lyneham and in due course told to collect a brand-new Beaufort (DE118) from the Bristol Aircraft Co., at Filton. After completing successful tests, we were sent to 1ADU at Portreath in Cornwall, and then the long haul to

1942 20 Beaufort Course

Palestine where we left the aircraft. I noted that this aircraft, whilst flying with 47 Squadron, struck high ground and the torpedo exploded, 50 miles SSW of Sidi Haneish. The date was 18 November 1942, only four months after leaving the factory.

We then converted to Glen Martin Baltimore aircraft for duties on a GR Squadron, No. 203. This was an interesting job, collecting, photographing and sending information back to base to allow the bombers to follow through. We worked in the Mediterranean, escorting convoys in the defence of Malta, also amongst the Aegean Islands searching for enemy shipping. We were then sent on six months rest, after November 1943, to 75OTU in the Egyptian delta on instructional duties.

Unfortunately, after this, the crew was split up and we went our separate ways. I was stationed briefly at Aden and finally with 209 Squadron, Mombasa, which flew Catalina flying boats. I was then posted back to the UK, where I did a course at RAF Cranwell, and with the swiftness with which the war ended due to the use of the Atom Bomb, we old hands were taken off flying duties and in the course of a few months, demobbed!

I will always have fond memories of my time at Turnberry and will never forget the warm welcome that we found there.

William Wilson

I was stationed for a short time at Turnberry in 1942 as ground crew, at that time under the Coastal Command Training Wing. The trainee crews were flying Bristol Beauforts and were engaged in air to surface torpedo attacking. This was initially done using concrete dummy torpedoes. Although I never witnessed it, this took place in the Firth of Clyde near Ailsa Craig.

The main entrance to the airfield was at the Turnberry/Girvan end, but most of the airmen used the road from Maidens. At the end of this road facing the sea, on the right were the living quarters, mess, NAAFI, etc. On the left were the airfield hangars. The road leading through the airfield ran along the side of a hill and had two stopping points on it, and nothing could proceed when an aircraft was coming in to land on one of the two short runways, the shortest running from the hill to the lighthouse. A very tricky approach to come down the face of the hill and level out to land. The pilots could plainly be seen from the stopping points. There were quite a few overshoots at the lighthouse! The main, much longer, runway ran parallel to the road and the sea.

The camp was quite a popular one having lots of west of Scotland men there. The living huts were long wooden affairs. Many a good night was spent at the pub in Kirkoswald, and a

1942 21 Beaufort Course

very unusual mode of transport back to base, was a taxi that was a converted hearse. If memory serves me right, we were locked in, so we couldn't scamper without paying our fare. It also had a back step that was used as a seat on many occasions when the inside of the hearse was full.

Turnberry was one of the few stations in my five-years' service that served chips at tea-time (or any time).

Peter Arden Hughes
My two grandfathers lived in Tarporley, Cheshire, in the early years of the twentieth century. The Rev. Walter Octavius Marsh Hughes was the Rector of St Helen's C. of E. Church and was the eighth child of the headmaster of Blundell's School, Tiverton. Doctor Forbes Fraser was a doctor in the town. My father Geoffrey was the second son of the rector, the eldest being Barty (later Rev. Assistant Chaplain in Chief to the Army) and the younger ones Jack (John Bernard) who retired as Captain RN and is buried at Meavy, Devon, and Robin. He also had one daughter, Muriel who had two sons and three daughters - Mark, Bernard, Mary, Elizabeth and Helen. Barty had one daughter who had five

Peter Arden Hughes

daughters, one of whom is Myrtle Kneen who lives at Whitbourne, near Worcester. Jack never married, and Robin had one son, John.

Doctor Fraser had two sons and three daughters the eldest of whom was my mother, Katherine, or Kit. The sons were Ted and Henry. Ted had no issue, but Henry had two children, Richard and Elizabeth. Richard had one son, Andrew but Elizabeth had no children. The three daughters were Kit, Alison ('Sass') and Margaret. Margaret was a Sister in the Army Nursing Service and died prematurely of sleeping sickness following a bite from a tsetse fly apparently imported in a fruit case at Avonweir, then the home of my parents. Alison had two daughters and a son. Margaret, who did not marry, William, who was awarded the sword of honour at Cranwell post war and retired from the RAF to become a test pilot with aircraft manufacturers Pilatus in Switzerland. He has two sons, David and Simeon, and Katherine, now Townsend, who has a son and two daughters.

After education at a preparatory school, Hildersham House (which he hated) and St Edwards School (Teddies), Oxford, my father read History at Magdalen College, Oxford, where he was appointed as aide to Edward, Prince of Wales, later King Edward VIII and Duke of Windsor. He did not graduate because of enlistment in the South Lancashire Regiment at the outbreak of war in 1914. He was commissioned and went to the Dardanelles with the regiment where in the rank of Captain he led his platoon in the assault from the Gallipoli beaches. A Turkish bullet hit him in the shoulder by his neck and exited his body from the small of his back. He was paralysed and left on the field for dead. The medical orderlies recovering bodies noticed that my father was able to twitch his nose and realised that he was alive. He spent two years in hospitals and was invalided from the army - without a pension!

He courted my mother, whose home was by now in Bath, her father having been in the RAMC on the Western Front throughout the war where he was awarded the CBE had moved to number 2 The Circus and later to number 5. He practised as a surgeon in Bath and a hospital was named after him, the Forbes Fraser Hospital. His wife Mary died following an operation and Fraser remarried and had a daughter, Mary. Forbes Fraser died in the twenties from poisoning from an infected needle while conducting an operation.

My parents' marriage was made against opposition from father's family who did not attend - a decision that they subsequently regretted - and I was born at the Walcott Street Nursing Home in Bath on 18 November 1918, just one week after the Armistice that ended the first World War and in good time to take part in the second one twenty-one years later.

The family moved from convalescence in Hereford to a clergy house in Cotebrook, near Tarporley where there are photographs of me as a baby. Father then trained as a nurseryman in the mistaken belief that this would be a non-strenuous occupation for a severely wounded ex-soldier and he bought an old John Bright water mill on the Bristol Avon at Christian Malford, in Wiltshire, which he renamed 'Avonweir' and built a greenhouse, later followed by six more. Initially he specialised in carnations but later diversified into mainly tomatoes and chrysanthemums with bedding plants in season - the perfect recipe for development into a modern garden centre, sadly not to be.

At the age of ten I was sent to a Preparatory School, Heddon Court, Cockfosters in Hertfordshire where the Headmaster was the son of a friend of my parents from Bath. Nicknamed 'Huffy' Hope by the boys he was a sadist typical of the Dickens schoolmasters, but the school and its grounds were beautiful and I made some good friends, the closest being John de Havilland, Geoffrey de Havilland's youngest son who was killed in a mid-air collision while testing a Mosquito during the war. The elder son, Geoffrey used to fly over the school in a Gypsy Moth which gave me my first interest in flying. One incident worth recording was the terrific noise of the R101 airship flying over London on its way to disaster at Beauvais in Northern France. We were in bed in the dormitory but rushed to the window to see its lights above us. At Heddon Court I

learned to live at ease with the aristocracy but my Mother never agreed with boarding education and with the slump, the fees (then about £80 a term I believe) were a severe strain and although Uncle Jack would have liked to meet them my parents decided they did not want to surrender their responsibility for me and I left Heddon Court to join my brother at the Chippenham Secondary School. This was a culture shock that I survived the stronger for. One had to bicycle to school, six miles, and back and there were girls at the school!

I soon became adjusted to the changed social environment and am grateful for the ability it gave me later to live with fellow senior officers but also to be completely at ease with the airmen and to understand them. I took a great interest in the Hawker Harts flying from the newly built Hullavington aerodrome (so called to the horror of the local gentry who pronounced it 'Hullington') and flew for the first time in an Airspeed Ferry tri-enged cabin aircraft of Sir Alan Cobham's circus at a fee of four shillings (40p) from a field by the Bristol Road outside Chippenham. I never had any doubt thereafter that I wanted to join the RAF as a pilot.

I became Head Boy at Chippenham School and on my sixteenth birthday I got my first motorcycle, a Zenith with a 250 cc OHV J.A.P engine which cost £32.10s. brand new. My father said that my idea of applying for a Short Service Commission was not good enough, if I wanted to join the RAF I should do it properly through the RAF College at Cranwell. This was some undertaking for a boy from a state school, this route being normally for those with a public school education and OTC (Officers Training Corps) experience. Such as me entered the Aircraft Apprentices Schools at Halton or Cranwell from where a very few were awarded cadetships at Cranwell. This was both too risky and too late, so I entered for the Civil Service Commission examination for all the three service colleges, the Indian Army, etc. specifying Cranwell as my first choice with the navy as an alternative considering the Fleet Air Arm as the fall-back option.

I continued schooling with this exam in mind and on the appointed days in the summer of 1936 rode on my motorcycle to Clifton College in Bristol where the papers were to be taken. There were about 400 names down for Cranwell and less than 30 places. I came nowhere! My Father gave me a sharp talking to. I simply didn't know what real work and study was. My outside interests came second to schoolwork. I sent off the application for the next exam in November 1936 and for the first time in my life gave my whole attention to the subject papers. I went to London for interviews and medical examinations at the Air Ministry and Admiralty and once more rode off to Clifton College on four or five days. Just before Christmas I received a brown envelope addressed to 'Flight Cadet P. A. Hughes'! I had got in by a close margin - 26th out of 28 accepted, although in the event there were 32 in my term. I believe I was the first cadet to enter directly from a state school.

On 7 January 1937 I arrived at Cranwell and was allocated a bed-sitting room in 'A' Squadron with a batman, 'Mr Speaks'. Kitted out in uniforms by Burberry's, which must have cost my father a substantial sum on top of the fees which were at that time payable each term. He never mentioned it, bless him. On arrival I contracted a terrible cold so missed the early settling in time by being in the hospital on the station, but on 25 January I was down at 'B' Flight and made my first flight in an Avro Tutor. Each day began with a Squadron Parade and inspection with rifles. Rifle drill was new to me, so I was soon in the awkward squad compared with those previously in OTCs or at Halton. However, I reached an acceptable standard although I never became very enthusiastic about foot drill with the result that I was never promoted to Cadet NCO status; this resulted in my never being practised in giving the orders in front of a parade, an omission in my training that was sorely missed later in my career when I became a Station Commander.

On 28 July 1939 my training was completed. Lord Gort VC, took the salute and exhorted us to serve our country in the military traditions that had brought it so much glory. I was gazetted to a permanent commission as a Pilot Officer and posted to 115 Squadron at Marham in Norfolk.

I then proceeded home on leave to Avonweir where I traded in my beloved Rudge Special motorcycle at Jack Williams's garage in Cheltenham for a new Morris 8, Series E, at a cost of £139. This was far outside my resources, so a hire purchase agreement was signed, soon to become of major concern to the finance company when they realised that I was an RAF pilot and that there was a war on; It proved a good investment as it sold seven years later for £345.

I served in various places across Britain for the next few years. The OTU was moved to Turnberry, Ayrshire, in May 1942. With that move I was sent to Upavon, where the Central Flying School, the school for flying instructors, had been retitled 1 FIS where I became qualified as a flying instructor on twin engined aircraft. On return to Turnberry I was given lhr:40 dual in a Beaufort and sent solo. Three days later I was giving dual instruction to Sgt Smit.

In July, I was given command of 'B' Flight, and later 'C' Flight, the night flying flight, with Hampdens and Beauforts. These two types were not very compatible in the days when there was no flying control. One joined the circuit and took ones turn at landing visually, although at night and sometimes by day there was some form of control by Aldis lamp from a caravan at the beginning of the runway. The Beaufort had a very steep approach and the Hampden a much more level one. I seem to recall an occasion when a Beaufort descended on to a Hampden in the approach. Active aircrew were kept away from crashes as far as possible for obvious reasons of morale.

A peculiarity of Turnberry was the high ground to the south which made the normal circuit height of 1,000 feet dangerous, so we used 1,500 feet when flying downwind for a landing towards the west. However, visiting aircraft could be caught out, especially in poor visibility or at night as there was no radio contact for us in those days (although air defence fighters were controlled thus).

I recall giving dual instruction to a crew and sending them solo at night. They crashed on attempting a landing from the west and all were killed. This aircraft had a non-standard Air Speed Indicator which I suspect may have misled the pupil pilot as to his approach speed and I felt some responsibility for not having demanded it be replaced. But this may not have been the cause. Either way, it was a great pity that I sent that crew solo and I have never ceased to regret this misjudgement.

In October, I was promoted to Squadron Leader (War Substantive) and in January 1943 the OTU moved to Long Kesh, in Northern Ireland, later the site of the 'H-Block' prison for terrorist prisoners. While there I considered that after three years of war if I was to justify my existence I ought to confront the enemy - and besides how else could I ever get a medal? I told the Chief Instructor, Willie Davis, that I wanted to 'go on ops'. I was soon posted to 489 Squadron, RNZAF, as Flight Commander of 'B' Flight at Wick. The Squadron was flying ex-Bomber Command Hampdens which carried a torpedo half in and half out of the bomb bay.

After a short special course at the Torpedo Training Unit at Turnberry I returned in April 1943 to Wick in the unenviable position of being a Flight Commander but without any real operational experience of any kind and let alone of torpedo operations.

On 6 May 1944 I led the Langham Wing in a strike on an Elbe-Ems convoy of eighteen vessels which we found off Borkum. I dropped my torpedo at a 3,500-ton merchant vessel and this was estimated as a hit; continuing my attack I fired 600 cannon rounds at two other ships. Too late I saw that the last of these was flying a balloon and I struck its cable with the wing root - the cable snapped, and I returned with the print of its winding on the paint. A fortunate escape and despite the considerable amount of flak we sustained no damage.

Soon after this attack I was awarded an immediate DFC (Distinguished Flying Cross). The citation for the award was, 'Hughes, Peter Arden, A/SL (33410). 489 (RNZAF) Sqn. *London Gazette* 26/5/1944. This officer has completed very many sorties, including several successful attacks on shipping. In May 1944, he

Torpedo School Instructors

flew the leading aircraft of a formation which attacked a large and heavily defended convoy off the Dutch coast. In the fight, Squadron Leader Hughes pressed home his attack with great skill and obtained a hit on a medium sized vessel which caught fire. His leadership was of a high order throughout and contributed materially to the success of this well executed operation. This officer has invariably set a fine example of courage and devotion to duty.'

David Reid

My arrival at Turnberry only happened after much thought. For a year previously, I had been stationed at an RAF Station near Edinburgh. It had been a pleasant posting, offering the opportunity to pay a visit home to Ayr every other weekend. However, the prospect of being stationed only eighteen miles from home helped me to make up my mind and volunteer to go to RAF Turnberry. Shortly after arrival, I discovered that the official 'living out' area encompassed Girvan and Maybole, but not Ayr. Although allocated a billet in a cluster of Nissen huts on the shore north of Maidens, I made little use of it, having decided to 'live out' unofficially! Getting in and out of the camp presented little problem, especially at the Culzean end. The Turnberry end was more difficult, due to the presence of a guard room. This necessitated a detour via a hole in the security fence on the Turnberry shore, a route I often used. The method of transport was a bigger problem, I had used a bicycle in the summer and the bus in winter. The need to be on the CO's parade at 8 am meant that I had to leave home at about 6 am and hope that my bike didn't have a breakdown or a puncture. My closest shave came in the winter when I used the 6.30 am bus from Ayr. The bus had almost reached the camp, when an airman suddenly leapt out of the hedge, into the path of the bus, to inform us that further down the road, the service police were instituting a check on all leave and living out passes. Never have I seen a bus empty more quickly, as the occupants took to fields and lanes to make it into camp, I was obviously not alone in living out unofficially.

Most evenings I didn't make it home until around 8 pm, but it was worth it to enjoy familiar surroundings, sleep in one's own bed and partake of home cooking, notwithstanding wartime rationing.

As a General Duties/Clerk, my work was of a humdrum nature, mainly in the orderly room and flight office, but it had its moments. One day as I was about to enter Station HQ, there was a loud explosion, and I was covered in dust and earth, along with several other airmen. When the air cleared, I could see our Commanding Officer, on the telephone in his office, in a very agitated state. It transpired that a naval vessel lying off the Arran coast had been firing blank shells at a sea target located between Maidens and Dunure. Their range and trajectory had gone somewhat askew. On another occasion, whilst working in the flight office near to the lighthouse, there was a terrific noise, rising to a crescendo and then silence. Curiosity got the better of me and when I opened the door to look out, there was a Hampden aircraft on its belly just outside! The pilot had been taking off when he experienced a loss of power, he chopped the throttle and applied the brakes, but could see that he would overshoot the runway. He then selected undercarriage up and the aircraft skidded to a halt just short of the flight office, another few yards and both flight office and I would have 'gone for a burton'. Fortunately, there was no fire and the pilot stepped out uninjured.

Another incident did involve fire, but again fortunately without fatality. A Beaufort had landed on the main runway and the pilot had inadvertently selected the 'undercarriage up' lever instead of the 'flaps up' lever. The aircraft flopped onto its belly and the pilot jumped out, but a small fire started. The airfield fire tender arrived at great speed, but from the wrong direction, with the result that the jet of water they aimed at the aircraft, blew back on them. Within a very short time the aircraft was totally consumed by fire. There were a few red faces and a subsequent court of inquiry.

It was at Turnberry that I made my first ever flight. The date was 1 July 1942, and the aircraft was an old Avro I. The duration was thirty minutes and the route was to RAF Heathfield, returning over my house. I thoroughly enjoyed it, but my enthusiasm was somewhat dampened with having to assist with the raising and lowering of the undercarriage, which in the Anson, was a manual operation and quite exhausting.

There were numerous accidents, but few involved fatalities, although I will never forget the three 'fatals' that occurred during my time at Turnberry. The first was when a Hampden and a Beaufort collided on landing approach on the same runway. The Beaufort which was higher and faster, apparently did not see the lower and slower Hampden. As the Beaufort flashed over the top of the Hampden, the pilot of the Hampden pulled up his aircraft and its propellers chewed off the Beaufort's tailplane. Both aircraft crashed to the ground in flames. This accident had interesting repercussions in the field of air traffic control. Because of a spate of similar accidents at various airfields throughout the country, it was decided to introduce a system whereby an air traffic controller was positioned in a mobile caravan at the landing end of the duty runway. The said controller was armed with a pistol and red and green flares. 'Red' was used to indicate 'not to land', and at Turnberry were often used to indicate to the pilot on landing approach, that the aircraft undercarriage wasn't down – a not infrequent occurrence with trainee pilots! 'Green' was 'clear to land' and was sometimes used in the event of radio failure.

Another fatal accident occurred when a Beaufort, taking off at night, exploded in mid-air above Turnberry Hotel (Hospital). In a period of two weeks a further two Beauforts disappeared without trace. there were rumours that Secret Service personnel had been sent up from London, suspecting the activity of a German Agent at the airfield. A more plausible explanation was that the Beaufort was prone to fuel leaks, with petrol vapour penetrating the cockpit, although any number of mechanical failures could have brought an aircraft down.

A further accident involved a trainee pilot who was performing local circuits of the airfield when an engine died on him, he struggled to return to the field on the remaining engine, but it soon became obvious that he wouldn't make it. He was too low to bail out and elected to ditch the aircraft in Maidenhead Bay. The plane sank immediately, three crew members managed to escape from the aircraft and were rescued by a local fishing boat.

The pilot was trapped because he had fixed a faulty harness pin by replacing it with a keyring, he was therefore unable to release his harness quickly before the aircraft sank.

One of the more interesting duties I had at Turnberry, was checking that the logbooks of flying personnel had been correctly compiled. One chap in particular stood out, a Squadron Leader Flying Instructor, whose name sadly eludes me. His logbook totalled thousands of hours: he had flown in both World Wars, old biplanes in WWI and here he was at Turnberry flying torpedo bombers. Ironically, this officer took off in an Anson to attend a conference in South Wales and lost his life in a mountain crash in North Wales. Weather forecasting was not so reliable in those days, neither were radio and navigation.

Although Turnberry was primarily a training airfield, it did have an occasional diversionary role. During my sojourn, I can recall seeing Liberators and other aircraft from Prestwick, Beaufighters from Heathfield and various bombers returning from air raids on Germany, diverted to Turnberry because their English bases were 'fogged in'.

Living out, I had little occasion to use my camp billet. I recall one hilarious fiasco when I spent the night in camp. There was an MT driver who worked shifts, and because he was a heavy sleeper, he used to ask to be woken up. On this night it was decided to play a joke on him. He duly came off shift at 20:00 hours, prepared to turn in, and asked to be woken at 06:30 hours. We let him sleep for about fifteen minutes, then everyone reset their watches at 06:30 and adopted various attitudes and dress of people just getting up. Apart from muttering that he didn't seem to have been asleep for long, he got dressed and lurched out of the billet for work. We were all keeping our fingers crossed, it had been a nice day and there was a beautiful sunset in the West, we felt sure that he must notice, but no, half an hour later he burst into the hut, and after some language, (unprintable here), he went back bed again. After a day or so, however, he saw the funny side of the affair and we were all forgiven!

When out at the flight office by the lighthouse, I often used to visit the adjacent War Memorial, to airmen who were killed training at Turnberry in WWI. It always gave me a sense of occasion, that I was treading in the footsteps of many WWI aces who had been here before me.

John Noble

I arrived at Turnberry on 3 July 1942 on Operational Rest after a tour of Ops on Hudsons. Our billets at Maidens were scattered all over the place and we were issued with bicycles to get around. 5(C)OTU was equipped with Blenheims, Beauforts and clapped out Hampdens from Bomber Command. As my experience was all on Hudsons I was allocated to the Anson Flight as a screen wireless operator. Morale at Turnberry was very low, as we heard from colleagues that accidents were prevalent, particularly on Beauforts, which were underpowered, and the gun turrets were unreliable. I soon saw proof of this when three aircraft were lost between 8 and 11 July.

On 13 July I was transferred to HQ Signals as a Supervisor. This involved direct wireless communication with the aircraft.

The following are events taken from my diary:

18 July	A Beaufort force landed in the Irish Republic and the crew interned.
19 July	A Hampden crashed at Girvan.
22 July	A Beaufort crashed.
24 July	2 Beauforts and a Hampden crashed.
25 July	A Blenheim crashed.

3 August	A Beaufort crew were killed, and a Hampden crashed.
5 August	A Beaufort crashed.
15 August	A Beaufort crashed.
25 August	A Beaufort crashed.
26 August	A Blenheim crashed.
31 August	A Beaufort crashed.
1 September	I personally witnessed the airborne collision over the runway between a Beaufort and a Hampden. Both aircraft caught fire and most of the crews were killed. At the spot where they collided the control tower was obscured so no warning signal could have been given or been seen.

All the crashes I recorded took place on or near the airfield – there were more on the many islands in the area, out at sea or further inland.

I was posted to Iceland from Turnberry and I wasn't sorry!

Robert Taylor Robertson

During the Second World War I was a pupil at Milngavie Public School. Three times our school travelled to Ayrshire to do our bit to help the war effort. I was on two of these trips which lasted for the whole two months of the summer holidays. Sixty boys went to the camps where we were split in to two squads. A squad would gather early Ayrshire potatoes at Forbes' farm at the Maidens while B Squad could take it easy and sightsee or be more energetic and help pick fruit on Culzean Estate. On my first trip we lived in the lower part of Culzean Castle and I recall two memorable occasions. The first we were allowed to view the wonderful armoury, the second we dined with the Marquis and Marchioness of Ailsa. The second time I was there we stayed in the Land Army Hostel at the Home Farm.

Kirkoswald Church comes into the story because every Sunday we would walk to church there, and my memory of it was of the history of the church, including Rabbie Burns and all of his friends.

Our week at the potatoes began when we were taken in a John Brown tractor and trailer lined with straw from Culzean Castle to Mr Forbes of the Maidens farm to pick the crop of early Ayrshire 'tatties'. We gathered them behind a chain digger, the first I had ever seen, but it was good for the soil of this area. Once you had picked your allotted space you had a few minutes to admire the view, which was impressive. We were on a hillside overlooking the Firth of Clyde and Arran. Below us was Turnberry airfield where many Commonwealth and allied airmen honed their flying skills. Many of them did not make it and we saw a few crashes especially on the hill behind us.

When our days 'tattie howkin' was over it was back to Culzean on our tractor and trailer, singing at the top of our voices all the favourites of the time.

The second week we could volunteer to pick varied fruits from the estate glasshouses and gardens – like peaches, tomatoes, strawberries, raspberries or blackcurrants, for which we received payment. Then it was down to the town of Maybole to spend our hard-earned cash. After the war my cousin and I and some of our friends returned to the Home Farm for a couple of years, cycling down on a holiday weekend and playing football from dawn till dusk.

Allin Hawkshaw

I joined the RCAF on 5 May 1941 and commenced my Elementary Flying Training at Stanley, Nova Scotia. I progressed to SFTS at Moreton, New Brunswick, flying Ansons and was awarded my wings between Christmas and New Year. I arrived in the UK aboard the SS *Vollendam* and was then sent to Bournemouth. The following months were filled with courses in vehicle recognition (from the air), advanced flying, general reconnaissance and astro navigation.

I arrived at the OTU at Turnberry for training on Mk I and Mk II Beauforts on 24 August 1942. The Mk I's had the Bristol sleeve valve engines while the Mk II's were fitted with the Pratt and Whitney engines. At least three Mk II's crashed on take-off at Turnberry during my stay there, with the loss of all crew. All accidents were blamed on pilot error and the opinion was voiced that we were all stupid, useless, no good pilots. I was enraged at these comments, and along with another guy, went for a walk in the countryside to let our anger cool. All the while we were cursing the training staff and the CO for their comments. I found out later that there had been a problem with one of the engines, it was thought that one of the propellers had gone into full fine pitch with disastrous results. I had the idea of getting the aircraft checked over and telephoned London to get the RCAF Chief Engineer to come and inspect them. He arrived early next morning and grounded all Mk II's. Later that day he found that the control units at the centre of the propellers were worn out and this would allow the props to go to full fine. In a few days all the aircraft were fixed.

My crew were Navigator Bert Pawsey from London, W Op Norman Brick from Wales, and W Op Frank Perry also from London. At Turnberry we would take our bicycles and go to dances in the village hall at Kirkoswald. All the local girls were young, and their mothers sat in the balcony to make sure nothing bad happened. Our barracks were Nissen huts and were freezing, we would walk along the railway track searching for bits of coal dropped by the trains, this helped feed the pot-bellied stove in the hut. Once, we couldn't believe our luck when a train was parked long enough for us to steal a whole pail full of coal - there was lots of heat that night!

I recall that at one stage in late 1942, the Beauforts were loaded with bombs, to take part in the thousand bomber raid, thankfully we were not called to do so.

An unfortunate mistake I made at Turnberry was writing in a letter home about the weather. As was usual all letters were censored before mailing. The CO called me in to his office and I was told that my promotion to F/Sgt was being withheld for six months due to my breach of regulations by mentioning the weather! I left Turnberry late October and went to Abbotsinch to complete my training. This done, in January

Allin Hawkshaw and Crew

1943, I collected a brand-new Beaufort Mk I at Filton, and along with my crew carried out many air tests at Lyneham before heading out to the Far East. We flew by Gibraltar, Algeria, Libya, Cairo, Palestine, Habbaniya, Bahrein, Sharjah, Karachi, Jodhpur, Bombay, Bangalore, Ratmalana, and finally to Vavuniya.

I always found the Beaufort a nice aircraft to fly, although the pilot's 'facilities' left a lot to be desired. Basically, they had a 'relief' tube on the floor in front of the pilot's seat. You had to pull it up and urinate into the small cup on the top, but it always blew back in my face. So, I used a tin can, jammed between my seat and the throttles. On the long flight to the Far East the tin can become rather full, so I slid open the side window and emptied it. Helped by the turbulence caused by the propellers, my wee went down the side of the aircraft and circled into the rear turret. Norm was the gunner that day who received an unexpected

shower! If you want to learn how to swear - try an angry Welshman!!! He wouldn't talk to me for a week.

We had at last arrived to join 22 Squadron.

On 8 October 1943, we set out from the Santa Cruz (Bombay) airport to do a 'creeping line ahead' convoy escort in Beaufort I, DW940. At about 10.00 am, 100 miles west of Bombay, we suffered engine failure.

So Sudden! I think I put the undercarriage down instead of opening the bomb bay doors to get rid of the depth charges. We 'landed' okay. The dinghy would not release, and Frank disappeared below the surface, still pulling the cord as the aircraft sank. He bobbed up again and we clustered on board the broken-off oleo leg, the tyre keeping it afloat (lucky mistake). In a long vigil overnight, we huddled together; luckily sharks chose not to attack.

Next morning a Beaufort search aircraft did

a port turn around us, so close we recognised the pilot, Martin Glynn. In our Mae Wests in the whitecaps we were invisible. Later in the morning we spotted a sail on the horizon. It tacked back and forth for what seemed like four hours, then came close enough to hail. All together boys, in a low-pitched voice: 'Ship Ahoy'.

Pandemonium broke loose on deck. One of the crew jumped overboard with a rope (sharks were around). We were too tired to be able to climb up. They hauled us on board with ropes. The ship was an Indian dhow commercial vessel with a load of grain on its way from Karachi to Cochin. They broke their journey to take us to Bombay.

Late afternoon, October 10, we arrived and were unloaded onto the harbour command ship, Bombay Harbour. A launch was summoned but they were going to leave us as they thought we were MI5 or some such underground (water?) security outfit. Finally, they took us and moored beside the third destroyer out, line abreast at dock. Major Strever (SAAF) came bounding over the decks, grabbed Norman who had a broken leg, carried him ashore and all of us were taken off to hospital for a few days.

Second time around, 22 January 1944, Beaufort Mk I DW872. The British Fleet was coming east for the Pacific War. The Squadron mounted a practice torpedo attack at first light. What a sight! Silhouetted against the morning sky were aircraft carriers, battleships, cruisers and destroyers. Probably the first time during the war that the classic 'A-K Line' could be assembled.

Then, engine failure, again, we made a good ditching, and the dinghy behaved itself. The navigator on this trip was Wilf Martin from Warrington, and we were also carrying a passenger, Bill Carrol, who had come along to see the fleet and to pick up his new uniform from his tailor in Colombo. There was a small tin lying on the dinghy floor, so Wilf threw it overboard in case someone stepped on it and ripped the fabric. Horrors!! Frank's cigarettes, waterproofed and lovingly packed as he had said, 'Never again will

I be without them like the last time'.

A couple of hours later an Albacore from one of the carriers spotted us. The brand-new Destroyer HMS *Petard* picked us up and at a speed of 42 knots we were in Colombo harbour for lunch. On arrival at the airport (Ratamalana) I was invited to the Officers' Mess. My commission had come through, maybe we should have ditched a third time, I might have made Chief of Air Staff!!

Murray Hyslop
In August 1942 I was ordered to 5(C)OTU. We flew Beauforts, Mk I and Mk II models. To me they were not much different from flying an Oxford; but they seemed war weary, almost a bunch of junk. From August 1941 to November 1943, 80 Beauforts were lost at 5(C)OTU. The Taurus engine that powered the Mk I must take some of the responsibility as it was unreliable and had a short life. But most of the losses were due to inexperienced, hurriedly trained pilots.

On 1 September 1942, while waiting at the end of the runway to take off, on only my sixth flying day at Turnberry, I saw a Hampden (P1199) and a Beaufort (L9932) making landing approaches together but unseen by each other – with the inevitable collision and crash. At the time it seemed to me like there was a crash every day at Turnberry. But, only five or six Beauforts were lost while I was there.

At 5(C)OTU we fired on towed targets and bombed fixed and moving targets. While my first month there went well, I recall that in October 1942 I had a few problems. My first incident occurred while I was wearing goggles to simulate night flying. While taxiing I cut a corner too sharply and hit a runway light with my tail wheel – not too serious but a portent of things to come. Shortly after that I lost an engine which caught fire while I was landing. The rest of my crew jumped out once the aircraft had slowed to about 30 mph – I got out of the burning aircraft at about 5 mph.

On 28 October the most serious incident occurred in Beaufort Mk IIA DD884 while flying at low level for air to sea firing. We had been higher to let my navigator get some practice, and when we came down to low level, I neglected to tighten the straps of my harness – with uncomfortable consequences. Suddenly, while barely off the water, both engines quit. The Beaufort quickly lost altitude and bounced off the sea. As my straps were loose, my head broke the windshield, shredding my helmet, and cutting my scalp. On the next impact the aircraft stayed on the water and rapidly sank – thankfully not before the entire crew was able to get out safely. I was later told that this was the first ditching without fatality at 5(C)OTU. It was also lucky for us that we were only about one mile offshore, near Girvan just south of Turnberry. Our crash into the water had been witnessed and we were soon rescued.

An investigation found that we had run out of fuel; but that the gauges were faulty – so I was cleared. In front of the board I was asked how I could have hit my head if my belts were on. 'It was a confused time,' was all I could stammer. That was, coincidentally, my last flight with 5(C)OTU. Among the Beauforts lost while I was there, I am including my own ditching. From the OTU some of us went to operational squadrons; but it was decided that I would become a torpedo bomber pilot. So, I was sent for specialised training to the Torpedo Training Unit at Abbotsinch. We studied the theory of

torpedo dropping in ground school, then practised with Beauforts against fixed and moving targets. HMS *Canterbury* and the *Isle of Guernsey* were our usual targets. They would steam along at various set speeds such as 8, 16, 24, or 32 knots while we would simulate attacks at various deflections. This all finished up on 10 December 1942, and I was assessed as an 'Average Plus Torpedo Bomber' pilot with a sound knowledge of Torpedo Bomber Theory. I now made my way to war. I was assigned to 39 (Torpedo) Squadron on the island of Malta.

Richard E. Wilson
After passing out on the Gunnery Course I was posted to 5 OTU Turnberry as a W Op/AG to meet up with other trainees including pilots and navigators, most of whom were New Zealanders. As we would be flying Hampdens, we had to form crews of four, a pilot, navigator and two W Op/AGs. I found another W Op/AG from Nottingham and we crewed with a New Zealander pilot and navigator. Our nicknames were Jack (pilot), Dick (navigator), Dick (myself) and Chick (W Op/AGs)

It was about one month before we were flying together as a crew, the W Op/AGs flying with staff pilots on other aircraft carrying out more training on the wireless and gunnery courses. The pilot and navigator in the meantime were on

1943 1TTU Maintenance Staff

familiarisation training on the Hampden. Our first flight together was on 8 November 1942 in an Oxford and we first flew in a Hampden as a crew on 20 November. After that it was Hampdens all the way – carrying out exercises in bombing, gunnery, wireless, navigation, night flying, etc.

The whole unit was transferred to RAF Maghabbery in Northern Ireland on 8 January 1943 where we remained for about ten days carrying out more local exercises. We then returned to Turnberry to join 1TTU to practise torpedo dropping.

My main recollections of Turnberry were the landmarks we were able to photograph from the air, Turnberry Lighthouse, Culzean Castle, Turnberry Hotel, Ailsa Craig and Girvan – I still have these pictures. I often think of the harsh conditions of the Nissen huts we lived in, as it was mid-winter and bitterly cold, we took it in turns to raid the coke compound for extra fuel for the stove. We would get the stove red-hot and make a round of toast with bread taken from the mess.

Our main enjoyment was to be found in Kirkoswald where there were two pubs, one called the Shanter. There was an upstairs room with a piano and as I was able to play, we spent many a happy hour singing and drinking. As well as Britishers there were the New Zealanders, Canadians and some Australians, and as long as I kept playing my drinks were free! My wife and Chick's wife were able to contact each other in Nottingham and arranged to come up to Kirkoswald for a week's holiday, staying at one of the pubs. We didn't dare let them join in our singing sessions because of the dirty songs being sung!

After training at Turnberry, we were posted to an operational squadron – 455 RAAF. We carried out several operations over the North Sea before being posted to 489 RNZAF Squadron at Wick. I completed my tour with Coastal Command and finished my service career at St Athan in South Wales as a Staff Signals Instructor with the rank of F/Lt.

Tony Lee

Tony Lee
I was born the youngest of seven children during July 1920, in Hove, Sussex. My Mother was fifty when I was born, consequently I was rather spoilt by my siblings. Father was the manager of a chain of grocery shops called Home and Colonial. My elder brother, Arthur, was killed at the Battle of the Somme during the First World War.

I left school in 1934 and got employment as a junior clerk at Hove YMCA, then a builder's merchants, before finally deciding on a career in local government at Lewes in Sussex.

I was conscripted in July 1940; I chose the RAF because I thought it would be more interesting and there would be less bullshit.

Basic training was undertaken on the sea front at Blackpool, after which I filled in time doing general duties at various other RAF stations, while waiting for a place on trade training.

Finally, I was posted back to Blackpool, to learn the basics of wireless procedures and the Morse Code along with ground experience at RAF station Whitby. From here it was on to Yatesbury, where I did my radio flying training. Pembrey Gunnery school came next – here we practised shooting at a drogue towed by a Lysander. All air gunners had the tips of their bullets painted with different colours; this paint left a trace on the target drogue. The drogue was recovered, and the holes counted, and colour matched to asses our score. After completing this course, I passed out with my Sergeant's stripes and left for RAF Prestwick. Here we studied the workings of ASV. I remember the damp, chilly Nissen huts under the trees. At the end of the course we were assembled in a classroom and the instructor held up his arm and told us that everyone to the left of his arm were to be posted to a Beaufort OTU and to the right a Sunderland OTU. It was as simple as that.

So, I arrived at Turnberry towards the end of 1942. We all knew that we would be posted either to a Beaufort OTU or a Sunderland Flying Boat OTU. We had hoped that it would be flying boats as we had heard of the appalling accident rate, particularly at Turnberry on Beauforts, mainly because of the weather and the fact that if one engine failed, it would be 'curtains'. We all accepted the risk of flying, but on Beauforts at Turnberry, the risk was unreasonable. I think everyone was in a state of shock. I hadn't even seen a Beaufort until I went to Turnberry, but I had heard plenty about their reputation!

I recall asking the station warrant officer how long we were likely to be at 5(C)OTU. I was told about five weeks and then we were for overseas. The aircrews mingled, with a view to getting to know everyone, and decide who we wished to fly with. The airmen mostly consisted of English, Canadians, New Zealanders and Australians. I made friends with another W Op/AG, Sgt Goddard. Our pilot was Sgt Murphy, a shepherd from New Zealand. He was a dour chap, who kept to himself, he never mixed with us socially. I think he was a bit of a loner.

A. Clark, Melville, Beatson, Lee and Thomas

After getting over the shock of being posted to a Beaufort OTU, we all proceeded to start training, and this was when there were so many accidents. We never knew what tomorrow would bring, survival was a matter of luck.

There were many accidents at Turnberry – I won't say that they went unnoticed, but they weren't discussed. The powers that be kept us away from crash sites, and funerals were so discreet that we never knew about them. The Station Sick Quarters and mortuary were situated well away from the airfield, so casualties were almost kept out of sight. I suppose it would have adversely affected morale if we had seen some of the things that happened.

We obviously heard things through the 'grapevine', missed a face in the mess, but we never dwelt on it. Looking back, our attitude seemed very hard and impersonal. We just prayed that it wouldn't be us next and took life a day at a time.

I vividly recall another W Op/AG on my course, Sgt McGuire who came from Brighton. He became fed up with it all, and wanted to pack it in. He tried to talk me into joining him, but I refused. I didn't want LMF on my record, nor did I fancy spending the war cleaning toilets. This is what happened to those that quit. He became very quiet and withdrawn, he refused to go out to the pub with us all week, preferring to be alone. His mood became even more melancholy as the week went on. He died on the Saturday whilst on a night flying exercise. All we knew of his passing, was that his bed hadn't been slept in. No one told us anything.

I used to stand and watch as Beauforts came in to land. I was in awe at the way they almost hung in the air, as if suspended by an invisible hand, then drop almost vertically, at great speed toward the runway. It was frightening to watch. One day, another W Op/AG, Bob Ashley, and I, stood for a while and chatted while we watched the Beauforts coming back to the airfield. Bob Ashley, an Anglo-Indian, had been a popular singer before the war. He was a regular entertainer on Pathe Pictorial. He could have had it easy by joining ENSA, he didn't have to

be aircrew, but he volunteered. I recall hearing him sing in the mess a couple of times. He had a marvellous voice. All the girls used to swoon over him. Tragically, he too was killed, when the Beaufort he was in crashed into the sea just off Turnberry Lighthouse.

We were enjoying a drink in the Sergeants' Mess one evening, when suddenly all the lights went out. Everyone went outside to find out what was happening, only to see a pall of smoke drifting across the hills towards us from the south. You could almost smell the smoke on the breeze. A Wellington had hit power cables on approach to Turnberry and crashed, killing all on board.

The closest shave I had was when our pilot forgot to remove the pitot head cover prior to take off. He managed to abort just in time, I still have a picture in my mind of the 'blood wagon' tearing along the perimeter track beside us.

I personally was always glad to be back on terra firma.

At the commencement of the war, most of the RAF torpedo squadrons flew the infamous Bristol Beauforts. These squadrons were known as the 'Suicide Squadrons'. More planes it seemed were lost in training than in action. The aircraft and its attached torpedo flew between fifty and one hundred feet above the water. The sunlight shining on the surface of the sea was enough to cause the loss of a plane. I must say though, I would far rather take my chances in a Beaufort, than fly with Bomber Command.

I was always happier when on my bicycle, on my way to Girvan to see my lady friend, Billie Colquhoun, who I met at a dance, along with her brother Bertie, who was ground crew in the RAF. I used to enjoy going to her home for meals, her father always seemed a bit stern, but her mother was a cheery soul who was a clippie on the buses. Sometimes she would let me ride for free.

We used to enjoy going to dances in the Catholic Hall in Girvan, or the Hospital (Turnberry Hotel), The patients and nursing staff would attend too. I still remember the steps leading up to the hotel. All the aircrew got on very well together (more so because of the dangerous flying we did), and there was wonderful comradeship, without which, life would have been intolerable.

Living at Turnberry, well, we were young and fit, we could cope with the cold and damp. It always rained for some part of the day, all the time I was there. If the SW wind blew up, we were sure of rain.

The CO at the time was a bit of a stickler for discipline. He seemed to enjoy the bullshit element of the service and had parades at the slightest excuse. There was a regular parade for the whole station on a Saturday morning which we all hated. Saturday was supposed to be a day off, so most of us skipped camp to avoid it if we could. There was no roll call, so we couldn't see the point of it. Mock battle exercises were another of his hobbies, all pretty useless when you are flying over the sea!

In the spring of 1943, when we had finished our Beaufort training, I and one other Englishman, Sgt Goddard, together with our N. Z. pilot, Sgt Murphy, and a Newfoundland navigator, Sgt Cooper, were ordered to collect a brand-new Beaufort from the factory at Filton and ferry it to India from an air base in Wales. All Beauforts that were bound for India had 'snake' written on the tail. For me, it was a trip full of adventure - I had never been abroad before. We flew to Morocco, where we stayed several days and managed to visit Fez. Then we flew across North Africa to Cairo. Leaving Cairo we flew over Palestine and on to the enormous RAF Base at Habbaniya in Iraq, finally arriving at our destination of Karachi. The only really scary moment on the trip was when our pilot forgot to switch on to the auxiliary fuel tanks and the engines cut out over the Bay of Biscay. I fervently prayed, as he tried to restart those silent engines. I have never felt more relieved than when I heard the roar as we were suddenly under power again.

From Karachi I was posted to an operational unit, flying Dakotas, dropping supplies to the 14th Army in Burma, as well as supplying the Naga hillmen, who were also fighting the Japs. Having completed my tour of five hundred hours,

I was then posted to 435 Squadron (Canadian), as an Intelligence Officer, this was my operational rest. While I was at Imphal, I was recommended for a commission, I didn't get it because, by this time the war was almost over.

I came home from India on a ship called the *Ormond* and had to get used to sleeping in a hammock. I was demobbed at Warton in Lancashire and went home and took three months holiday, before returning to work. My job had been held open for me, so I knew I had a future. I returned to work for local government, although, at first it seemed very dull after the RAF. I moved to a post in Chiswick, after twenty years in Lewes, and found things much different in a London Borough. I then transferred to Lambeth Borough and worked with 'Red' Ken Livingstone. The political backstabbing was awful and made a difficult job almost impossible.

It was during this period, that I made my first trip to Dunure Cemetery. I had started to think more and more of the lads who had died and felt I must return to pay my respects. I travelled overnight by coach and hitched or walked the rest of the way to Dunure. I have made this journey many times over the years, wondering what it was all about, couldn't there have been a better way, avoiding all these senseless deaths. I am glad I survived, but heartily saddened at the futility of war.

I retired from local government at the age of fifty-nine, and then worked for the Royal Airforce Association at their Head Office. On my first day, I walked into my office and was surprised to discover that the seat at my desk was the Captain's chair from the *Ormond*.

I shall always remember the pals we left at Turnberry.

Jack 'Harry' Mihell and Norman Robert Mihell

My father, born in 1911, was part of a family of four boys and one girl from the town of Sault Ste. Marie, Ontario, Canada. My father, Jack 'Harry' Mihell, and my uncle, Norman Robert Mihell, joined the WWII war effort in Ontario and trained under the British Commonwealth

Jack 'Harry' Mihell

Air Training Plan. They trained in places such as Ontario, Manitoba, Alberta and British Columbia. Ronald, another brother was also in training as a W Op/AG and Bob, the youngest, as he became of age joined the navy. He was sponsored by his parents, who 'adjusted' his age on the forms. So, even the baby of the family headed overseas at only 17. Sister Kathleen met and married a training pilot of the BCATP stationed in Canada, before he was sent on operations overseas. Their poor parents had their five children and a son-in-law directly involved in the war, and such a distance from home. There was no chance of any of them being able to go home on leave, and there was always the fear that if one - or more-should be wounded, it would not be practical or possible for the parents to visit, and the dread that if they were killed that they may never come home but be buried where they fell. The family was to be apart for over four years.

Both Jack and Norman were reunited in 1942 when they arrived to further their operational training at Turnberry. Dad and uncle Norm were both training as W Op/AGs on Hampdens. By this time the siblings were scattered around the globe, newspapers from home and family mail was a vital link which the boys eagerly awaited. News of each other and their hometown was only obtained through the parents. The boys would write home, with censored mail, giving clues to their whereabouts and activities to their English born father, who would then deduce where they were stationed and what they were up to. He would then in turn pass this information amongst the family and as incoming mail wasn't censored this information enabled the brothers to get together occasionally.

Socialising wasn't limited to family gatherings though...

It was while he was at Turnberry that my

father met my mother, Mary 'Betty' Rodger, from Maidens. She was born in 1917 and when she left secondary school, war with Germany was a very real probability. With this in mind, mom decided that she didn't want to end up working in a factory, she wanted to something more direct for the war effort. Consequently, she joined the Royal Navy and trained in torpedo testing. She was stationed at Loch Long and during the summer of 1942 went home on leave to Maidens to visit her family who lived in Ailsa Buildings. She and her sister Jean attended a dance at Kirkoswald, at which my dad and Uncle Norm were also present, and they met at about half past midnight when they all realised they had missed the last bus back to Maidens and Turnberry. Jean was the only one with a torch so the four of them walked home together. Thus, started the courtship between mum and dad, and Uncle Norm and Aunt Jean for the rest of 1942. Mum and dad met

up whenever their jobs allowed.

My father had always been a keen hunter, and with his bicycle and service revolver would travel in search of a rabbit for the pot, although Maidens being a fishing village was never short of fish, meat was a welcome addition to the diet during wartime. He would present his 'catch' to mom's family and they would all enjoy it for supper. I'm sure the RCAF wouldn't have sanctioned this, but being a long-time Canadian hunter, it would have seemed perfectly natural to dad. He would have field dressed his kill and slipped the wrapped meat inside his jacket. The result of this was that he was able to prove himself to his prospective in-laws as a capable breadwinner in time of need! His eyes used to twinkle when he recounted his tales of Turnberry, he recalled a time when he and Norm were walking back to their billet, probably after visiting the sisters, when they were challenged

Jack Mihell and Mary Rodger wedding day at Kirkoswald

Jack Mihell in a Hampden Gun Turret

by a policeman because Norm was smoking a cigarette. They were given a severe talking to, after Norm was instructed to extinguish his smoke, they were warned that Britain was in black-out and they glow from Norman's cigarette could have been seen by a German bomber. The brothers thought this fanatical and considered that the policeman was suffering from stress.

Mom and dad got married on 29 December 1942 at Kirkoswald Parish Church. Many telegrams of congratulations were received, the most memorable one being from mom's pals at Loch Long. Mom tested two sizes of torpedo, the Mk 9 and the smaller Mk 6. The greeting on the telegram was that her friend, 'hoped that she would get a Mk 9 on her trestle.' This was read out at the reception in the Kirkoswald pub causing much hilarity.... Years later mom explained to me that even though Britain was at its darkest hour, with the threat of invasion still very real and the hardships of war all around, that the people never lost their sense of humour,

and especially cherished the companionship of working together with a common purpose. They laughed together, cried together and had good times and bad times together. The war brought people closer. All of dad's buddies from his training course attended the wedding reception, and for the rest of his time at Turnberry, he was never allowed to forget that telegram. Both my parents were sober, God-fearing people, but their eyes always sparkled whenever this telegram was referred to!

At some point after this event, the powers that be realised that two brothers, dad and Norman, were in the same unit. Family members were not permitted to serve together, and so they were split up. Years later, Norm told of how badly that separation had affected him. He had always relied on his older brother, throughout his early years, and suddenly he was on his own, facing a high probability of death, thousands of miles from home.

After training, dad was posted to an operational squadron in southern England, returning to Turnberry as an instructor in 1945. He stayed with his in-laws and his new bride at 4 Ailsa Buildings.

With the war ending, dad returned to Canada with his 'war bride'. All the Mihell siblings arrived home on the same train. Those from overseas came into Halifax harbour aboard the *Louis Pasteur*, there they met the youngest (still only 18), Bob, and they then all took the trans-continental train for the two-day journey across Canada. All five of them disembarked at Sault Ste. Marie, to be met by their parents. Grandma didn't know who to kiss first - and kept going from one son to the other, then to her new (first) daughter-in-law - and back again! The incredible stress of waiting while her four sons were away at war was over. They were all home safe, including her son-in-law. Uncle Ron was to say that the only time he ever saw his father cry was on the first night they all had supper together again after four long years apart.

When mum and dad first decided to marry, they had already planned to live in Canada after the war. They saved as much money as they could, mom worked her off-duty days on the telephone switchboard at Loch Long, they both bought war bonds and hoarded every penny. When they returned to Canada, they had almost enough money for a deposit on a house, they planned to work for a few more years, and put off buying their house. Fate intervened, I was born and within a short time became very ill. I required one and a half years of hospitalisation, and with no government sick plan in those days, I used up all the war savings and put them so deep in debt that they were unable to buy their own home until I was at university, in the 1960s.

I grew up hearing dad talk so much about his time at Turnberry. These were very happy times for him, and he reminisced about them often. He wouldn't talk about his time spent on operational bombing. Being born in 1946, I grew up with the war still fresh in people memories and dad's attitude of only talking of the good times was a pattern also displayed by many of my friend's fathers.

I am sorry I never asked dad more about his war duty, but he was always so reticent, and especially after he suffered a nervous breakdown shortly after his return, the subject of the war was rarely raised.

Dad passed way in 1988, followed within three months by Norman. Mom passed away in 1997 and it was while I was tending to her belongings that I found their mementos of Turnberry, including the infamous telegram!

That's their story as I know it, I'm sorry I don't know more. I am sure mom and dad could have talked for hours on end. I so wish that they were still here so that I could now hear them with a maturity that perhaps would enable me to appreciate what they said, and better document these memories of a people and a time long gone, of whom mom and dad were just a couple out of thousands of others. God Bless them all.
[With thanks to Jack Mihell Jnr]

Leslie Scott

Sgt Leslie Scott (RCAF attached to RAF) was posted to Turnberry for a Hampden long range navigation course in the fall of 1942. Travelling with Phyllis, they made their way by train through London and north to Ayr. They arrived late in the day and stopped at a teashop for a bite to eat. They inquired of the owner where they could get a place to sleep for the night and were told they would not be able to find anything at that time of night. The owner of the teashop

23 Beaufort Course

kindly offered to let them sleep on the floor of the teashop - and that is what they did. This was in September of 1942 by which time Phyllis was about six months pregnant!

The next morning, they took a bus to Girvan, and Les, in his inimitable way, started walking down the street and knocking on doors to inquire if anyone could accommodate them. Les, being military and on course, would have been provided with accommodation on the base. However, no provisions or allowances were provided for wives travelling with their servicemen husbands during wartime; indeed, it was not supposed to happen! He was eventually able to find a person who would let them have an upstairs room, on Vicarton Street, the main road leading north from Girvan along the coast towards Turnberry.

With accommodation secured, the next important item was pay. Fortunately, shortly after arriving, a pay parade was held. In those days' personnel received their pay in cash from a pay officer. Those to be paid were assembled on a parade square, perhaps organized by course or in some other way. As each person's name was called out, he would march up to the paymaster, salute and then his pay would be counted out to him. He would count the money (I never saw anyone with the nerve to do that!) and sign in receipt of it. Then, one step back, salute again, right about face and march back to position. At this particular pay parade, Leslie waited for his name, 'Sgt Scott', to be called. It wasn't. This could be and would have been a major problem. Living from pay to pay was the way they lived at that time. To not be paid would be a serious difficulty, particularly since, although he could eat on the base, how would Phyllis eat and how would the rent be paid (it was probably overdue, as it was!). So, at the end of the pay parade Leslie marched up to the Flight Sergeant in charge of the parade, a crusty older, regular RAF Flight Sergeant and enquired as to where his pay was, stating that his name had not been called. The Flight Sergeant looked down at his list, and informed him that it had been called -- 'Flight Sergeant Scott' -- and why was he out of

Leslie and Phyllis Scott

uniform? (Leslie was wearing Sergeant's stripes without the Flight Sergeant's crown) And didn't he read orders? And he had better get down to stores and get his crown on his uniform and he could be paid. And in this manner Les learned that he was now a Flight Sergeant. He did get his pay and the Flight Sergeant was much more civil to him, now that he was properly of equal rank.

With accommodation and pay sorted out, attention could be turned to the training. A course consisted of several pilots, navigators and wireless/air gunners (W Op/AGs). From these people crews were created, each with a pilot, navigator and two W Op/AGs. These crews would train together and then go on to operations together. Les's pilot was Spackman, a Canadian from Lethbridge, but he does not remember the names of the W Op/AGs on this crew. Each course had a number and a photo was taken.

The purpose of the course was to train crews' operations on the Beaufort and Hampden bombers. On completion of training the crews would be posted to Coastal Command, which was generally considered to be safer than being transferred to Bomber Command.

The course consisted of classroom work and flying assignments. The flying assignments could be dangerous, and it appears that most courses would count at least one aircraft and crew lost during training. There were also instances of aircraft colliding with each other on the ground, with resulting casualties

At some point during the training, probably in early November, Leslie did not return to Girvan after work. He returned from a training mission and complained of a pain in his abdomen. He was sent to the base hospital, which admitted him. Phyllis became anxious and was eventually able to determine that Leslie was in the hospital but could get no other details. This was in fact Turnberry Hotel which had been converted into a hospital for the duration. Obviously worried because she knew he had been flying and feared there had been an accident, Phyllis decided to make her way to the hospital to see what was wrong. By now it was late in the day and dark. It was overcast or at least moonless when Phyllis, now 7½ months pregnant, took the bus up to Maidens and was let off outside the hotel/hospital. In the blackout she had to make her way up the long flight of steps. With relief, some level of exhaustion and skinned shins she learned that Leslie had been diagnosed with appendicitis, had been operated on and was fine.

However, the operation meant that Leslie could not complete his training. After recovering from the operation, he was assigned as a supernumerary to base operations to await the next group, thus being separated from Spackman and the rest of the crew, who had a tragic future, being posted to Malta with Beauforts to strengthen the air defences during what proved to be the last days of the German occupation of North Africa (Operation Torch, the landing of American troops in North Africa, occurred in November of 1942 and the Germans capitulated there in May 1943). When the Beauforts arrived in Malta the W Op/AGs were declared surplus

and were ordered back to Britain. They were sent back by air transport and the aircraft involved were shot down over France with the loss of all crews and passengers, thereby killing all of the W Op/AGs who had been on the course with Les. The pilots and navigators stayed at Malta for their tour. Les ran into Spackman in 1945 when they were both on the same ship back to Canada.

During this period, probably about the beginning of December, Les and Phyllis moved to Maidens, on the edge of the airfield. Les had managed to find a family who would rent him their front room, which was on the main floor and only three steps up, as a new home for his soon to be expanded family. The view from the room looked out to the harbour at Maidens on the southern edge of the bay, near the jetty.

The month passed, and it became evident that the birth was near. It was decided that Phyllis should go to Seafield Hospital in Ayr for the delivery. Phyllis entered Seafield about 26 December. Now, Ayr was some distance from Maidens, in those days what little telephone service was available was mostly tied up by the war effort. In any case, airmen were not supposed to have their wives with them. Because of this, it was a day or two after Phyllis delivered Les' first son, to be named Kenneth, that he found out about it. He then made arrangements to go to the hospital and pick Phyllis up and return her to Maidens. He did not have a car and could not afford a taxi, so he took a bus up to the hospital and brought them back by bus -- but only to Kirkoswald the next village about 2 miles away.

Maidens

From there they took a taxi so as to create the right image in the village! It was only in later years that Les realized that everyone in Maidens would have recognised the taxi as being from Kirkoswald.

Sometime about mid-January, Les and Phil went for a walk along the beach at Maidens. They probably often did this, but this walk was different because some scenes from the walk were recorded as photographs. Of course, this being wartime, photographs could not include military facilities and even photographic materials were rationed, so the photos stand out for that reason. The walk was made with a sergeant who was an American serving with the RCAF (his shoulder flash says 'CANADA' and below the propeller is 'USA', which was worn by Americans who had joined the RCAF). His first name was Johnny, but his surname is forgotten. Les remembers trying to locate him a year later and learning that he had been killed on operations.

Sometime after New Year another course assembled at Turnberry. This was 23 Beaufort and Hampden course. Crews were assembled from those assigned, after the group had been together for a while. This allowed some opportunity for the course members to create their own crews. One of the airmen, Sgt 'Bim' Verrell RAF, was a bit older than the others and was thinking forward to the extent that he wanted to survive his operational posting and the war. He apparently checked out the pilots and decided that Sgt Frank Molloy, a New Zealander who

Johnny the Yank

Johnny the Yank and Mrs Scott

had been in a flying club before the war, was the best pilot on the course and he convinced Molloy to have him as a W Op/AG. (Malloy refused his commission later in the war and left the forces as a WO1. He is deceased and never married.) Bim then decided that Les was the best navigator and approached him with the proposition that he should join Molloy's crew. In this manner the crew was put together, the fourth person being Sgt Ben Lyons, RAF.

Shortly afterwards the whole course was relocated to Northern Ireland for training, thus bringing an end to Les's tour at Turnberry. *[With thanks to Ken Scott]*

Reg Faber
I joined the RAF in July 1940 at Edgeware in Middlesex. Being unemployed, I volunteered, thinking that, as a volunteer, it would be easier for me to have some choice in what I wanted to do. When the recruitment officer asked me if I wanted to fly, I couldn't say yes quickly enough. At this time there was a need for W Op/AGs.

WWII airmen, 1TTU

After attestation I was sent to RAF Uxbridge where we lived under canvas. Training was a lengthy business and after two years I passed my gunnery course and was sent to an OTU and be 'crewed up'. I was first sent to 3RDFS at Prestwick to train on an ASV Radar Course, and from there, on to Turnberry, to 5(C)OTU, No. 21 Beaufort Course, A Squad. I flew in Ansons, Blenheims and Hampdens, piloted by staff pilots, on wireless and gunnery exercises. It was at Turnberry that I crewed up with Sgt Roy Sutton, pilot, Sgt 'Rusty' Ruscoe, navigator, and Sgt Robert Farnon and myself as W Op/AGs. Roy Sutton had been a staff pilot at Turnberry on Ansons. These pilots would fly Ansons and Blenheims for specific W Op and gunnery practices before we crewed up.

Roy Sutton, through being at Turnberry for some time, had made friends with the Johnstone family who lived at Millbank in Maybole. I believe that Mr Johnstone was the Town Clerk, and the family consisted of three very attractive daughters, Mayne, Vera and Stella. Mrs Johnstone was connected to the St John's Ambulance and Red Cross. They were all well known in the town and a very popular family and I was privileged to become one of their close circle of friends. We had some wonderful times, and whenever we could get away from the airfield we would head straight for Millbank. The family had a sort of 'billet' erected in the garden, which was a hut containing six beds, and we all loved staying there.

1942 : 21 Beaufort Course A Squad

Reg Faber and crew at Millbank, Maybole.

Sgt Farnon and I quickly became firm friends. I soon discovered that he was none other than Bob Ashley, of radio and film fame, and was a vocalist with Jack Payne's Band from 1937 until 1939. He volunteered when war broke out. He had made a series of films for Pathe Pictorial and regularly sang on the BBC; he had a beautiful baritone voice. Bob was a lovely chap, an absolute gentleman. As was common practice during the war, entertainers and celebrities of Bob's ilk were usually given 'cushy' postings, as drill instructors and the like, and sent to places such as Blackpool where they would be out of danger. But, as a measure of his character, Bob volunteered for aircrew. While we were at Turnberry, Bob often sang in the Sergeants' Mess, NAAFI, and hospital as well as taking leave to sing at a special RAF Concert at the Royal Albert Hall in London, where he sang two popular songs, 'You Walk By' and 'I've got Spurs that Jingle Jangle'. We spent a lot of time together – one evening we went into Ayr for a meal, Bob discovered that a friend of his, George Elrick was performing with his band at the Gaiety Theatre, On the spur of the moment, he decided to pay a visit and was asked to go on stage as a surprise guest-artist, he asked to borrow my jacket, as my 'best blue' was better than his, so he entertained the crowded theatre as I stood in the wings with his jacket over my shoulders. He also borrowed

my jacket for his broadcast from the Royal Albert Hall. He was supposed to be on exercise the night he left to perform in London, but we swopped duties, at this stage of our training the navigator and one W Op/AG were not needed.

The pilot was doing night circuits and bumps, so a full crew was unnecessary. When he returned, he took my turn on night flying exercises.

I was at the nurse's dance at Turnberry Hotel (Hospital) when Roy Sutton and Bob were killed when their aircraft crashed into the sea and was only told about it when I returned to my billet. Rusty and myself were sent on fourteen days leave, which enabled me to tell my parents before the news of Bob's death was announced on radio and in the press. My parents, knowing Bob was my friend and fellow crewman, would have feared the worst if they had heard the news of the loss of our aircraft before I was able to tell them. Rusty and I were deeply shocked and saddened by the loss of our friends, we wrote to Bob's father and sent a photograph of us all for him to keep. He sent the following reply, which I have kept throughout the years

My Dear Rusty and Reg,
Thank you, boys, for your kind sympathy and for the touching manner in which you have presented it. The sending of the photograph was a particularly charming thought on your part.

Reg Faber at Kallang (Singapore).

When Bob was home last, he spoke enthusiastically of his crew and what splendid fellows you all were, including poor Roy; and he looked to a long continuance of the pleasant association - God has willed otherwise. His Holy will be done.

You boys know what a great lad Bob could be. Well, from his boyhood up, we were always pals; and I have lost not merely my only son, but my best pal.

However, let us remember him as he appears in the photograph, gay and bright and happy: the very embodiment of clean-living youth; and let us thank God for such as Bob.

Goodbye boys. Here's safe flying and happy landings. Thank you again for your great kindness and God Bless and Guard you.

J. A. St John Farnon

Upon returning to Turnberry, I moved with 5(C) OTU to Northern Ireland, as a W Op instructor, not to Long Kesh with the Beauforts, but to Maghabbery with the Hampden Flight. I was there until March 1943, when I was posted to Cranwell, 3(C)OTU, to make up a Wellington crew. Here I teamed up with Sgt J. Briggs, pilot, F/O Vic Dale RCAF, 2nd pilot, Sgt J. Rose RAAF, navigator, Sgt R. Davies, Sgt N. Evans and myself as the three W Op/AGs. Wellingtons were being transferred to torpedo bombing, and you can imagine my feelings when I learned that my new crew and I were to be posted to 1TTU, Turnberry!

On one day we were on a bombing exercise where the target ship was HMS *Cardiff*. Visibility was patchy, with areas of mist in parts, and we suddenly came upon a ship where we had hoped to find the *Cardiff* - only this one seemed huge and was certainly not our target vessel. We were flying low over the sea and had to take avoiding action to avoid a collision, the ship turned out to be the *Queen Mary*, bringing US Forces to Europe!

Reg Faber

During my second visit to Turnberry, I was summoned to Admin and told that a body had been recovered from the sea and was being buried locally, and as I had been on the same course as the casualty, would I attend the funeral. This was Sgt Harrild, a W Op/AG, who had been on my Beaufort course. His family were present, and when I attended the funeral service, I was surprised to see that Sgt Harrild had a twin sister, the likeness was obvious.

After our course, we left Turnberry at the end of June 1943 for Talbenny in South Wales, where we were to collect a new Wellington and fly it to North Africa. However, for some reason, our skipper John Briggs was taken off the crew. We never saw him again or were given any explanation as to why he left. He was replaced by Sgt George King who flew us from RAF Hurn to Ras-el-ma, then French Morocco, in 9 hours and 10 minutes, then on to Cairo and the transit camp at Almaza. I flew with George for the rest of my time in the RAF, from July 1943 until January 1946. While in the transit camp awaiting posting, our main occupation was playing football. Finally, via the Norwegian ship

Unknown Airmen

Bergensfjord, we sailed from Kazfareet, Suez to Greenock, the journey took nineteen days and we had corned beef and dried apricots for Christmas Dinner seven decks below, on the original hell-ship! On arrival in the United Kingdom, we were posted to 271 Squadron (Transport Command) at Doncaster. This squadron was in the process of converting from Handley Page Harrows to Dakotas for the forthcoming European Campaign through D Day, Arnhem and the crossing of the Rhine. Our main purpose was glider towing, re-supplying by airdrops, and of course flying in troops. We were also fitted out with twelve stretcher racks for the badly wounded and carried a WAAF flying medical orderly. Just before August 1945, we were sent on a glider-snatching course at Ibsley, snatching gliders off the ground was not our idea of fun. Very dicey! As George and I were both newly married we were not much enamoured of this idea, having no desire to make young widows of our new brides! In February

we moved to a new airfield at Down Ampney in Gloucestershire. Here we transferred to our sister squadron, No. 48 and on V. J. day flew in stages to Patenga, (Chittagong-Bengal), being not long married, to Ina, a lovely Scots WAAF, this was not an ideal posting and while I was stationed at Chittagong, my daughter Janice was born.

Whilst with 48 Squadron and flying duties permitted, I often played football, as I was the squadron football team captain. At one match the ball was kicked into a herd of water buffalo nearby. I retrieved the ball and was walking back to the pitch when one of the buffalo took a dislike to me and proceeded to charge. As the 48 Squadron magazine 'News and Views' records it, 'The Funniest Sight of the Week … the Senior NCO who was chased by a water buffalo on the Burma Oil Co's football pitch and broke the World Sprint Record for the 100 yards!' 48 Squadron were operational until January 1946. When the squadron was disbanded, I was posted separately from my crew, to be Signals Briefing Officer at Kallang (Singapore). Here I was CMC Sgts Mess for three months before embarking on the *Capetown Castle* and arriving in Liverpool fifteen days later, Easter 1946.

On demob at the end of March 1946, I returned to London where I was born, and went into the tea and coffee business, firstly with Lipton Ltd., then Tetley Tea Co., and on to Lyons Tetley as a sales and marketing manager. My family had been lucky enough to get a flat in London, and my wife and I placed a tribute to Bob on his grave at Brookwood Cemetery in Surrey. Every year, up to and including 1985, when I retired and moved to Devon, we religiously carried out this pilgrimage every Armistice Day.

Robert St John Farnon (Bob Ashley)
Having a wonderful baritone voice Bob Ashley worked hard to achieve recognition. He auditioned at the BBC in 1932 but was unsuccessful. It wasn't until 1936 that by pure chance he happened to be at the HMV studios when the comedy star Anne Zeigler failed to turn up to record 'Moon for Sale'. Bob was invited to step in and performed so well he was offered a contract then and there.

Later he came to the attention of the BBC's Eric Maschwitz, which brought more work his

Reg Faber and Wellington Crew

way, including a recording session with Carroll Gibbons and the Savoy Hotel Orpheans, when he sang 'You were There' and 'Play, Orchestra, Play'.

Louis Levy and his Gaumont-British Symphony Orchestra specialised in recording 'Music from the films' and he employed Bob at recording sessions. In 1936 Bob recorded two titles with the orchestra - 'You're Sweeter Than I thought You Were' and 'She Shall Have Music'. In 1937 he also recorded 'The Eyes of the World are On You'.

Bob started working as a singer with Jack Payne's Band in 1938 but volunteered for active service with the RAF as soon as war broke out. He recorded a few records with Jack Payne's Band including, in 1939, 'There'll Always be an England' and 'Lords of The Air'. In 1940 he released 'The Grandest Song of them All', 'The Navy's Here', 'Out of The Blue' and 'I'll be Waiting for You'.

Before mass ownership of television, people would go to the cinema to watch the news produced by Pathe. This was usually followed by a short 'feature' Pathe Pictorial before the main film. Bob Ashley was the star of many of these including:

1936	Love Remembers Everything
1937	Moon for Sale
1938	Without a Song
1938	My Heart Will Never Sing Again
1938	County Clare
1939	Deep Dream River
1939	Killarney
1740	A Garden in Grenada
1940	Lights of London
1940	Absent Friends
1941	Room 504
1942	The London I Love
1942	I'll Walk Beside You
1942	The First Lullaby

He always finished these songs with a boyish grin.

Bob made a brief film appearance in 'Such is Life', made at Shepperton Studios in 1936,

and he frequently broadcast on radio. His last broadcast was in 'The RAF Takes the Air' on 25 October 1942. His final stage appearance was as a surprise guest with George Elrick at the Gaiety Theatre in Ayr.

Sandy Ferguson
During the war we children were 'voluntary evacuees' and were sent to my granny's at Gamesloup, between Lendalfoot and Ballantrae. As we lay in our attic room beds at night, we could hear the German planes flying up the coast towards the ship-building Clyde, near Glasgow, causing a heavy droning sound with the weight of the bombs. The planes were much quieter on their return as they were a lot lighter. Grandpa and granny used to worry a lot over these bombing trips as they still had their own children in Glasgow, including our own mother still working there.

During the day we would often see Lysander planes towing long tubular targets which were fired upon by Spitfires as the pilots practised their skills. These targets would then be released over a ground marker which was situated on Troax farmland about a mile east of Gamesloup. They would then be collected by two airmen who lived in a big bell-tent in the same field. They would count the holes in the target and report the score by radio. There were times when the target missed the ground marker, and in fact many were lost out at sea. On one famous occasion while we were at school a target was dropped about 5 miles off course, and landed up a hill about three miles north of Lendalfoot School in a place we knew as Hillend.

It was lunch time at school, and I could see the target just lying there. It didn't look very far away, and as we were very friendly with the RAF men at the dropping zone, I thought that I would be doing them a favour if I collected the target for them. I conscripted Rab Falls to assist, and like me, he was only too willing to be away from school. We set forth with my bike which was an 18" BSA. At long last we reached the bottom of the hill and left the bike by the roadside before proceeding on foot or rather knees and feet as we struggled upwards toward our goal. Eventually we reached it just as the rain came on, so in order to shelter we crawled inside the said target tube, which was made of nylon, and we could almost stand up inside it. When the weather settled, we rolled up the huge target and nylon lines which took two of us all our time. We then dragged it and lifted it down to where the bike was parked. We heaved it on to the handlebars. We were now walking, pushing a loaded bike, and we returned to school well after 3.30 pm when school closed. I was not flavour of the month with the dreaded Mrs Cunningham, although I did get some thanks from the airmen. Thinking back, they were probably not bothered whether they ever saw it again, as they would then have to count the holes, which is a tedious job, I guess.

My sister Maireen came running down to the house one day, having seen a Lysander crash in a field, just over a mile south on the east side of the main road. It was on farmland called Little Bennane, owned at that time by a farmer called Shankland. It turned out that Maireen had helped the pilot out of the plane and along to the nearby farm to get treatment for his scratches, which luckily were minor.

Eric Forte
As a lad of about seventeen I worked with my father in our cafe and confectioners by Girvan harbour. There were three launches of the Marine Craft Section based here, I can recall the names of Sergeants Smith and Spring as well as LACs Brooks, Wellock and Pheobe – they were regular customers at 'Fortes Café'. They were billeted in a hotel called 'Oxenfoord' in Louisa Drive in Girvan.

The Cafe was always busy with servicemen and women, despite the rationing. Girvan was always busy too. I was also a member of the Air Training Corps and we were often taken to Turnberry for short flights in Anson aircraft, usually just around the bay. I remember the Australian Air and Ground Crew at Turnberry, they stood out because of their dark blue uniforms.

Pauline Clarke
I was only sixteen when I applied to join the WAAF. My father was in the RAF and I wanted to join up too. My father had to sign my papers, which he did, but saying at the same time, 'Don't ever come to me and complain that you don't like it'. I didn't! I never had any cause to. I had put my birth date as two years older than I was, no-one ever questioned it and when asked for my birth certificate I said that I couldn't find it, so I became 2064476 ACW 2 P. Clarke.

After training I was posted to RAF Waddington, which was a bomber operational station. I was put in to the WAAF Admin Office and given an old Oliver typewriter which produced stencils when it shouldn't have, there were more holes than letters. In the SHQ there was another WAAF called Doris May, whom I only met when we were posted to RAF Turnberry. A posting had come in for two

Pauline Clarke

Pauline Clark and friends

WAAFs to be sent to RAF Turnberry, so they selected two they wanted rid of. Doris was one because she had overstayed her leave last time and I was the other because they did like my typewriting on that awful typewriter.

We set off for Turnberry having found out it was in Scotland. It took us all night in the train with so many stops and starts and we were both shattered when we eventually arrived at the 'drome. Transport took us out to the lighthouse, where there were a few Nissen huts and we realised that this was where we were going to stay. We discovered that we and another eighteen girls were the first arrivals and that the aerodrome was not yet functional. There was one WAAF Officer whose name strangely enough was the same as mine, P. Clarke. It looked very strange on a leave pass! A small camp had been set up and we were taken there by lorry and issued with bicycles, fortunately most of us could ride. These bikes were our only means of transport both on and off duty, whatever the weather. In fact, it was very bleak when we arrived in the winter of 1941, and I think that cycling back and forth to the main camp caused me to have pneumonia. I was very ill and taken to Turnberry Hotel which had been turned into a hospital. I was there for about five weeks and then began to get better, and noticed it was a rather palatial hotel - the main thing I remember was its sunken baths. I then went to convalesce at the Duke of Hamilton's house in Dungavel, which was wonderful, such food and service! The Duke had very kindly lent his house for an RAF convalescent home.

When I eventually got back to the camp, I went to work for the Adjutant of the Flying Training Wing, F/Lt Henderson. He had the awful job of writing to the parents and families of those killed and he felt it very keenly, as did we all. F/Lt Henderson called me into his office one day and said he had something to tell me, as he was rather hesitant, I wondered whatever it could be. He then said that he had received a memo from the Air Ministry to say that all WAAF were to be called by their surnames. He finished up by saying, 'I am so sorry, Miss Clarke!'

In 1943 the entire unit 5(C)OTU moved to Long Kesh in Northern Ireland, where we lived in purpose-built huts, but that's another story. By the time the war ended I had reached the exalted rank of LACW.

I met my future husband, James McLachlan, at Turnberry, and we were married in 1944. He was a Corporal Airframe Fitter, and our best man Joe Lawrence, was also a fitter at Turnberry. James, or Jimmy as he was called, changed his vocation after he left the RAF, going on to university and becoming a Church of Scotland minister. We spent a few years at Bannockburn before we moved to Edinburgh, where sadly Jimmy died in 1965.

David Withey

I was born on the same day as HRH Prince Philip, 10 June 1921, in Liverpool.

When a Territorial Army Unit was raised in my hometown of Crosby I signed up in May 1939. I was at annual camp when the war began on 3 September 1939. We had gone to south Wales on 19 August and from that date on I was in uniform, until demob in January 1946.

In April 1941 I transferred to the RAF, I had already tried to join 611 (County of Lancaster) Squadron, RAuxAF, but as I was only eighteen years old, I was not accepted.

I had been part of the Torpedo Training Course at Abbotsinch, before assisting in the move of the TTU to Turnberry, prior to joining an operational squadron. I recall there was a

David Withey

great mix of aircrew, from the UK, the British territories overseas plus men from other parts of Europe.

The aircraft I flew in at Turnberry between 4 October and 19 November 1942, as a navigator, was the Handley Page Hampden, which after service with Bomber Command had been modified to a torpedo bomber. The Hampden had an extremely narrow fuselage, but it was an excellent flier, the only drawback was that it had insufficient defensive armament. There were stories that the other types of aircraft there were 'well past their sell-by date' and consequently not too reliable. One aircraft type, the Blackburn Botha, was not considered a safe aeroplane.

Our torpedo training consisted of making imitation attacks on naval ships off the Ayrshire coast. Torpedoes were carried and dropped as in a real attack, but they were fitted with dummy warheads, painted red and filled with water. When the torpedo had lost power and stopped running, this water was blown out by compressed air with the result that the torpedo floated to the surface and was recovered by a launch standing by for that purpose. As part of our training we visited the Royal Navy torpedo factory at Alexandria, near Dumbarton, to learn how these very intricate weapons were manufactured.

All of our training was vital to our ability to carry out attacks against shipping with a reasonable chance of success. At the same time, it was clearly illustrated that the torpedo bomber was very vulnerable during the attack as tactics required a low-level approach and crossing the ships bows!

After completing the course at Turnberry, I was posted to RAF Leuchars.

George Wilkie Newman

'Bill' Newman and his firm friend Sgt Hancock, known as Taffy, had passed out as Sergeants W Op/AG for Bomber Command. The powers that be decided to sperate them from their pilot and navigator, over which Bill and Taffy were most upset, so upset that they decided to become ground crew. The faith and friendship of the crew was one of the most important things to them. However, they had a change of heart and decided to take another course, this time on torpedo bombing at Turnberry.

They had been grounded for two weeks due to inclement weather, before taking off on a night navigation exercise. The aircraft became lost and although tried to contact base for a position, they could not receive a reply through faulty wireless equipment. The crew of Beaufort DD881 perished after running out of fuel while lost over the sea.

Cecil LeRoy Heide

When I arrived at Turnberry as a Wireless Operator Air Gunner, I was unaware of its fame as a golf course. Had I been a golfer I would have been distraught to find that the runway for the RAF airfield cut right through the middle of the course with the accompanying taxiways, aprons and hangars destroying what was left.

There were forty airmen on our course, a few officers among the pilots and navigators, but all the wireless operators and air gunners

were NCOs, mostly British, apart from a few Canadians and one Australian.

For the first few familiarisation flights in the Beaufort we had a different crew each time, this was before we 'crewed up', a process of 'natural selection' that was left to us, usually over a few beers in the mess!

It was on one of these occasions in the mess that I met Harry Deacon, an RAF pilot from London with three years' experience and over 1,000 flying hours. Thinking him an ably suitable candidate, I quickly made friends and it was soon agreed that I would be his navigator. Our wireless operator, Fred Jones, a tall, red headed Yorkshireman, was discovered playing darts, with incredible accuracy and cheerfulness and when I explained that we were looking for someone to join our crew (having a pilot with over 1,000 hours) 'Ginger' as he became known, quickly agreed to join us.

Harry found our other air gunner, a Welshman, Stafford Evans also known as 'Taffy'. He was only nineteen and had joined the RAF straight from school.

So, now we were a crew and we could get on with the job in hand, namely, getting to grips with our aircraft, the Bristol Beaufort Mk I. A sturdy all-metal aircraft, the Beaufort could take a lot of punishment and remain airborne, its weakness was it was almost impossible to fly on one engine, especially when loaded, it 'flew like a brick'.

The fuselage was broad, and the crew had room to move about easily, the pilot and navigator sat 'up front' on take-off and landing, the navigator's position normally being in the perspex nose. The wireless operator sat immediately behind the pilot, with an armour plate between them. The rear air gunner had a power-operated turret, protected by armour plating. And we even had an Elsan toilet!

Our training at Turnberry entailed converting to the Beaufort and initial bombing and gunnery practice. Torpedo exercises would follow.

I soon discovered that what I had learned at gunnery school before was not wholly compatible with the machine guns and bombsights in the Beaufort, they were totally different. At Turnberry we practised bombing both fixed and moving targets, fired the machine guns at ground targets and at a drogue towed by another aircraft. I found the guns very hard to handle, they threw me around so much when I pulled the trigger that I was never very accurate. The clattering noise was deafening.

I also had to learn how to use a large handheld 'Camera Gun', which was supposed to record the accuracy of a strike on operations.

We continued our training as a crew, bombing, gunnery and navigation. We were all kept busy except Ginger, who with nothing much to do spent his time in the air listening to the BBC. Our course at Turnberry ended just before Christmas 1942, when we were granted ten days' leave and informed that we were to be posted to RAF Abbotsinch for torpedo training.

So, we arrived at Abbotsinch to find that out of the ten crews that left Turnberry, we were down to eight – we never did find out what happened to the others. The base at Abbotsinch was small and could easily be reached using the trolley bus from Glasgow. Living quarters were quite spartan, wooden huts with inadequate heating.

Our course began with four days of classroom theory on torpedo mechanics followed by flying training on alternate days, involving formation flying and dropping torpedoes against fixed and moving targets.

Classroom work done, we moved on to flying operations for which we were put into the hands of Instructor, Flight Lieutenant Inglis, who had completed a tour on Beauforts with 22 Squadron at Chivenor. He informed us that he would teach us how to drop a torpedo, practising 'dummy drops' on a merchant vessel off the coast which was fitted with photographic and measuring equipment to assess our accuracy or lack thereof!

F/Lt Inglis told us, 'Torpedo operations are done at low level, at a height of about 50 feet. The reasons for this are obvious: the camouflaged aircraft are hard to spot against the sea by the enemy ships and fighters; and most heavy cannon on large ships cannot be lowered to this

elevation. The torpedo must be dropped flying straight and level at an altitude of about 75 feet and a speed of 140 knots at a range of between 1,000 and 1,500 yards.'

A couple of days later we were to put this theory into practice. Taking off from Abbotsinch our navigator set course for an island in the Clyde called Ailsa Craig. There we would find our target ship. We quickly found the small freighter and did a couple of practise runs before dropping our dummy torpedo. Our final run looked well placed and Taffy reported the torpedo to be running well. The target ship would analyse our run and radio our results back to base. As we flew back to Abbotsinch, in my mind I couldn't help but shudder at the thought of torpedo dropping in action against the real enemy, having to fly straight and level for over a thousand or so yards at 50-75 feet above sea level with every gun in the convoy firing at us. We were informed later that our height and speed were okay, but our range was out. Disappointed, we put our poor results down to the inaccuracy of the sighting device we had been issued with and all agreed that we would get better with practice.

We only visited the city of Glasgow once during our stay at Abbotsinch, and found it to be drab and dirty, badly suffering from bombing. The Germans had targeted the Rolls Royce Aero Engine works and several Luftwaffe squadrons were equipped with a new radio direction finder aimed solely at the complete destruction of this important site. Although badly hit, the factories never closed. The docks at Greenock, where many Atlantic Convoys arrived, were also getting a lot of attention from German bombers and were heavily damaged.

We practised several more torpedo drops, this time with the ship moving and taking evasive action, we were getting better and maybe had a hit score of perhaps two out of three. We were all acutely aware that operational targets would not be anything like this little ship and they would be firing back at us ….
[From his book, 'Whispering Death.']

Bob Yorston

My mother was from the Isle of Harris, the daughter of a Harris Tweed merchant. My father came from Montrose and was the son of the rector of Montrose Academy. They met at Edinburgh University where my mother was a medical student and my father studied forestry. I was born in March 1920, and when I was eighteen months old my mother and I went out to Kenya to re-join my father who was at that time a coffee farmer. In 1928 we came home from Kenya so that we children could go to school. Apart from a spell in Laurencekirk, we stayed in my mother's old home in Harris, so ended up fluent in the Gaelic tongue.

Bob Yorston

In the summer of 1939, I was granted a place at St Andrews University to study medicine. When WWII broke out, young men immediately became liable for conscription into the armed forces. Unless, of course, they were in a reserved occupation. Medical students were among those classed as being in a reserved occupation, but they had to have reached a certain stage in their studies to avoid being conscripted. Because I had taken a year out between leaving school and going up to university, it was quite possible that I might not have reached that stage in my studies before my age group became liable for call-up.

I have often been asked why I volunteered for aircrew, and the usual trite answer is, 'Because it seemed a good idea at the time.' Which begs the question, 'Why was it a good idea?' Perhaps a word or two about the relevant factors would be in order.

Having heard my father's tales of the horrors of the trenches in WWI, I had no desire to be conscripted into the army. Likewise, the navy did not appeal, because during schooldays in the Hebrides I had heard from old salts too many grim stories about life at sea. So that left the RAF.

Volunteering for aircrew was one of the very few ways to circumvent the regulations governing reserved occupations. Besides, it offered the slight element of choice which conscription did not. So, to avoid the possibility of being conscripted into the army or navy, I volunteered for aircrew. I was placed on deferred service, told to go back to my studies and await further orders. In March 1940 I was summoned to Dundee for attestation and was selected for training as an observer.

Then came the medical examination, and I was found to have a slight weakness in my left eye. The Medical Officer said, 'Not good enough for an observer. Got to have 6/6, 6/6 vision for observer, and he's only got 6/6, 6/9. He'd be all right for a pilot though.' And so, I was remustered for pilot training.

This pleased me considerably, because from a very early age I had had two ambitions; one was to be a pilot and the other was to be a doctor. Before I had left school, my father had asked me what I wanted to do, and I told him my ambitions.

'Well, son,' he said, 'whichever you choose, you're going to have to do it on scholarships or bursaries. But I would just point out to you that by the time you're 45 you'll probably be finished with active flying; whereas in medicine you'll have come nicely to your peak with another good twenty or more years ahead of you.'

So that's why I started in medicine. But when the opportunity came to be a pilot, I took it. I was called up in June 1940. After doing my initial training in Cambridge, I was sent to what was then Southern Rhodesia for my flying training and got my pilot's wings on Christmas Eve, 1940. From there I was sent to Blackpool to the School of General Reconnaissance, and then to Silloth in Cumbria to learn how to fly the Lockheed Hudson. In October 1941 I was posted to 269 Squadron at Kaldaoarnes in Iceland. After a tour of operational flying from Iceland, I was posted to 5(C)OTU at Turnberry.

As was my custom both at school and university, I wrote home to my mother every week. She, bless her heart, kept practically all the letters I wrote to her from the RAF, and these are they, my 'RAF Years, the Turnberry Chapter.' I have for the most part omitted things like greetings, and the 'thank you for your letter,' and the 'give my love to' bits, and the mundane replies to my mother's news of local arrivals, departures and casualties. But otherwise the letters are as I wrote them, grammatical errors and all. The language is often a bit cringe-making - but that's the way we spoke and wrote in those days. Oh, and I've omitted all the more scurrilous, nay, even libellous bits! I'm no fool - I've no desire to make lawyers rich!

Non-Operational Flying, UK
Turnberry, Ayrshire, Friday 20 November 1942.
The journey down to Glasgow was quite uneventful, and I got here OK on Friday, despite it's being a Friday the 13th, and settled in fairly quickly. This is a 'Beaufort' torpedo-bomber OTU, and I'm a bit puzzled by the posting because I've never flown Beauforts and I've had no torpedo training. So how the hell I can train sprogs in the art beats me. However, that's academic, because I've been assigned to the stooge flight flying Avro Ansons, which is rather droll, because I've never flown Ansons before either. But that problem was quickly solved. The Flight Commander took me up for about half-an-hours dual, and when I made a couple of quite good landings in a strong crosswind, he was happy enough to send me off for my '1st solo on type.'

The Anson is a nice aircraft to fly and is powered by the same Armstrong-Siddely Cheetah engines as the Airspeed Oxfords that we flew in Kumalo, so it's an engine I know and trust. But of course, there is always a snag somewhere, and in the Anson it's the undercart. It's raised and lowered manually by a crank beside the pilot's seat, and it requires over 150 turns to raise it. So whenever possible, I get one of the sprogs to do the job.

My job is to fly u/t Navigators and W Op/ AGs around on navigation, W/T, gunnery and bombing exercises. Each flight lasts between one and a half and one and three-quarter hours,

and I've been flying two or three details nearly every day since I came here. I keep an eye on the navigators and see that they are coping, and help them out with any problems, which is where my Kalda experience comes in useful. I'm getting to know the Firth of Clyde and the west coast quite well now. I usually have a screened (staff) W Op on each flight and listening on the intercom to him teaching the sprogs is fair improving my knowledge of signals.

The airfield has been built on what was a golf course, and it's a dispersed station, which means that instead of all the buildings being concentrated in one place as on the pre-war airfields, they are widely dispersed around the countryside. It makes the station less vulnerable to air attacks. This was a lesson learned in 1940. A couple of bombers with a few well-placed bombs, especially when mixed with incendiaries, could completely disable an entire 'regular' station. Not so easy with a dispersed station. Of course, there is a price to pay. It's nearly two miles from the Flight Office to my billet, which is almost in the village of Maidens, right across the other side of the airfield. And the mess is about a mile away in the other direction, sort of at the apex of a triangle. Fortunately, push-bikes are available on loan, and I'm making full use of mine.

So far, I haven't had much time to explore the hinterland. The airfield is about 6 miles from Girvan and about 15 miles from Ayr. Tomorrow one of the other staff pilots and I intend going into Ayr to see what the town offers in the way of entertainment and grub, particularly grub.

By the way, I was rather tickled to find that Shanter, where Tam o' came from, is within the airfield circuit. Of course, Ayrshire is Burns country, so it's not surprising that a lot of the local names crop up in his poems. I must read a bit more Burns poetry and absorb some local colour.

Turnberry, Saturday 28 November 1942
I've been waiting for my books to roll up, because I've arrived in a camp where there's an acute shortage of reading material. I've

managed to borrow from one of the boys a book by a fellow called Frank S. Slaughter. It was all about a young American surgeon and gave some interesting sidelights on American medicine.

My own news isn't very plentiful. A week ago, tonight, I went into Ayr with another staff pilot, and went to have something to eat. We tried to get into a flick, but the queues were too damn long, so I came home.

Last Wednesday was my day off, so I went to Ballochmyle (hospital) to see Les Goodson. Les was the only survivor of a very nasty crash in which five people were killed when the Beaufort Les was flying was barged into by a Hampden on the end of the runway. Les was taken to Ballochmyle suffering from extensive burns and

Les Goodson

wasn't expected to live. That was three months ago. I didn't recognise him when I went in to see him. His right eyelid is twisted, his nose is badly burned, and the right side of his face is pretty badly messed up. They've managed to save the thumbs on his hands, and the fingers of his left hand up to the first joint. They don't know what is going to happen to his right hand yet. His legs are quite serviceable, though fairly badly burned in places, and Les told me he expected to be up and around fairly soon. I was really quite shocked at the change in him. Where before he was well built, he's now as thin as a rake. It was really his strong constitution that pulled him through. He's the prize exhibit of Ballochmyle now, he told me. He also gave me some gen. on the rest of the boys. Eric Rawes went missing on a raid on Hamburg, and Ken Wright disappeared on the first thousand bomber raid over Bremen. Allan Marriot is still going strong and so are Willys, Roxburgh and John Owen. Willys is now a Flight Lieut. On my way back from Ballochmyle I

stopped in at Ayr to have some grub and did a picture show, which I enjoyed quite well, in spite of it being a Bing Crosby effort.

Last night we had a dance in the mess to which only WAAF were invited. It was quite a cheery little show, and I got off my mark with a WAAF in stores and one in accounts. Useful contacts! Tonight, there is an ENSA show in camp, and I'm going along to inspect it.

Turnberry, Sunday 6 December 1942
Your letter of 25 October arrived here from Iceland last night. Obviously, there's nothing now in that one that needs an answer.

My books duly arrived on Wednesday, and I was very glad to get them because I'd been reduced to reading sauce bottle labels. It was very good to get my teeth into a Kipling again!

I don't think Les is going to have any mental scars after his ordeal. He was certainly very chirpy when I went to see him.

The ENSA show last Saturday night was quite a success. I enjoyed it immensely. Some of the wisecracks I'd never heard before, and the comedian kept his jokes fairly pure; he had to, for we'd lots of WAAFs present. Not that they'd have been worried, but still, it wouldn't look good.

On Monday night we had a film show in the mess. 'The Man Who Could Work Miracles' by H. G. Wells. We all thoroughly enjoyed it and decided that the first miracle that we'd work would be to end this blasted war!

Apart from flying and work generally, there was nothing much doing until Thursday, which was my day off. I went into Ayr to do some shopping; had a haircut and a wizard lunch which included a dirty great hunk of sauté chicken. After lunch, I went to see 'Eagle Squadron' and didn't think much of it. Had some tea, went to the ice rink, decided it was no go, and came back to camp.

Friday, Saturday and today have been very quiet. The weather rather leaked on us and I only got in one trip yesterday afternoon.

I had quite a thrill today when a Hudson from Silloth landed with engine trouble. I went over to inspect it and found it was the self-same machine in which I'd had my ropiest do in Silloth; the time when the 20-inch camera went through the astro-hatch. It was a pupil crew on a nav. trip, but they were able to give me some gen. on the 269 crowd at Silloth. I told the pilot to give all of them my regards. He didn't accept my offer to air-test his machine for him when it was serviceable again. I'd loved to have had a crack at the old 'Hudswine' again.

Turnberry, 13 December 1942.
Nothing very much to report from here. We've been a bit at the mercy of the weather. There were four consecutive days this last week when we couldn't do any flying. Then when we did eventually get airborne, we were only half way round the detail when we were recalled by control. Just as well really, because I landed in very heavy rain which was a wee bit dicey.

Then the following morning I had to cut short a trip because of an oil leak on the starboard engine. Just to show off, I did a single engine landing using the port engine only. For the afternoon flight that day, I had two navigators, one of them being a staff navigator. So, while the pupil navigator was doing the navving, I gave the staff lad some dual on straight and level flying and gentle turns.

You never know, it might come in useful for him later, even if only to give the pilot a chance to nip to the back to spend a penny.

As I said, we had four days last week when the weather put paid to any flying. That gave me the chance to catch up on the latest intelligence reports. After checking the Coastal Command reports, I read up on what's been happening in other parts of the war. The 8th Army seems to be doing well in North Africa, don't they? More power to their elbows, I say.

I also took the opportunity on one of the no flying days to go around to the hangar and watch the fitters do an engine overhaul on a Cheetah engine. I was gratified to find that I hadn't forgotten all the stuff about the engine that I'd learnt at Kumalo (where we had the same engine in our Oxfords), and it all paid off when I'd had

that oil leak a couple of days later. One thing I did notice, however, was that there wasn't the same atmosphere of camaraderie in the hangar as there had been at Kaldaoarnes, nor the same urgency. The difference between an operational and non-operational station, I suppose. It's understandable that an operational squadron generates more of an esprit de corps than a non-operational unit.

On Thursday I went into Ayr with one of the other staff pilots, a lad called John Bradford, for a meal and a flick. We only just caught the last bus back to camp. But that's about the limit of one's social life.

I nearly fell out with the armoury one day last week. All our practice bombs failed to explode, and I was about to give the armoury a roasting when something made me check the bomb racks. Which was just as well, because I found that the sprog navigator had omitted to arm the bombs before we dropped them. One red face saved. Mind you, I should have suspected what had happened, because although you can very occasionally get one bomb failing to explode, it's extremely rare for the whole lot to be duff.

Turnberry, Sunday 20 December 1942
There was too much flying to be caught up with because of the time lost through bad weather. The flying is a bit of a bind just now, because as well as it being very bumpy because of the high winds we've been getting, there's the strong, low sun to contend with when we're coming into land on the west facing runway, especially just around sunset. It can be almost blinding, and it makes it very difficult to judge your height above the runway at the point of touchdown.

Mind you, I was very glad of the high winds the other day. I had a slight case of the twitch when I found that my air speed indicator was u/s. Fortunately, the strong wind allowed me to come in with a fair amount of throttle and so keep my air speed above the stalling point.

The station is getting all geared up for the Christmas festivities. I can't remember there being such a palaver about it in Iceland, but then we didn't have the WAAF up there, and they're the ones who seem to do most of the organisation.

Turnberry, Sunday 27 December 1942.
Thanks very much indeed for your letter which arrived on Tuesday, for the greetings telegram, and for the parcel of eggs which arrived last night. The eggs are especially welcome, and only one of them got damaged in transport; so that I have ten whole ones left. They're beautifully fresh, too.

On Monday we had our Christmas dance in the mess. I had a grand time, but for once I was caught napping. I forgot about the grub, and only got in in time to nab the last of the best bits, instead of being in the forefront of the battle.

On Tuesday afternoon I went into Ayr to do the odd spot of shopping and managed to get another couple of Kipling's to add to the ones I bought earlier this month. One of them is his autobiography, *Something of Myself*, which I'm very much looking forward to reading. Then I had a bite of food and went to the odd flick. A nice quiet evening, which ended up by my getting my feet absolutely soaked in a puddle.

Wednesday night was spent at a dance in Girvan, quite a posh affair at seven and six a go. I had a very good time, and this time made sure of some eats by sitting out the two dances before the buffet was due to open. After the dance I was out of luck for transport; but one of the men who drove the launch for practice bombing came to the rescue, and I spent the night in the Marine Craft Section. They woke me up in the morning with a cup of tea and some hot buttered toast. I got a bus to the 'drome and spent Christmas Eve quietly in my bunk.

I slept in on Christmas Day. It was 09:20 before I moved. I went up to the mess and had an al fresco breakfast of grapefruit, spamwiches, dates, sultanas and bread and marmalade. Then I went over to help serve the Airmen's Christmas Dinner. Old service custom.

Our own Christmas dinner consisted of brown soup, turkey, pork, Christmas pud, dates, sultanas and almonds, cheese, coffee, and a bottle of beer. You may be sure I did full justice to it.

After dinner I went into Girvan to pay a visit to the Bairds. After tea, Ellice and I came to Turnberry Hospital to see their little concert (Ellice worked in the hospital last summer). After the show I put Ellice on the Girvan bus and retired to bed.

Yesterday morning, I went in and fetched Ellice for a jaunt to Ayr. We had lunch chez Ellice first, then caught a train to Ayr. I did my odd shopping, and then we went to a flick. After the flick some tea. Tea was a wizard meal of roast duck and all the trimmings. After tea we looked in at a dance and then caught a train for Girvan at 21:20. I had a strupag in the Baird's and got a bus back to camp, in which I was lucky, because it was a special bus running some people back to camp from a dance. Today has been very quiet. I went to have tea with the Bairds and made a provisional date to go and see Ellice tomorrow.

And that's just about all my news. On the whole it's been a very quiet Christmas, and yet I've managed to have quite a bit of fun.

Only one thing more - Bliadhna Mhath Ur [Happy New Year] to you all in Tobair Mairi [Mary's Well – name of their home on Harris]. I just wish I could be up there with you.

Turnberry, Sunday 3 January 1943.
Now that the Bacchanalia have finished, it's back to work as usual. In the Flight Office on Tuesday I said, 'Ah well! Back to the daily round, the common task!' and was promptly corrected by one of the other pilots, who said it was 'the trivial round, the common task'. He said it was from the works of a clergyman called John Keble. I said I thought it was from the Bible. He said no; and that the Bible one that I was confusing it with was the one from Exodus, 'your works, your daily tasks'. We had an interesting five minutes swopping quotation. It turns out that this bloke was a lay reader in the Church of England.

The flying has been the main activity this past week, because we're busy having to make up for the time we lost over Christmas. The weather has been quite blustery for the past few days, and on one trip I had one of the sprogs being airsick, and also an ATC cadet who had scrounged a trip with us. We also had to curtail one trip because the W Op reported severe jamming on the W/T.

Yesterday I had a bit of engine trouble over Girvan and had loss of power from the port engine. When I got down, I reported it to the Chiefie fitter and said that I suspected one of the cylinders was not working properly. He sought me out later, and said I was quite right, and that there had been a broken rocker arm on number 1 port. He said, 'You're wasted being a pilot. You'd be far more use in the hangar than flying bloody sprogs around.' High praise, indeed! I was quite bucked. Mind you, the Chiefie himself looked quite pleased when I told him I found it quite interesting and rewarding to see what went on in the workshops, and that I made a point about learning as much as I could about the engines that I was flying, and the best place to do that was in the hangar.

The social life has been much quieter this past week. I met Ellice on Monday as arranged, and we went up to Ayr for the afternoon. She had some shopping she wanted to do, and afterwards we went and saw a flick. It was so damn boring that we came out before the end and went and had some tea. We got a train back to Girvan, and after a cuppa with the Bairds I headed back to camp.

Since then, it's been a case of 'scorning delights and leading laborious days.' There was a little impromptu party in the mess on Thursday night, but we did little more than see in the New Year. I was in bed before 1 am. These wartime New Year's don't match up to the ones of yore!

Turnberry, Sunday 10 January 1943.
There's not really very much to report this week. The flying has been the main activity, and that's been pretty much routine. The weather has been lousy; not bad enough to stop us flying, but bad enough to stop us enjoying it. The other day I had my altimeter and ASI freeze up after an air firing exercise, so I had to get down to deck level to thaw them out before landing. Yesterday the W Op tuned in to the BBC after he'd finished his exercise, and we landed to 'Music While You Work,' which we thought was appropriate.

The social side of my life is in the doldrums at present. Not that I'm complaining, because it's giving me a chance to catch up with my reading. Those Kipling books which I bought last month have been a good thing. I'm now reading his autobiography and finding it very interesting.

The Anson flight is probably going to be moving to another base sometime later this month, but I'll let you know as soon as I get the pukka gen.

Turnberry, Sunday 17 January 1943.
Not a great deal has been happening since I wrote last week. In fact, I was very nearly going to send this communiqué on a postcard. Apart from a visit to Ayr with Bob Garnett and John Bradford, the only other activity has been the flying. Bob, Brad and I have taken to doing part of our trips in formation to break up the monotony. I had a nice little trip the other day, though. I had to take a load of homing pigeons down to Castle Kennedy. They were to be released there as part of their training. I couldn't help thinking of Johnnie Howitt's 'brainchild' of crossing a homing pigeon with a parrot so that it could tell you the course for home.

I was talking about that trip in the mess that evening, and one of the other lads told me that when he had been at Castle Kennedy, the boys were nattering about night flying when one of the fellows said, 'Pheasants don't have nav. lights, do they?' it was agreed that they didn't. 'And they don't have flare paths either.' That was also agreed. 'So, they can't fly at night, can they?' 'True enough,' was the reply. 'So why don't we go out tonight and knock a few off while they're not flying?'

Still no news about our move, but I gather it will be somewhere in Wales. I'll let you know as soon as I hear anything definite.

On 28 January 1943 the Anson Flight of No.5(C)OTU was posted to 10 Radio School at Carew Cheriton in Pembrokeshire, Wales.

Just as the Lord works 'in mysterious ways his wonders to perform,' so does the RAF. And so, having completed a tour of anti-submarine duty with a Hudson Squadron in Iceland, I fully expected to be posted to the Hudson OTU in Silloth. So, imagine my surprise to find myself in a Beaufort torpedo-bomber OTU. I was assigned to the Anson flight which was posted as a complete unit to Carew Cheriton. After a few months I began to find the routine somewhat tedious - just flying u/t W Ops around the same old triangular courses without any practice bombing or air gunnery to add to the odd bit of variety that we had at Turnberry. So, when I saw an Air Ministry circular offering NCO pilots the opportunity to be seconded to fly with Scottish Airways, I volunteered. This entailed a visit to the Air Ministry, where I was seen by a Squadron Leader who, after asking me lots of questions enquired whether I was really keen to go to Scottish Airways, or whether I would prefer to go on ops. 'Oh, ops., sir,' said I. By this time the Hudson had been phased out from the anti-submarine work for which I had been trained, but the Leigh Light had come into service, and they wanted pilots for that job. So, I was posted to a Leigh Light OTU and from there to 179 Squadron flying Wellingtons from Gibraltar, and later from Cornwall. For a few months in 1943 I flew with Bob McGill (see page 443)

After serving six years in the RAF I returned to St Andrews and received my Medical Degree on Christmas Eve 1950, which by coincidence was exactly ten years to the day after I was awarded my pilot's wings. Following the usual round of junior hospital posts I became an ENT specialist, and by another coincidence received my specialist diploma on Christmas Eve 1960, exactly ten years after my Medical Degree. I retired in 1985.

Pierre Le Brocq

'Pete' Le Brocq was born on the Island of Jersey in September 1924. He attended Victoria College where he received an excellent education and also excelled at most sports. He was evacuated to England prior to the German invasion of the Channel Islands, along with his mother, brother and sister.

He was tall, good looking, charming and the life and soul of any party. As one former member of the RAF once said, 'if you went into a room full of strangers, the one you would remember when you left would be Peter Le Broc'.

Joining the RAF at just fifteen years old, Pete had gained his wings by the time he was sixteen, he was posted to 38 and then 40 Squadron, completing a tour of operations in the desert as a 17-year-old F/Sgt flying Wellington Mk Ic's.

Promoted to P/O and later F/O he was a Torpedo Flying Instructor at 7 OTU Limavady and 1TTU Turnberry at eighteen.

Posted to Turnberry at the same time as Pete was S/Ldr Albert Lloyd Wiggins DSO, (Wiggy), who was O/C No. 2 Squadron 1TTU, where he remained until 1944 when he joined 455 Squadron RAAF at Langham. Squadron Leader Wiggins had been awarded his DSO in October 1942 for torpedoing a ship in the Mediterranean

1945 : 1TTU Ground and Flying Instructional Staff

Piere Le Brocq cartoon

Piere Le Brocq

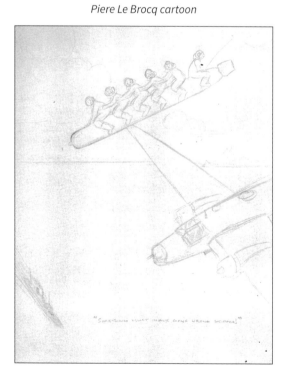

Piere Le Brocq cartoon

Albert Wiggins and friend

S/Ldr Albert Wiggins

Sea in a Wellington of 38 Squadron on an operation he recalls as 'The night I handed over to God'.

Another famous Australian at Turnberry during this period was Wing Commander Grant Lindeman, of the wine producing family, at one time Commanding Officer of 455 Squadron, but for a while was Chief Instructor at 1TTU.

Training was carried out against nominated target ships. Where there exist rivalries between Scotland and England, likewise between Jersey and Guernsey. This can be traced all the way back to the English civil war (Jersey for the King and Guernsey for Cromwell) F/O Le Brocq must go down in history as the only Jerseyman to torpedo the *Isle of Guernsey*!

Returning to operational flying he was killed while flying with 144 Squadron, at the young age of nineteen.
[With thanks to Mr Phil Le Brocq, nephew of the late Pierre Le Brocq]

Bob Spells
Making our way to Turnberry we arrived at Ayr Railway Station on a Sunday after travelling from London. There were four of us, and whilst waiting for the connection to Girvan, we decided

Wiggins and personnel

Bob Spells

Bob Spells and crew

to walk into town. The place was deserted as it was so early in the morning. We saw some people going into a baker's shop and followed them in. We asked the lady behind the counter if it was possible for us to get a cup of tea anywhere, and she told us to wait there, while she disappeared through to the back shop. When she came out, she instructed us to follow her, and this we did. To our surprise, she didn't take us to a tea shop, but took us to her home where she woke her family and then proceeded to prepare breakfast for us all. She had some lovely bread rolls that she had bought at the bakery. I have never experienced hospitality like it, before or since!

So, I arrived at 5(C)OTU, Turnberry in 1942 to undergo training on Bristol Beaufort aircraft, as a W Op/AG. The other type of aircraft used for training at Turnberry was the Handley Page Hampden. There were also Avro Ansons and a couple of Mk IV Bristol Blenheim. The permanent staff at RAF Turnberry seemed a

Target Ship and Ailsa Craig

Bob Spells and crew

Bob Spells and crew

sullen lot, officers of no distinction, half trained as we were, morale was very low. The TTU was a lot better endowed, efficiency reigned supreme.

At first, I and another chap, Bill Raper, were attached to the Hampden Course, but we agreed to swap with two other W Op/AGs, because they were married men, and they had heard that Beaufort crews were being sent overseas on completion of training (both had their wives with them at Turnberry). Sadly, these airmen were later both killed whilst on their first mine laying mission over Norway.

We then met our pilot and navigator, Bill

Bill Carter

Bill Carter and crew

Carter and Bob Bird, both Canadians. We completed our conversion onto the Beaufort and transferred to 1TTU to learn torpedo dropping, this entailed a lot of low flying over the water, which was very frightening.

I had an unforgettable experience in an Anson during one navigation exercise. Most of the Perspex was missing from the windows; I got absolutely drenched when we flew into some rain clouds!

The torpedoes and servicing unit were housed in garages belonging to Turnberry Hotel (then a hospital). The torpedoes we dropped had a dummy warhead, which was hollow and set at

a depth to run under the target ship. When the torpedo reached the end of its run, it would come to the surface and emit a smoke signal, this to enable the recovery ship to locate it and retrieve the torpedo for further use.

The ships that were used as targets were HMS *Malahne,* HMS *Isle of Guernsey,* HMS *Canton,* HMS *Cilicia,* HMS *Heliopolis* and HMS *Canterbury.* All our practise runs on these ships were recorded on special cameras which were used to attest the accuracy of the attack. These cameras were the responsibility of a lovely WREN Officer who as you can imagine was very popular, especially among the Colonial types!

We enjoyed dinghy drill in the swimming pool in Turnberry Hotel, very posh it was too, particularly the showers, with hot and cold water, which sprayed at you sideways, all around you, you didn't even get your hair wet!

Sometimes we went ice skating in Ayr or to the cinema in Girvan, I have had to walk back to the airfield a few times when I missed the last bus.

Bill Raper and I met two girls in Maidens, which was right in the middle of things at this airfield; it was a grim place in the winter of 1942/43. Originally, I and another chap had 'chatted up' these girls at the camp cinema, which was a gymnasium cum chapel on the base; tragically the other chap was killed on a night exercise. Anyway, they invited us to meet their relations, the main ones I remember were Mr and Mrs John Andrew, who lived in a bungalow called 'Tigh na Mara', right on the waterfront. Come Christmas, they laid on a special Christmas dinner for all our crew, brilliant when one considers that the Scots don't celebrate Christmas as much as New Year. We also celebrated New Year with them; it seemed to go on forever with the whole village visiting one another, what a smashing people!

I remember that the Andrew family had an interest in the Ailsa Craig, where the granite for making curling stones was quarried. One could see Ailsa Craig very well from the Maidens; it was a good landmark for locating the airfield.

Prior to meeting the Andrew family, Bill and I had been waiting at Ayr for a train to Girvan, when a RN Petty Officer approached and asked us if we were going back to Turnberry. We said we were, and he then kindly offered us a lift. His car was an open top Sunbeam Talbot. We were to learn later that he had been killed when his car crashed after he was stung by a bee. He was a relative of the Andrews.

On completion of our training we had an end of course party in a pub in Kirkoswald called the Shanter where there were scratch marks made by Robert Burns on one of the small windowpanes.

My memories of Turnberry are always tinged with great sadness; we lost a lot of good friends there. The Hampdens were a lot of clapped out ex-Bomber Command aircraft. The Beauforts had an inherent design fault that made it uncontrollable on one engine. The main problem on the Mk I (Bristol Taurus engines) was that the main propeller pin would snap which could cause the propeller to windmill and then break away complete with reduction gear. Servicing left a lot to be desired, the airmen of all trades

were deficient in training and expertise which didn't help. This was not so with the Torpedo Training Unit. The Mk II Beaufort, which had Pratt and Whitney engines, was more reliable and came from a family of very good engine manufacturers. The propeller on the Mk II could be feathered [turning the angle of the propeller blade to a point where there is the least drag caused by 'wind milling'], this was not so on the Mk I. Although, the aircraft was still difficult to control on one engine. Despite the nasty reputation of the Beaufort, it was amazing how some pilots became enamoured of it.

The Hampden crews suffered some hairy experiences as well. One crew survived after a crash in the Outer Hebrides. They had hit the top of high ground on a small island after suffering engine failure; they had then proceeded to crash onto another island. Although they survived the crash, it took a long time before they found someone to help them; it was a couple of weeks before news of their survival filtered through to the outside world! Another Hampden ditched in the Clyde the navigator and two W Op/AGs survived. The navigator, being a big New Foundlander, kicked his way out of the perspex nose. The pilot died when he could not unfasten his harness, the trident shaped locking pin was missing so he used a key ring instead, this he could not detach.

The method by which the three crew entered the Beaufort was via a hatch on the starboard side, which opened in two parts. The top half was metal and was hinged upwards and the bottom half was wood and slotted into the hatch housing. When not in use, this part was stowed inside the fuselage, secured by a cord which was supposed to prevent the hatch being swept overboard by the slipstream. As was usual, this cord was missing, and there were quite a lot of these covers lost this way. Bill Raper, my co-W Op/AG, lost ours on one trip and was told he would be charged to sign a F664B, which meant he would have to pay for a new one, plus one to replace it in the equipment store. When informed of this we all searched the foreshore, where we hoped it may have been washed up. We walked all the way to Culzean

Castle and back and had almost given up, when we found it almost as we reached Maidens, what luck! Whether it was the one we lost or some other, it didn't matter, everything was okay, and no charge was made.

21 Course lost over one third of the crews that arrived at Turnberry, they went out on exercise and just didn't come back. I especially recall a Mexican chap, Joe Dukick. He had already lost two crews. He appears in the course photo with his third and final crew. Alas! they also failed to return after a night exercise and this time Joe perished with them. The last crew of our course to be lost was that of Sgt Wood. They went on a fuel consumption test in a brand-new Mk II Beaufort – they were never heard of again.

It still makes me sad to remember the lads who were lost at Turnberry and the families that suffered so.

Jack Egger

I came into this world on 1 November 1920 at Nanaimo, a small coal mining town on Vancouver Island, British Columbia. My mother hailed from Buxton in Derbyshire and my father was from Grindlevald in Switzerland. Father had trained as a Maître De in Paris, London and Buxton and was subsequently moved to a hotel in Vancouver, The Wigwam Inn. My mother followed him to Canada and eventually they were married in the early 1900s. My father was a difficult man, and I grew up in a shadow of fear, always feeling that whatever I achieved, it was never good enough for him, longing for the day when I could get out on my own. In his own way, he was a good man; he always found work during difficult times and managed to put food on the table. He instilled in me a strong sense of duty and responsibility, but whatever I attempted in the way of gaining my father's approval, seemed bound to fail. I could

Jack Egger

never please him, and forever conscious of his criticism, I became a bit of a loner, it was safer that way.

As soon as I was old enough, I got a job as a caddy at a local golf course and was greatly helped and befriended by the Scottish pro and green keeper. After graduating from high school with much rejoicing, at the age of seventeen, I finally was able to leave home by getting a job at Qualicum Beach Golf Club, which was about thirty miles north of Nanaimo. I did alright on my own and was happy to be away from the oppressive atmosphere at home.

1939 arrived, and I knew I must go into one of the services and do my bit; conscription was in force, but not for serving overseas. A friend and I went to join the army, but when I saw the chaps marching with full back packs, in seemingly pointless circles, I decided that it wasn't for me. The sea made me unwell, so the navy was out of the question, so as a last resort, I applied to join the RCAF. I was accepted for pilot training and in December 1941 was sent to 15 EFTS Regina. The first aircraft I flew was a Tiger Moth, and along with classroom exercises covering the technical stuff, I graduated on to the Cessna Crane, a twin-engine trainer. Training included some lovely flights over places such as Melfort, Moosejaw and Dauphin. By June 1942, I was assessed as an 'average pilot' with my 'single engine approaches rather weak', although as a pilot/navigator I was considered above average! By July I had moved up to Ansons, and spent just over 40 hours on this type, finally graduating as a Sergeant Pilot.

I was posted to Halifax, Nova Scotia, to await transport overseas. When our ship came in, the *Louis Pasteur,* our commanding officer refused to let us sail in her, because of the filthy conditions. Having been a cargo ship carrying grain, she was full of bugs, and deemed unfit for human accommodation. So, we were promptly marched back down the gangplank to await further orders. Now there was the problem of what to do with us while we were waiting for another convoy.

Command, in their wisdom sent eight of us to Summerside, Prince Edward Island, to do a Reconnaissance Course, covering all types of navigation, including astro-navigation. I thoroughly enjoyed the challenge of this course and passed with flying colours.

We eventually boarded the *Athlone Castle*, and after fifteen days at sea, we arrived at Bournemouth on the south coast of England.

There was much milling about at the reception office, Command picked ten of the Canadian pilots, including myself, who had the GR ticket, to be sent to Coastal Command OTU at Turnberry. Blissfully unaware, we asked around as to where Turnberry was, only to be informed that it was considered a most dangerous posting. The general opinion of the staff at Bournemouth was that it was the worst airfield for losses during training, and the gossipmongers advised us not to go! As if we had a choice in the matter! Any refusal would have been considered LMF (lack of moral fibre). So, it was with great concern and wariness we boarded the train for Scotland.

The journey was awful and seemed to take days. We thought it was normal for this country; the trains were so slow and dirty, compared to the railways in Canada, with their sleeping compartments with fresh crisp linen on the beds. Everyone was speculating what we would find at Turnberry, the imagination running wild.

We arrived at Maidens Station, tired and hungry, on a wet and windy winter's day, and were marched to the accommodation huts with our kit. We hadn't been issued with much, so thankfully our load was quite light.

After settling in, I became part of 22 Beaufort Course, 5(C)OTU. Although the Beauforts we flew were seemingly not the safest of aircraft, I must have been very fortunate or very stupid because I enjoyed flying them just above the sea. I lost several friends in accidents at Turnberry. I was blessed with the help and advice of Squadron Leader Hughes (see page 353), who passed on to me all his experience and the 'trade secrets' of flying the Beaufort. I found the aircraft comfy to fly – everything was within easy reach and visibility was relatively good, the main drawback was that they were under powered. Entry to the Beaufort was by climbing up the walkway on the

port wing and then sliding through the top hatch into the pilot's seat. The other crew members got in through the waist hatch on the port side, and then the navigator made his way forward, past the pilot, to the navigator's compartment in the nose. The wireless operator sat behind the pilot; an armour-plated partition separated them. The air gunner climbed into his dorsal turret and checked his gun and ammo before taking up his take-off position, by the waist hatch, facing towards the rear of the plane, sitting on the closed lid of the lavatory! Before entering the cockpit, I would do a 'walk-round' and check the exterior of the kite, I never once had to abort a take-off or return to base due to mechanical problems, I must have had luck on my side. Before I climbed into the cockpit, I would make sure that the weather clips on the ailerons had been removed, the same with the elevators and rudder.

Then I always checked that the Pitot Head Cover was off, and static port for the air pressure instruments was clear, so that affected instruments would function correctly. A false reading, giving me the wrong air speed, could mean the difference between life and death.

Once in the driving seat I would check the undercarriage lever was in the down position and that the hydraulic system indicator showed 'undercarriage and bomb doors closed'.

Once I had turned on all the main switches, I would plug in my intercom and check that the rest of the crew was also plugged in, and that the system was working. Then it was on to GCP, controls and fuel - carburettor heat to cold, gills fully open, pitch fully fine, controls all moving correctly and fuel tanks full, with levers set to inner tanks ready for starting. When I was satisfied that all was ok on board, I would poke my head out of the window and shout to the waiting ground crew, 'prepare to start!'

The ground crew would prime the induction system and I would then press the electric starter to rotate each engine twice, then the chaps on the tarmac would switch on the magnetos in both engines and after a slight pause and checking that the ground crew was clear of the propellers I would yell out 'contact'.

I would switch on the main magnetos and

press the port starter button, then the starboard. With both engines running I would check all the gauges, move the flaps up and down to check that the hydraulic pump was working, then I would ensure that the propeller control governor was working, and with the brakes applied would run up each engine in turn to full throttle. The noise was incredible, and the aircraft would vibrate and rattle, the crew sat patiently throughout these checks, letting me get on with the job in hand. Finally, I would check the air pressure for the brakes and when I was satisfied that all was okay, I would give the signal to the ground crew that I was ready to taxi out. This was by raising my arms, and palms facing outwards, wave my hands crosswise above my head. This meant 'chocks away'.

Taking off in the Beaufort, which was a 'tail dragger', was an experience that stood out on its own. It was essential to keep the aircraft straight by the applying asymmetric power to the engines, this meant increasing the power to one engine to prevent yawing, until sufficient airspeed was gained to allow the tail up, then you could guide the plane by using the rudder.

We were taught to land using the three-point system, all wheels to touch down at the same time. This was a little difficult in a crosswind!

I vividly recall my first night circuit at Turnberry, Pilot Officer Shute accompanied me on the first circuits, a training session lasting one hour and ten minutes, and upon landing I was instructed to take-off and do a couple of circuits myself. I can remember turning to Sgt Weatherley who was aboard and asking him if he wished to leave the aircraft, I was concerned for his safety.

He smiled and said, 'I'm staying, let's get it over with.' I am happy to report that I brought both Beaufort 219 and Sgt Weatherley back to terra firma without incident, although I thought I'd better not push my luck and only did one circuit!

Once we had reached a certain stage in our training we were brought together as a crew, pilot, navigator and two wireless operators/air-gunners. Training continued as a team, we practised navigation, night flying and torpedo runs. My crew was Sgt McGuire, Canadian, navigator, Sgt Weatherley, W Op/AG, and Sgt Standfield, W Op/AG, who were both English. Once the whole crew became familiar with their roles, we all enjoyed the opportunities to low fly over the seas and practice dummy attacks on target ships such as the *Heliopolis*. Alternatively, we practised mock attacks using a camera to record our 'hits'. Flying at such low level had its problems, especially if the sea was calm; it was difficult to judge the horizon, essential in keeping the aircraft flying straight and level.

Another hazard we had to watch out for was, if flying an aeroplane with a wingspan of 58 feet, there wasn't much room for error when banking. Not only did you risk the wing tip catching the waves, but the stall speed of the aircraft was increased by the angle of the turn.

This was even more dangerous when formation flying, if the chap in front messed up, you were likely to be caught up in his mistake!

Squadron Leader Hughes advised me to always hold my aircraft two or three feet above the lead aeroplane when flying in a 'Vic' formation, advice I always followed! Another little trick we learned was to use the radio antennae to set our height. Altimeter readings were not always too accurate, so the wireless operator would unwind the trailing aerial, which was about sixty feet long, and when it touched the water, I would pick up static on my headset and set my altitude accordingly.

Training was not always serious, we had our light-hearted moments when boyish humour took over, my crew were bemused when I would open my window and stick my arm out, when rounding the Ailsa Craig, preparatory to landing approach. When asked what I was doing, I informed them that 'I was turning left, wasn't I?'

I had become pals with Sgt Pollock from Darvel in Ayrshire and spent many happy hours at his home; in fact, I was almost adopted as one of the family. We took great delight in 'beating up' the small town, much to the concern of the locals, especially flying down the main street, between the two church steeples which were quite close together. This was in complete breach of regulations, but what the hell, we were told to fly low, it still makes me chuckle to think that during my pilot training I was required to undergo altitude tolerance tests, to see how I could cope flying at 35,000 feet, whilst at Turnberry I was rarely over 100 feet!

Living conditions at Turnberry were horrendous, the food was awful and the general attitude of the staff almost unbearable.

There seemed to be in place an outdated hierarchy based on the class system which was not evident in Canada or any of the other 'colonies', this caused a lot of resentment amongst the aircrews. For example, we had to queue in the canteen for our food but were constantly pushed back in the line-up by commissioned RAF types who just marched up to the head of the queue and demanded service. There was a distinct discrimination between Pilot Officers and Flight Sergeants; we were treated as second class citizens, very badly indeed. The CO at the time even referred to us as 'those Canadian pigs'. He took great delight in humiliating us whenever he could, setting up mock battle exercises, having us march with full kit through smoke etc., seemingly for the amusement of his civilian friends. We became thoroughly fed up with the situation, I felt I had come over to Britain to 'do my bit', and I thought I deserved better treatment than this, so a couple of us wrote to the Canadian Embassy in London and complained. We never heard of the outcome, but there seemed to be an awful lot of inspections of the station by top brass after that!

Not being a drinker or a dancer, my social life was rather limited. We used to go to the local pubs in Kirkoswald and enjoy the chat.

I remember a pilot called Quale, an Englishman, whom I flew with on occasions. At first glance, you wouldn't think he was a pilot, he didn't seem the type. He had a receding chin and just looked totally out of place. I soon discovered what a solid man he was, the sort you could rely on, a brilliant pilot. We had wonderful times in the pub listening to his 'yarns', he was most

amusing, he used to make up ditties and was always singing and telling jokes.

Sgt Quale was to lose his life ferrying a Beaufort to India, for some reason he decided not to stop at Gibraltar, but to fly on, to Port Lyotti on the Moroccan coast. It was thought that he ran out of fuel and came down in the sea a couple of miles from land. This was a tragic loss of an excellent pilot.

Another pilot I recall was 'little Eddie', he had to have extra cushions issued because he was so short, he couldn't reach the rudder pedal!

A most memorable flight was when we were flying over Turnberry carrying out training exercises, we were contacted by radio, and told to head up toward the Holy Isle and have a look-see as to what was going on up there. Reports of towering black smoke had been received, and it was thought that either an aircraft had crashed, or a U-boat had torpedoed something. It took only minutes to reach the scene, we were greeted by the sight of a ship sinking and watched as men were jumping overboard to escape the flames of the burning vessel, even though the sea itself was on fire, the leaking fuel from the ship having ignited. We felt awful as there was nothing we could do, we circled until the ship sank beneath the waves, I watched as a man launched himself from the rail at the stern of the ship, almost eighty feet up, only to land in the burning sea. Years later I met this very chap, who had survived his ordeal, at a memorial service in Ardrossan, where a plaque was unveiled in memory of the men who lost their lives in HMS *Dasher* on 27 March 1943. I vividly remember the plume of black smoke billowing into the sky.

Ultimately, our training complete, my crew and I were ordered to pick up a 'kite' from the factory at Filton, near Bristol, with the intention of ferrying it to India. I watched as the aircraft was brought from the hangar and was flown a circuit by the test pilot. Upon landing, the pilot walked up to me and said, 'there you go son, she's all yours, take her away'. The way he spoke to me was almost challenging, insinuating that perhaps I wasn't man enough. So, with the grand total of fourteen hours and ten minutes flying time on Beauforts, I proceeded to climb aboard my new charge. As I prepared for take-off, the words of the test pilot still ringing in my ears, and aware that I was being watched, I was anxious that I didn't mess up and make a fool of myself in front of an audience. Thankfully, that didn't happen and once airborne I set course for Templeton in South Wales to undertake air testing of the aircraft and fuel consumption tests prior to setting out on the long haul to India.

During these tests we had some engine problems, so we took the aircraft to Redruth in Cornwall, and then went on leave to Scotland, while awaiting the engine replacement. Finally, we were able to restart our trip to the Middle East, so, on 23 June 1943; I took off from 1O ADU from Portreath, Cornwall, flying Beaufort DW977.

After a tour with Bomber Command, returned to Canada and spent the rest of the war as a pilot instructor at 5 OTU Abbotsford, flying Liberators and then on to 1 CFS OTU at Trenton, flying the Mitchell B25. During this period I was introduced to the pleasure of practice flights in a Harvard. I really enjoyed flying this aircraft; it was fast and responsive, in all, a joy to fly. I needed (?) lots of practice on this type, and just had to fly it almost every other day!

I made very many good friends during the war, and I kept regular contact with my navigator, Mickey McGuire, a friendship that has spanned over fifty years. There was a special relationship that developed between a 'skipper' and his crew, you relied on each other completely, and you had to.

I suppose in reflection I was very fortunate to have come through it all unscathed, you never allowed yourself to think that you might die, we were young, and we almost felt it was tempting fate to think that way.

We were sad for the pals that we lost, but we never spoke of it, talking about it wouldn't bring them back. We had no option but to get on with the job in hand. It was as though we built a protective wall around ourselves, keeping our personal feelings under guard; it was the only way we could cope with what was happening

all around us. I think me being a bit of a loner helped me through; I certainly grew up very quickly. I can't say that I had fun, no war is fun, but I was able to justify my place in society, I had fought for something that I believed in, and for better or worse, it made a man of me.

My association with the Scottish aircrews, along with the other Commonwealth lads, who were nestled in the same Nissen huts, gave me a tremendous education of the differing cultures of the world. I seemed to have an especial affinity with the Scots and was made most welcome in their homes in the nearby villages. I repeatedly returned to visit them whenever the opportunity arose, and I was able to get leave from the various stations to which I had been posted.

Some forty-three years later, I met again a Scottish lass, Madge, whom I had known during the war. We have been married now for almost eleven years and live in Ayr, overlooking the Firth of Clyde and toward Ailsa Craig. I must confess that I am very happy with my final 'posting'.

As a postscript, towards the end of the war, I was able to demonstrate to my father that I didn't need his recognition or approval. I had grown to do things that he couldn't – I was a confident man in my own right. Whilst an instructor at Abbotsford, I was able to telephone home and tell my father that I would be flying over the house and to watch out for me, and that I would 'waggle' my wings for him. So, with an RAF crew, in a Liberator, I fulfilled my boast. As I approached the house, I flew low and slow, looking out for the old man. Suddenly and unexpectedly, there he was, standing on the veranda, excitedly waving a white sheet. Was this finally the long-awaited sign of approval?

Ronald W. (Ron) Butcher
In early 1943, I was stationed at 4 Advanced Flying Unit, West Freugh, south-east of Stranraer, on a navigation course preparatory to going to an operational training unit.

Soon after take-off one night in an Anson on a cross country exercise, our radio became unserviceable. We proceeded on our briefed exercise but, at the end of the exercise we had, obviously, to descend and land. The sky was cloud covered and there had been no success in repairing the radio. Our only clue as to where we were was our dead reckoning position, which was only as good as two student navigator's computations.

It was decided to let down in a north westerly direction over the Irish Sea and hope that we were in the confines of its location. If we were correct in our assumptions, the only obstacle would be Ailsa Craig. This bore heavily on our minds as we let down through cloud. The cloud broke at about 800 feet and the first thing we saw was the coded pundit flasher at Turnberry.

I went on to finish a tour on Lancasters. The other navigator was killed on the Nuremburg raid of 30 March 1944.

Eric Cameron
For some six months in 1943, I was a young RCAF F/Sgt Wireless Operator/Air Gunner posted to RAF 1TTU as a wireless instructor. I had earlier flown a complete shipping strike tour in Malta based Wellingtons during the Islands 1942 blitz, surviving several close calls, then posted to Egypt to ferry bombers from West Africa to Cairo with 3ADU. I returned to the UK in April 1943 on the SS *Mauretania*.

I have located a 1943 daily diary in which I had at that time jotted down my impressions of that RAF station. Generally, rather negative I must say, probably influenced by the dreadfully wet, chilly summer and the loss of camaraderie one experienced on an operational squadron. The Turnberry administrative officers and NCOs seemed to view NCO aircrew as an inconvenience to be endured and all too often, harassed with nagging etc., especially the fledgling trainees. Now I can understand why the station staff personnel felt sour at being stuck at Turnberry for an indefinite period, whereas instructors and trainees moved on to other things elsewhere.

My positive and pleasurable memories of Turnberry relate to the two friendly farm families that Dennis Cooke and I (and others) became

Eric Cameron and Dennis Cooke

acquainted with that summer: the Craigs of Barlaugh farm and the Templetons of Chapelton farm. The Craig farm had Italian POWs working in the fields and I wagered that they were likely glad to be 'out of the battle' and well fed.

There were three well-bred and attractive Craig girls (Cecilia, Mary and Jenny) and Jean Templeton, a personable lass that I dated quite often. We really enjoyed the hospitality of those families who truly made us welcome. Mary Craig later married one of our wireless instructors, P/O Ralph Cope, RCAF, from Drumheller, Alberta.

When I was at Turnberry there were Beauforts, Hampdens and Wellingtons. Each aircraft type was in a separate unit with its F/Lt or Sqn/Ldr Commander, flying instructors (commissioned NCOs), wireless instructors and ground maintenance personnel. Our No. 2 Wellington Flight had its few buildings close to the seashore and was a long walk or cycle from our billets. As I don't recall being involved with the 5(C)OTU, my feeling is that the TTU operated independently and received trainees from other units to instruct the pilots in the fine art of accurately dropping a torpedo from an aircraft flying fifty feet above the water – something that required skill and good judgement, especially at night. Particularly during mid-summer 'calms', when the surface of the sea became very 'glassy' and the pilot seldom had distinct visual horizon to guide him and the altimeter was not sensitive enough. The result was the chance of flying into the sea itself.

The duty assignments for the wireless instructors usually lasted for eight hours, sometimes with a short break for a snack at the mess. However, weather changes often cut short the shift when flying was cancelled due to rain, mist, fog or turbulent winds.

If memory serves, the Wellingtons tended to concentrate on night-flying torpedo work while Beauforts and Hampdens on actual operations often flew daylight strikes. Beaufort Squadrons made many shipping strikes from Malta during my stay there. The units would arrive from Egypt for a week or so of ops and then go back due to the chronic shortage of aviation fuel, food and many other items. (During my six months on Malta in 1942 my weight dropped by some sixteen pounds because of our near starvation diet.)

The food in the Sergeants' Mess at Turnberry was far from appetising, although it was plentiful. Our Nissen huts were heated by a small iron stove, but coal or coke limited to one small bucket per week – consequently, we huddled under blankets that often felt damp from the chilly, humid air off the sea. And it was impossible to keep a uniform pressed in a presentable form. The huts had no running water, so we kept a towel, soap and shaving gear in the washroom of the mess. Scrounging fuel for the little stoves became something of a lark at night. Two aircrew would hoist another over the tall fence topped by barbed wire. He in turn would fling coal or coke over the fence to be stuffed into spare canvas kit bags or whatever was handy. The 'fuel raids' usually were done late at night and the fuel hidden under the beds, because there was an administrative S/Ldr whose mission in life

seemed to be to inspect the aircrew billets and punish anyone whose possessions were untidy. The Group Captain/Station Commander also conducted regular inspections. The officer types, of course, had batmen to look after them.

Given that we aircrew instructors were young men in the prime of life, I suppose it's not surprising that we found life at Turnberry by and large dreary and uninspiring. The frequent mess dances certainly helped to brighten things and usually were well attended by local girls.

Altogether I spent five years in the RCAF as a W Op/AG, four of which I was overseas and for most of those 'attached' to the RAF. I flew on operations with RAF 221 and 69 Squadron from Malta, RNZAF 489 Squadron from Wick and Leuchars and finally in 1944/45 with RCAF 407 Squadron from Chivenor and Wick. During those years, numerous friends and acquaintances fell by the wayside due to aircraft crashes for the most part, including several lads that I went to school with pre-war. My pilot on 489 Squadron, S/Ldr Stanley Kellow DFC, was killed a few months into 1944 when training pilots to convert from Hampdens to the more formidable Beaufighters at Langham.

The following are excerpts from my personal diary written in 1943:

April 26: Arrived Ayr 6:30 am after very long, tiring train journey from London. Then local train from Ayr to Girvan. Phoned to Turnberry for some transport, without success. Took a taxi due to having suitcases and kitbags. Later, the RAF refused to reimburse the taxi fare. Were we supposed to walk, we wondered? Weather cold and blustery. NCO mess about half a mile from our Nissen hut billet where Dennis Cooke and I share a damp, chilly room. The Wellington Flight near seashore seems about one and a half miles from the mess. First meal was not inspiring.

April 28: Caught 2 pm bus to Ayr from village. Town absolutely dead. (Because it's Wednesday, somebody said.) All we could get to eat was beans and chips. Then a beer. Found the ice-rink, which was crammed with schoolgirls trying to skate on artificial ice that had nearly an inch of water over it. Back to Turnberry.

May 1: A Hampden crashed into the sea from about 500 feet up and the four aircrew aboard killed. The wireless instructors who've been here a while before us say things are quite 'dicey'. They fly only enough in each month to qualify for their flying pay.

May 11: Explored Kirkoswald village. Drinks at the Shanter Inn and Kirkton Arms. Then to dance at the local church hall, there I met a pleasant young Scottish girl from Dumfries who is with a Land Army Unit billeted in the big hospital near the airfield. It used to be a posh hotel and the airfield a golf course that was commandeered by the RAF for an OTU and the TTU.

May 24: Took two weeks to get issued with a bicycle – a girl's model! Cycled to the RAF Regiment location for a week long 'commando' or battle training course. Managed to persuade the sergeant to let us skip all activities like route marching, except for the essential weapons lectures; Sten, Browning, Vickers and hand grenades. A couple of nights ago three Wellingtons flown by trainee crews crashed into the sea. Walking the nice beach this afternoon, Dennis and I came across wing ribs, petrol tanks and pieces of fabric. Concluded that the seabed near Turnberry must be littered with aircraft debris and human remains. Chilling thought if one is expected to fly with trainee pilots.

June 14...Was supposed to go up for a flight with F/Sgt 'Curly' Withers, ex-Malta pilot and one of the instructors. Had a couple of ops trips with him in Malta late 1942. Steady pilot. However, the assigned Wellington was not mechanically 'flyable'. This morning a Beaufort crashed into the sea and, being all-metal, sank. No survivors picked up.

June 17: Flew on an air test with F/Sgt Withers to work the wireless if need be. Airborne again after lunch, but an engine cowling came loose so we hurriedly landed and got the ground crew types to fasten it down again. These Wimpys seem to be on their last legs. Took off again and no further problems.

June 20: On a flight today an engine cowling of the Wimpy tore right off and narrowly missed slicing into the tail assembly. Pilot made a very hasty emergency landing. It wasn't the same plane as the previous incident with a loose cowling.

July 23: Sergeants' Mess dance and booze-up. Afterwards, walking Mary back to her billets in the big hospital, she asked what the inside of a Wellington looked like. There was one parked not far away so got her up the ladder under the nose and inside the kite. She was sitting in the pilot's seat when two RAF Service Policemen shouted at us to come out with our hands up! They had seen my small flashlight beam inside the aircraft and wondered if it might be saboteurs (There was a rumour going around that a Hampden's metal control rods had been partly cut with a hacksaw). Mary was terrified the SPs would charge her with trespassing and was in tears. We got a stern warning after they marched us to their nearby guardhouse.

August 6: One of the training Wellingtons had both engines conk out at about 1,000 feet but managed to get down again without damage -- an amazing bit of luck some instructor pilots said. However, a Beaufort of No. 4 Flight flew into the sea and two of our Wellingtons were sent out to look for survivors around 6:00 am. Apparently all they sighted was some wreckage, an empty dinghy and part of a tailplane. No survivors found.

August 16: Very cold tonight. Feels like November in Canada!

August 23: Dennis Cooke, Ralph Cope and I cycled up to Barlaugh, the Craig farm, where Mrs Craig insisted upon serving a very nice tea with fresh scones, home-made jam, etc. With her three daughters and Jean Templeton from down the road, we cycled to visit a pub in Kirkmichael village. Mr Craig employs several Italian POWs to work his land and they did a lot of whistling at the girls,

August 26: Dennis and I have been posted back on to operational flying, since we had responded to the Adjutant's appeal for volunteers. We're to join 489 RNZAF Squadron, Wick, flying old torpedo-Hampdens (handed over from

Bomber Command) to do shipping strikes in daylight off the Norwegian coast without fighter escort. Sounds like it could be quite dicey. Heard that Wick is 'dry' -- no pubs. The other guys think we were nuts to volunteer. But when we did, we didn't expect to be posted to a torpedo-Hampden outfit! They say there are lots of enemy fighter 'dromes near the Norway coast. It rained all day today. Will my battle dress ever get dry?

August 29: Rain let up around 3:30 pm so Dennis and cycled up to the Craig farm in 35 minutes. I diverted to the Templetons' and had a super tea with Jean and her parents, who then drove off to visit friends somewhere. As it looks like more rain, they invited me to sleep over. Late afternoon watched Jean milking the cows in the barn. Gave it a try but can't get the hang of it. She's an expert and doesn't make the cows nervous the way I seem to.

August 30: Eggs and bacon for breakfast with muffins fresh from the oven. I helped Jean to dig some potatoes. About 11:00 am Mr Templeton took us for a drive to Mauchline. Then back to Chapelton in time for lunch. Later, Dennis and I saw the three Craig girls off at the railway station as they're going to visit relatives.

August 31: Tried to coax some leave out of the adjutant before we return to ops flying with 489 Squadron, but no luck. He says that 489 badly needs 'experienced' aircrew. After the lovely food at the Craig and Templeton farms the stuff in the Turnberry NCO mess looks and tastes awful. The officers likely are getting much nicer meals. Let's hope Wick will be better.

September 1: Craig girls are back home again. Took them to the dance at Maybole. Cycled to the King's Arms by 7:30 pm, then to the town hall to meet the girls who had been driven there by Mr Craig. It was a fun evening. After the dance, Mr Craig arrived to pick up the girls and drove them home. Cycling back to Turnberry in the dark, Dave fell off his bike a few times but wasn't hurt. One of the hills on that road is quite steep and winding.

September 3: F/O Branton sent a message saying we're flying up to Wick tomorrow in a Beaufort, so Dennis and I rushed around getting 'clearance chits' signed off by various sections. At 4:00 pm we cycled up to Chapelton farm for tea and, later had supper at the Craigs. Cycled to the Kirkoswald dance and the girls came along by taxi – reserved by Mr Craig to take them home after the dance. Chatted with Mary at the dance and afterwards gave her a lift back to the hospital near the aerodrome, sitting on my bike's handlebars. She was scared going down the big hill, because the bike's brakes are not very good, especially with two bodies aboard. Sadly, it was a round of goodbyes for us as Wick is such a long distance from Turnberry.

September 5: Wick. Second day on 489 Squadron. Dennis and I have been selected for S/Ldr Kellow's crew, the 'A' Flight Commander, and an experienced pilot. Seems like a quiet, decent fellow. Pre-war, he was a journalist in London. After lunch we examined a torpedo-Hampden. The fuselage interior certainly is narrow and cramped compared to the Wellington. Armament consists of only pan-fed Vickers K guns of the type used in World War I! RAF Wick strikes me as a fairly well-organized unit and all sections seem to want to do their best to help new arrivals, unlike Turnberry which often seemed chaotic. Met some ex-Beaufort, ex-Malta fellows who are assigned to F/O Branton's crew. Billets here are steam-heated! But Dennis and I are sharing a dormitory with about two dozen other aircrew NCOs – a mix of RAF, Canadian and other Commonwealth countries. S/Ldr Kellow told us we probably could get some leave in a couple of weeks or so.

Dennis Cooke

In the spring of 1943, in the close company of my chums, Eric and Ralph, plus a few thousand tour-expired troops, I sailed in the SS *Mauretania* from Freetown, British West Africa, to Liverpool. For me it was a homecoming, but not so for my chums, Eric Cameron from Three Rivers, Quebec, and Ralph Cope, from Drumheller, Alberta. They had joined the RCAF in Canada, trained in the UK, were sent overseas for active service, and were now very far from home.

After a short leave in London we were posted

Dennis Cooke and Cecilia Craig

charms and into the arms of Cecilia, Mary and Jean who loved to dance. We forgot about any dance steps we vaguely knew and found ourselves dancing dizzily to 'The Gay Gordons' and most energetically to the 'Highland Schottische'.

Our partners Cecilia and Mary Craig and their friend Jean Templeton were daughters of local farmers, and very soon we found ourselves the happy recipients of genuine Scottish hospitality. Mrs Craig looked upon us as the sons she never had, and took a delight in seeing that we were all well fed. Her potato scones were mouth-watering, and breakfast time was a dream come true in the days when food was scarce. Mr Craig, a man of intense strength whom I first saw guiding a horse-drawn plough, would sit quietly in his comfortable chair smoking his favourite pipe. He was by nature a gentle man who kept his own wise counsel in a home where he was outnumbered by the fairer sex. Aunt Bella was a kindly soul and guardian of our morals. She accepted us 'men of action' with some reservation as companions for her nieces but would have liked us a lot better if we had been teetotallers. It was not long before loving friendships blossomed as naturally as the flowers that grew in the surrounding fields. The friendship which Ralph and Mary enjoyed indicated to everyone how admirably suited they were to each other, as later events proved.

Eric, on his off-duty days, whether it was wet or fine, would cycle the five miles to the Templeton family's farm where he and Jean spent many happy hours in each other's company.

Cecilia and I enjoyed a very close liaison, but she could not return my love since she had given her word to await the return of another RAF fellow who was pining for her in some God-forsaken outpost of the British Empire. Cecilia was an honourable young woman who wouldn't break her word. (I must admit I often hoped she would, but she didn't)

to 1 Torpedo Training Unit, RAF Turnberry, as Wireless Operator/Air Gunner instructors, following our recent action in torpedo aircraft in the Mediterranean. Trying to impart our knowledge to others was a dangerous occupation and we didn't relish the thought of 'getting the chop' in the hands of semi-skilled pilots.

To relax we sought refuge in Scotland's finest beverage and played games of snooker to pass the hours away – that was until to our delight we met three of Ayrshire's most bewitching lassies at the Maybole Town Hall Saturday night dance. We three fearless latter-day musketeers emboldened with an intake of highland dew fell for the

All the girls lived to a high moral code which was rather amusingly put at risk when one Sunday evening the six of us mounted our bikes and cycled to the only pub in Kirkoswald. The landlord thought it prudent to accommodate us in a private room away from the prying eyes of local busybodies. The girls drank lemonade, while we quaffed pints of good Scottish ale. The girls giggled a lot knowing that in future they would probably be known as the 'brazen hussies of Barlaugh'. It was harmless fun, but sadly the last occasion when we all met together.

It has been said that 'love knows no bounds', but we had dared to ignore the danger signals that

had been a chapter in our lives that we were not likely to forget.

Eric and I were transferred to RNZAF 489 Squadron at Wick and took part in operations off the coast of Norway. Early in 1944 Eric and Ralph were repatriated by ship to Canada for home leave. They were welcomed by their provincial newspapers as 'War Veterans', and by their families as heroes.

After a month of luxuriating in the love and comfort of their family home, Eric and Ralph were summoned to return to Europe, they travelled rough in a troopship and on arrival in the UK were posted to RAF squadrons stationed

Craig family - Aunt Bella, Mr Craig, Cecilia and Mrs Craig

loomed. Wartime love was always a transitory affair, especially for those sworn to serve 'King George VI, his heirs and successors'. The day came when unceremoniously and with little notice we were whisked back into the fracas of war with no assured guarantee of ever returning to those we loved. Turnberry and Barlaugh farm

in distant parts of the United Kingdom. They were elevated to officer status and flew on active service for the remainder of the war. I joined RAF 524 Squadron flying in Wellington aircraft patrolling the sea areas around Britain until May 1945 when peace was restored. Then I was posted to Burma to complete my war service as a

Signals Officer. I enjoyed the closing years of my service which gave me the opportunity to travel in Burma, India, Siam, French Indo China and Hong Kong. I sailed from Rangoon in March 1946 for Southampton and resumed my former life as a civilian.

Ralph became one of the 14,000 Canadian servicemen who fell in love and married British girls. He returned to Maybole to court and marry Mary Craig. They began their blessed married life in Drumheller, in the province of Alberta.

Eric returned to his hometown in Quebec, but he did not settle easily after his world travels. He wrote to me enclosing photos of the log-cabin home where he lived an outdoor life in a remote area of Canada. Then his life course changed suddenly and unforgettably. His travels took him to the Caribbean where he met Grace and fell in love. Eric and Grace were married and lived for a time in Montreal where their daughter Gail was born.

Jean Templeton, Mary and Cecilia Craig and Mrs Craig

In late April 1946 I returned to my family home in London and was accepted for training as a schoolteacher. This new occupation and the promise of financial security led me to think seriously about marriage. I wrote to Cecilia on the off chance that she was still a single young woman and I hoped that we could meet again.

Cecilia replied with a cheerful letter which included the news that she had married Robin, a local farmer. It was a disappointment to me, but I realised that Cecilia would be happier following the family tradition, farming was in her blood, as it had been her parents. She would have hated London had we been married and lived there. I concluded erroneously that was the end of our story. It wasn't.

In 1984 after my second marriage, I retired after thirty-eight years teaching in London's primary schools. Around ten years later I decided to write an autobiography for my family. I began with chapters of family history and concluded with my war memoirs I mentioned the kindness I met from the people of Ayrshire. I wrote with a genuine feeling of admiration for the Craig family, and my girlfriend and dancing partner Cecilia Craig.

When I heard that Margaret Morrell, resident of Ayrshire, was engaged in writing a book about RAF Turnberry as a tribute to the men who served there, I wrote to her telling of my time there.

One day Margaret telephoned me with the incredible news that she had traced Cecilia Craig who was now Mrs Wilson, a widow and living in Kilwinning, Ayrshire. At first, I found it hard to believe. My instant reaction was to telephone Cecilia, but on second thoughts I considered a letter would be a much gentler approach. I wrote immediately, enclosing photographs of some happy occasions we spent at Barlaugh farm, plus a recent photo of myself.

Cecilia by this time had met with Margaret Morrell and knew of my existence. She replied to my letter saying she was delighted to hear from me again and ended with an invitation to me to visit her if ever I was in Scotland so that we could talk about old times.

It was a joy to receive her letter so long after our last correspondence. Having already broken the ice I decided to telephone her. It took a while to overcome our feelings of surprise at meeting again after sixty-two years, then we began conversing easily with each other as though it was still 1943. In that respect nothing had changed. It was a joy to hear Cecilia's voice again, which sounded exactly like her mother speaking. I sent Cecilia a bouquet of flowers to mark the occasion and shortly afterwards telephoned again to enquire about the events which had happened to us since 1943.

Cecilia told me she had been to Canada to be reunited with Mary and Ralph. Sadly, Ralph had since died following a sudden heart attack. Cecilia was grieved to hear about Eric who had suffered a severe stroke which left him needing constant nursing care. I asked Cecilia about Eric's friend Jean – she thought the Templeton family had moved away from Maybole. We talked about our families, our grandchildren and the depressed state of farming etc. We could have gone on talking for hour after hour.

Celia sent me a photograph of herself taken by her daughter in the garden. I showed the picture to my young wife Phyllis who thinks it is quaintly amusing that an old man in his eighties has been reconnected with his wartime sweetheart.

I am deeply moved by this extraordinary and lovely event which has happened in my departing years. It has confirmed my belief beyond doubt that real friendship and true love, which differs in many ways, are both timeless. And timeless means forever.

Bill Lucas

At the outbreak of war, I was an engineering student at University College London, and immediately volunteered to join the RAF.

I was at 1TTU at Turnberry from May 1943 until August 1943, at that time Flying Officer and pilot, having just converted from Hudsons to Beauforts at Long Kesh in Northern Ireland. I found the Beaufort, both Mk I and Mk II, nice to fly. My navigator was Sid Davies from South Wales and the W Op/AGs were Canadians Keith Southwood and Johnny Bunce.

I became good friends with a chap called Raymond Rootes – he was later killed on his second operational flight when his aircraft hit a 'slice' cable that the Japs had stretched across the Irrawaddy just north of Rangoon in Burma.

At Turnberry, we practised torpedo dropping, singly and later in formation with other aircraft, a Channel Island steamer, the *Isle of Sark*, was used as a target vessel. We had to work on ship recognition using 'Jane's Fighting Ships' as a textbook. The pilots also had to use a Torpedo Attack Trainer, which was a form of 'Link

Trainer' set in the centre of a sphere. An image of a warship was projected onto the inside of the sphere, and it could be made to take avoiding action. When the torpedo was dropped, everything stopped at that point, and a beam of light indicated the likely path of the torpedo. Today, I suppose it would be called a flight simulator.

We also took part in joint exercises with the Royal Navy who were flying Swordfish aircraft.

Whilst I was at Turnberry, I believe that there was something very hush-hush going on with the Dambuster Squadron, who were lodged in hangars on the perimeter of the airfield. We were not allowed near the hangars or aircraft. I think it was something to do with the bouncing bomb.

I recall that on one training exercise, one of the aircrews were late, so we took their place. By the time we returned, the other crew had turned up and took over the aircraft. Shortly after they had taken off, the aircraft nose-dived from about three hundred feet, straight into the Clyde, just a short way from the lighthouse. There were no survivors.

Bill Lucas

A memorable trip was when I was flying a Beaufort on a navigation exercise near the islands of St Kilda in the Outer Hebrides, when some electrical gadget in the rear of the aircraft caught fire. Johnny Bunce used the fire extinguisher, but the liquid evaporated, and the fire kept burning. So, in desperation, he grabbed the emergency axe and hacked through the cables connected to the problem gadget, yanked the offending part from its moorings and chucked it overboard. This left us with no electrical systems, but the fire was out, and we managed to get back to Turnberry safely.

On Saturday nights we would go to the dance in the village hall in Kirkoswald, after 'tanking up' at the 'Shanter Inn'. I remember that you couldn't buy a drink there on a Sunday, unless you had a meal.

Finally, the day came when we were sent to Stranraer for a Yellow Fever injection. This gave us a clue as to where we were going. The medical orderly there was a lady, I think some chap must have treated her badly, because she really seemed to enjoy jabbing that needle in!

Bill Lucas

Bill Lucas course

Bill Lucas and crew

After our time at Turnberry, most of us were sent to the factory at Filton to collect a new Beaufort. This we flew to South Wales to have radar fitted, then on to Maghabbery in Northern Ireland where we undertook compass tests and fuel consumption tests before flying to Portreath in Cornwall. From here we left for India, arriving at Karachi, where we left the aircraft (JM508) before reporting to an aircrew pool at Poona to await further orders. After Poona I filled in time at 319MU at Jodhpur, test flying aircraft after various overhauls.

I flew many types and always thought the Beaufort compared quite favourably. However, it was slow, and I don't think I would have liked to fly it so much on 'ops'.

In 1945 I was in Calcutta, having delivered a Mosquito there for an overhaul, when I contracted polio and had to be sent home. I was discharged and then finished my studies at Manchester, taking a degree in Chemical Engineering.

Ian Masson
I started the war in the army but found life rather dull and uninteresting, so I transferred to the RAF as aircrew in 1941. I did the EFTS course in the UK and then was posted to Canada for further training. This consisted of SFTS in Alberta, General Reconnaissance course at Charlottetown, Prince Edward Island, the OTU at Patricia Bay, Vancouver Island. On completion of this I returned to the UK in April/May 1943 and was sent on the 12th course to the TTU at Turnberry. My crew were P/O John Skidmore RAF, Navigator, P/O 'Snowy' Lennox, RCAF, W Op/AG and P/O 'Ty' Tyler RCAF, W Op/AG.

The Chief Instructor at Turnberry was W/C G. M. Lindeman DFC, who was an Australian serving with the RAF. Prior to taking up this post he had been the CO of 455 RAAF Squadron and was largely responsible for the development of torpedo tactics which became the RAF standard.

The Officers' Mess was at Maidens and we were billeted in quarters on the hill behind the village. I remember the wonderful views from there over the Firth of Clyde. Turnberry Hotel had been taken over as a hospital. The doctor in charge at that time was Capt. Alan Brown, who had been a contemporary of mine at school (George Watson's, Edinburgh) and I paid regular visits to their mess and the hotel swimming pool.

The object of the course was to train aircrew in flying low-level formation and the tactical use of torpedoes. I recall that the following ships acted as targets to assist our training: *Isle of Guernsey, Malanne, Isle of Sark and the Heliopolis.*

The aircraft we were to train on was the Hampden Torpedo Bomber. This aircraft had its plus and minus points. The main bad feature of the Hampden was its lack of performance. As a torpedo bomber, ships had to be located and attacked in daylight and its slow speed and poor defensive armament made the aircraft very vulnerable, particularly as operations were often beyond the reach of fighter escorts. The propellers were non-feathering and in the event of an engine failure, height could not be maintained. Defensive armament consisted of a single rear firing gun. The second rear gun position could not be used in the TB conversion because the torpedo extended into the gun position. Another problem was that certain manoeuvres would cause the rudders to lock, resulting in what was referred to as a 'stabilised yaw'. This was a form of spin and could not be recovered from at low-level.

In its favour the aircraft had the following good points – it was pleasant to fly. The pilot was centrally placed and by sliding back the canopy he had all the advantages of an open cockpit. For its size the Hampden was reasonably manoeuvrable.

I can recall several Hampden accidents at Turnberry, most were the result of inexperience with low flying and with formation flying. Instruction given to pilots at the start of their course was to 'always fly at 50 feet or below', only to rise above when joining the circuit to land. I seem to remember the short runway caused some embarrassment to the inexperienced at times as it caught them unawares.

There was a navigator on our course of

Ian Masson and crew

Russian extraction who constantly quoted extracts from Omar Kyan, regrettably he was later killed in an air accident.

At the end of my training at Turnberry I was posted, along with my crew, to 455 (RAAF) Squadron at RAF Leuchars. We operated on Hampdens from there (also Tain and Wick) until the end of 1943 when the squadron converted to Beaufighters. In 1944 we were posted to RAF Langham in Norfolk where 455 joined 489 RNZAF to form the ANZAC wing. This operated along the European coastline on anti-shipping strikes and also covered the D-Day landings.

I was tour expired in June 1944 and was then posted to the OTU at RAF East Fortune, instructing on Beaufighters. Subsequent postings included the RAF College at Cranwell and the Central Gunnery School at RAF Catfoss where I flew Beaufighters and Mosquitos.

Fred Eyre
I was stationed at Turnberry from 15 May until 8 June 1943, one of a Wellington crew undergoing training for night torpedo attacking. The weather at that time is the first memory I recall: it was beautiful, that and the scenery, I have never forgotten. The second is that we lost three aircraft and crews during the time I was there, the fine weather being the primary cause. The sea was like a millpond and judging 60-80 feet above the sea on a moonlit night, doing 120-140 knots asked a lot from the pilots. The Royal Navy ships that we used to exercise with, heard an aircraft approaching, then a thud and all was quiet, this happened twice.

Although I had been at other stations where accidents has occurred, Cottesmore comes to mind where Whitley's and Hampden's used to fall out of the sky regularly, here at Turnberry

Mr and Mrs Fred Eyre

the W Op in one of the Wellingtons that crashed had come through training with me, I found that hard to come to terms with, that he had gone so quickly forever.

I know for many years people didn't believe me when I recalled these accidents, the weather being so perfect, it's hard to take in. But the sea was like a sheet of glass, probably causing the misjudgement. No computer operated instruments in those days. I have tried hard to remember the name of the lad who died in one of the Wellingtons, but no, the name evades me, yet I can recall him as plain as day, a little chap with a Geordie accent. He was a bit of a novelty to me, we didn't have much dealings with 'up country' people in those days, nobody strayed far from their roots.

Tony Spooner
From early 1943, until September 1943, I was Flight Commander of the torpedo-carrying Wellington Flight at 1TTU, Turnberry. We trained pilots, most of whom came from 7

OTU Limavady, to drop torpedoes at night from Wellington VIIIs. These were ordinary Wellington Is (Bristol Pegasus Engines) with ASV added. No real torpedoes were used, but once or twice each crew was allowed to drop a real but unarmed torpedo which was recoverable. I held the rank of Squadron Leader. I have many pleasant memories of Turnberry. The station commander was Group Captain Elton and his deputy was RAAF Wing Commander Lindeman of the Australian wine producing family.

I regard 1TTU at Turnberry as the most efficient RAF Station which I served throughout the war (and I flew from the first day, 3 September 1939, until the very last one and was in the air when news of the Japanese surrender was received.)

From my point of view, as a S/Leader Instructor in charge of the Wellington (Night) Torpedo Training, at an airfield far distanced from any possible enemy interruption and living out of camp at a seaside resort with a young wife, Turnberry was 'all gold'. Those above me at Turnberry never once bothered me and I was free to run my own show as I liked it. I also had with me a faithful co-pilot who had appeared 'from nowhere' when in Malta 1941, my long-standing one was killed. I had later helped Norman Lightowler to get commissioned and he was, as usual, a tower of unobtrusive support, as were the several other instructors who comprised our torpedo dropping flight. We seemed a happy team. The pupils who had come to learn to drop torpedoes at night had otherwise completed all their training. At that stage of the war, this usually meant at least one year and a period in Canada, USA, South Africa or Rhodesia. They came with their crews with which they had already formed a happy team. A Wellington crew comprised a skipper (either a Sergeant or a P/O), a co-pilot, an observer who was the navigator and three W Op/AGs. These were persons, virtually all Sergeants, who had been trained as wireless operators and air gunners and, only in Coastal Command, of which we were all a part, had to operate our early airborne radar. This was ASV (Air to Surface Vessel) Mk II, crude but in trained hands, fairly effective. With ASV, given relatively calm seas, a good operator could detect a ship at sea at about five miles and follow it down to about one mile, whereupon the 'blip' became lost in the sea-return clutter which was due to radar picking up the waves. It was therefore useful for locating ships at night, but not effective for accurate torpedo attacks as these required the aircraft to drop their 'fish'

Tony Spooner (third from left)

at a range of about 600-800 yards if they were to have the least chance of a hit. (A torpedo dropped from the air, immediately dives deep, then surfaces and porpoises for approximately 200-400 yards before settling down to a steady track at the prescribed depth of about 20 feet (or so) below the surface. Drops within say 400-500 yards are thus unlikely therefore to score a hit). Beyond 800 yards, the time that a torpedo, which travels at about 40 knots, takes to reach its target is generally so long that the moving ship will probably have either seen its track in the water (it released bubbles as it went along) and taken avoiding action or have gone ahead of the danger.)

Later radar marks (1943 onwards) were much more efficient with a PPI, (Planned Position Indicator) and a Centimetric frequency (5 cm or 10 cm) which, to a large degree, cut out the sea-return clutter. With ASV Mk II, it was necessary, once having found a 'blip' which looked as if it might be a ship (it might also be a whale, a friendly ship, a life-boat full of survivors, a rock, a sandbank, a U-Boat etc.) to turn the aircraft, as directed by the W Op/AG on the ASV set, until the aircraft is lined up with the 'blip' dead ahead. With a PPI an aerial picture of the scene was always on the screen, with the aircraft at the centre of the circle.

As the pupil crews went from my flight straight to an operational squadron, they were as keen as mustard and had by then become a disciplined crew; not that the Canadians were ever truly disciplined (in contrast to the UK and most RAAF and all RNZAF crews.) Up to their arrival at Turnberry, it had been a court martial offence to indulge in low-flying. This is always fraught with danger, but pupils at times felt the urge to let off steam, or to show off, by beating up their own or friends' houses or airfields in low-flying passes. If caught, the pilot was 'out'!

At Turnberry, by contrast, we had to teach the pilots to become thoroughly familiar with low-flying and to comprehend its dangers. As soon as they were airborne with wheels and flaps retracted, day or night (we obviously taught them to fly low by day first), they had to drop down to 100 feet or less and stay there. With the large wingspan of a Wimpy, turns at low level placed the inside wingtip perilously close to the sea.

Two things led to what I regard as our ability to turn out effective night torpedo bomber crews – the splendid co-operation of the navy allied to the amazing Torpedo Attack Trainer (TAT). The former loaned us a light cruiser, HMS *Cardiff*, which was fitted with 'black', i.e. infra-red lights, which gave away its length and height. As, whenever we pressed the bomb-tit to drop a theoretical torpedo we automatically took a camera shot, this meant that, judged by the 'black-lights' on the cruiser, which showed up on our film, we could calculate the distance from the ship at the moment of release and by comparing the cruiser length (in the film) to its height, we could also calculate the angle between the aircraft's track and the cruiser. The aircraft's exact height and heading also appeared on the film, along with its speed. This enabled us to check that the drop was being correctly made. (From a 48-year-old memory the height had to be 60 feet, the distance from target 800 yards and the aircraft speed 140 knots). All the parameters taken by the camera were fed into a machine, which in a large dome, akin to a planetarium, instructors and pupils could later watch the ship's progress (its track and speed curve were also added to the equation), the torpedo drop, its initial dive and gyrations, the its steady run towards the target (in a dotted line) and to see whether or not it had (in theory) hit the moving cruiser. This was the fascinating, very advanced TAT.

We did not have, but much needed, low level radio altimeters, only the pressure sensitive ones which were marked in hundreds of feet and lacked the detail we required. Low-flying, therefore, had to be judged solely by the pilot's eye. This was okay for as long as the sea had visible ripples, waves, white caps etc., but in a still calm sea it was deceptive. The sea looked the same from 500 feet to 0 feet.

This led to our only losses. One night when I had left Norman Lightowler in charge, a rare, dead-flat calm persisted. Consequently, one of the 'Wimpys', flown by a crew towards the end of a course, flew into the water. Trying to be helpful, when it became known that an aircraft was 'in the drink', the other crews put on their landing lights, flew low and looked for survivors. As a result, due to the flat and deceptive calm sea, providing no definition of height, two other aircraft also flew into the sea. Of the eighteen men aboard the three aircraft, only seven were picked up that night or next morning.

It speaks well of the high morale, that the next morning the skipper of the aircraft which had initially flown into the sea, came to report. He had spent about six hours in the freezing water, been fished out early that morning near the Mull of Kintyre, had found a flight back to Turnberry from Machrihanish, and after discussing the night's losses with the other survivors, had realised that between them they could make up a complete crew. 'Would it be alright, sir, if we formed up as a crew and finished the course?' He had already obtained the consent of the survivors from the other lost crews. What could I do in the face of such determination but say yes.

On another course, one pilot fell ill and missed about 75% of the exercises. His name was Del Rosso, of Italian extraction. He begged to be allowed to 'fly all night and day, if need be' to catch up, rather than be put back to the next course (about six weeks). Somehow, we managed it. The pleasant outcome of this was that he let me know that his father was the manager of the restaurant of the Cumberland Hotel in London. Accordingly, when my wife and I next had the chance to visit London we went there and asked for Manager Del Rosso, when he appeared, and we told him who we were and of his son's determination to get into the war, we enjoyed the best 'rationed' meal possible!

A curious incident was when, for a punishment, I detailed a pilot to spend the morning keeping up to date the large coloured wall charts, which I had devised so that, at a glance, I could see how each crew were progressing. To my annoyance, since these charts were my pride and joy and did away with the need to keep any files etc., the colour coding (self-evident) had been mixed up. As this pupil was normally a conscientious one (in the air), I was puzzled as well as annoyed. I did not think that he was the type that would purposely rile me. To my surprise and horror, it dawned on me that he was colour-blind! Somehow, he had got through the year-long previous courses un-detected. I felt obliged, as he had five other lives to preserve, to report my findings and was pleased that the outcome was that he would not be grounded but would be switched to train on single-engined fighters, where he would only have his own life in his hands.

One nice surprise for me was to discover that 'across the water' at Machrihanish, the Fleet Air Arm was also training pupils to drop torpedoes: in their case from Barracudas, Albacores and Swordfish. Their CO was Lt Cdr Edgar Bibby RNVR. Now, Bibby like myself (RAVR), had been a pre-war civilian pilot. I had been a (junior) flying-instructor at the Liverpool and District Flying Club at Speke. Bibby, of the well-known Liverpool corn and cattle cake family, was the best private pilot club member with his own small plane. We became friends and to my joy, he turned up in Malta 1941 in the FAA as one of the leading pilots of 828 Squadron, flying the wonderful obsolete Swordfish torpedo bomber. Both of us, because we could fly well, thanks to our pre-war experiences, had been rated 'exceptional' as pilots and had received rapid promotion in our respective services. (I was only commissioned in about July 1940). Neither Edgar nor I knew much or cared over much about regular RAF/RN established (peacetime) procedures. When Edgar was being frustrated about the inability to obtain some vital spare parts, he flew himself to London and personally called upon the First Sea Lord! As this High Admiral, in charge of all FAA, was known to Edgar's family, they ended up dining together - and Edgar promptly got his spare parts.

In my case, when similarly frustrated, and after having filled in (in triplicate of course) the necessary RAF requisition forms with no results, I personally flew to a maintenance unit and arrived back with the parts required, much to the annoyance of our station engineer officer. He was an RAF regular.

This attitude paid off handsomely, when lamenting that our ASV was inadequate in that its shortest and most sensitive range scale, was one which went from 8-10 miles to zero (others went from 100 miles to zero and 40 miles to zero), a bright Canadian Corporal (virtually all ASV mechanics were Canadians as they had ransacked their technical schools and universities for such persons) came and said that he could modify our ASV sets with an additional 4th range scale which would read in hundreds (or less) yards! This was music to my ears as our ASV operators, using this short-range scale, could now easily report to the pilots when they were approaching, and when actually at, 800 yards, which was our aim. We promptly modified all our sets and thereafter TAT showed that we were dropping our 'theoretical' fish at a precise 800 yards.

Realising that the modification was of great value to many others, I reported what we had done, only to be ordered by some distant HQ, to put the sets back to what they had been as there was an added risk of 'fire and explosion'. I failed to see how monkeying around with valves and condensers, as my corporal had done, could add to such risks – I paid no attention to the order. Later I was delighted to be able to get my RCAF corporal a well-deserved BEM (or was it an MBE?). As I now remember it, his name was Few.

Living out in a bungalow at nearby Girvan - slap opposite the Ailsa Craig, which I could see from my window – was a great joy as:

1. We were introduced to a local poacher who had been taking salmon from a local river for 'donkeys' years. His wife used to deliver to us about once a week, a whole fish. She brought it round in the bottom of the baby's pram. Whether this added to the flavour or not we were never too sure, but I have never tasted better salmon since.

2. Nearby was Culzean Castle and I got to know its gardener. As my father was an excellent (amateur) fruit grower, I was able to demonstrate to him that I knew how and when to pick fruit. As a result, I was given free rein of the enormous

walled kitchen garden. About once a week, I came home loaded with fruits of the season – figs, peaches, nectarines, pears, apples, various berries etc., plus superb vegetables, as my wife was sharing the kitchen with the bungalow's sole owner/occupant, an elderly lady. 'How much?' I would ask my gardener friend, 'Would five shillings be too much, Sir?' was always the reply. Remember, this was in food starved Britain of 1942-43!

There was also a double summertime (supposed to be good for the cows!) This meant that at Ayrshire latitudes, in mid-summer (1943) it was not dark until really late. We had teamed up with an Admin Officer and his wife, the Beatties, and the four of us would feast on salmon and fruit before repairing to the garden, where in a still evening we could play bridge up to midnight without lights!

Although seldom in the Officers' Mess, except for lunch, I attended a party which was held in a converted hut (all buildings at RAF Turnberry were wartime Nissen or other ugly huts. The magnificent Turnberry Hotel was merely used as a hospital. Shame!) to celebrate the fact that the Officers' Mess had been repainted and 'tarted up' to look slightly more comfortable, unfortunately, someone who shall be nameless, (not me), introduced a mysterious, open at the top, large paper bag. The curious were invited to peer inside. Whereupon the bottom of the bag would be smartly tapped, and the soot inside would rise up and adorn the face of the viewer! In no time, with the bag of soot being fought over, the Officers' Mess was much in need of another redecoration!

Perhaps the one sadness about the Wellington Torpedo Flight at Turnberry, was that the idea of using the Wellington to drop torpedoes (it carried two), came too late in the war, and, except for one or two crews from my first and second courses, the well-trained happy crews arrived in the Mediterranean (to where all were destined) too late to be as effective as they might have been, as by mid-1943, the Med, thanks to the Battle of Alamein at one end and the successful 'Husky' Operations at the other, had been almost

swept clear of enemy shipping. But, F/O 'Hank' Donkersley, a RCAF pilot, sank several enemy ships and earned himself a rapid DFC and bar before the enemy finally surrendered in Tunisia. The idea of dropping torpedoes from 'Wimpys' had been conceived by 38 Squadron in the desert quite early in 1942 and, after my ASV equipped special duties flight at Malta, I had written to my AOC Malta and thence to London to suggest that their torpedoes and my ASV be combined in one aircraft. This was done, and the torpedo Wellingtons then became 'Fishingtons'!

As Turnberry also had flights which trained Beaufort and Hampden crews to drop torpedoes, as well as Beaufighters, the idea, (probably Bibby's), of planning massed attacks was born. As Bibby from Machrihanish could contribute Barracudas and Albacores, by the time that the giant battleship, HMS *Nelson*, came steaming up the Clyde, we managed between us to arrange for it to be attacked, happily in daylight, by no less than 79 aircraft simultaneously! As Beauforts, Hampdens and my 'Wimpys' all dropped from 40 feet, as also did the aircraft of the FAA, the chaos of the simultaneous attack can be imagined! I rate it my (I led the six or seven Wellingtons) most dangerous five minutes of a hectic war. By then we had adopted a Zulu formation – in this we flew in a vast semi-circle in fluid pairs (I could raise six or seven of these) which allowed the target no chance of being able to feather i.e. turn into or away from a sighted torpedo track , as to do so, exposed the ship to others of the formation. How no aircraft collided remains a (thankful) mystery!

Brian H. Shelley
I was in the Terriers (Territorial Army) before the war, Duke of Lancaster's Own Yeomanry, mounted cavalry. I was very badly injured from a horse kick in November 1939 and subsequently graded to 'B2'. In early 1941 the RAF asked for volunteers for aircrew training and I got our Regimental MO to upgrade me, and I was then accepted for pilot training at RAF, HQ in Belfast. My home is Manchester, but I married a girl from Winnipeg during my training with the BCATP so

Brian H. Shelley

have lived in Toronto for over 50 years.

I attended 28B Torpedo Course at Turnberry from 8 to 26 August 1943, as second pilot in the Wellington crew, captained by F/O Jock Nicol. Also, on our course was F/O Jimmy Edwards who later became a well-known comedian of radio and stage, who even in those days entertained us with his trombone playing!

We were trained as night torpedo bombers, operating in threes, one high up to drop flares over the target for the other two to come in from port and starboard and drop torpedoes from 50 feet above the water at about, if I remember correctly, 500 yards from the target. Sounds simple, but, at night over a calm sea, with no accurate altimeters at that height, it was very easy to go straight into the water and be wiped out, as many were during their training. We had Beauforts and Hampdens who operated by day and who also lost quite a few aircraft during their courses.

There was good morale among students, and I remember several great parties on evenings when we were not flying. Our living quarters

were down by the seashore, and I remember one or two crew members at times were brought home on planks being unable to navigate for themselves. I remember going ice-skating at a rink in Ayr, and on a weekend in Glasgow we found a pub with three floors of bars!

Our crew consisted of two pilots, a navigator and three W Op/AGs, one on the radio, one in the rear turret and one on the SE (a very primitive form of Radar to find our targets). We also spent a fair amount of time on a contraption which taught us to fire a torpedo at a moving ship, judging the correct offset for target speed. You were shown what happened with each shot, mostly misses! We ended the course by dropping two 'runners' at the target ship. This ship sailed all day and all night, up and down the Firth of Clyde, from the Island of Arran, being attacked by Hampdens, Beauforts and Wellingtons.

Our crew were all RAF, but most others had at least one either Canadian, Australian or New Zealander on board, so we were a mixed lot, but all young and full of fun and playing tricks on each other! We had been crewed up at 7OTU at Limavady in Northern Ireland and after Turnberry we were given new 'Wimpies' and sent out to the Middle East. We ended up at Cairo, where they took our new aircraft and told us that we were no longer required in the Mediterranean! We spent three months in a transit camp in the desert outside Cairo, before finally going back to the UK by ship to train on Dakotas for the 'D' Day op.

I joined 271 Squadron in February 1944 and was promoted to Captain in March. We did the D Day op. on 6 June, at night, resupplying the paratroopers which had been dropped the night before. Then for the next three months we flew to into France with supplies and returning each trip with eighteen stretcher cases and three sitting wounded. Just before the Arnhem op. we were posted to 194 Squadron stationed at Argatarla in Assam, to support the 14th (or forgotten) Army in the Burma Campaign. In June 1945 I finished my tour of 700 hours and flew Expeditors carrying VIPs and on scheduled services until my release in November 1945.

Lawrence Evans

Lawrence Evans

Although I now live in Cornwall, I was born in the delightful county of Surrey, in the small village of Dulwich, which has now been absorbed by London. My grandfather was the village blacksmith until he sold out in 1919 when the gentry and tradespeople sold their horses with the advent of motorised transport.

As an air-crazy schoolboy, I can recall cycling to Biggin Hill Aerodrome in Kent to visit one of the Empire Air Days, which were begun in the late thirties to encourage interest in the RAF during the 'expansion period', when lost time was to be recovered once it was realised that Herr Hitler did not mean to use all of his lovely new aircraft for Lufthansa Airways! At Biggin Hill I sat in the cockpit of a Gloucester Gladiator fighter and remember thinking that I would never

be able to get it off the ground, let alone fight the enemy in it. I was filled with awe at the skills of the RAF pilots in their planes.

War began when I was eighteen, and all men over the age of twenty-one were to be enlisted in one of the services. But I didn't intend to wait to be called-up, to be placed in the PBI, (poor bloody infantry). So, I discussed the situation with my cousin, who had been brought up in the same school as myself, and after much deliberation we decided that we would 'try' for the RAF.

I joined in spring 1940 at Cardington, Bedfordshire, the site of the giant hangars that housed the pre-war airships, R100 and R101. The medical board were satisfied with my physical condition, and despite the interview board's efforts to change my mind, I elected to train as a W Op/AG (aircrew) and not as an observer, as my mathematics weren't too brilliant. My introduction to service life was 'square-bashing' on the prom at Blackpool. We u/t aircrew were temporarily sent to operational 'dromes to carry out typical erk's duties, (night guards on 'drome perimeters, cookhouse 'essentials', etc.). We subsequently went through 10 (Signals) School at Blackpool, and then to 2 (Signals) School at Yatesbury, Wiltshire, where one was introduced to flying, on W/T exercises, in Percival Proctors and De Havilland Dominies. In my personal case, when we should have gone on to Gunnery School, it was revealed that because all GSs were fully booked with 'common or garden' air gunners (i.e. without wireless training) for Bomber Command. I fancy that someone in authority in Coastal Command put his foot down and decided that 'Bomber Command be blowed' - and arrangements were made that proposed crews for Coastal Command had better be sent to OTUs direct, for completion of training. I was therefore, sent to 3(C)OTU at Chivenor for a gunnery course, and then to be crewed with an observer and a pilot, and thenceforth trained as a crew on the aircraft stationed at that unit - in my case, the Coastal Command's Bristol Beaufort torpedo bombers, in preparation for the full OTU course. This was a perfectly delightful situation

- on the North Devon coast at the confluence of the Taw and Torridge rivers, opposite Appledore, and close by the shore at Woolacombe Sands, Saunton Sands, Braunton Burroughs, Mortehoe, Ilfracombe, etc. Who could ask for more, in the Summer too! After passing-out as a crew, we went on to the TTU, then at Abbotsinch, and subsequently to North Coates Fitties 'drome - 86 Squadron.

The Beaufort squadrons were frankly admitted, officially, to have the highest rate of losses in action of any RAF aircraft. (As an ex-Beaufort crewman, I have always contended that more losses occurred to Beauforts on training units, than ever happened on operational squadrons in flying against the enemy.) As for flying the Beaufort, what can I comment? When loaded, it had a tail down attitude as if it was 'hanging on the props', as we used to say. The complete tail unit, (when viewed from the turret), used to vibrate from the vertical - at least three to four inches each side of the vertical axis. I once watched the complete tail empennage of an accompanying Beaufort (AW350) sliced of by the propellers of an escorting Beaufighter, which has exceeded his duty as escort cover. The Beaufighter managed to return to Malta, but the Beaufort didn't!

The aircraft had an Elsan toilet, which was where we sat, on the closed seat flap, to have a crafty cigarette on occasion - and to flick the ash out of the open access hatch. It also came in very useful on really long flights! The Beaufort was not an easy aircraft to handle, although many pilots grew to like them, once they had overcome all the idiosyncratic habits which the aircraft possessed. Like all Bristol-made aeroplanes, it was all metal and therefore a very strongly constructed machine, capable of absorbing a great deal of punishment. Many are the stories of 'narrow squeaks' of all varieties, told by RAF aircrews about flights in Beauforts. Our crew managed to reach Malta, despite flying on only one engine after the other had been put out of action by cannon fire, during a torpedo strike on an Axis Convoy. The main problem that Skipper Marshall had on this occasion, was in gaining enough height with only one engine, to land at Luqa, where the runway was roughly at 300-400 feet, and the aircraft was almost below sea level! (I jest).

I survived a total of seven crashes, including a forced landing in Staffordshire with wheels down and which occurred due to lack of fuel after a seven-hour consumption test, whilst air-testing the Beaufort we were later to fly out to Egypt. This was on a day when the weather had deteriorated into a complete 'clag': cloud base was down to about two hundred feet all over England and Wales. Add to the scenario a u/s radio set, and you have the ingredients of a prang!

I arrived in Egypt in January 1942, in a Mk I Beaufort, Serial No. N1182, flown by Sgt R. Marshall, RAAF, and crewed by Observer Sgt T. Thompson, RNZAF; W Op/AGs Sgt B. Doggett and myself. We flew from the United Kingdom via Gibraltar and Malta, and by night - through some horrendous electrical storms, which hadn't been forecast by the Met men! (as usual).

We were posted to 39 Squadron, who were struggling to build up their numbers of aircraft, having disposed of their Martin Marylands in 1941. The Squadron's first successful operation took place on January 23, when a large convoy of ships was sighted en route to Tripoli. It included the MV *Victoria*, known as ' the pearl of the Italian fleet'. A strike was set up from Benghazi, which included three Beauforts from 39 Squadron. These aircraft dropped torpedoes which crippled the *Victoria*, leaving her to be finished-off by some Albacores of the Fleet Air Arm.

Rommel's Afrika Korps began their 'big push' soon after this, and the most advanced landing ground then available to the RAF was Sidi Barrani. Our crew was sent up to stand by for a strike, with a take-off planned before dawn on March 26. Our aircraft, N1170, crashed soon after take-off, with an engine failure. The aircraft was wrecked, but fortunately it didn't catch fire, and the torpedo thankfully remained inactive!

It was about this time that the 'wizardry' of ASV was suddenly introduced to us poor

W Ops, when the Mk II Beauforts began to arrive in the desert. It was all rather 'hush-hush', and instructors appeared as if by magic to demonstrate this new contraption. How the ground-staff laddies were taught to effect repairs and find faults never became apparent. In any event, I believed it was never a great help to us as torpedo bomber crews, since we always flew at such low heights over the sea, that the limited range of this so-called aid was non-helpful. I simply left it to the other W Op/AG on the crew! In any case, the pesky contraption was always going wrong!

My introduction to torpedo strikes did not occur until mid-June. Again, we were sent forward to Sidi Barrani, with torpedoes loaded. The squadron had managed to get about fourteen aircraft serviceable and the 'grapevine' promised a maximum strike effort against the Italian naval fleet, which was expected to sail from Taranto to attack an Allied convoy, code named 'Vigorous'; and which had sailed from Alexandria on a relief mission to attempt to supply Malta. The Italian battle fleet left Taranto on June 14. The next day at 06:15 hours, twelve Beauforts of 39 Squadron took-off and set course to intercept the enemy ships, which were by this time about 205 miles NNW of Benghazi. We flew in four 'vics' of three, but because the distances involved were too great for the squadron to return to Egypt, the plan was to fly on to Malta after engaging the enemy. On the way to intercept the Italian fleet, we were suddenly under attack from ME 109s. We were hopelessly outclassed, and even though our pilot made a valiant effort to evade our attackers, we were hit. Marshall lost rudder controls and hydraulics, and we were forced to drop the torpedo too early. Our navigator sustained shrapnel wounds to his ankles, and Skipper Marshall was forced to concentrate 100% on simply keeping the aircraft airborne. Despite wallowing alarmingly, a course to Malta was set and maintained. We arrived over Luqa to find the airfield free from any attack by the Luftwaffe, which at this period of the three-year siege of Malta, went on throughout each twenty-four-hour period! Marshall succeeded in lowering the under-carriage by use of the emergency cartridges. But with virtually no controls to the rudder, and little hydraulic pressure for braking, we veered of the runway on landing - straight into another Beaufort (from 217 Squadron), which had returned from the very same 'op' against the Italian fleet, to its base at Luqa, but which had been left on the runway after running out of fuel. Our aircraft caught fire and was destroyed.

Thus, ended our initiation into torpedo-strike warfare. 39 Squadron had lost half of its task force of a dozen aircraft, with eight airmen killed and at least two hospitalised. The convoy from Alexandria never managed to get the supplies to Malta and had to return to port. The Italian Fleet turned tail and returned to Taranto, and, having encountered the RAF and its torpedo bombers, never ventured out of port again!

On March 3, 1943 an enemy convoy was reported sailing north of Palermo. Three reconnaissance Wellingtons of 221 Squadron were sent out at 18:25 hours, followed shortly before midnight by six torpedo carrying Wellingtons of 221 and five Beauforts of 39 Squadron who were then based at Luqa. It was my 'turn' for turret duty on that flight - I was by then the only original member of Marshall's crew. 'Skipper' Marshall was by then a Squadron Leader, and Flight Commander of 'B' Flight (this is, of course, an ample demonstration of how one could be promoted on 39 Squadron, purely because of the high rate of casualties among established crews. Experience and competent flying (and luck) counted very much in lasting long enough to complete your 'tour'.)

All that I remember of the flight was Marshall setting course for western Sicily at a height of about 800 feet and then - recovering consciousness while floating in the sea in my Mae West, the remains of the Beaufort on fire nearby. I shouted for the rest of the crew, but there was no reply. Then, by good fortune, the aircraft dinghy floated by, only partially inflated. Somehow, despite the pain of a broken ankle and forearm, I managed to pull myself aboard and spent the night in bitter cold, drifting in and out of consciousness, with the sea washing over me. At one stage I found the packet of fluorescent marker powder, but most of this I spilt over myself, and my skin was dyed bright yellow. As dawn broke, I could see land receding in the distance, but then a fishing boat appeared, and I was rescued. The Maltese crew thought at first that I was German, but then thankfully took me to a nunnery on Gozo, where I was treated for hypothermia and numerous injuries. The next day, an ASR launch ferried me to hospital in Malta, where I was visited by S/Ldr Milsom RAAF, the other Flight Commander. I have no answer as to why Marshall's Beaufort crashed into the sea, probably engine failures. I recall a continuous stream of red-hot sparks flying from both engine exhausts, which I think may have been the result of full boost being left on longer than usual after take-off. I am fairly sure that the cupola of my gun turret was sheared off as the Beaufort hit the water, and I was catapulted out.

I was in hospital in Luqa for nineteen days and then told that I would be repatriated for further treatment under better medical conditions. In any event, my tour of operations was almost complete, with just one more 'mission' required.

So, on 20 March 1943, I boarded a Hudson of 24 Squadron (then based at RAF Hendon), along with a time-expired Chief Petty Officer of the Royal Navy, and we left Luqa bound for Gibraltar. After flying for about four and a half hours the CPO noticed that the starboard propeller had stopped! At this, the W Op opened the cabin door and told us to prepare for a crash landing: the starboard motor had failed, and the port engine was overheating. The pilot made a good-ish belly landing in the undulating sandhills of the Algerian desert, although the Hudson almost finished up on its nose, and everyone got out without delay!

No one was seriously hurt, but the navigator was slightly injured, and my wounds were not improved much! We could see aircraft transiting to the south and the navigator fired off Very cartridges to which there was no response. However, a jeep-type vehicle appeared some while later, manned by French Foreign Legionnaires, who took us to their 'fort'.

The nearest habitation turned out to be Touggourt, 430 miles west of Tripoli. We were taken to this town, where a few of the local 'dignitaries' insisted on entertaining us for the evening. By this time my injuries were causing me great discomfort, and the navigator was suffering from slight concussion. The two of us were taken to the local hospital - which looked like a Victorian workhouse! There was little by way of medication, and most of the patients seemed to be native Algerians awaiting their end.

After a couple of days an ancient twin-engined Potez of the French Air Force transported us to Biskra, which was a slightly larger town, 125 miles to the north, and from there a DC3 of the USAAF flew us the 200 miles to Maison Blanche aerodrome at Algiers. From here another DC3 took us, finally, to Gibraltar, where I was glad to be under RAF control once more. I reported to the MO and asked if I could now go home by sea! This request was granted, and I boarded an armed merchant vessel which was part of a convoy bound for Liverpool. By amazing coincidence, the CPO who survived the Hudson crash with me was on the same ship! The accommodation in the sick quarters was superb and the home journey took four days, by a circuitous route. On arrival in Liverpool, I was met by an ambulance which took me to a convalescent home for the RAF at Hoylake in the Wirral.

I returned to active service in August, after spending five months in hospitals and convalescent depots. The powers that be decided to send me, being tour-expired, to RAF Turnberry as an instructor, as I was no longer classified as active aircrew, and rather 'out of contention' for any flying duties. So, I passed the time giving the occasional lecture to the W Ops and navigators of 5(C)OTU.

Turnberry was a good 'drome, in a beautiful situation. I was much impressed by the grandeur of the hotel. This was one of the few RAF Stations I had been posted too at which I was provided with a bicycle - which, of course, was inevitably stolen. The showers were sheer luxury to an ex-battlefront wallah from the desert, an early visit, soon after working hours was essential, before all the hot water ran out.

I fondly recall visits to some of the houses in Maidens, treated to meals and hospitality most kindly provided by the parents of girlfriends(!), and which usually featured home-made Scotch pancakes, which I did not experience again until many years later while on holiday in Oban. I used to enjoy a trip on the double-decker bus, via Maybole, to Ayr and Troon, and the municipal golf courses, which seemed amazingly cheap to us southerners.

The weather at Turnberry as I recall, was fairly amenable for the time of year. Far better than in Northern Ireland, where I spent the following winter, stationed at Killadeas at the OTU for Sunderland and Catalina flying boats, and where I seemed never to be out of 'wellies'. So much water was always involved!

Vern McDougall

I was at Turnberry in January 1944 doing a course on torpedo attack. There were six of us RCAF pilots with our RAF navigators. We flew Bristol Beaufighters, an excellent aircraft. We enjoyed many cups of tea at a tearoom in Maybole. We thoroughly enjoyed our stay at Turnberry, in spite of the constant rain.

When we finished our course, we were posted to 254 Squadron at North Coates airfield near Grimsby. We returned to Turnberry for a week in the summer for a refresher course. After leaving 254 Squadron we were sent to East Fortune to instruct. In June we were sent back to Turnberry to ferry back three Beaufighters to EFT. The weather turned bad and we had to turn back for Turnberry, I almost hit the Ailsa Craig coming in to land. We spent the night at Turnberry and left early next morning, and that was my last flight in Scotland.

Out of the six RCAF pilots who trained at Turnberry, only four made it home to Canada, two were lost on 'Ops'.

In 1986 my wife and I joined a golf tour and visited Turnberry. We spent three nights in the beautiful hotel and golfed on the old aerodrome.

Our guide said I was the only one he had met that had flown from there.

Hugh Gwyn Jones (Taff 103)

I was born in the tiny hamlet of Gorad, one and a half miles north of Valley in Anglesey, in 1921. After a disappointing year at university, in 1941, I met one of my friends from school, George Hamley Williams, who was home on leave having just gained his wings as a pilot in the RAF. I made up my mind to volunteer to become a pilot, and I told George of my decision. So, he gave me the 'low-down' about joining. One of the things he said was 'Don't have a pee on the morning of your medical. It is very embarrassing to stand naked holding a beaker and you can't do it.'

In due course I was instructed to attend for a medical examination in Chester. When the day came (in November) I was to catch the 07:05 train and I remembered what my friend had told me. 'Don't pee'. I did not - not after breakfast nor on the train. Later that morning I was standing naked in a cubicle in the medical centre being examined by different doctors - one for the heart and blood pressure, one for reflex actions, one for eyes, ear and throat etc., until eventually I was given a beaker and told to produce a sample - there and then on the spot! By now I was bursting and because of the long delay I could not start to urinate. Eventually I did, but could I stop? The beaker quickly filled and overflowed! There I stood, naked, in everyone's view producing a Niagara Falls of urine all over the coconut matting upon which I stood! I simply could not stop, and the attendants and doctors were all smiling and chuckling at me!

At the end of the medical I was seen by the Senior MO, who said 'I cannot pass you A1. Your bladder would be an embarrassment on long flights! I am therefore passing you Grade B2. You must be A1. to fly, but you can still join the Air Force. I suggest you become a wireless operator.' That is what I did, and after signing up I received the King's shilling.

On reflection, I think my failure to hold my water at that medical probably saved my life, since the survival rate of pilots was extremely low. Had I become a pilot my chances of survival would have been very slim indeed.

Hugh Gwyn Jones

My friend, George, did not survive. Shortly after I joined up, he was on a training flight when his plane crashed killing both him and his instructor.

I was called up in the first week of January 1942 and instructed to report to Padgate. Dad saw me off on the early train and being an ex-serviceman, he gave me some sound advice as we waited for the train to depart.

There was snow and ice when I arrived at Padgate – all we did that day was stand around in freezing conditions. We had to fill in some forms and we were allocated huts for the night. We would be woken up at 06:00 for breakfast at 07:00. There were about thirty recruits in our hut in which there were two stoves to keep us warm. I went to bed having placed what little money I possessed inside one of my socks which I kept on my feet. This had been part of the advice that dad had given me. A sergeant came into the hut and in an authoritative voice said that he was collecting valuables for safe keeping. 'There are thieves about,' he said 'and they take advantage of rookies.'

Most of the lads dropped their wallets into the sack, which the sergeant was carrying, on the understanding that their belongings would be returned to them in the morning. The sergeant disappeared into thin air and from the distinct lack of enthusiasm shown by the staff the following day to follow up complaints from several huts of raw recruits I would guess this 'con' trick was a regular occurrence with each new intake at Padgate.

As a W Op/DF, at the end of November 1942, I was on my way to 5(C)OTU Coastal Command, at Long Kesh, Northern Ireland. There aircrews did their final training before going on active service. It was also a base for a squadron that flew sorties over the Atlantic searching for U-Boats. Our call sign in the DF station was 1BM. I was to be there for about fourteen months.

Apart from giving signals concerning weather, altitude, signal strength and courses to steer, I became adept at an exercise known as QGH. This was a planned way of getting an aircraft to descend through cloud so that the pilot could land by visual means at the last moment. I would guide the aircraft by means of QDMs until it within sight of the airfield. It was then up to the pilot to use his own vision. I had a great advantage in that I could send on the Morse key with my left hand whilst virtually writing in my logbook at the same time with my right hand.

Thus, the pilot was receiving more instructions in a given time via his W Op. I had the opportunity to accompany pilots on QGH exercises because this procedure was then a very important one. I also went up to test the transmitters and receivers on the planes while the pilots were practising 'circuits and bumps'. The aircraft most in use was the Beaufort, but I also flew in Ansons and Hampdens.

On one occasion we had a narrow escape. Beauforts were nose-heavy and consequently had to do a steep approach before levelling out at the last minute to land. A Hampden, however, did a long low approach to the runway. Having been given No. 1 to land by ground control, I shut down the wireless and went forward to the navigator/bomb aimer seat just to the right and below the pilot in our Beaufort. As we began our steep approach a Hampden glided in below us quite oblivious of our presence and proceeded to land. My skipper (a Scotsman who could swear profusely) quickly opened the throttles and made another circuit before landing!

We were not allowed to use 'plain language' when we were 'on air', but at the end of an exchange in code we would sometimes add our initials and TU (thank you) or GL (good luck) to which the usual reply was STU (same to you).

I intercepted an SOS call whilst at Long Kesh. It was a very weak signal and I could not get an accurate bearing on it - just enough to see that it was from the south. I reported it to Flying Control and about two weeks later I was told that our signals officer had been informed that the SOS had come from an RAF Short Sunderland plane that had been shot down in the Bay of Biscay.

One night I received a call on my wireless requesting a QDM, i.e. a course to steer to our aerodrome, but the aircraft W Op did not give me his call sign. I insisted on having his call sign, but he had either forgotten it (a very unlikely event) or he was in an enemy plane trying to fool me. Flying control confirmed my decision and I sent a further code signal that meant 'Cannot Comply'. He replied 'GS' which was the unofficial sign for 'Get Stuffed!' Was he a fool who had forgotten his call sign or was he a 'Jerry' trying to outsmart me in order to bomb Long Kesh? I will never know! Come to think of it the skipper on that plane would have known the call sign anyway!

I was posted from Long Kesh in February 1944. The journey turned out to be a nightmare. When our transport lorries arrived in Larne for the ferry to Stranraer it was blowing such a gale that sailing was out of the question. We were taken to some Nissen huts and given sacks with instructions to fill them with straw from a near-by barn. Those sacks were to be our beds for the night, and we found that the floor in the huts was bare earth! The rats did not help either.

We boarded the ship, *Princess Victoria*, at 08:00 the following morning having had just a mug of tea for breakfast. The ship carried cars and lorries in its hold. It had two large doors in its bows which were opened at the docks to allow the transport vehicles to enter or leave the ship.

At 09:00 we sailed in very rough seas and it was bitterly cold. Strangely I was not seasick. When we arrived outside Stranraer an Aldis lamp was flashing in Morse code and I could read the message. I broke the news to my colleagues - the sea was too rough for the ship to berth. They would not believe me, and they thought that I was pulling their legs! As time went by, they realised that I had been telling the truth. It was so cold on deck I made my way down to the hold and found a lorry, the doors of which were not locked. I climbed in and managed to sleep for a while. Eventually there was an announcement that hot tea and biscuits would be served. The biscuits turned out to be hard as rocks, but that was all we had to eat. At least the tea was warm! At 11:30 the following morning the Aldis lamp was flashing again and the message this time was that ship could now berth. The crossing which normally took two hours had taken twenty-six hours! Years later after the war that same ship sank when it's bow doors were battered in by the waves of another storm! [*Princess Victoria was the first purpose-built ferry of her kind to operate in British coastal waters, she sank during a fierce storm in the North Channel with the loss of 133 lives on 31 January 1953.*]

We travelled in our lorries to Turnberry and eventually settled into our huts close to the village of Maidens on the seashore.

The DF station where I was to work was situated on top of a hill that overlooked the airfield and was close to the village of Kirkoswald. I was issued with a bicycle as a means of transport to and from the DF station. I walked most of the way up there and free-wheeled all the way back. The bicycle came in handy because it enabled me to cycle across the golf course - on a very straight road, to the seaside town of Girvan where there was more 'life' than in Maidens.

I also visited Ayr quite often by bus, and Maybole on Sundays where I as a guest (with others) of the local people for chapel service and supper. I was surprised to find that the local fish and chip shop was open every Sunday evening.

Our call sign at Turnberry was 3VA. With regard to the radio messages, these could be prefixed with priority letters to indicate the level of priority. From memory the letters were: P, OP, UP, and SOS, in that order, the P being the lowest priority, but taking preference over non-priority messages, progressing to the obvious highest priority - the SOS.

I was on duty one evening when the weather was poor with low cloud and rain. There was no 'traffic' on the air as we called it, just 'mush' in my earphones. I still had to 'listen in', although our own aircraft were grounded. Suddenly I heard the 3VA call sign followed by a QDM request with UP priority. In other words, the aircraft was in urgent need of a course to steer towards Turnberry. I had experienced plenty of practice runs at Long Kesh, but this was the real thing! I coped, and the following day flying control let me know that the plane I had brought in the previous evening was that of the Air Vice Marshal of Coastal Command. It was a good thing that I had not dozed off as I sometimes did!

On another occasion I was on night duty and no one had told me that Turnberry was 'number one diversion' aerodrome that night. That meant if England was covered in fog, the homecoming bombers were to use Turnberry as their first

alternative base. Luckily for me, I had not fallen asleep and at about 05:00 I heard a faint call of 3VA. Before I knew it, I had several aircraft calling me requesting QDMs (course to steer to reach Turnberry). I kept my head, sent them all 'AS' (the signal to wait) and then proceeded to deal with each one in turn giving them the QDMs they required. It took me an hour to bring in 21 aircraft that night. They were bombers returning from missions over Germany. When I went into the mess for breakfast after coming off duty, I saw the air crews eating their bacon and eggs. I was given baked beans, but I made no complaint! For me this incident was the highlight of my service at Turnberry. For once our base had not been merely a training unit!

I am reminded of a fatality at Turnberry - an airman (deliberately it was rumoured) walked into the revolving propeller of a plane in dispersal. Very little was said about it and I have no details.

We had our own dance band on the station and dances were held twice a week. A Land Army girl named Agnes formed a strictly dancing-only liaison with me and she taught me to dance properly. We never went out with each other, just danced, and I became quite proficient. I did kiss her goodbye when I was posted from Turnberry, and all she said was, 'You are different, you are!'

I enjoyed going for tea and scones in a cafe on the seashore at Maidens and also the fair at Girvan. In Ayr I went to the Bobby Jones dance hall, but I never enjoyed it since there were too many Yanks 'jitterbugging' and that was definitely not my style! I cycled here and there to enjoy the countryside and I had the experience of freewheeling up hill and pedalling hard downhill. The hill was called 'The Electric Brae' and I think the slopes were optical illusions that gave the hill its name.

The invasion of Europe took place in June and all leave was cancelled. I had not been home since I was at Long Kesh, so Mam decided to spend a week at Girvan and Roy came with her. I booked lodgings for them and later I met them at Glasgow. Mam and Roy could not understand a word when their landlady, Mrs MacClachie, was speaking to them in her broad Scottish accent and life was difficult for them. But of course, when I was off duty, they enjoyed the time I spent with them.

Later that summer I met a girl in Girvan, she was beautiful and was a good dancer. She was on a short holiday from Glasgow and was returning home in two days. She agreed to see me in Glasgow, but as soon as this happened, I was posted to Chigwell in Essex. However, she promised to write to me, and she did!
[from his book 'A Welsh Journey', an autobiography]

Donald Stewart McNeil

Turnberry has a warm place in my heart, a great country with great people. I was there for five to six weeks, from March to May 1944, on No. 3 Course (Ventura).

On 16 April I had a night flying check out and did circuits and bumps for two hours in a Lockheed Ventura. On my last circuit the fog rolled in and closed the airfield. I was instructed to orbit the searchlight we could see through the fog. After the control tower found an airfield still open, we were ordered to proceed to Tiree, so we set course, managed to find it and landed about midnight. I had a great crew, and although we were all scared, tired, hungry and short of fuel, not one complained. The island of Tiree sure looked good. This was my first experience of night flying in the UK, and I was taught that black-out meant black-out!

On arrival, we found that there was no tow-bar to fit the Ventura, so we improvised, and managed to wrap a cable around the wheel and brake drum - which seemed a good idea at the time - anyway, after four days a replacement brake drum arrived, and we were able to return to Turnberry!

After leaving Turnberry I spent the next year as a Coastal Command pilot flying Hudsons, Venturas and Flying Fortresses out of Skitten and Wick.

The area around Turnberry was very nice. I remember visiting Ayr and Girvan where the Houston Sisters, Billie and Renee used

Donald McNeil

to perform. *[Author: Sisters Billie and Renee Houston were very popular Scottish comedy and revue artistes]* We used to drink at White's Bar in Maybole with the commandos who were practising cliff assaults.

I knew a lot of the Beaufort Boys (22, 42, 217 Squadrons) having worked with them on Ops over Norway, Denmark and France, and I didn't envy them their jobs. On Hudsons we had options of how to attack shipping, but on the Beauforts they had to maintain a set height and speed to launch their torpedoes. As such they were sitting targets. What a reward for the trauma they had to suffer after training at Turnberry.

Eric Robinson

I attended a course at RAF Turnberry from 20 June until 6 September 1944 after which I was posted to 280 Squadron, an Air Sea Rescue Unit at that time based at Langham in Norfolk.

At Turnberry I crewed up with skipper F/Sgt George Borlau, navigator Sgt Jack Brownlee and wireless operators P/O Andy Bell

Don McNeil and crew, B. Smith, B. Porter and M. Brunello

Eric Robinson

and Sgt Sammy Bilton, the air gunners were Sgt Jimmy Cook, and myself.

Unfortunately, all records of my time at Turnberry were lost, as my Logbook was in my kit bag which was stolen whilst I was travelling from Turnberry to my home in Worcestershire by train.

One thing I can remember very well was when we had to land at Turnberry with no undercarriage, which resulted in the aircraft catching fire and was a complete right-off, but happily, no one was hurt.

John Allen Bateson
John Bateson was a Flying Instructor at 5(C)OTU. With this unit he had trained Beaufort crews, at Chivenor, Turnberry and Long Kesh, before taking on ASR training (with Vickers Warwicks) and torpedo training (with Bristol Beaufighters) at RAF Turnberry from 1944.

On 30 June 1944, John Bateson was flying with Kaptein Løytnant Carl Johan August le Maire Stansberg in Beaufighter VI JL835. Late that morning, their aircraft suffered engine failure and crashed into the sea off Turnberry. Stansberg survived, seriously injured, but Squadron Leader John Allen Bateson was killed. He was 33 years old. Elder son of George Roland and Florence Maud Bateson, husband of Doreen Joan of Toat Hill, Sussex. Bateson lies in the Commonwealth War Graves Commission section at Dunure (Fisherton) Cemetery.

Both men were highly experienced pilots.

Kaptein Løytnant Carl Johan August le Maire Stansberg
Before coming to Turnberry Stansberg had by 1936 completed five years training at the Royal Norwegian Navy Academy. He was posted to the RNoN Flying School in 1937, and by 1939 he was a pilot serving at various Naval Air Stations, including Stavanger, Bergen and Tromsø, until the fall of Norway in the Summer of 1940.

He spent a period instructing in Canada with RNoAF, and from 1941 Stansberg was a First Lieutenant and Flight Commander with No 330

(Norwegian) Squadron RAF, based at Reykjavik, Iceland.

After further training duties in Canada, Stansberg was posted to the UK in 1943, for operational training to 132(C)OTU East Fortune and, then in June 1944, to 5(C)OTU Turnberry.

On the morning of 30 June 1944, he was flying in Beaufighter VI JL835 with Squadron Leader John Bateson when their aircraft suffered engine failure and crashed into the sea. Stansberg survived, although seriously injured, and spent the next six months in hospital. Stansberg held the Norwegian Naval rank of Kaptein Løytnant: the equivalent of a British Lt Commander RN and hence so recorded in the 5(C)OTU entry.

Stansberg had been awarded the British DFC in 1943, and the Norwegian Order of St Olav with Oak Leaves in 1944. Fit to return to duty in January 1945, senior staff and liaison appointments followed and by the end of the war he was at the RAF Staff College. More senior appointments in the post-war Royal Norwegian Air Force followed. In 1959 Oberst CJ August Stansberg was appointed Inspector General commanding the RNoAF. He retired, highly decorated, in 1970.

Harold Blampied
I was born in Bayswater, Auckland, on 12 September 1920, the son of Martin and Elsie Blampied, both of whom served in WWI, Martin as an infantry officer in France (wounded and awarded the MC), and Elsie, who travelled to the UK as a VAD to nurse wounded soldiers. Being I suppose, something of a military family, on 4 September 1939, both my father and I were mobilised. My father volunteering for any task, for which his WWI experience may have been useful, and finished up as a major, commanding an AA battery. I joined the Territorials through a rugby football club recruiting scheme at the time of 'Munich', as a result I spent over three years as a Coastal Gunner, but after the Japanese threat died down, I resigned my Army Commission and transferred to the RNZAF to begin flying training, which had been my main ambition. I trained in New Zealand and Canada

before Turnberry. I arrived at the Torpedo Training Unit flying Beaufighters mid-1944 and there was of course a fully established RAF station and personnel, such as WAAF and Officer Assessors, who interpreted the results of our training exercises, both in the air and in the elaborate training device which was like a cinema and I think was called a TAT (Torpedo Attack Trainer). There were instructors both for flying and for the unit, and the WAAF officers were glamorous. My main memory is of the runways, the one in general use, due to the prevailing wind, stopped just on the edge of a high cliff, and as we were encouraged to fly low (attack height being 150 feet, and range of 1,000 yards at a speed of 180 knots) we did not climb after take-off, but simply went down to about 50 feet and immediately practised our low flying over the sea which was tricky. This only applied to the one runway.

Turnberry Hotel was unoccupied at the time, but we used the pool for dinghy training. We also used to make diving attacks on the hotel before landing and it was a miracle no one crashed into it. We were young and reckless. We had no casualties on my course, but many 'graduates' were later lost on operations as shipping attacks were hazardous.

We enjoyed our time at Turnberry immensely, the training was exciting, and the locals were friendly.

The course lasted about six weeks, from memory, after which we were posted to the Beaufighter Strike Wings - Langham, Dallachy and 489 (NZ) Squadrons in my case, in August

Harold Blampied

1944, just in time to 'have a crack at the Hun'. I survived nine months of anti-shipping strikes with 489 and then converted to Mosquitos at Banff, in preparation for the Japanese theatre of war. Before I could get there, the Japs surrendered, which was a much better idea! I was demobilised at Banff with the rank of Flight Lieutenant and only one 'gong'.

Archie Clark and crew

Archie Clark

Altogether I suppose I was at Turnberry for more than three months in 1944. This included ground as well as flying training. We were being trained to fly in Hudsons for posting to one of several meteorological flights made from various parts of the country. Our five crew members were: Graham Melville, pilot from Bristol, Ron Lee, navigator from Sheffield, Alex Beatson, W Op/AG from Glasgow, Norman Thomas, met. observer from Caerphilly and myself, Archie (Nobby) Clark, W Op/AG from Glasgow. I see from my logbook that I flew in Hudsons HM843, AM666, AM700, and AM562 among others. Wing Commander Sandeman was our Chief Instructor.

Unfortunately, Alex Beatson became ill during training and was not able to complete the course as one of our crew, but he did complete later and joined 519 Met. Squadron flying out of Wick. Sadly, Alex was killed when his Fortress aircraft crashed near Wick on 1 February 1945. At Turnberry our replacement W Op/AG was Charles Walker from London. Once we were fully trained, we were posted to 251 Squadron on met. flights called 'Magnums' out of Reykjavik in Iceland.

Memories of our stay at Turnberry are now rather dim, but from my logbook I see that we struck bad weather on one of our sorties and were diverted to the Isle of Islay. I had relatives in Port Charlotte who used to keep the post office, but as we were in Port Ellen there was no chance of seeing them. I can't even remember if we talked on the telephone!

Another memory of Turnberry was watching two Hudsons being brought in from an operational squadron and being flown by experienced pilots. One landed safely, but the other didn't, it 'ground looped' - the landing legs twisted round on landing, probably due to a sudden gust of cross wind. I believe that is the reason why my skipper, Graham Melville, decided to land with wheels up when we were completely frozen over, and with no radio contact, on our first Magnum met. flight from Reykjavik.

One of the training exercises was photography and was much enjoyed. Who could fail to enjoy flying up and down the Clyde coast taking pictures of various landmarks including the Cloch lighthouse and Ailsa Craig.

On one navigation exercise we were sent to find Rockall, that small projection in the Atlantic, north-west of Northern Ireland, and only visible when the tide was out and relatively calm seas running. We were in luck and found it amidst a patch of disturbed sea, surrounded by reasonable calm, and took our photographs!

Life outside the station was the usual round of visits to Ayr and surrounding towns and villages in search of entertainment.

Charles Goldspink

I had the pleasure of a very enjoyable two and a half month stay at Turnberry when I undertook my operational training at 5(C)OTU. We arrived mid-July and departed at the end of September

Charles Goldspink and crew

Charles Goldspink and crew Warwick BV518

1944. I travelled up with a colleague, Sgt Ken Jarrett, from South Wales, having successfully completed an advanced course for coastal command wireless operators. My friend and I were to serve together for the rest of the war. As was usual, at Turnberry we met up with the other branches of air crew and eventually formed ourselves into a crew of six.

We had become friendly with a Canadian pilot, Pilot Officer Bruce Lee, who was to be our skipper, an Australian navigator joined us, Flight Sergeant Mark Matthews, and completing the crew were two air gunners, Sergeant Jimmy Bassett and Sergeant David ?, who were from the north east of England.

Having established ourselves as a crew, our next step was to become acquainted with the Vickers Armstrong Warwick, the aircraft in which we would carry out operational sorties when with a squadron. At first, our initial training flights were under the supervision of a Flying Instructor, two of whom were F/O Mossley and F/O Taylor. There was also S/Ldr Brotherstone

Charles Goldspink

Ken Jarrett

who was the Chief Flying Instructor (he was later killed at Prestwick with pupil pilot F/Sgt Holmes on 29/12/44). *[see Warwick BV 340 - Author]*

After the completion of these tests we went into an extensive programme of operational exercises during which time we covered a large area of the Atlantic and the west coast of Scotland. Several times when returning from an Atlantic exercise we would 'home in' on Ailsa Craig with our radar. We also had classroom studies, which involved lots of note taking.

An important part of the training was local night flying. Our skipper was required to practise take-off and landing 'circuits and bumps'. This training did not need the whole crew to be present, only a single wireless operator. One night while negotiating a landing, a throttle linkage broke and caused a runaway engine, spinning the aircraft into the ground where it burst into flames. Fortunately, both he and the W Op managed to get clear before the whole thing became a fireball. I recall a similar incident with another aircraft catching fire on landing, this time with the whole crew aboard, again all were

fortunate in making an escape before it too was engulfed in flames.

We had a very good Sergeants' Mess where we were never short of entertainment, and on most Sunday evenings we would have a social event to which we could invite our friends from the local community. I can well remember two ladies who came from Maybole and Kirkoswald, their names were Elizabeth and Mary. On our off-duty days we would go to Ayr where we would visit our favourite restaurant for a meal of braised rabbit and all the trimmings. In the village of Maidens was a quaint little tea shop, well patronised by all – here we had tea and toast with homemade jam. Turnberry will always have a place in my heart, a lovely area with a very friendly and welcoming people.

On the completion of twenty-four exercises we were allocated to 280 (ASR) Squadron based at Langham on the north Norfolk Coast.

Don Hamar

I was stationed at 5(C)OTU Turnberry during the months of August and September 1944, while 'on

Don Hamar

rest' (so called!) between operational tours as a pilot in 144 Squadron.

We compromised a group of instructors whose job was to train newly qualified Beaufighter pilots in the art of dropping torpedoes. We occupied a group of wooden huts at the southern end of the airfield, next to the sand dunes, and had both visual and radio contact with our pupils while they carried out dummy torpedo attacks on the target ships, HMS *Heliopolis* and HMS *Rosemary*. Cameras were mounted in the nose of each aircraft to operate when the pilot pressed the 'drop' button and the resulting photograph would be assessed to establish whether the target would have been hit. The attacks could be by single aircraft, or by six or more aircraft converging on the target, at the same time, but from different directions, at a height of 100 feet.

When the weather was warm, we could pop into the sea for a swim while the Beaufighters were being refuelled.

With us was a lady, Miss Hogarth, who was a scientist from the Royal Aircraft Establishment, Farnborough, working on the development of torpedo sights.

On 17 September I flew an Airspeed Oxford to Edinburgh, loaded with flowers and produce from the Culzean Castle greenhouses for a VIP wedding. I am sorry to say that Warrant Officer Thomas and I helped ourselves to some of the grapes on the way!

William 'Bill' Harris

I had originally been categorized as PNB (Pilot/ Navigator/Bomb-Aimer) and was a member of the Southampton University Air Squadron (2AEF). At that time, 1943, there was a long delay before beginning training for aircrew, and it was suggested to me that if I wished to fly, the easiest way was to train as a meteorologist and volunteer for aircrew duties. This I did and after basic training was posted to RAF Blakehill Farm near Swindon. I spent several impatient months kicking my heels before I was finally posted to RAF Bridgnorth, and then at last, RAF Halfpenny Green for flying training in Avro

Ansons. We did navigational exercises, learning to use a 'Dalton Computer'. At the end of this course of training I was awarded a 'N' brevet, much to the disgust of real navigators. I was then posted to 5(C)OTU, RAF Turnberry, where we did training flights to Rockall in Lockheed Hudsons until we were considered proficient enough to be let loose on a squadron.

At Turnberry we were engaged on practice weather flights, learning our trade as Meteorological Air Reconnaissance Observers. The main purpose of this was to learn the code to be able to transmit the weather we encountered on our flights back to base. There were also some aircrew duties we had to learn at the same time. I did some air to air firing on one occasion and I was later told by the gunnery officer that I had scored 5% hits. I thought that was very poor, but he assured me that it was a good score for my first attempt.

One incident I do remember was on the instance when only a few minutes after take-off it was discovered that our gyro compass was faulty, so we turned back to base which was still visible. However, we were refused permission to land as our fuel tanks were still full, so for what seemed like days, but were of course only hours, we had to circle Ailsa Craig until we had used up enough fuel to allow us to land.

On my last trip at Turnberry we broke through cloud in order to measure the pressure at sea level when we spotted a German submarine being escorted by a destroyer. There had been an order from Germany for all submarines to surface and surrender.

After completion of my training at Turnberry, I was posted to 517 Squadron at RAF Brawdy. There were more exercises, called Dry Swims, to be done using the met. flight code watched over by F/Lt Aldridge who was I/C the met. observers. At last I was considered capable to fly alone with a crew. I had no standard crew and flew with several different aircrews.

In April 1945 we were awarded our own brevet and the unique 'M' brevet was seen for the first time. I moved with the Squadron to Chivenor in November 1945 and stayed with them until it was disbanded at the end of June 1946.

I was then posted to RAF Ballykelly I/C the Met. Office until August when I was posted to MO4 at Harrow to work on the calibration of met. instruments. I finally ended up in Plymouth where I remained as the Senior Assistant until I was demobilised with the rank of F/Sgt.

Arnold 'Barney' Whitehead

I had joined the RAF by volunteering for aircrew, because I was in a reserved occupation when the war began, and I could only join the forces if I volunteered for a job that required extra personnel. After the usual drilling etc., I trained as a pilot in Canada on Tiger Moths, and Oxfords, then on a navigational course on Ansons, then over Vancouver Island on Hampdens where the emphasis was on shipping attacks.

32 Beaufort Course

Returning to Britain, we were posted to Northern Ireland where we converted to Beauforts and continued to strengthen our skills at coastal navigation and shipping attacks. I was not very good at flying Beauforts, as, in contrast to other planes I had flown, the Beaufort had to

be brought in to land with the tail up, a technique I was never happy with.

On leaving Long Kesh in the late summer 1944, I was sent to Turnberry, 32 Beaufort Course. My crew consisted of Navigator Gordon Cameron, and two W Op/AGs, Bob Wallin and Jim Wright, both from Alberta. I record here that I had just been married and my wife joined me for a week at Turnberry, staying in a hotel in Girvan. Turnberry seemed to be a cold bleak place when we arrived and found our accommodation in the usual Nissen huts. There were four crews on our torpedo training course, two RAF crews and two Canadian, though between us we included British, Canadian, Australian, New Zealanders and Americans. We had two instructors (whose names I can't remember) but we understood that they had been in attacks on German pocket battleships, so we respected them as brave and experienced flyers.

There was very little instruction on the ground, though I remember a link trainer which simulated attacks on enemy harbours and gave what we would these days call a virtual reality experience. Most of the training was in the air. The two instructors would each take two crews and fly with them over the sea in mock attacks. Probably I was rather a cautious pilot, and the first time we flew with the instructors, I got more and more alarmed as we took the planes lower and lower. When we reached operational height, I don't think we could have got any lower, we were literally leaving a wake behind us as the propellers took off the tops of the waves. Perhaps the scariest part of the exercise was when we were flying in an arrowhead of three and the instructor took us into a steep turn. The plane on the inside had to reduce speed and do the tightest turn in order to act as the fulcrum of the formation. Very white knuckle flying! Then when the exercise was over, we had to fly down the hillside to the runway.

Looking back, I always said that I learned to fly at Turnberry, before that I was licensed to fly, but always felt that the plane partly dictated what happened. At Turnberry I learnt that it was possible to control the plane and make it do what you wanted. Our final exercise was a mock attack on a ship on the Clyde, but this time with a whole flock of Beaufighters, and perhaps the odd Mosquito. A taste of the future of shipping attacks. Now I was trained to the ultimate degree in torpedo bombing, so what was my next posting - to an Air Sea Rescue Squadron! Then to India and Burma and long searches over the Bay of Bengal. What a contrast!

For years after the war, my navigator, Gordon Cameron, a plumber from Aboyne, Aberdeenshire, used to reminisce about the sweet shop in the Maidens where we obtained sweets off the ration.

Ken Abbot

While at Brighton Personnel Depot in March 1944, several RAAF Wireless Air Gunners were selected for attachment to Coastal Command and sent for further radio courses at Yatesbury in Wiltshire and Carew Cheriton in Wales. At the end of the second course, members were paired off and postings to operational training units were decided - some to Ireland for training on Catalinas, some to Turnberry for training on Air Sea Rescue Warwick's, and Stuart Wright and I to Alness in Scotland for training on Sunderlands.

While on leave in Edinburgh, our posting was cancelled, and we were told to report to 5(C)OTU Turnberry to train on Hudsons for meteorological and anti-submarine work. To say the least, we were both very disappointed at this change.

On arrival on 12 September, the course intake of sixteen aircrew was told to form into crews consisting of a pilot, a navigator, and two W Op/AGs. Stu and I were the only RAAF, W Op/AGs and the only other Aussie on the intake was a very short navigator named Bill Wong Hoy - a full Chinese from Cairns in North Queensland. We three Aussies settled on Ben Jackson from

Ken Abbott and crew

Derby as our pilot and our newly formed crew was then allotted a met. air observer, Harry Jones.

In just over two months on a series of training exercises we made twenty flights for a total of 65 hours in all kinds of weather. Stu and I became proficient at photographing Culzean Castle on our pilot's call of 3...2...1...now! as he flew low over the trees. We took turns operating the guns in the turret and using the radar set as an aid to get back to base in bad weather, but our main training was in getting up to speed in the transmission of meteorological data compiled by the met. observer. The training exercises were designed to cover all the possible problems we might face when we were on operations. The final exercise was to locate and photograph Rockall in the Atlantic - a rocky outcrop far to the west of Scotland. It was reckoned to be a good test of a crew's capability - we were one of the majority of crews over the years which did not find it.

At the end of our course two crews were to go to Iceland and the other two to Wick at the northern tip of Scotland - we were one of the two lucky enough to be posted to Wick.

We quickly found that Wee Willy Wong, as Bill Wong Hoy was soon labelled, was indeed a cunning fellow who seemed to be able to organize things to make life more bearable. In no time he was able to arrange for the crew to be welcomed into the home of a Mr and Mrs Sloan who lived on the edge of the aerodrome. Some of the crew visited the Sloan's fairly often, and as Mr Sloan had a fishing vessel, they soon introduced us to fresh herrings, beautifully cooked by Mrs Sloan. Jan, who was the only child of the family, was a bright lass of sixteen, with several friends of her own age. Bill, Stu and I sometimes made trips into Ayr with the girls to go to the pictures.

The fact that the area around Turnberry was the real heart of Robert Burns country, did not have much significance for the Aussies, but Uncle Jimmy (as we called Mr Sloan) introduced us to the Bard's verses. Stu Wright was a stutterer, and to hear him trying to recite Burns poetry with the correct accent, coached by Mr Sloan, was one of the most enduring and amusing memories of my time in Maidens.

Turnberry was certainly not an ideal location for an aerodrome and we always thought that the only reason for its establishment was because it had been one in the First World War. It was lucky there weren't more accidents and casualties over the years. There were plenty of opportunities for rookie crews to make mistakes because of inexperience and the local hazards - a hospital building (a hotel in peacetime) not far from the end of a runway, a lighthouse at the side of another runway, Ailsa Craig not far offshore and often shrouded in cloud, and wind that always seemed to be blowing strongly. The nearest our crew came to an accident was on our first night flight when we almost hit the hospital building because the aircraft was sluggish on take-off. That was our only flight in that aircraft while on the course, and we were quite happy not to have to fly in it again - its number ended with 13, and having had one lucky escape, we didn't want to chance our luck again. On another occasion Ben had great difficulty landing in driving rain and strong wind and we all swore that he bounced us as high as the lighthouse just off the runway!

We had to practice dinghy drill in the hotel swimming pool. Stu was a lifesaver from Sydney, which was just as well because we almost drowned Bill. He wasn't a good swimmer and was handicapped by a shoulder that had been injured and operated on in Canada.

My only recollection of a serious accident while we were on course was of a Warwick of the Air Sea Rescue Course catching fire just as it touched down at the end of the runway. Two RAAF W Op/AGs who had been at Yatesbury and Carew Cheriton with us had to jump from the astrodome and both sustained burns. One had the outline of the shape of his helmet marked on his forehead as the leather protected his scalp, but it was not too serious. However, the other lost the use of his left hand as it was almost reduced to a claw. One of the crew was called Holmes, but I'm afraid my memory has let me down regarding the other crew members.

One of the houses on the edge of the

aerodrome was that of Mr Sloan Senior (Jan's grandfather). His house was in line with a runway and just below the level at which aircraft touched down. If a W Op/AG forgot to wind up the trailing aerial before landing, the old gentleman lost slates off his roof!

Bill Wong Hoy was, I believe, the first Chinese commissioned officer in the RAAF and he lived in the officers' quarters at Turnberry, apart from the rest of the crew. Bill did part of his training in Canada and, being a canny fellow, acquired some items that proved very useful when he arrived in the UK. One was a Kodak Retina camera, with which he took almost all the photos of our crew at Turnberry and later of the Fortress crew at Wick after the changeover from Hudsons.

In the NCO quarters huts, at times there were lively arguments at night between the English and Australian lads, sometimes culminating in boots being thrown in the dark. Usually the arguments were caused by the English lads, who had trained in South Africa, trying to convince everyone else that England was the finest country in the world. The Australians who had other ideas and had spent some time in South Africa on the way over, countered by telling them that they were an ignorant lot whose South African experience was not enough on which to base their contention. Jock Beatson, a Glasgow boy, did not take sides and thoroughly enjoyed the fun as only a true Scot would when someone took on the Sassenachs!

Doug Whittaker

I joined the RAF in June 1941 and began my basic training at Newquay. I was then sent to the USA to learn to fly with the US Army Air Corps. I did my initial training at Albany, but it soon became evident that I'd never make a pilot so was sent to Winnipeg to train as an Observer (Navigator/Bomb Aimer) The training was successful, so I was then sent back to the UK and Squires Gate at Blackpool to crew up. I was posted to 279 Squadron and completed 71 Ops with Coastal Command, our job was to attack U-Boats and look for ditched bomber crews

returning from Germany. The job was quite hairy at times!

I was then posted to RAF Turnberry on 1 November 1944 as a Navigator Instructor. I remember that I learnt to swim in the pool at Turnberry Hotel. We enjoyed our time off with outings to Kirkoswald, Maybole and Culzean Castle. Memories of Bobby Jones Dance Hall features prominently!

From my Diary:

1 November 1944 Arrived at Turnberry
25 December 1944 To Bobby Jones Dance Hall
1 January 1945 Visit to Burns Monument in Ayr
24 January 1945 Very Cold!
11 February 1945 F/O Beagley and crew crashed in sea – all lost.
20 February 1945 Warmest February day ever.
24 February 1945 Visit to Astoria Ballroom in Beresford Hotel Glasgow.
25 February 1945 Enjoyed a good meal at the Beach House in Ayr.
2 March 1945 To Ballochmyle Hospital for check-up.
7 March 1945 Dance at Turnberry Hospital.
13 March 1945 Gave 5 Shillings to Batman!
3 April 1945 W/O Vear got lost in bad weather and landed at Wick!
10 April 1945 To RAF Hospital Lochnaw near Stranraer for another medical.
14 April 1945 To Hampden Park with 140,000 others to watch England play.
18 April 1945 Bad tonsillitis for ten Days.
27 May 1945 Cricket: Turnberry v Kilmarnock
13 June 1945 Instructors beat Pupils at Cricket. Took three wickets.
15 June 1945 Cricket Turnberry v I.C.I. Girvan.
20 June 1945 Left RAF Turnberry for RAF Aldergrove.

I crashed on 26 March 1944 and had to have over fifty stitches in my head and was under the care of the famous surgeon, Archibald McIndoe. Whilst at Turnberry I had to go to various hospitals for check-ups.

My posting as an instructor was cut short due to a bout of appendicitis and I was rushed to the hospital at Lochnaw near Stranraer. I left Turnberry on 21 June 1945.

Robert (Bob) McGill

One of the corporal drill instructors on my basic training was the comedian Max Wall - I remember him best for the turns he did at the concerts held in one of the theatres to entertain recruits on Sunday evenings. One of his jokes involved bringing on stage various vegetables and attributing them to parts of Hitler's body - finally tossing two potatoes in the air - as the laughter subsided, he shouted, 'You're all wrong, they're King Edwards!'

I arrived at RAF Turnberry, 5(C)OTU at the end of 1944 to train on ASR Warwick's.

The most memorable flights I recall were an air sea rescue search for two Royal Navy Corsairs lost over the Irish Sea and a special trip to take photos of a mini-sub that was beached off one of the Western Isles. I seem to remember being told that this was very 'hush-hush'. We used the photo camera to take pictures of our 'targets' as well as local landmarks. I recall taking a picture of Turnberry Lighthouse, which showed it an angle of 45 degrees - I think it was displayed as an example of what not to do! I am still a rotten photographer.

One of the staff pilots at Turnberry was F/Lt Rickaby, one of the famous pre-war amateur jockeys.

The early part of 1945 was very cold, and we searched the seashore for wood to burn on the stove in our Nissen hut, we also raided the coal dump and tried to extract a few lumps of coke from under the wire fence.

We went to Turnberry Hotel, which was in use as a hospital, to use the swimming pool for wet dinghy drill. I recall jumping into the pool in full flying kit and Mae West and having to right an upturned six-man dinghy. Coastal Command had not been issued with Irving Jackets, at least not when I joined, so full kit consisted of a Sidcot suit and a life jacket. For dinghy drill the suits were passed from one crew to the next and were disgustingly cold wet and dirty. I'm glad I never had to do it for real: it was quite difficult

in a swimming pool, never mind on the open sea. When flying on Warwicks and Wellingtons you kept very low over the sea, so most of us just wore battledress and our Mae Wests. I think that many thought that wearing too much gear might hamper one's escape if you 'ditched' in the sea. Likewise, parachutes were not worn. I must admit that because the rough battledress trousers chaffed my legs, I cheated by sometimes wearing my pyjama trousers underneath. We also wore whistles attached to the collar of our tunic; we were supposed to blow the whistle to attract attention whilst bobbing about in a dinghy in mid-Atlantic – no comment!

Bob McGill

There were visits to Ayr and Girvan, mostly using the overcrowded bus services, one of the conductresses was known as fat Elsie, and her struggle through the packed crowd of airmen to collect fares was always accompanied by much banter and merriment. On the journeys between Turnberry and Girvan I recall seeing horse drawn

Bob McGill and crew

Bob McGill and Warwick

carts collecting seaweed from the shore. I was told that this was taken to a factory for processing into iodine.

There was still a small railway line behind Turnberry Hotel, but no passenger trains, just the occasional very short goods train.

Our diversion airfield was RAF Prestwick. This was a major transport command base as it had an excellent weather record. We never actually had to divert, but were in no doubt that Prestwick being so near was good news to us!

On our time off, our crew used to take a walk along a footpath through the grounds of Culzean Castle. In the spring of 1945, on one of our walks, we picked some daffodils to present to the girls in the WAAF Admin Office. Shortly afterwards a notice appeared on Station Routine Orders warning us that the practice was to stop forthwith - obviously someone had seen us pick the daffs!

My memory of training flights seems to involve a tiny lump of rock off the Western Isles - I would hardly call it an island - this was the favourite target for navigation exercises. This could be picked up on Radar quite easily in calm seas but not when the sea was rough, it was easily missed.

Neville Beale

My trade was a little-known one: meteorological air observer, recruited from ground staff met. personnel who had volunteered for flying duties after some experience in their trade. Because weather plays a vital part in air operations and, indeed, other wartime activities, the RAF had formed several squadrons in Coastal Command based all around the UK and in Iceland and Gibraltar, to undertake long patrols over the Atlantic and the North Sea, with the main task of scientific weather reporting. That was analogous to the land based meteorological stations, but with obvious differences imposed by being airborne.

In my case, I had been called up in November 1943 at the age of 18, after having spent a year in the Observer Corps in the London area. I joined

Neville Beale and crew

the RAF at Lords cricket ground which, until the V-weapon attacks led to its being moved to South Devon, was the aircrew receiving centre where aircrew volunteers got their first experience of military life; medicals, fatigues, 'square-bashing', Morse code, injections etc. etc. Unfortunately, it turned out that my eyesight was not good enough for the aircrew trades then being trained and I was sent at the end of the course to Eastchurch on the Isle of Sheppey (now a prison like other former RAF bases) for remustering.

You had to choose from a list of ground trades. I picked radar operator and, though I cannot now remember why, meteorologist. The RAF must have been shorter of meteorologists than of radar operators, for the former was to be my lot - but not yet. Until there was a vacancy on the met. course, I was sent up to Staffordshire, where, until April 1944, I spent long hours guarding and then helping to service three-ton Bedford lorries being readied for the Normandy landings.

Then finally came a posting to the met. course, held at a school in Kilburn, north London and we billeted in what had been the St Regus Hotel on Cork Street, know famous for its art galleries. But we were delighted to be in London, and since my home was then in Ruislip, I was able to get home at weekends. Weekdays, we marched to Piccadilly Circus and took the tube to Kilburn. Lectures and practical experience with the instruments which are the tools of the

meteorologist's trade led to my qualification by the 10 May. What now I thought?

My sister, who is four and a half years older than me, had been for some time a WAAF staff car driver. She told me that the atmosphere seemed very good when she had visited the bases where 2nd Tactical Air Force was preparing for the Normandy landings. An uncle of mine was a senior officer in one of the 2nd TAF groups and he was able to arrange my posting to 39 RCAF wing at Odiham in Hants. There I joined a small detachment of a sergeant (Alan Kiss) and two corporals (Churchill and Pease), already living and working in tents set up in a field at the southern end of the airfield.

Most of the other personnel of 39 Wing were Canadians, both French and English speaking. Among them, the atmosphere was quite relaxed among different ranks, compared with an RAF unit staffed solely with my fellow-countrymen. Perhaps that was because they were all volunteers, since no Canadian serviceman could be sent overseas against his will.

Neville Beale

This is not the occasion to recount my experiences in the Northwest Europe campaign, suffice to say that we landed on a Normandy beach in June and had reached Eindhoven in Holland by October 1944, when I was told that I was going back to the UK for flying duties in a trade where my short sight could be corrected by special goggles: namely, meteorological air observer.

So, it was back to the Aircrew Receiving Centre, now at Torquay and Paignton in Devon, where one of the same corporals I had known at Lords greeted me, and I had the strange experience of virtually repeating the earlier course. Then we all sewed on sergeant's stripes and took the train to Millom in Cumbria (also now a prison) where we underwent a course in navigation for about two weeks.

Our flying experience was gained in Avro Ansons, whose undercarriage had to be wheeled up by hand. Flights around the Irish Sea and elsewhere over land gave us a taste of life to come. Then, at the beginning of December, I arrived at RAF Turnberry. Now one learned about airborne meteorology, as well as the more martial arts such as air gunnery. We started flying what were called 'operational flying exercises' - both by day and by night - which lasted for six or seven hours over the North Atlantic, west of Scotland, and were a taste of the longer patrols carried out by the operational met. squadrons.

My readings of air temperatures and pressures, cloud formations and amounts, etc. were recorded in code and cipher and written down on message pads which I gave to the wireless operator to send back to Turnberry by Morse code. I also had my chance to practice with a Lewis gun, both on the ground at the 'butts' and air to air against a drogue target.

Over the sea there was no restriction on our flying height and we sometimes flew at low-level to practice aerial photography, by taking photos of such targets as Ailsa Craig, Culzean Castle, Turnberry Lighthouse etc. The Turnberry Hotel was used as an Officers' Mess, but the NCOs were billeted in Nissen huts, which were extremely cold in the winter weather, despite small stoves.

Local people were most friendly and hospitable. It was not unusual for a call to come to the Sergeants' Mess, asking for several men to come to a party where there would otherwise be too many girls on their own!

On one occasion I made a date with a girl who operated the local telephone exchange (literally a blind date!) and took her to the cinema in nearby Maybole. On the way back by bus, at night, she laid her head on my shoulder, which was spotted by the gunnery instructor. Next morning, at his lecture, he announced to the class, 'Sergeant Beale seems to have got his feet under the table' - a graphic phrase which I had, in my innocence, not heard before, and I actually looked down at my feet, much to the amusement of my classmates!

At Christmas the usual practice was for officers to serve the Sergeants their Christmas lunch and I shall never forget sitting next to a sergeant pilot who announced that he was going to be killed. We all told him not to be so stupid, but he did indeed crash not long afterwards at Prestwick, killing himself and his instructor, during a training flight. (Warwick BV340, 29/12/1944) Was this premonition, or what we used to so unkindly call 'the twitch', which my dictionary describes as a 'state of nervousness' which could have affected his flying ability?

One of the runways at Turnberry was quite short and ended at the sea. We had a difficult landing on that runway, during a snowstorm. I mentioned this to a Scottish warrant officer pilot who shared our hut. He had just returned from Canada, where he had been a flying instructor, but he pooh-poohed it. He muttered, 'It's all in Pilot's Notes'. Pilot's Notes was a small booklet giving the pilot all he needed to know about the aircraft and does include landing instructions, but it does not really help a pilot wanting to land on the short runway at Turnberry, even in good weather! As the runway was short and ended almost at the sea, it was essential to land as near to the other end as possible and have the whole length of it in which to come to a stop. But just beyond the start of the runway, there is some high ground which must be avoided in the descent.

My recollection is that our pilot had difficulty in spotting that high ground in the snowstorm, so a very sharp descent was required when he finally saw the runway. (This may have been the cause of my subsequent ear trouble).

After I left Turnberry on leave at the end of our course, the WO pilot crashed into the sea at the end of that runway on take-off and the whole crew were killed. The Met. Air Observer was Sergeant Joseph, who had started his training with us but had been delayed after ACRC when his medical fitness had been questioned. I think it was his hearing that was in doubt, so he himself sought a second opinion, at his own expense and was found fit. Poor Jo, he paid the price for that!

The weather over the Atlantic was very bad that day in January and one of the crews from 518 Squadron, based on the Isle of Tiree in the Hebrides, couldn't land back at its base and was diverted into Turnberry. As they prepared to land, they saw flames coming out of the sea where Jo's Hudson had crashed. As I was on leave at home, I only heard about this later as it was to 518 Squadron on Tiree that I was now posted.

Geoff Lawson and his crew were posted to Gibraltar, while 'Pete' Dutfield and I were the two met. air observers from Turnberry who went to Tiree. The others went to Cornwall, Wales and other bases on the UK mainland.

On one flight at Turnberry, I had a bad cold which affected my inner ear. On arrival at Tiree, I was put straight into sick quarters for a couple of weeks. On coming out and reporting to the senior met. air observer officer, known as the 'Met Leader', a position similar to 'Navigator Leader', 'Signals Leader', etc., on the squadron, I was told that I was flying on ops. the next day. So, at the aircrew briefing I met for the first time, the six other men with whom I was to spend the next few hours over the Atlantic in an aircraft (a four-engined Handley Page Halifax) which I had never been in before! How I was then airsick for the whole flight, (but never again, I'm glad to say) is another story!

The squadron commander and several other senior officers had recently been lost over the Atlantic in unknown circumstances, so maybe the acting met. leader would have to fly himself if he hadn't sent me instead! At that time, until the cause of the previous loss had been investigated, we were flying for only about seven hours, but soon we returned to patrols of over ten hours, which took us halfway to Newfoundland and back, thanks to special fuel tanks in the main bomb-bay. We also carried depth charges in the wings, for use against any snorkelling U-boat we might spot - a rare occasion. We flew most of the time at about 1,500 feet, every 150 nautical miles, descending as close as possible to sea level and for several hours at 18,000 feet, with 'stepped' climb and descent. On the Hudson, the met. air observer sat next to the pilot but on the Halifax (Mark V and Mark III) sat in the Perspex nose.

The loss of aircraft and crew mentioned above actually happened on the same day that the other crew was diverted to Turnberry. The diverted crew had experienced carburettor icing over the Atlantic, so perhaps this could also account for this disappearance. Another possibility was that they had been losing height owing to engine trouble and had tried, with only partial success to jettison the long-range fuel tanks and suffered catastrophic 'drag' as a result.

The crew I joined on Tiree also suffered engine trouble on two occasions which nearly forced us down into the sea.

I still have the folder with all the conversion tables etc. which I had to use for various learned calculations, which I have now totally forgotten. I also have the log in which I recorded the VE day flight my crew and I did which lasted ten hours and twenty-five minutes.

I was demobbed from the Meteorological Office at RAF Waddington in 1947, although met. flights continued into the 1960s, when satellites took over.

Elizabeth Murray nee Harkins
Following a tearful farewell to my parents, brothers and sister, I travelled by train to the coastal town of Ayr to report to the Office of Agriculture at Burns Statue Square. A few weeks previously, I had received my call up papers

and had to choose between munitions work, the Timber Corps, or the Women's Land Army. I chose the WLA and have had no regrets since.

That first morning I met up with other girls, all 'rookies' like myself, and we were told to report to the hostel at Culzean Castle – my home for the next four years. We were taken in the back of one of the Land Army trucks, the journey lasting about thirty minutes. We were met on our arrival by the leader of the unit and were provided with the following items of clothing:

2 Milking Jackets
2 Pairs Dungarees
1 Pair Boots
1 Pair Gaiters
2 Pairs Socks
2 Green Pullovers
2 Short Sleeved Cream Blouses
1 Pair Shoes
1 Pair Wellington Boots
1 Felt Hat
1 Heavy Coat
1 Raincoat
1 Waterproof Hat

To supplement this, we bought ourselves khaki shirts, a brown leather belt and smart green ties.

The staff present at the hostel were:

Mrs Glen	Warden
Miss Anderson	Assistant Warden
Miss Swan	Cook
Miss McLellan	Leader
Miss Paton	Assistant Leader

Some of the Land Girls at Culzean:

Georgie Hamilton	Isabel Cross
Margaret McLelland	Maisie Marshall
Helen Zeleski	Neagle Twins
Jean O'Byrne	Jean Aitken
Margaret McCann	Eva Masterton
Janet Craig	Nan Sandilands
Mary Platt	Anne Boyle
Chrissie Sinclair	Isa Cunningham
Joan (Jewish Girl)	Margaret Mitchell

Sally Campbell	Rose Oats
Dora Glendinning	Madge Murdoch
Margaret Harris	Winnie Dolan
Irene Bovill	Frankie Donovan
Marie Paton	Helen Penman
Jessie Stevenson	Betty Parker

Our truck driver was Georgie Hamilton who turned out to be originally from my hometown of Coatbridge – we were to become lifelong friends.

The accommodation in the Hostel consisted of:

Downstairs – A lounge with a big open fireplace with lots of chairs around it.
The dining area was at the other end of the room and the kitchen was off that.
A piano was available which provided us with endless entertainment.

Upstairs - A large room with about 40 beds.
Each girl had a bed, chair, a chest of drawers and a small wardrobe.

All the girls were from different walks of life but over the next few years we had much to share.

Our day started about 6:30 am each morning when we would rise, wash and dress. Our washrooms and toilets were downstairs off the entrance hall. Breakfast was at 7:00 am and by 7:30 am we would be on our way to work. Farmers phoned the hostel with the number of girls they required for that day and we were allocated to various farms throughout the area. South Ayrshire was well known for its rich farmlands and its rich farmers too!

We were given different tasks dependent on the season. During January and February, we graded potatoes, cleared out barns and scaled the dung (manure). Sometimes we were out in the fields before daybreak and wouldn't start work until daylight arrived. Late February into March was potato planting time – the drills seemed as if they were miles long and our backs ached as we were not used to the work. During April and May, the potato fields were hoed, and the turnips were lifted and shawed. The potatoes were harvested between June and July, another back-breaking job. August and September involved harvesting, which most of the girls enjoyed as we were well looked after by the farmers' wives and the work wasn't so tiring. We were so well-fed during harvest time we didn't need to eat the contents of our 'piece' tins.

The work was hard, but we had our laughs. I remember once during harvesting at Enoch farm, I was busy collecting the seed into bags at the side of the threshing machine, when this lad – the farmer's son – put something in my pocket. I reached in and to my shock and horror pulled out four or five small pink baby mice, they must have been only days old. Well, I dropped them through sheer fright but chased that young 'devil' all around the field. We all laughed knowing his mischievous reputation.

The first time I donned my Land Army uniform, I was pleasantly surprised. The breeches were most comfortable (especially the corduroy ones) and very warm which was fortunate during the winter months. The thick socks came up the legs to the breeches and a thick sweater worn under our pullovers kept us warm.

Boots and gaiters finally kept out the chill and we all felt very smart. When I first went home for a visit after a month, my parents and family were very impressed and proud of me in my smart uniform.

On very cold winter days, the farmers must have wondered what on earth was arriving on their doorstep. On these occasions, on top of gloves and lots of woollies, we wore headscarves and balaclavas.

About three miles south of Culzean Castle was Turnberry Airfield where RAF personnel were stationed during the war. We were often invited along to their social events in the NAAFI (in the Sergeants' and Officers' Mess). We were also allowed to use the YMCA; the manageress was Mrs Thomson and she was affectionately known as 'Arsenic and Old Lace' because of her dress, although she was very kind to the 'Land Girls'.

I met my future husband at Turnberry towards the end of 1944 when he was stationed there. We were subsequently married in 1946 during his demob leave.

The Land Army girls always looked forward to the various invitations from the airfield as the delicious food was so much better than at the hostel. The music of the Big Band sound contributed to the many super dances we attended, despite the jealous competition with the WAAFs. It really was a very happy time, despite the sadness of war all around.

We also organised our own social events and we could invite up to three friends. These were mainly RAF personnel, and occasions included Burns' Suppers. The Marchioness of Ailsa was invited to our parties. She was a fine old lady who dressed in tartan – she was very friendly and allowed the Land girls access to the beautiful grounds of Culzean.

We were still working at Culzean when General Eisenhower was gifted a flat in the castle in recognition of his great command during the war years.

I am proud to have been connected with Culzean Castle as it is visited by many people from all over the world.

My days in the Women's Land Army were very happy days even though the work was very hard. I made lifelong friends and have many happy memories which, to this day I recall often.

Peter Gibson

My time in the RAF began by serving an apprenticeship in 'electrics and radio' at Cranwell, after which I spent some time at RAF Station Dyce, and then with 612 Squadron. and their Avro Ansons. At the age of eighteen I sought training as aircrew, and initially I undertook a course in aerial gunnery and the basics of navigation at Dumfries and West Freugh. I was then posted to the first of the Radar Courses at Adamton House in Prestwick, where we received instruction on how to use an early type of radar system that was to be fitted to the Beaufort. This was supposed to be capable of locating and homing in on ships from fifty miles away, dependent on the aircraft height. Having learned the theory, it was then on to the OTU at

Peter Gibson and crew (top left)

Turnberry to put it into practice.

Aircrew arrived to join a particular course, usually consisting of eighteen to twenty crews of four. Having come from all over the British Isles and the Commonwealth, we were mostly unknown to each other. We were left to mix in the mess halls and crew rooms, and by some magic process, pilots, navigators, wireless operators/air gunners formed into fours. So, we emerged as a crew, I joined with Sgt Bill Ratlee who was from Moosejaw in Saskatchewan, P/O Dennis Forman, an English navigator, and Sgt Frank Heywood W Op/AG, another Englishman.

We very quickly formed a good relationship and tackled the training with enthusiasm. Dennis was slightly apart, ate and lived in the officers' quarters, and he was also several years older than we were. We 'lads' appreciated his more mature approach and outlook.

We had one bad crash on landing at Turnberry, after a night exercise in marginal weather. We were all shaken, but thankfully unhurt.

The station living and eating quarters were widely dispersed, for the obvious reason of spreading the likely target areas for enemy bombing. Everyone had a bicycle to get around the airfield, the distances between the various dispersal points and accommodation sites were quite large. We were accommodated on the hillside, beyond the railway bridge, on the road to Culzean.

There is no disputing that Turnberry was not a good location for operating Beaufort and Hampden aircraft. Familiarising young, inexperienced pilots with these aircraft required, ideally, flat terrain and hazard free approach paths, parameters one can see Turnberry clearly does not meet. For torpedo training you need water, this Turnberry did have. The alternative stations were not viable. Lincolnshire was over extended, the sheer volume of aircraft and personnel was too much for Chivenor, as it was already an operational station, and of course the south coast was far too vulnerable to enemy action.

Pilots and crewmen were arriving into the Clyde ports from training grounds in the Commonwealth. Command were under considerable pressure to provide crews for the Beauforts coming off the production lines at Filton, and to get them out to Malta and the Middle East. Coastal Command had to train these crews quickly, and I am sure that the commanding officer and training staff at Turnberry would have preferred a more measured approach to the introduction of young pilots, who had very few twin-engine flying hours and these mostly on benign training aircraft, to the vagaries and limitations of the Beaufort Mk I. Add to this the hazards of night flying at a difficult location, there were bound to be problems. All commands had the same difficulties, training times were being slashed, therefore, attrition rates went up, but, vitally, pilots and crews were getting to squadrons in sufficient numbers to make a difference in the course of the war. So, it was at Turnberry, with all its drawbacks that these crews were formed, trained on the basics, before collecting their new charge from Filton and flying to the Middle East. Turnberry was no different to any other operational training unit at the time, pressure was on, right the way down the chain of command.

Having completed our training at Turnberry, we were sent to Abbotsinch and spent about two weeks practicing dummy attacks, without torpedoes, I can't recall now, but we may have actually got to drop a torpedo.

Then it was post haste to Filton to collect our new Mk II Beaufort, we spent a few days getting it tested and sorting out the differences between the old and new models.

Having a short time to get used to our new aircraft, we were posted to the Middle East. Our trip to Gibraltar was uneventful, there we refuelled, spent the night and the next morning set course for Malta in the company of two other Beauforts. During the flight, one of the other aircraft inexplicably peeled away from the formation, lost height and went into the sea. We circled once, reported the position to control and then were forced to leave the area, and continue

on or journey as our fuel load was only enough to get us to Malta, with very little in reserve.

We arrived on Malta and stayed for two weeks. Some crews were held in Malta to join 39 Sqdn. We did two sorties to lay mines at the entrance of Souse and Gabes harbours. It was said that these mines were difficult to sweep.

Command then relieved us of our Mk II Beaufort, gave us a beat-up Mk I instead, and sent us to Shalufa, near Suez for further training with torpedoes before joining 47 Squadron.

We operated out of a base near Alexandria, attacking shipping that was supplying Rommel's army in North Africa, this was prior to the battle of El Alamien, and we continued our offensive as Montgomery advanced with the 8th Army to Tobruk and beyond, thus Rommel was chased out of Egypt. As Montgomery pursued the enemy westward, landing grounds became available and we occasionally moved up to them with a detachment of aircraft, to get closer to the shipping routes, and keep battering Rommel's supply lines.

Eventually, we returned to be based in Malta.

The ace reconnaissance commander, W/Cdr. Adrian Warburton, had 69 Squadron operating a mixed bag of aircraft, including Spitfires, Beaufighters, Marylands and Baltimores. When we arrived back in Malta, we had the opportunity to join 69 and fly Baltimores and I spent my time in Malta with them up to the conclusion of the Sicilian Campaign. Unfortunately, Bill and Frank were killed soon after joining 69, I was shattered, but I was quickly placed with another crew and carried on.

On finishing my tour in Malta, I went with 325 Wing to Palermo in Sicily and from there to Salermo. I had my 21st birthday near Trapani in Sicily. In the spring of 1944, I was repatriated by troopship from Naples to Liverpool.

I had come full circle and found myself back at Cranwell for further technical training and it was whilst I was here that I received my Commission. I joined 206 Squadron at RAF Leuchars, who were flying Liberators on anti-submarine operations, until VE Day. After VE Day, 206 moved to Cambridge where the

Liberators were modified to carry passengers. We then ran a service to Burma and India, to repatriate troops who had been POWs in Japanese hands and urgently needed medical assistance. I was de-mobbed and joined Scottish Airlines at Prestwick until 1951 and then moved to Edinburgh and spent twenty-five years in avionics and aerospace before moving to the United States of America for a further fifteen years in the same business.

I retired in 1990 and returned to Edinburgh, where I was employed on a consultative basis for a few years, finally coming home to Prestwick, native heath of my wife and myself, for total retirement.

George Fox

After a twelve-hour journey from 11AFU, Calveley, Cheshire, I arrived at 5(C)OTU Turnberry at 8 pm on 19 December 1944. A good supper (for wartime) was provided, of two eggs, bacon and a sausage. Apart from its being an operational training unit, I knew nothing about

George Fox

Turnberry, what aircraft I would be flying, how many would be in my crew, how long the course would last, just where I, and presumably my new crew, would be going after completion of the course (Far East?), and so on. In other words, everything was normal, and we'd be finding the answers in due course. My diary entry for the following day at Turnberry is illuminating: 'Most disorganised station I've been on. No one knows we're here or why.' All I knew was that, having been allocated to Coastal Command when I had gained my flying badge in December 1942, I could expect to continue flying much more over water than land, and this proved to be the case. Next day's diary entry (21/12/44) showed some progress: 'Lectures all day. 16 pages of notes.' The following day, there were more lectures, and a welcome issue of Raleigh bicycles for getting around the airfield, and for relaxation when off duty.

Later I met my new crew. The navigator was Derick Wright from Wealdstone (near Harrow), the two wireless operators/air gunners were Les Chambers from Birmingham and Jake Burford from Castleford, Yorkshire. To complete the crew was the Met. Air Observer, a Scot, Jock Jolly. These were all pleasant young men and we all got on well together. However, in another crew was a W Op/AG from London, Harry Mariner, who soon became my best friend, and I happily accepted his invitation to be his best man when the war was over. Because of the great spirit of camaraderie among aircrews, I regret not having aimed at an all London crew, when we might have been able to arrange periodic reunions in London after we had all gone back to civvy street. I did in fact keep in touch and exchange visits with Derick Wright, becoming his best man. He and his wife, Jean, then emigrated to Australia.

Coming back to Turnberry, the weather there was pretty awful and often bitterly cold. We were billeted in huts which were poorly heated. Entries in my diary record: '8 January 1945 - woke at ungodly hour, though had 5 blankets (9 thickness) and greatcoat on bed, and woollen long-sleeved pullover over pyjamas. 10 January

- electric fire in hut confiscated. 12 January - no coal, so no fire in hut. Brrr. 17 January - played football in freezing rain and high wind. 27 January - hot water frozen in washroom.'

I made my first flight in a Hudson on 23/12/44, accompanied by another Hudson trainee, F/Sgt Sedgwick, and the necessary instructor, P/O Passlow. Other instructors I flew with during the sixteen-week course were F/Lt (later S/Ldr) Sherwood, F/Lt Pullum, F/Lt Hanson, F/Lt McNeill, F/Lt Cronin and S/Ldr Allsopp. During the course several training flights were curtailed or cancelled due to bad weather or aircraft unserviceability. For example, the weather prevented any flying on 16 and 20 Jan. On 18 January my diary records: 'flew for an hour until weather closed in.' On 26 January, after flying for only 25 minutes with an instructor on my first night flight in the Hudson, we had to land because of deteriorating weather. Earlier the same day I had taken off with my crew for an air to air firing exercise (Aircraft AM681), but this had to be abandoned because neither turret gun would fire. Repeating the exercise on 11 February (Aircraft FK502), all went well. But a similar exercise on 13 February (AM681 again) was jinxed - this time the target drogue broke away from the towing aircraft! However, repeating the exercise later that day in AM681, was successful.

The course included several Operational Flight Exercises (OFEs) which took us over the Atlantic for a few hours and were mainly navigational exercises. These were plagued with similar weather and serviceability problems, as evidenced by my Logbook:

The 3 April flight was my last main flight of the course and was not uneventful. Because neither the W Op's wireless telegraphy equipment nor my radio telephony was working, we could not communicate with the control tower. As I came in to land, a red Very light was fired, warning me not to land. As I climbed away, I could see no obstruction on the runway, and so circled the airfield and made another approach to land, only to have another red flare fired at me. There was a slight crosswind on the runway, but not sufficient to cause me any concern, so I was very puzzled at the red lights. I made a further two attempts to land, merely attracting more red lights. And then a procedure came into force (which I think had the codename 'Darky'). A searchlight suddenly appeared, circled the sky a couple of times, and then settled into a steady beam indicating the direction I was to follow. So, I did this, picking up other searchlights on the way, as we fell out of range of earlier ones, until the lights of an airfield appeared ahead. So, with no red lights this time, we promptly landed. Having turned off the runway and taxied to the parking area, I was surprised to be guided into a parking space, not by the usual uniformed RAF man, but by a sailor brandishing Ping-Pong bats, exactly as you see in aircraft carrier films! I have no record of where we landed, but it was my first and only experience of landing at a Fleet Air Arm base. We were made very welcome and given a meal. Soon after, transport came for us from Turnberry, and we were back there at 2.30am. The reason for the diversion turned out to be the crosswind, and

3 Feb.	OFE1	3hrs20min	Day	Returned early - bad weather
7 Feb.	OFE2	- 15min	Day	Abandoned - aircraft fault
9Feb.	OFE2	6hrs10min	Day	Recalled by base
23Feb.	OFE3	6hrs20min	Day	Completed normally
1 Mar.	OFE5	5hrs15min	Night	Completed normally
11Mar.	OFE4	5hrs35min	Day	Recalled by base
16Mar	OFE1	4hrs40min	Day	Completed normally
25Mar.	OFE7	1hr55min	Night	Aborted-Radio contact lost
1Apr.	OFE6	2hrs30min	Night	Aborted-Radio contact lost
2Apr.	OFE7	2hrs-	Night	Aborted-Radio contact lost
3Apr.	OFE6	2hrs45min	Night	Aborted-Radio contact lost

the-powers-that-be's lack of faith in my ability to cope with it.

To complete the course, one had to do a minimum amount of night flying. The weather and serviceability problems meant I was falling behind the requisite hours. I was about to get airborne on 28 Jan. but had a burst tyre just before take-off. With no other plane available, my night flying had to be abandoned. On 1 Feb., I managed to get in my first hour of night flying without an instructor aboard. My next night flying was scheduled for 4 Feb., when my diary shows that I waited from 6pm until10.30pm, when firstly no serviceable aircraft was available, and then, when a plane did become available, the weather closed in and all flying was cancelled. On 5,6,7 and 10 Feb., night flying was also cancelled because of the bad weather.

11 Feb. proved a fateful day. I was due to make two-day flights and was also down for night flying. The first flight was the air-to-air firing exercise mentioned earlier (in FK502). The second day flight was in a different Hudson (AM681), in which we did a "W/T QGH" exercise. This is a controlled descent through cloud, where the aircraft is tracked from the ground, and instructions given by radio to the pilot on height and course to fly to get him safely down through the cloud. (Even today there are cases of pilots coming down through cloud, unaware of obscured high ground below, leading to a high risk of fatal consequences). On the evening of 11 Feb., it was very dark, raining, and there was low cloud around the airfield. However, this was no worse than I had experienced before coming to Turnberry, and I had no qualms about flying. I still had only one hour's solo night flying in the Hudson and was required to bump up my hours by getting airborne as quickly as possible in the nearest aircraft. This was FK502, parked very near to the crew room. So, after a brief word with my close friend, Harry Mariner, I and my crew got aboard the Hudson, stowed our parachutes and awaited the ground staff to bring their accumulator trolley

to plug in, so that we could get the engines started. Meantime, I completed my preliminary cockpit check. We then had a visit from the maintenance Flight Sergeant, who told me that the plane would not be ready for another twenty minutes. When I explained my need to get airborne without delay, he directed us to another plane quite some distance away, which he said would be the first to go. So, we collected up our gear, and trudged off to the other plane. Having got settled in, I was astounded to see, after a few minutes, that the accumulator trolley, instead of coming to us, was going to the other plane, which proceeded to get its engines started. It was then our turn, but the opportunity to be first in the air was gone. By the time I was ready to turn on to the runway, the other plane had already taken off. Just as I was about to call the control tower for permission to turn onto the runway and take off, a red glow appeared in the night sky and quickly disappeared. The tower then repeatedly called the first aircraft, getting no response. Knowing that my friend Harry was in that plane, and that only a miracle could save him, I called the tower, seeking permission to take off to see if I could spot a dinghy, or any sign of life, although I realised this was very much a forlorn hope. I was told to stand by, and then told to return to the parking apron, as further flying was now cancelled. Sadly, there were no survivors from the four crew, and as far as I know, no bodies were ever recovered. As I had flown the crashed aircraft previously that day, I was asked at the subsequent enquiry whether I could shed any light on why the aircraft had crashed. I could only report that the plane had performed perfectly normally earlier.

Nine days before the crash, we had a day off, and Harry invited me to join him and his fellow W Op/AG to a late-night party in Girvan. Harry was a very likeable person, and he had somehow got an invitation from some local girls for him and friends to join their party. As I was scheduled to fly the next morning, I was far from keen on having a late night. However, we learnt that widespread fog was expected, and that no flying would be possible. So, it seemed well worth taking a chance. I can recall no details of the party, but my diary tells me that it finished at 1.30am, and that we "arrived back at camp at 6am after two hours sleep". I have no idea where we slept - possibly at someone's home or perhaps in a bus or rail station if Girvan was so equipped. But far from being foggy, the day started bright, and I and crew were soon in the air on OFE1. I remember finding it difficult to avoid dropping off and was greatly relieved to be recalled to base, because of deteriorating weather. We flew for 3hrs 20min, but without the recall, I would have had to struggle to keep awake for at least another hour.

The weather continued to be bad, and night flying was cancelled on 12, 13, 14, (just as we were ready to taxi towards the runway), 17, 19, 20, 22, 25, 27 and 28 Feb., 4, 6, 9, 18, 26 and 27 March. Throughout this period, I was airborne at night on just three occasions, totalling only 9hrs 10min. (However, with better weather in April, - but still suffering curtailed flights due to loss of radio contact - I was able to increase my solo night flying for the whole course to 17 hrs 25 min).

Three weeks before the February crash, on 21 Jan., another Hudson (AM562) had violently fallen into the sea shortly after take-off, again with no survivors. My diary records that the bodies of three of the five crew were recovered. This crash occurred in daylight in good visibility, and with a very experienced pilot at the controls. At the time the accident seemed inexplicable, and I do not know if the later enquiry was able to establish a likely cause.

We had one crash on the airfield itself, fortunately with no casualties. This was on 12 Jan. when a Swordfish with spluttering engine came in at a steep angle, hit the ground and stopped dead with its tail high in the air. We were some distance away and feared the plane might catch fire. It didn't, possibly because it had run out of fuel. We waited and after a few minutes we were relieved to see the pilot climb out of the plane apparently

uninjured. If memory serves me right, the pilot was a lady on ferrying duties. *[Cdt. Miss P. M. Provis, ATA. Flying Swordfish NF369 of 4FP, force landed, undershot airfield, hit embankment, avoiding hangar and parked aircraft – Author].*

I referred above to the QGH descent-through-cloud procedure. There was one flight from Turnberry when I would have liked to use this but couldn't because our radio wasn't working. This was on 30 Jan. The exercise was a "Dual Climb" (i.e. an instructor, F/Lt McNeill, was aboard, as well as my crew). The climb entailed my taking the plane to 18,000ft., turning the plane periodically through 90 degrees, so that we kept roughly above the airfield area all the time. Systematically the plane would be levelled out to enable the instruments to settle down, and for the Met. Observer to take temperature readings etc. at the different heights. We were soon above cloud, but during the exercise, we lost all radio contact, which made accurate navigation impossible. Theoretically, it is possible to get a bearing on the sun, using a sextant, the precise time, and complicated mathematical tables. Sailors appear to have mastered this process, but I could never get a sufficiently accurate result, and I think my navigator had similar difficulty with his astro-navigation. Eventually, as we descended (Derick being reasonably optimistic that his navigation had brought us over open sea) we broke through the cloud, and indeed were over water, with land on our left, and open sea to our right. However, there was a complication. We were then on a northerly course so, if we were near Turnberry, the land should have been on our right, not our left. So, we appeared to be off the east coast of either Ireland or Scotland (or perhaps England). With the radio still dud, we had no-one to ask nor, without some idea of where we were, could we begin to map read. Fortunately, in wartime, there were many airfields equipped with "pundits". These flashed a two-letter code in Morse, and we carried a list of the pundits with us. It wasn't long before we came to an airfield, whose pundit was flashing AK, which we promptly identified as Acklington (near Newcastle). So, I turned west, while the navigator worked out a more precise course for Turnberry. Luckily, the cloud base was high enough for us to fly below it all the way back. The exercise normally took about two and a half hours. The way we did it, we took four hours and ten minutes, including an unscheduled forty minutes night flying. Back at Turnberry, because nothing had been heard from us and we were so long overdue, we had been all but written off!

Because of the poor weather, much of our off-duty time was spent on the "camp". We had various ENSA shows of varying quality, and there was a well-attended camp cinema. Once, we had a Fitness Parade, which turned out to be a route march to the nearest pub about a mile away. Other activities included swimming, cycling, (several times into Girvan about 6 miles away), and walking along the beach to Culzean Castle. Twice, on days off, we went into Ayr. On the first occasion the bus was too full to take us, but we managed to get a lift in an RAF lorry (which also had some Italian Prisoners of War aboard), with the lorry soon overtaking the bus.

One other, very important, activity was writing to my home, to my 3 brothers (2 of whom were serving abroad) and to various friends. My diary makes only one reference to incoming mail. This was a letter from my married brother in Enfield (who was rejected for military service because of heart trouble), in which he mentioned his fire having been blown out by the blast from an exploding V2 rocket. I kept a record of letters I sent, to help make sure that I didn't fall behind with my correspondence. In my 106 days at Turnberry, I managed to write and send 71 letters. My navigator, Derick, was (and is) a prolific letter writer, and his letters, which he claimed were to his sisters, would have far exceeded my total!

On the evening of 11 April (having completed as much of the course as the bad weather and poor aircraft serviceability had allowed), I was glad to go on leave from Turnberry, arriving home

at Enfield next day at 8.30am. On 23 April, I left London at 4pm and arrived in Belfast next day at 12.30pm (after a tedious journey via the Stranraer-Larne ferry) on my way to RAF Aldergrove. Here a 7-week conversion course to four-engined aircraft awaited me. I also acquired a further three crew members (second pilot, flight engineer and wireless operator/mechanic). After the trials and tribulations, we had suffered at Turnberry, this course seemed quite idyllic. During it, on 8 May 1945, we had the highlight of going into Belfast to join in the VE Day celebrations, now that the war in Europe was at last over.

Alex Smith

When I was stationed at Corsewall Point, a notice appeared informing us that a Confirmation Class was commencing, and would those interested please contact the padre. At the initial discussion/talk, I was surprised, but previously unaware, that the only other candidate was a very pretty, petite WAAF, the 'apple of many of the unit airmen's eye'.

After attending the required number of 'lessons' we were given a day's pass and requested to be at headquarters, RAF Turnberry on a certain day and time.

Travelling on the Stranraer to Glasgow bus, we arrived at the Turnberry guardroom, where identities checked, we were directed to the large headquarters building, where we joined candidates from other units and met the Bishop of Glasgow, who 'performed' the Confirmation and issued us with certificates to that effect.

After joining with the Turnberry airmen for lunch, we made our way through what seemed to be a typical wartime airfield and caught a return bus to Stranraer, where we went to the cinema and saw 'New Moon', before we returned to camp and I walked the young lady back to the WAAF quarters before returning to my billet, where all my mates on being informed of my fellow traveller, in unison said, 'You lucky B-----, Smithy!'

Three weeks or so later, whilst crewing HSL152 in company of another HSL on training exercises off Arran, heavy weather was encountered and being training, not operational, it was deemed prudent to shelter in Ayr harbour overnight, aiming to return to base early next day. With the early start in mind, the engines were being run up, when a radio message was received informing us that an aircraft had crashed, just off Turnberry. Within minutes both craft were leaving the harbour, line abreast, flat out, almost leaving the water as they met the still running, heavy seas, and the scene still vivid in my mind is looking across those stormy seas to Arran where snow-capped Goat Fell glittered in the wintry sunshine.

152 soon outstripped her companion launch and covered the short distance to Turnberry quickly, so within a very short period of the aircraft crashing, rescue craft were on the scene. The sister launch soon arrived and an intensive search commenced, crossing and criss-crossing the indicated areas. The search continued for a long period, with all crew acting as lookouts, but there was no sign of any wreckage or leaking fuel. As the Hudson crashed on take-off, it would appear that the 'kite' went straight in, with no chance to try for a forced sea landing. After a long period, the search was abandoned, and we reluctantly returned to base.

Watching Pro-Celebrity Golf from Turnberry on TV, with shots of the hotel, Ailsa Craig and Arran, with the sea and sky so blue and calm, thoughts of those brave men out there still come to mind, whose age would not be so different to mine. Invariably triggering memories of those waters, (usually much rougher!), of passing through those long lines of ships in convoy, with their flanking destroyers; of the guard ship near the Cumbraes, flashing out with its signal Aldis lamp, 'who are you and where bound?, as we came through at night. Being told that U-boats would hide beneath the Ailsa Craig, where apparently it was difficult to detect them with sonar.

Those waters are the home of many, many memories. Not much later I was destined to pass close to the Heads of Ayr on the troop ship *Mea Hellas* (Greek), on the way to the Japanese War,

finally finishing up in North Borneo, still with the flying boats.

Claude Henry Turner
Henry (only his parents and blood family called him Claude) came from a yeomanry family of Church of England priests and military officers.

A sickly child, he was sent to Malvern College in the hills as it was felt that this was the best place for a child with chest problems. This childhood weakness possibly influenced his attitude to life: he was a fanatical sportsman, allegedly representing the RAF in more sports than any other person, while he always enjoyed the company of beautiful women.

After training at Cranwell, he returned there as an instructor with his wife, Margery. They had met at a country house party: coming to the last fence of a friendly point-to-point, Henry realised that the young Margery Lazenby was going to beat him: he drove his horse into hers and went on to win!

The young officer was transferred to Northern India, (now Pakistan) where there was a constant guerrilla war against the hill tribesmen. Henry and Margery put their skill with horses to good use. She broke-in imported Australian ponies, which Henry rode in polo matches and then sold. Unfortunately, the Quetta earthquake interrupted this idyllic lifestyle shortly after the birth of their son, when thousands died, and they had to rescue the young Tony who was buried in the rubble.

The family finally returned home just before World War II. Henry becoming an instructor at RAF Halton, his task was to give the young ground crew apprentices the experience of flying. One youngster got his feet caught up in the controls and Henry's Tiger Moth turned turtle shortly after take-off: no one was hurt, but what Henry did not know was the spectators had included his wife and son. Typically, on returning home for lunch, Henry said nothing of his escapade.

Later in life, Henry often said that the Second World War came too late for him. By which he meant that in his thirties he was considered too old for active service as a pilot, although he spent some months in 1940 flying Ansons on anti-submarine patrol.

After spells in charge of barrage balloons in North London and at Wick in northern Scotland, Henry was sent to Malta as Second-in-Command of the RAF base there. RAF Malta played a crucial role in protecting Allied fleets supplying the Eighth Army in North Africa and then Sicily, but he had a far more personal memory of his time in Malta: having succeeded in getting submarine captains to bring squash balls to

travel north to Troon for golf (when petrol was available), so he established some practice holes on the aerodrome, which enabled him to earn a place in the RAF team to play other services.

An appearance before the CO was not looked forward to as he sat with crossed revolvers in front of him (he practised every day). It is rumoured that he had quite a reputation among the local ladies.

In 1946, Henry was transferred to Transport Command as CO of RAF Pershore in

Group Captain Claude Henry Turner (centre)

Valletta, the Germans then destroyed the back wall of the court during a bombing raid, so he had to learn to play three-walled squash.

Returning to England, Henry was stationed at Coastal Command Headquarters before taking command of the Torpedo Training Unit at Turnberry. He and Margery lived in two Nissen huts at the Maidens end of the airfield.

As the war approached its conclusion, Henry was able at Turnberry to lead his men in successful sporting campaigns including hockey and cricket. He found it irritating that he had to

Worcestershire. It was here that he claimed to be the first RAF Officer to replace his batman with a batwoman (WAAF).

Henry's final posting in the RAF was as Air Attaché to the British Embassy in communist Poland. While many of his embassy colleagues spent their time meeting with their western allies; Henry believed his task was to gather intelligence and to this end he employed a beautiful Polish girl to teach him the language: his affair with his tutor, was but one of many, however his attempts to rescue her and bring her to England after his embassy posting, was to lead to a show trial and imprisonment by the communists. [He recounts in his book *International Incident,* published by Alan Wingate 1956]

His wife Margery had stood by him during all his adventures, but on his return from Poland in 1951, he demanded that she divorce him.

Having tried his hand at selling life insurance: a disaster - his method was to play his clients at golf in the hope of trying to then sell them a policy - the trouble was that Henry always insisted on winning.

Inheriting a small sum of money, he went into a partnership to establish the original 'Painting by Numbers': unfortunately, his partner disappeared and so much of the inheritance was spent paying over creditors. However, when the partnership launched its product at the Ideal Homes Exhibition, they had employed an attractive young demonstrator, Sara Travis: Henry and Sara married in 1956 and had a son, Richard.

The family settled at Cleeve House near Malmesbury in Wiltshire, where for many years they prepared broiler chickens for the market.

As he grew older, Henry was unable to play any of his beloved sports, though he did try fishing. His finely tuned, athletic body was no more and so Henry's other great passion - attractive women - was no longer an option.

He died in 1980, near Woodstock, Oxfordshire.

[With thanks to his son, Tony Turner]

William (Bill) Williams

William Williams was born in Liverpool and enlisted in the RAF on 29 August 1939, at the tender age of 15. With living in Liverpool, it was a close call as to whether he joined the navy, as some of the men in the family had.

His first flights were in Defiants at the Air Gunnery School at Walney Island, and then on to wireless operator training at Hawarden. His first operational posting was to Invergordon in 1942 to fly in Sunderlands, this lasted until 1944 and he experienced many 'run-ins' with JU88s.

After this it was to Turnberry and 5(C)OTU training on the Beaufighter TFX. On the morning of 13 April 1945, with F/O Payne at the controls of Beaufighter RD426, Bill Williams walked away from a crash on the beach at Turnberry. The only comment in his logbook records, 'nearly had my dish'. It was a miracle that both crewmen survived. Bill Williams was flying again later that same day.

There followed a short stint with Sunderlands again in the Middle East and then on to Dakotas at Wymeswold, Dishforth, Syerston and Kabrit again, in the Middle East. The next posting was on Operation Plainfare - the Berlin Airlift - operating between Fassburg and Gatow, completing over sixty sorties.

It was then back to Sunderlands at Pembroke Dock, and Ansons at various locations before being posted to Kuala Lumpur to take part in the Malayan Emergency from 1956 to 1958. He was with the 'Voice Flight' on Dakotas doing propaganda broadcasts and leaflet drops to the Guerrillas.

On leaving Malaya in 1958 he joined Coastal Command at St Mawgan and Kinloss, flying in Shackletons. His final posting was to Brize

Bill Williams

Norton with Air Support Command flying in Belfasts, Britannias and VC10'.

He left the RAF in 1969 and joined Hawker Siddeley, later to become British Aerospace, working as an electrical inspector on Harriers and Hawks.

My Father never spoke much about his time in the services, I know it is my fault for not asking, I just wish I could now.

F/O Payne had already been to Turnberry in September 1943. Crewing with 'Mac' McGaw's crew flying Wellington Mk I's and Mk II's at 7 OTU Limavady, McGaw volunteered his crew for torpedo work. They were then posted to 1TTU for a seven-week intensive course. They started with low flying over the sea by day and night (with the assistance of flares) The object was to keep the plane at fifty feet above the water. They practised simulated attacks on target ships, using first a camera that operated when the bomb doors opened, and later, a torpedo with a dummy warhead.

'At night the method was to first find the ship using ASV then, if it was a moonlit night, the skipper would lose height and make his attack, using the moon to silhouette the target. If there was no moon, two sticks of 4,100 candlepower flares were dropped at right angles to each other so that which-ever way the ship turned it would still be silhouetted'.

'There were many incidents; the fifty feet above sea at night being one of the main problems. Too high, the torpedo bounced (as one of the target ships found out), too low usually resulted in propeller damage, or, as happened on more than one occasion, the Wellington hit the sea. When we were lumbering down the short runway at Turnberry in a Wellington Mk I with a 2,000 lb torpedo on board, we wondered more than once if we were going to make it'.
[Thanks to Chris Williams]

John Entwistle
I was an RAF Navigator and, prior to commencing my second tour, I was cross trained as a Meteorological Observer and was en route to a Met. Squadron when I spent April and May

1945 at 5(C)OTU at Turnberry, flying with F/Lt Applebee in Hudsons. I was there on VE Day and vaguely remember the celebration, but naturally after all this time, the exact events are a bit hazy!

One of my fondest memories is of an elderly lady in Maybole who served us egg and chips for a very reasonable price, and it became a habit of some of us to cycle to her house as often as possible. I also remember there was always a huge urn of real milk in the Sergeants' Mess and we were encouraged to drink a glass a day. I sometimes thought that I was the only occupant who liked milk, but I recall drinking gallons of it.

We did navigation exercises and always used Ailsa Craig as a navigation point. We all enjoyed regular excursions to Ayr and exploring the surrounding countryside.

From Turnberry I went to Tiree with 518 Squadron and from there to Aldergrove and 202 Squadron. We flew Halifax's on meteorological flights over the Atlantic, which I continued until my release in 1948.

I emigrated to Canada the same year and in 1951 I joined the Royal Canadian Mounted Police, serving with them until I retired in 1982.

I now live in Nova Scotia, many areas of which resemble Scotland, I guess that's why they call it 'New Scotland!'

Cledwyn A. Evans
On 18 April 1945 I received a posting to No.5 (Coastal) Operational Training Unit at RAF Turnberry as an instructor.

I cannot say that I left my mark on any subject. While there I was sent to RAF East Fortune, on an instructor's course, I forget how many weeks that lasted, anyway I knew more about things when I left there than I did before. Then back to Turnberry, shortly afterwards six of us including two officers were invited by the navy to pay a visit on board an aircraft carrier at Greenock. Off we went and there was the usual formality of senior rank leading, followed by the rest of us, up the gangway, and a salute given to the senior officer and the quarter deck. It was a very interesting visit and hospitality was great. We were taken all over the ship which was originally a banana boat which had been given a flight deck over the original deck so converting to a small aircraft carrier. I forget the name. Things went on as usual at Turnberry until 10 October. I should have said that I was on 1 Torpedo Training Unit flying Bristol Beaufighters (not me personally). Two aircraft went out one morning on a torpedo exercise, I don't know what had happened but they both collided, one managed to struggle back but the other went into the sea and both occupants were lost, one a pupil and the other an instructor. Flying went on as usual. At one stage, I think it might have been during September, I had some leave.

While at Turnberry, I, along with others was issued with a bicycle, a very useful machine. I used to cycle miles on this thing, anyway when going on leave or leaving the camp for any reason we were supposed to hand the cycle in at the cycle store to the sergeant in charge. On this occasion I decided that while on leave I would leave my machine in my billet; I had my own billet. So off I went on leave, and when I eventually returned, I found that my bike had gone.

So, very red faced had to go and see the sergeant in charge, fully expecting to have to pay for it. But living in the same mess and having the rank of warrant officer by now, the whole thing was forgotten about. Had I been an airman no doubt things would have had a different ending. I liked being stationed at Turnberry. I cycled many times to Maybole and the surrounding areas. We aircrew could use the indoor swimming pool at Turnberry Hotel. I was mainly involved with ground training of radio navigators. I distinctly remember a building that was spherical inside where the pilots and navigators carried out exercises. Inside this dome all kinds of lighting effects could be used, crews could simulate flying, and the lighting could be changed to suggest different conditions or times of day or night. A lot cheaper and safer than the real thing!

I recall one summer evening, a Wellington aircraft coming in to land but its wheels would not come down, so they had to circle for a long time, using up their fuel before they could land.

Eventually the pilot managed to get one wheel down and attempted a landing like that. He managed to keep it level for as long as possible, before it sank on to its belly and slid to a stop. Luckily there was no fire, and no-one was hurt.

Then came 10 October and 1 Torpedo Training Unit was moved up to RAF Tain, still in Scotland. Right, up at Tain training still took place, still with Beaufighters and still with practice attacks on a Royal Navy destroyer called HMS *Havelock*. When these aircraft of ours attacked the target ship, the people on board used to take all kinds of bearings and different readings during the attack and eventually these readings were passed to me in Morse by radio who in turn passed them on to people who worked out exactly how the attack took place and all the errors and whatever was not done properly.

Margie Fraser

I was stationed at Turnberry 1945-1946 after completing my course as a Radar/Wireless Mechanic. I was one of the first batch of WAAFs to be trained in this, allowing the release of servicemen for active duty.

Apart from the year I spent training in London, I really saw very little action, and after May 1945, time was mostly given over to social activities.

At Turnberry we had a mixed hockey team and I instigated a sports day which was a great success. My only claim to fame was that I played for the WAAF hockey team, at both Scottish and British level.

Albert Anderson

My RAF career started at the recruiting office in Bothwell Street, Glasgow, during September 1938. I passed the educational and medical examinations, although the latter caused the doctor to summon his female assistant, saying, 'Come and see this, he's like a skin't rabbit, but he's warm, so he's in!' A group of us were then sworn in and told to be at the station to catch the midnight train to the RAF Reception Depot at West Drayton.

After initial training, I had completed a tour of operations in Bomber Command with 420 Squadron, RCAF and then took an admin course at the RAF School of Administration and Accountancy, Hereford, whereupon, the Air Ministry posted me to RAF Turnberry. I served there for a short period, 3 July 1945 to 5 December 1945. My final release date from the RAF being 31 January 1946.

On arrival at Turnberry, I was to take charge of the technical library for 2/3 weeks and fill the assistant adjutant's chair until my demob papers arrived. The officers in charge of the various sections were very peeved, when I, a total stranger, being a new broom in charge of the Technical Library, issued a letter to each department requesting the return of long outstanding library books. They expressed their chagrin at an officer commanding's 'In night' and had arranged with the bar staff to have my soft drink 'spiked'. I was given an extremely strong Pimm's No. 1. However, as a non-smoker and non-drinker, after the first sip of the drink my head started to spin, and I drank no more!

During this period, I was to find a Flight Sergeant Sandy Davidson in charge of the Salvage Department. Sandy and I were both born and brought up in Grantown-on-Spey. Sandy was older than I and had joined the RAF in the early 1930s. When he was on leave his blethers about the RAF and his service abroad in Singapore gave me the notion, and I joined in 1938. At Turnberry, Sandy and I did our Orderly Officer duties together on a few occasions. Counting the Italian prisoners of war at 'lights out' was a problem. Any POW not standing by his bedside was always 'in the toilet'. A prisoner from the hut previously inspected had not been quick enough to fill the vacant space. They would be covering up for a chum who was in town visiting a girlfriend. Fortunately for me, Sandy knew all their tricks and how to deal with them!

By coincidence I saw another familiar face in the Officers' Mess. It was Norman Shaw, a Glaswegian, we had served together in the Shetland islands at Sullom Voe during 1941 and 1942. Norman as a clerk and me as a driver,

Motor Transport. Now Norman was a pilot and still flying the Halifax from Tiree, and I had just finished a tour as a Flight Engineer on Halifax Mk IIIs.

I had requested this posting to Turnberry as my wife, Mary, was in Robroyston Hospital, Glasgow, having artificial Pneumothorax treatment for tuberculosis. Our two children were in a Glasgow Corporation home. The terms were fixed and were paid monthly through my RAF bank account. Mary had been unwell for quite some time but had refused a bed in Robroyston as she was caring for her aged and bedridden mother. Her mother died on 21 May 1945. With her mother at rest in Craigton Cemetery, Mary first ensured that there was a bed for her in Robroyston. We had no relations that could take on the responsibility of looking after two young children, they were all elderly and unable to cope. After considerable argument with the Glasgow authorities, Mary finally had the bairns placed in a home and herself in hospital. She had been under considerable strain for quite some time. She was an ex-nurse from Mearnskirk Hospital and had probably caught the tuberculosis germ there. Part of a nurse's duties was to clean out sputum flasks, bed pans etc. masks were seldom or if ever used even when available. With her own illness steadily worsening, her elderly mother's health deteriorating, her aged father not able to give a hand and two young children to care for, two houses to look after, provisions to get and laundry to be done for both households, Mary's tasks were endless. The worry about me on operational flying didn't help much.

Thankfully my request for a home posting was accepted and from Turnberry I was able to visit at weekends, providing I was not on duty. I would visit the children in the morning and Mary in the afternoons. Sunday evening would see me back at Turnberry. Due to my family circumstances I didn't socialise much, I had always been a bit of a loner anyway, I never got to the stage of knowing the names of the other officers. Although there was one Flight Lieutenant, from Fraserburgh, who befriended me.

Mary had many a month in hospital, going through all the curative treatments known for TB – fresh air, sleeping on an open veranda, the bed placed there and sheltered from winds, rain and snow by a waterproof sheet. Artificial Pneumothorax, an operation whereby air is injected into the space between the lung and the chest wall, known as the pleural cavity. The air causes the lung to collapse and prevents air entering it during breathing. This is the most effective way known of resting the lung. It was a very trying period for Mary, unable to see the bairns, or able to visit them.

Out of hospital, and now back at home with the children she loved, she was soon back to full household duties, provisions to get, meals to prepare, two children in need of care and attention. All this and still to attend the hospital for treatment.

I had hoped to continue my service in the RAF, however, circumstances put paid to that. So, my release date as confirmed as 31 January 1945. One of the final things I did at Turnberry was to wangle a 'flip' in a Warwick and enjoyed it very much, it would be the last flight in an RAF aircraft.

It was marvellous to be at home with Mary and our two children Anne and Andrew. There was one great problem though, there were no jobs. The few jobs that there were had thousands of servicemen and women filling them on demob. Mary's brother, Tom, walked right back into his old job he left to join the Army with his TA Unit. Tom was a driver and managed to get a position for me. I tried very hard to make a success of the job, but there was no job satisfaction and the pay was very poor.

I had tried the employment exchange and the ex-officers exchange, but there were no jobs available. Mary agreed that I should go back to my hometown of Grantown-on-Spey, but again, the same story, no work. Eventually, a job as a driver with a timber merchant came up, and I grabbed it. Soon I was working alongside a German prisoner of war. We became good friends and worked extremely hard for our employer, both of us for a pittance, he as a POW and me because of circumstances, i.e. no other employment. Then I left the driving and took a job woodcutting which involved cutting down the trees and sawing them into lengths for the sawmill. It was hard work, but enjoyable.

By this time Mary and the weans had moved to Grantown, we had the shell of a house built on my mother's property, and soon Mary was sawing wood and thumping in nails to fashion her wee bungalow. Building materials were still on permit and Mary had quite a few arguments with the authorities, but she always got her permits. No one and no authority daunted Mary! I took over after work and at weekends, soon we had the house up to a standard to enable us to move in and live in reasonable comfort. The pay I was earning was still very poor. Mary had to make the best of it, and I can assure you she did! A storeman's job became vacant in the 26 Command Workshop REME at Grantown. I applied and was successful, although the pay was still small. I remember the words Mary said when we first got married on an airman's pay of seven shillings a week, 'Give me four walls and a wee bit money and I'll make a house we can live in.' And she did just that. From house to garden, flowers, fruit, vegetables poultry. She had roses round the door and climbing on every fence. Eggs and a cockerel for the pot, everything was at hand. Mary was talked into having her father stay with us, but he was no problem.

Unfortunately, Mary's TB started to flare up again. Soon it was back to treatment involving trips to hospital in Elgin, 36 miles away. Mary spent a horrific few months undergoing surgery, and was in dreadful pain, the drugs she was given seemed powerless to help.

After a long spell away from home, Mary was over the moon the day she was released, back home to her two bairns, Anne was now 11 years old and Andrew 10.

From my storeman's job with the REME, I took charge of the stores section as the foreman. With the closure of the depot, I requested a posting to the Command Ordinance Depot at Stirling. Mary wanted to be close to our daughter Anne, who was now married and living in Falkirk.

I finished my service with the MOD, being the OIC Workshop and Transport section, with three Queen's Honours. No monetary value, just a thank you for services rendered. They were: Her Majesty's Silver Jubilee Medal; The Imperial Service Medal, and Certificate of Commendation, awarded by the Commander in Chief, United Kingdom Land Forces, General Frank Kitson.

Mary and I were able to enjoy many good years together, in all we had 63. We celebrated our Diamond Wedding and enjoyed many holidays at home and abroad. We were inseparable, she truly was a wonderful lassie. During my retirement years I have written about our life together, simply to pass my time and for my own reading. I also wrote about my sorties with 420 Squadron, our pilot and our bomb-aimer both winning the DFC. My executors will have quite a bonfire! I feel I have nothing much of interest to recall about RAF Turnberry, but as I hope the reader will understand, I had other things on my mind.

William Aitkinhead

I arrived at Turnberry in August 1945 after a tour overseas. I was billeted in the last hut on the shore at the north of Maidens. My unit was in a hangar at the furthest end south of Maidens and I had to cycle there and back.

Whilst at Turnberry I met Forrie Reid, an ex Allan Glen's rugby player who soon talked me in to playing for RAF Turnberry against the Naval Commandos at Heathfield.

One Saturday night I went with a few pals to see Kilmarnock FC at Rugby Park. I can't remember much about the game, although I do remember thinking that I was daft going to see the game, when I was only seventeen miles from home and could have gone there instead.

I was told to report to Station Headquarters to make a claim for my medals and was interviewed by a WAAF officer – I had never seen one before! I gave her the dates and approximate durations I spent at various RAF stations overseas. I made a claim for the Africa Star, but was told I had arrived too late in Egypt. So, I withdrew my claim for any medals at all, I was very

disillusioned with my time in the service and feeling very cheesed off!

I made an application for a weekend pass for the Glasgow September holiday. We used to spend this in Whiting Bay on Arran before the war and fancied a long weekend there. I arranged for a friend to take my 'civvies' from Glasgow to Arran. On the Saturday morning I had a breakfast of fish in the dining hall. Then I found there was no transport available but managed to get a lift in a lorry to Ayr bus station, then an onward bus to Ardrossan only to find that the ferry had been moved to Wemyss Bay because of stormy weather. So, another bus managed to get me to the ferry for about 2 pm. I made for the dining salon and found that there was only fish for lunch, so I had to make do. We set sail to Brodick, Lamlash and finally Whiting Bay.

While sailing between Holy Isle and just off Lamlash I saw a biplane landing on a floating landing strip, I think this was possibly the Fleet Air Arm.

I arrived in Whiting Bay, met up with my pal and headed up the hill to the Clifton Hotel, where I changed into my 'civvies' and was pleased to hear that the kitchen had kept a meal for me – more fish!

I had an awful thirst after eating so much fish, so we went to the Whiting Bay Hotel, which was the only place with a drinks licence at that time, only to be told that they had run out of beer!

It had taken me almost twelve hours to get from Turnberry to Whiting Bay, yet they are not even twenty miles apart. I don't remember how I got back to Turnberry, but I was in a foul mood.

Shortly after this I was posted to Chivenor in Devon. This was in October 1945. Chivenor was just like Turnberry, no activity and lots of 'bods' hanging about doing nothing. I ended up playing Rugby again at Chivenor and lost my two front teeth, thankfully the Station Dentist made me up a denture.

One of the officers discovered that I was an only son and that my father was in business for himself in the building trade. He suggested that I try for a compassionate release. I made the application, which required a doctor's letter confirming that my father's health was not good, and I needed a signed declaration from a business associate to confirm that I was needed back to help him run the business. I sent letters home by registered post requesting this information and received the required papers by return of post. The RAF moved fast, and I was cleared from Chivenor the next day!

I arrived home at Hogmanay, and after two days' holiday, started back at work on 3 January 1946.

Alan King

I was sent to Turnberry for re-crewing following a rest tour 1945/46. I had been in line for posting to the Far East, but unfortunately due to a re-occurrence of Malaria which I had contacted on an 'ops' tour in Africa made me unfit for overseas duty.

I was promoted to W/O around this time and was given all the most important jobs, like I/C salvage, I/C Fire Piquets, I/C Italian POW Camp, CMC Mess etc. The responsibility was frightening!!

My stay at Turnberry was very enjoyable, the locals from the villages were the friendliest people I have ever met (especially at Hogmanay). The fishermen's cottages along the front all opened their doors and made us most welcome.

Douglas and Margaret Fairweather

Of all the names commemorated on Turnberry War Memorial, perhaps the most poignant are those of husband and wife Douglas and Margaret Fairweather. Neither trained at Turnberry, but uniquely are buried together in a Commonwealth War Grave at Dunure (Fisherton) Cemetery. It is believed that this is the only known instance world wide of a joint war grave.

Captain Douglas Keith Fairweather was the son of Sir Wallace Fairweather and Dame Margaret Eureka Fairweather of Glasgow. He served as a Chief Motor Mechanic in the RNVR during WWI. He received his Royal Aero Club Aviators' Certificate at the Scottish Flying Club on 14 December 1928. His father, Sir Wallace Fairweather, kept an Avro Avian plane in a field at King Henry's Knowe, Newton Mearns, just after WWI.

Douglas Fairweather was one of the first to sign on with the Air Transport Auxiliary. [The Air Transport Auxiliary (A.T.A.) was a civilian organisation which ferried aircraft from the factories and maintenance units to the squadrons of the R.A.F. during World War II. It also transported service personnel and undertook air ambulance flights.] He set up the Air Movements Flight at White Waltham in 1942. Known as 'Poppa' Fairweather, gregarious and eccentric, it was reported that he chain-smoked, (smoking was against regulations anywhere near an aircraft) measuring his journeys by the number of cigarettes he smoked. A journey from Belfast to White Waltham took 23 cigarettes, or 2 hours 40 minutes! He even had an ashtray installed in his cockpit. Another regulation he paid lip service to, was the ATA instruction that each pilot had to carry their own maps, his was about three inches square, nowhere in the directive did it mention scale!

He had married female pilot, the Hon. Margaret Runciman, in 1938. Margaret was the Daughter of the Rt. Hon. Walter Runciman, 1st Viscount Runciman of Doxford, and of the Viscountess Hilda Runciman of Doxford, Northumberland. Prior to the war she was an instructor with the Civil Air Guard. She also taught at Glasgow Flying Club and had logged over 1,000 hours of civilian flying. Joining the Air Transport Auxiliary, she was one of the first women pilots at Hatfield, starting in January 1940 and was reputedly the first woman to fly a Spitfire. She and her husband Captain Douglas Fairweather, were both posted to Prestwick in November 1941.

Margaret, being heavily pregnant, took a leave of absence from flying in early 1944, when her husband Captain Douglas Fairweather lost his life on an ambulance flight, somewhere in the Irish Sea, in Anson N4875, on 4 April 1944, while flying in atrocious weather, en route to pick up a patient with serious injuries. His body was washed up on the shore at Maidens. Margaret gave birth to a baby girl a few days after Douglas was killed. She returned to work on 15 June 1944.

Flight Captain the Hon. Margaret Fairweather was killed on 4 August, flying a Percival Proctor from Heston bound for Hawarden, near Liverpool, with passengers Louis Kendrick of the Ministry of Aircraft Production and her own sister, the Hon. Mrs Kitty Farrer, who was Adjutant to the Commandant of the ATA's women's section. Almost twenty miles from her destination, the engine of the aircraft stopped suddenly, she tried switching fuel tanks in an attempt to restart it, to no avail. She picked a ploughed field for a forced landing, the aircraft's wheels stuck in a furrow and its nose tipped forward. Her sister Kitty was thrown clear but injured and Louis Kendrick broke his thumb. Margaret suffered extensive skull fractures when her head was banged forward on to the instrument panel with the impact of the crash. She died later that night at Chester Royal Infirmary without regaining consciousness. At the subsequent inquiry in to the cause of the

accident, investigators found that the port fuel tank's vent pipe was blocked by a membrane of weatherproof paint, causing the petrol gauge to indicate that the tank still contained fuel, when in fact it was empty.

Andy Young
Prior to training on the last course of 5(C)OTU at Chivenor I had completed a radio course at Prestwick, billeted in Dankeith House, before being sent south where we formed into crews. There were eight crews on our course (No. 15 Beaufort), and on 1 May 1942 we were sent to Turnberry. We arrived flying the unit's Beauforts and found that the runways were still in the process of being finished, so we had a few days rest before carrying on with our training.

Eventually things became more organised and we combined the end of the OTU course with the torpedo course using the target ship *The Isle of Guernsey*, which we had to find and then torpedo. During my time at Turnberry, the powers that be decided to have the first 1,000 bomber raid, and

15 Beaufort Course

as they appeared to be scraping around to make up the numbers our crew were seconded. On 26 May we flew to North Coates in Lincolnshire, were briefed, refuelled and bombed up. Just before taking off it was realised that we were fitted with secret equipment and they couldn't risk it falling into enemy hands, so our trip was aborted. We had mixed feelings about being stood down. We wanted to go and take part and felt that being withdrawn was an anti-climax, but also realising that we had been lucky to escape the bombing raid, all because we were carrying ASV.

Returning to Turnberry, we finished our training and were signed off as a competent

a new aircraft. We did the usual consumption and W/T tests, then on to Portreath. We took off for Gibraltar, and the next day arrived in Malta, then on to Egypt and finally to Palestine. A test pilot then decided to check out our Beaufort and promptly pranged it! It was written off, but thankfully he was okay, so much for the brand-new aircraft!

We then travelled by train to Cairo, and on to Shallufa where we joined 221 Squadron, converting to Wellingtons carrying two torpedoes. We also had to do a night torpedo dropping course. In the beginning we just went on patrols looking for enemy shipping in the Mediterranean, and guiding attack aircraft in to

L to R - J. Kendall, R. Westcott, R. Wood and Lt. Stewart. Course 15

torpedo crew. We were to be sent on operations to the Middle East, based in Malta, attacking shipping bound for the Afrika Corps. The first stage of our journey was to Lyneham, then to the Bristol Aviation Company at Filton to collect

the targets. We were then posted to 69 Squadron on Malta where we did night searches and attacks.

Once I had completed my tour of operations in April 1943, I was posted back to 1TTU at

Turnberry as a W Op instructor. In the early hours of 21 May I was the duty signals instructor and F/Lt Alexander was duty pilot in charge, when we briefed three Wellington crews for a low flying exercise during the night. The weather was very good with bright moonlight reflecting off a very calm sea. Everything was proceeding as normal, a listening watch was kept on their activities by W/T, until all communications stopped for some time. We waited a while, and

Andy Young and Beaufort crew

then F/Lt Alexander and I took off to see if we could locate them. The weather was still good, but nothing was found. We took off again at first light to continue our search. The weather was still clear, peaceful and dead calm, the only thing we saw was a school of basking sharks close in to the coast. No sign of debris of any sort, the aircraft just seemed to have vanished. To this day I still remember that morning vividly. There was never any satisfactory explanation to why they had disappeared. I had flown with Sgt Forsyth-Johnson and crew the day before; they had probably spent the following day in the classrooms and had some time off prior to the fateful night-time exercise. They were a fairly well experienced and capable crew – I just couldn't understand what had happened.

I left Turnberry towards the end of September 1943 and was posted to Cranwell then eventually 240 Squadron in the Far East flying Catalinas until the end of hostilities. I then joined Transport Command in East Africa until demob.

11
WWII Aircraft

War is a great driver of technology, forcing the rapid evolution of military aircraft. Long gone were the wood-and-fabric biplanes as seen at Turnberry during the Great War, replaced by high-strength, all metal, construction airframes and supercharged engines. There are not many pictures of WWII aircraft at Turnberry, probably because servicemen were not permitted to take cameras on base and the taking of photographs was prohibited, although it must be mentioned that some of the ones I have been given of aircraft and personnel have been printed on MOD photographic paper with 'Top Secret' printed on the reverse. A brief description of the main aircraft on charge at Turnberry during WWII follows.

Blackburn Botha

One of the first aircraft to arrive, and crash, at Turnberry during the Second World War was the Blackburn Botha, probably one of the most unsuccessful aircraft to come off the drawing board. The Botha was designed to meet Air Ministry specification M15/35, for a land-based general reconnaissance/torpedo bomber, intended to replace the Vickers Wildebeest.

A total of 382 aircraft were built at Blackburn's factory at Brough and 200 at a new factory in Dumbarton, Scotland, powered by the Bristol Perseus engine.

Entering service in May 1940 with 608 Squadron and a few serving with 502 Squadron, it was soon found to be seriously underpowered and unstable - the type was universally unpopular. When initially tested the aircraft was found to have many inherent issues, it had poor elevator control, a high stall speed, a critical lack of longitudinal stability and exhaust fumes entering the cockpit.

After only five months, in spite of its bad reputation, the Botha was relegated to North Sea reconnaissance patrols, convoy escort, training and communications duties, with a few being used as target tugs. It had a very high accident rate, almost one-third of all training aircraft crashed, aircrew had scornfully nicknamed the

Blackburn Botha

aircraft 'Why Bother!'

The Botha continued in service with units such as 3 School of General Reconnaissance and 11 Radio School as a navigation and gunnery trainer until 1944 when it was declared obsolete.

Handley Page Hampden

The Handley Page Hampden, or HP52, was one of the most important twin-engine medium bombers of the RAF, along with the Armstrong Whitworth Whitley and Vickers Wellington.

Powered by Bristol Pegasus radial engines, the prototype first flew in June 1936, entering service in 1938, and with the Armstrong Whitworth Whitley and Vickers Wellington was the mainstay of the RAF in the early stages of the war. The Hampden had an impressive speed, manoeuvrability and capability for carrying a large bomb load. With a crew of four, it could carry a payload of 4,000 lb conventional drop bombs and took part in the heavy bombing over Europe with RAF Bomber Command, such as the first 1,000-bomber raid on Cologne and the first night raid on Berlin.

The Hampden had a distinctive appearance, almost like a fighter, with a narrow tapering fuselage, (it was only 3 feet wide) and a large tail boom, carrying twin tail fins. In fact, Handley Page dubbed it a 'fighting bomber'. The look of the aircraft led the aircrews give it the nicknames 'Flying Tadpole', 'Flying Panhandle' and 'Flying Suitcase' – some went as far as to say that it handled like a suitcase too!

Underpowered and slow, the Hampden was retired from Bomber Command in 1942, was converted to a long-range torpedo bomber, and transferred to Coastal Command and became the Hampden TB Mk I with a Mk XII torpedo, carried in a slightly enlarged open bomb bay, supplemented with a 500-pound bomb under each wing. Some were used as a maritime reconnaissance aircraft. The Hampden also operated as an ASR aircraft carrying the Lindholme dinghy.

The Hampden had a distinct weakness for low-level strikes against shipping. Height could not be maintained on one engine, due to the minimal altitude required on torpedo drops. Losing an engine was fatal. Another issue was the aircraft was known to develop a flat spin, or 'stabilised yaw', from which recovery was difficult and required height.

The Hampden achieved great successes in the Coastal Command role.

Bristol Beaufort

Competing with the Blackburn factory to meet Air Ministry specification M.15/35, the Bristol Aeroplane Company came up with the Type 152, or the Beaufort, a somewhat larger and heavier variant of the Bristol Blenheim, designed by Frank Barnwell, who had designed many of the WWI fighter aircraft. The RAF urgently needed a new land-based torpedo bomber, and both designs for the Botha and Type 152, were ordered straight off the drawing board.

Problems with the prototype delayed the introduction to service, the intended Bristol Perseus engines could not provide enough power for the heavier aircraft and the replacement, more powerful Bristol Taurus engines, were rushed into production, although having not been fully developed and tested. They were prone to continually overheating during ground testing, delaying manufacture. It soon became apparent that the Taurus engine had an even more serious flaw, the frequent failure of the Maneton clamp bolts in the crankshaft, which lead to the connecting rods slipping and destroying the engine, sometimes catching fire, causing a sudden loss of control. Engine failure at low altitude was disastrous with poor single-engine performance. In theory it was possible for the pilot to fly the Beaufort on one engine, but in practice this was not always successful.

With a crew of four, the Type 152 could carry a single 18-inch Mk XII aerial torpedo. It was Coastal Command's main torpedo bomber from 1940 to 1943, seeing action against the German fleet. The Beaufort was particularly successful operating from Malta where it played a vital role in denying Rommel's Afrika Korps desperately needed supplies.

More Beauforts were lost through accidents

Bristol Beaufort

Bristol Beaufighter

and mechanical failures than were lost to enemy action, flying more hours in training than on operational missions.

Bristol Beaufighter

As the Bristol Beaufort was development of the Blenheim, so the Bristol Beaufighter was a derivative of the Beaufort torpedo-bomber. The prospective heavy fighter/strike aircraft was designed to share the same jigs and tooling as the Beaufort so that manufacture could easily be switched from one aircraft to the other and so increase the speed of production, by re-using Beaufort components, but the fuselage required further modification and had to be redesigned. Intended to be a twin-engine, two-seat long range day and night fighter, powered by Bristol Hercules engines giving a maximum speed of 320 mph, the Beaufighter entered service with Coastal Command in mid-1942, the type used at Turnberry being the TF Mk X or more commonly known as the 'Torbeau'.

Bristol Buckmaster

The Bristol Buckmaster, or Type 166, was an advanced British training aircraft designed to meet Air Ministry Specification T.13/43. It was an advanced dual control, unarmed twin engine trainer, able to carry a crew of three, student pilot, instructor pilot and radio operator. When introduced in 1945 it was considered the highest performance trainer in the RAF - only 110 entered service.

Avro Anson

In May 1934, the Air Ministry asked the Avro Company to design a twin engine landplane for coastal reconnaissance tasks. Avro came up with the Anson, a small low-wing, twin engine monoplane, a highly liked, reliable aircraft which earned the nickname 'Faithful Annie'.

Ansons were mostly used as trainers and light transport aircraft, with over 11,000 built. 'Faithful Annie' continued in service with the RAF until 1968.

Vickers-Armstrong Wellington VIII

First produced as a twin-engined, long-range, general reconnaissance bomber in 1937, the Wellington was the main aircraft of Bomber Command in the early years of the war. It was noteworthy for the amount of damage it could suffer in action and remain airborne, due, mainly

Bristol Buckmaster

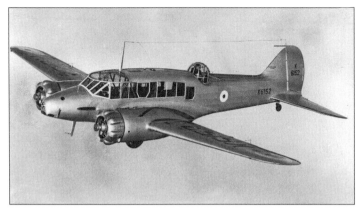

Avro Anson

to the geodetic form of construction designed by Barnes Wallis (of bouncing bomb fame), a lattice-work of steel which gave the airframe great strength and rigidity. With the arrival of the new larger four-engined aircraft, such as the Avro Lancaster, it was converted for use in Coastal Command, Transport Command, and Flying Training Command duties.

The Wellington Mk VIII was converted for Coastal Reconnaissance, fitted with ASV radar, Leigh Lights and a bomb bay housing two 18-inch torpedoes or anti-submarine ordnance.

The aircraft was very popular with aircrew and was nicknamed 'The Wimpey', after the character 'J. Wellington Wimpey' from the Popeye cartoons.

Over 11,000 of all variants were produced and were retired from service in 1953.

Vickers-Armstrong Warwick
Though similar to the Wellington, being a fabric covered geodetic structure, the Warwick was designed and built to a different specification. It had a larger wingspan at 96 feet 9 inches, was longer, and its weight was increased by 20,000 lbs. Developed at the same time as the Wellington, the two aircraft shared around 85% of their structural components. Originally designed as a twin-engine bomber, by the time it entered production in 1942 it was already outdated as required by the original specification and was adapted to a number of duties, including Coastal Command anti-submarine

reconnaissance, Transport Command and air-sea rescue.

The Warwicks at Turnberry had been converted to air sea rescue duties, carrying an airborne lifeboat which was powered by two 4 hp marine engines. Sails and oars were also provided in case of engine failure and it was equipped with a survival kit, wireless and rations. This was suspended under the aircraft from an attachment in the bomb bay between two auxiliary fuel tanks. The lifeboat was dropped from a height of 700-800 feet using six 32-feet diameter parachutes at a speed of 115-140 mph.

The Warwick had a crew of six, namely, pilot, navigator three W Op/AGs and an air gunner. The ASR Mk I carried an armament of eight .303 machine guns in three turrets, two in the nose turret, two in the mid upper turret and four in the tail.

A total of 850 were built. It was the largest British twin-engined aircraft used during the Second World War.

Vickers-Armstrong Warwick ASR

Airspeed Oxford
The Oxford was a twin-engine advanced trainer that was developed from the Envoy, a commercial passenger aircraft, during the 1930s

Vickers-Armstrong Warwick ASR at Turnberry

in response to the requirement for training aircrews in navigation, radio-operating, bombing and gunnery. In addition to training duties, Oxfords were used in communications and anti-submarine roles and as air ambulances.

A total 8751 were built, and at the end of the war some surplus ex-RAF 'Ox Boxes' were converted for civilian transport becoming the Airspeed Consul.

Westland Lysander

The Lysander or 'Lizzie' was introduced in 1938 as an army co-operation and liaison aircraft used for communications, artillery spotting and as light bombers. The high winged Lysander was powered by a Bristol Mercury air-cooled radial engine, in spite of its strange appearance, the Lysander was very aerodynamic with a low stall speed of 65 mph. Serving in France, they made easy targets for the Luftwaffe, and it was decided that they were 'quite unsuited to the task; a faster, less vulnerable aircraft was required.' The Lysander then found its forte flying for the for the Special Operations Executive on clandestine missions. Fitted with a fixed ladder over the port side, the 'Lizzie' could drop off and collect agents in occupied France, landing and taking off in fields or short makeshift airstrips.

Operational Training Units, such as Turnberry, used the Lysander in a less glamorous role - for target towing duties.

Over 1,700 Lysanders were built and were withdrawn from service in 1946.

Today, airworthy 'Lizzie' V9552 can be seen flying as part of The Shuttleworth Collection at Old Warden in Bedfordshire.

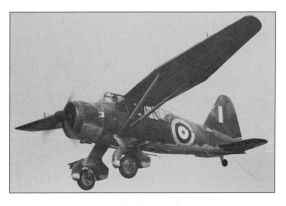

Westland Lysander

Bristol Blenheim

The Bristol Blenheim, designed and built by the Bristol Aeroplane Company, was initially developed as a civil airliner. Its excellent performance drew the attention of the Air Ministry when, during tests, the aircraft proved that it was faster than any fighter in service with the Royal Air Force at the time, having a top speed of 307 mph.

A military version was ordered and Type 142M (M for military) entered service in 1937. The Blenheim light bomber was one of the first British aircraft with an all-metal stressed-skin construction, retractable landing gear, flaps, a powered gun turret and variable-pitch propellers.

Bristol Blenheim

Rapidly outclassed by the vast advances in technology and the increase in the performance of the enemy fighters, the Blenheim was relegated to the home front. Some Blenheim squadrons took part in the Battle of Britain, suffering heavy casualties but never receiving the same publicity as the Spitfire and Hurricanes.

The Turnberry Blenheims were used as bomber/gunnery trainers and were retired in 1944.

De Havilland DH.98 Mosquito

Designed by Geoffrey de Havilland and nicknamed the 'Wooden Wonder', the Mosquito was a British, twin-engined, multi-role combat aircraft, unusually having a frame constructed mostly of wood.

De Havilland DH98 Mosquito dropping a 'Highball' Bouncing Bomb.

With a crew of two, in 1941 the Mosquito was one of the fastest aircraft in the world.

The type seen at Turnberry were Mosquito Mk IVs, which were modified by Vickers-Armstrong to carry the Highball 'bouncing bomb', becoming Type 465. In all a total of 33 Mosquito IVs were converted.

De Havilland DH.95 Flamingo R2766 'Lady of Glamis'

The Flamingo was a twin-engined civil airliner powered by two Bristol Perseus air-cooled radial engines. With two pilots it was capable of carrying twelve passengers. The Flamingo first flew in December 1938 but during WWII the type was pressed into service by the Air Ministry for use by the Air Council and as the King's Flight. The King's Flight aircraft, 'Lady of Glamis', was to be used in the event of the Royal Family having to flee the country in advance of invasion by enemy forces. This aircraft visited Turnberry on 15 September 1942. Only 14 DH.95s were ever built.

Lockheed Hudson

Initially designed as a civilian aircraft, the Lockheed Model 14 Super Electra airliner, and later converted to a light bomber, the American built Lockheed Corporation Hudson was ordered by the RAF to be used as a maritime

De Havilland DH95 Flamingo R2766

patrol aircraft supporting the Avro Anson prior to WWII. By the outbreak of war a total of 78 aircraft were already in service with 224 Squadron at Leuchars.

Production of the new twin engine light bomber for the RAF continued, but due to the United States' neutrality in the war, could not be seen to be providing direct support to Britain, this was overcome by flying the aircraft from the factory to the Canadian/US border where they landed and were then towed over the border into Canada by tractors or teams of horses before being flown on to Canadian airfields where they were prepared for transportation as deck cargo, by ship to Liverpool. On arrival in the United Kingdom the Hudson was fitted with the Boulton Paul dorsal turret and mostly went on to serve with Coastal Command Units.

The Lockheed Hudson was nicknamed the "Old Boomerang" due to its ability to withstand heavy enemy fire, safely returning home and was the first aircraft of American design operating from the British Isles to shoot down an enemy aircraft.

At Turnberry, the Hudson was used mainly in Navigator and Meteorological crew training.

Lockheed Hudson AM562 flying over Rathlin Island from Turnberry.

12
WWII Training

Before the outbreak of WWII, aircrew undertook their operational training on their squadron, but as hostilities increased and operations began, it soon became apparent that this method of training couldn't be successfully carried out by units and personnel who were on active service. It was therefore decided to temporarily remove some squadrons from operational duties and assign to them the role of training new aircrew. This became a permanent arrangement and these 'training squadrons' were re-designated 'Operational Training Units'.

At these OTUs the aircrews were taught the theory and methods of high-speed torpedo dropping. In the Beaufort this was to be done after a standard diving approach, at a speed of between 300 and 350 mph. With the improved performance of aircraft such as the Beaufort, it was realised that there was no British air-launched torpedo suitable for this purpose – astonishingly, the only design available was the WWI Mk XII torpedo, and its ideal drop speed was a mere 160 mph to avoid the torpedo breaking up when hitting the water.

This problem was overcome by abandoning the diving approach method of dropping, instead, training the pilot in the skill and precision required to fly low, slow, and straight and level, allowing the torpedo to enter the water smoothly. Dropping too high or too low and the torpedo could bounce or dive to the seabed. The optimum height was 68 feet and had to be judged by the pilot, being too low to register on an altimeter, and in calm conditions miscalculation was easy, with disastrous consequences. The range of release was 670 yards.

The trainee pilot had his hands full, flying the aircraft to these strict parameters. Later, it would take immense bravery to fly straight into defensive enemy anti-aircraft fire during the run-in against the target. Once the torpedo had been dropped, the aircraft had to pull up at full throttle, over the ship and turn away, usually below mast height. This whole procedure was fraught with difficulty and tragedy. The skill and precision could be taught, the courage they brought with them.

All airborne torpedo training for both the naval and air services was undertaken by the RAF, which, at the outbreak of war, was performed at the Torpedo Training Unit at Gosport using Fairey Swordfish aircraft. This area of the Solent was vulnerable to enemy air attack, which led to the transfer of training to the Clyde in March 1940, the

Beaufort Formation on the Clyde

unit being accommodated at the somewhat inadequate and boggy grass landing ground at Abbotsinch.

The practice torpedoes used were eighteen inches in diameter, slightly smaller than the twenty-one-inch operational weapons and were set to pass under the keel of the target vessel. They were designed to float to the surface at the end of their run enabling their recovery and reset for further use. The Admiralty provided small target ships such as the *Malahne* and the *Heliopolis* but also some larger warships, including the battleship *Ramillies* the cruiser *Dido* and the *Cardiff*. The latter was based on the Clyde to give courses to naval gunners. The RAF used the *Adastral* (RAFA *Adastral* was formerly the Royal Navy trawler *William Gillet*, sold to the Air Ministry on 9 March 1921. She was then renamed *Adastral* and was employed by the RAF as a weapon recovery vessel, based at Ayr). The ranges were in the general area of Ailsa Craig and standard exercises included 'A.L.T.' (Attack Light Torpedo) 'F.A.L.T.' (Formation Attack Light Torpedo) and 'A.R.T.' (Attack Running Torpedo)

Beaufort dropping a torpedo (Maidens in the background)

The Fleet Air Arm established its own torpedo training school at Crail on the Fife coast, leaving the TTU at Abbotsinch with a lightened workload. The aircraft used for this training was the ageing Vildebeest torpedo bomber. It was planned to replace these with the Bristol Beaufort, but production was running late, and the arrival of Beauforts at the training units was accordingly rather slow.

Before they began their specialised torpedo training on the Beaufort, the four-man crew had to be trained to fly, navigate, and defend their aircraft. Pilot training was undertaken all over the world, South Africa, Rhodesia and Canada, in areas far removed from the conflict and likely disruption by the enemy. A complete aircrew did not normally come together until their arrival at the OTU. In November 1940, Beaufort training was temporarily transferred to the new 3OTU at Chivenor in Devon, in anticipation of the opening of 5(C)OTU at Turnberry. Considerable delays in the planned construction of Turnberry meant that the newly formed 5(C)OTU and its Beauforts remained at Chivenor until May 1942. Here the new Beaufort crews completed their eight-week OTU course before moving on to Abbotsinch for their final four weeks training at the TTU which was now equipped with Beauforts, complemented for a while with a few Blackburn Bothas.

When Turnberry was completed, and finally became available in May 1942, 5(C)OTU moved in to carry out training courses, consisting of 56 four-man crews at any one time. The Handley Page Hampden and the Vickers Wellington had also been chosen to serve as additional torpedo bomber aircraft and a number of these types had been transferred from Bomber to Coastal Command, both to augment the training force and take the place of the Beaufort Squadrons which had been sent overseas.

Beaufort gun camera image

In November 1942, it was decided to greatly expand the RAF Torpedo Force by increasing the total number of torpedo squadrons to thirty-one. It was also decided to add torpedo armed Beaufighters ('Torbeaus') to the other three types of aircraft and to expand both the OTU and the TTU accordingly. As a result of these changes, the TTU was split into two groups. At Turnberry, 1TTU was formed to handle the training of Wellington and Hampden crews, 40 at any one time, and it took over part of 5(C) OTU (Hampdens at Turnberry) and 7OTU (Wellingtons at Limavady) in the process. The Wellington crews, unlike the Hampden crews, who were much fewer in number, were all intended to be posted overseas. This resulted in a surplus of Wellington torpedo crews in 1943, who were then posted to man the rapidly increasing Dakota troop carrying flights or to ASR Warwick Squadrons instead. The Hampden crews were mostly posted to Coastal Command Units.

As 1942 gave way to 1943, training was being progressed as quickly as possible in order to meet the new requirements, but even as this was happening, the military situation in the Mediterranean was changing rapidly, starting with the successful Allied invasion of North Africa in November 1942 and culminating in the retreat of German forces, first to Sicily and then to Italy. A major consequence of this Allied success was a greatly reduced need to attack enemy supply line shipping across the Mediterranean, whilst by now, in any case there were torpedo training units in both Middle East and Far East theatres of war - 5METS at Shallufa in Egypt and 3TRS at Ratmalana in Ceylon.

Back in the UK, it was accordingly possible to make substantial cuts in the expanded torpedo training organisation. So, in September 1943, 2TTU was closed and its remaining responsibilities transferred from Castle Kennedy to Turnberry. Here too reductions were made, and all Wellington training ceased. Further cuts came the following spring when 7OTU (Wellingtons) at Limavady was closed completely whilst 5(C)OTU ceased training on Hampdens and reduced its Beaufort commitment by two thirds. Spare capacity in the unit was now taken up by the addition of Hudson and Ventura training for Air Sea Rescue and Meteorological Squadrons. The intention was to concentrate many different types of training, including specialised and refresher courses within one unit.

During May and June 1943, a detachment of 618 Squadron with a few Mosquitos appeared at Turnberry from their base at Skitten. This unit had been formed on 1 April 1943 for the sole purpose of sinking the German battleship *Tirpitz* in its lair in the fjords of Norway. To this end, it was proposed to use a smaller version of Barnes Wallis's bouncing bomb, called 'Highball'. At Turnberry, the squadron was issued with 'practice rounds' which they used on training attacks against ships moored in Loch Striven. Problems soon developed with this new technology and the training program was temporarily suspended. As further development of the 'Highballs' allowed, training recommenced on 11 September 1943 and this continued right through to 12 October 1944 when the unit left for Australia, never having the chance to make its originally intended contribution to the war effort.

Mosquito

In May 1944, 5(C)OTU returned to Turnberry, whereupon it took over what was left of 1TTU and at the same time absorbed the role of the Air Sea Rescue training unit from Thornaby. For this it received a substantial fleet of Warwicks, many of which were fitted out for dropping lifeboats. Warwick's predominated as preparations were made for the expansion of ASR units both at home and overseas.

Six months later the Venturas had gone, but a handful of Wellingtons had been added, whilst the unit also received its own Spitfire for fighter affiliation practice. Coastal Command Flying Instructors' School was moved to Turnberry from St Angelo on 8 June 1945, with a detachment

based at Alness for flying boat training and had Wellingtons and Beauforts on strength. On 1 August 1945 5(C)OTU was disbanded, but 1TTU seems to have reappeared under the wing of CCFIS and started to train with renewed vigour as if they were expecting another war! The aircraft types on charge at Turnberry were: Ventura, Hudson, Beaufighter, Warwick, Oxford, Anson, Martinet, Lysander, Mosquito, Sunderland, Catalina, Liberator, Buckmaster, and Master.

This was the status of RAF Turnberry towards the end of 1945, but like most other training units at this time, it was then rapidly closed down. Both CCFIS, renamed Coastal Command Instructors School on 29 October 1945, and 1TTU moved to Tain and the ASR training was transferred to 6OTU at RAF Kinloss. This brought to a close all military flying at Turnberry. The airfield was now only used by No. 10 Gliding School of the ATC until January 1948 when the station was totally abandoned by the RAF.

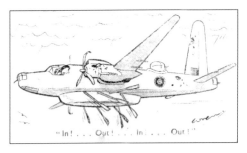
Vickers Warwick cartoon

13
WWII Accidents

The first casualty at Turnberry was not an airman, but a general labourer, James McIntyre from Girvan, who died on 25 November 1941. He was killed in a motor lorry accident on the southern extension to 'A' runway which was in the course of construction.

As with many wartime training units throughout Britain, Turnberry was not unique for the high accident rate among the trainee crews. Charterhall in the Scottish borders was dubbed 'Slaughterall'. It is estimated that over 8,000 aircrew were killed in training accidents during the war before ever seeing active service.

Unable to comprehend the carnage that was taking place around them, local rumours as to the cause of the accidents flourished, such as 'clapped out' aircraft, poor maintenance, wind shear off the hills, inadequate runway length and even sabotage. Official Accident Record Cards (Air Ministry Form 1180) were filed for each training incident, and an official inquiry held, allowing the cause to be analysed and the resulting information used in future accident prevention. A summary of the details given on these include the date, place, aircraft type and serial number, pilot's name and the verdict of the enquiry.

An example of a statement given by the Group Captain Commanding, RAF Station Turnberry, regarding the loss of Wellington's LB237, LB187 and LB193 follows:

21 May 1943
1. Four Wellington Aircraft (403 Sgt Forsyth-Johnson, 404 P/O Wells, 405 P/O Hollom, and 406 F/O Griffin) took off at approximately 02.00 hours on 21 May 1943, to carry out ALTs by flares and moon against target ship HELIOPOLIS. Weather was fine,

visibility 12-15 miles with slight haze. Sea was calm, wind light and cloud base 3,000 feet. Moon obscured.
2. On this type of practice, one aircraft drops stick of flares from 3,000 feet whilst the remainder make their attacks. They then change places so that each pilot has practice in flare positioning and ALTs.
3. Although the target vessel was only about 5 miles off Turnberry Aerodrome, it appears that three only of the four aircraft which were airborne arrived over the target vessel, since neither the target vessel or the other aircraft observed more than three aircraft at any time.
4. Aircraft 405 (P/O Hollom) was the first to crash (02.45 hours) whilst making his approach on a steady course to the ship. The pilot stated that he was quite happy about estimating his height under the conditions prevailing as he had already successfully carried out ALTs earlier on that evening on the first detail. However, he misjudged his height and struck the sea at the moment when his second pilot had appreciated the danger and was warning him to fly higher. P/O Hollom was the only survivor from this crash.
5. The pilot of the second aircraft, F/O Griffin, having appreciated that an aircraft had ditched, in view of the fact that *Heliopolis* was using her searchlights and subsequently through intercepting a W/T message, switched on his landing light and started hunting for survivors at a height of about 150 feet. He states that whilst flying at this height he hit a violent bump which put his wing into the sea. He did not misjudge his height in his opinion.
6. The Captain of the *Heliopolis* did not see this

Beaufighter RD426 crash

Beaufighter RD426 crash

Beaufighter RD426 crash (Turnberry Lighthouse in the background)

Beaufighter RD426 crash

Beaufighter RD426 crash

aircraft crash and it was quite by chance that he heard shouts coming from some distance away whilst still hunting for survivors of the first crash.

7. Four survivors of F/O Griffin's crew were then recovered by the safety drifter *Animation* at 04.00 hours, approximately 45 minutes after ditching.

8. No news of aircraft 403 (Sgt Forsyth-Johnson) was heard other than W/T contact just after take-off. It is suggested that he might possibly have collided with F/O Griffin, as it would take an inordinately big bump to put a Wellington into the sea from 150 feet without actual contact.

9. All aircraft were burning navigation lights, but it can be expected that a pilot would fail to notice these if searching the water with his landing lamp.

10. The fourth aircraft (404 P/O Wells) sensibly remained at 2,000 feet dropping his remaining flares to assist the search for survivors.

(Signed) J. G. Elton
Group Captain Commanding
RAF Station Turnberry

In conclusion, the tragic loss of life at Turnberry, was a result of inexperience, mechanical failure and every pilot's dread, poor weather conditions.

Turnberry War Memorial with the additional panels commemorating the WWII Casualties.

14
WWII Memorial

In November 1990 four sections were added to the base of the monument bearing the names of those who died at Turnberry during the Second World War. It is thought that these names were listed from the gravestones of the casualties that are buried at Dunure (Fisherton) cemetery. Those who were taken home for burial, and those who have no marked grave, having been lost at sea, were therefore not included. A total of 83 names were commemorated for WWII, a further 89 which had been omitted have since been added.

The Inscription on the memorial reads:

Their Name Liveth For Evermore

as Also those Commemorated below
who Died in the
1939 -1945 War

Royal Air Force

F. H. Aldous Sgt	G. K. Gardner AC1	B. Lunn Sgt
J. A. Bateson Sqdn/Ldr	W. F.Gray F/O	G. Maguire Sgt
D. H. Battrick P/O	E. Greaves P/O	R. F. Miles Sgt
J. C. Campbell LAC	J. B. Hall Sgt	W. Monkhouse Sgt
W. V. Chambers Sgt	E. J. Harrild Sgt	K. W. P. Noble F/Lt
T. B. Clark Sgt	H. H. C.Harris Sgt	E. L. Padfield Sgt
P. R. Comley Sgt	F. Helliwell Sgt	S. Park F/Lt
T. E. V. Docherty Sgt	E. W. Hirst Sgt	H. Pike Sgt
E. Everall AC2	J. L. Hiscock Sgt	R. W. A. Quin Sgt
D. Fairweather Capt.	A. N. Hunter Sgt	D. N. Smallie Sgt
Mrs M. Fairweather F/Capt.	H. F. Hunter Sgt	F. B. Smith Sgt
F. S. Fisher Sgt	C. Jones Sgt	D. R. Sorensen F/Sgt
J. F. Forsyth-Johnson F/Sgt	B. H. Lowe Sgt	H. W. Walpole Sgt
S. Weaver Sgt	G. C. Whitehurst F/O	K. M. Kershaw 2nd/Off
J. L. Webber Sgt	G. J. Wilcox F/O	J. Whitehouse Sgt
J. A. Wild LAC	G. Duncan Sgt	Unknown Airman

Royal Canadian Air Force

D. B. Barnie Sgt	F. J. H. Highmoor Sgt	J. G. Newman Sgt
F. G. Brezina Sgt	G. H. Lacerete U/Lt	J. P. O'Neill Sgt
J. A. Calder Sgt	J. W. Leyland F/Sgt	G. K. Shaw F/Sgt
W. T. Gale Sgt	J. O. Maclean F/Sgt	D. G. Taylor F/Sgt
L. O. Glay F/Sgt	D. B. S. Malcolm Sgt	W. M .Vacheresse Sgt
W. H. Harris WO2		

Royal Australian Air Force

B. S. Burns Sgt	J. H. W. Johnstone	F/O E. V. Roy F/O
E. D. Lee F/Sgt	N. W. Huggins F/Sgt	D. S. McDonald Sgt
A. D. Riordan F/Sgt	T. S. Tamblyn F/Sgt	J. V. Jenkins F/Sgt
J. A. Kingsmill Sgt	J. E. McGrath Sgt	F. J. Wright F/Sgt

Royal New Zealand Air Force

L. N. Buttimore F/O	J. E. Jones P/OT	D. K. H. McDonald F/Sgt
S. R. Grier F/Sgt	B. Moore Sgt	A. C. Smart F/Sgt
M. C. Jolly F/O		

Another anomaly is the commemoration of air crew who were not based at Turnberry, or even crashed in the vicinity. This can also be explained by the copying of the names of those buried at Dunure. There follows a List of Remembrance, with brief notes giving details of the accident and casualties. This information has been gathered from the ORBs, Crash Cards, CWGC and eye-witness accounts. I have also included details of aircraft accidents attended by the Medical Officer from RAF Turnberry.

◆

IN MEMORY OF THE CREW OF
BLACKBURN BOTHA I L6211
3RS

who died on 24 February 1942

Sergeant William Tyndall Sheppard 407731
Pilot
Royal Australian Air Force
of Mount Gambier, South Australia,
who died aged 28.

and

Sergeant Esmond Elliott Brown 1188027
Royal Air Force Volunteer Reserve
of Wokingham, Berkshire,
who died aged 22.

William Tyndall Sheppard

Blackburn Botha L6211 of 3RS Prestwick crashed on the foreshore at Turnberry whilst on a training flight. Starboard engine failed.

◆

IN MEMORY OF THE CREW OF
AVRO ANSON I DG861
3RS

who died on 19 April 1942

Pilot Officer Henry Harding Want 119874
Pilot
Royal Air Force Volunteer Reserve
of Holloway, London,
who died aged 24.

and

Pilot Officer Herbert Kenneth Daniel 114733
Observer
Royal Air Force Volunteer Reserve
of Minehead, Somerset,
who died aged 32.

and

**Leading Aircraftman
Anthony George Hawes 1395105**
Observer U/T
Royal Air Force Volunteer Reserve
of Northwood, Middlesex,
who died aged 19.

**Leading Aircraftman Francis Owen Hogg
1465762**
Observer U/T
Royal Air Force Volunteer Reserve
who died aged 33.

and

**Leading Aircraftman the Revd. William
Godfrey Gorton 1412875**
Wireless Operator/ Observer U/T
Royal Air Force Volunteer Reserve
of Cheltenham, Gloucestershire,
who died aged 28.

*Anson DG861 broke up in the air
south of Maybole.*

◆

IN MEMORY OF THE CREW OF
BRISTOL BEAUFORT II AW357
5(C)OTU
who died on Monday 11 May 1942

Sergeant James Whitehouse 563725
Pilot
Royal Air Force
of Ashton-Under-Lyne, Lancashire
who died aged 28.

and

Sergeant Reginald Frederick Miles 1375370
Royal Air Force Volunteer Reserve
of Clench Common, Wiltshire
who died aged 26.

*Beaufort AW357 crashed into sea west of
Girvan after engine failure and sank
immediately. Two other aircrew were saved
by the crew of a fishing boat.*

◆

IN MEMORY OF THE CREW OF
BRISTOL BEAUFORT I N1023
5(C)OTU

who died on Wednesday 27 May 1942

Sergeant Edgar Charles O'Donnell 655939
Pilot
Royal Air Force Volunteer Reserve
of Ward End, Birmingham,
who died aged 22.

*Beaufort N1023 was on a single engine flying
exercise when on approach to Turnberry with
undercarriage down, the aircraft was seen to
swing at about 200 feet and dive into the sea
off Maidens village.
The possible cause was given as engine failure.*

IN MEMORY OF THE CREW OF
BRISTOL BEAUFORT II AW306
5(C)OTU

who died on Saturday 6 June 1942

Sergeant Horatio Herbert Morris 656486
Pilot
Royal Air Force
who died aged 26

and

Sergeant Frederick Payne 1293749
Air Observer
Royal Air Force Volunteer Reserve
of Norton, Kent,
who died aged 27.

and

Sergeant Donald Campbell Smith 640963
WOp/AG
Royal Air Force

and

Sergeant Samuel Hunter 1070098
WOp/AG
Royal Air Force Volunteer Reserve
of Glasgow,
who died aged 29.

*Beaufort AW306 went missing while on a
cross-country flight over the sea. The last
wireless message received by the station W/T
was a request for a weather report and was
almost impossible to decipher. Return message
was sent as a matter of routine. There was a
certain amount of cloud and also sea fog.
The weather deteriorated in the area of the
exercise and the inexperienced pilot may
have decided to descend through cloud and
in so doing hit the sea.*

IN MEMORY OF THE PILOT OF
BRISTOL BEAUFORT I L4515
5(C)OTU

who died on Thursday 18 June 1942

Sergeant Howard Roy Fidgin 1382134
Royal Air Force Volunteer Reserve
of Eastbourne, Sussex,
aged 30.

*Aircraft failed to return from training flight.
Search unsuccessful, except patch of oil on sea.
Government transport saw a Beaufort crash in
sea 33 mins after L4515 took off.*

IN MEMORY OF THE CREW OF
BLACKBURN BOTHA I
L6418
3RDFS PRESTWICK

who died on Friday 19 June 1942

Sergeant Arthur Richard Dipple 655658
WOp/AG
Royal Air Force
of St Austell, Cornwall,
aged 26.

*Blackburn Botha L6418 flown by
Pilot Officer J. Haydock, force landed in sea
after engine failure due to defective cylinder
in one engine and subsequent overheating of
remaining engine by overloading.
Pilot did everything possible.*

*The body of Sgt Dipple was washed ashore at
Maidens Bay on 22 June 1942.*

IN MEMORY OF THE PILOT OF
BRISTOL BEAUFORT L4459
5(C)OTU

who died on 24 June 1942

Sergeant Jacob Letterstied Smit 778643
Royal Air Force Volunteer Reserve
of Gwelo, Southern Rhodesia,

who died aged 22.

*Whilst on a local training flight aircraft
was seen by the captain of a ship, to spin into
the sea, ten miles south of Girvan.
Cause Obscure.*

IN MEMORY OF THE CREW OF
Avro Anson N5297
2OAFU

who died on Thursday 2 July 1942

**Flight Sergeant
William Thomas Gale R/84287**
Pilot
Royal Canadian Air Force
of London, Ontario, Canada,
who died aged 22.

and

Sergeant John Benson Hall 1354005
WOp/AG
Royal Air Force Volunteer Reserve
of Milton Bridge, Midlothian,
who died aged 22.

and

**Leading Aircraftman James Cameron
Campbell**
of Perth,
who died aged 30.

and

Aircraftman 2nd Class Ernest Everall 1126654
WOp/AG U/T
Royal Air Force Volunteer Reserve
of Higher Openshaw, Manchester,
who died aged 21.

and

**Leading Aircraftman Joseph Arthur Wild
1576055**
Air Bomber U/T
Royal Air Force Volunteer Reserve
of Burton-Upon-Trent, Staffordshire,
who died aged 31.

*Anson N5297 was on a navigational training
exercise from RAF Millom.
Crashed on Shalloch-on-Minnoch*

William Thomas Gale

◆

IN MEMORY OF THE PILOT OF
HANDLEY PAGE HAMPDEN I AE137*
5(C)OTU

who died on Thursday 9 July 1942

Sergeant Terence McCormick 590892
Pilot
Royal Air Force
of Larne, County Antrim, Northern Ireland.

*After failure of the port engine, the pilot
lowered the undercarriage too early on approach, the aircraft stalled and spun in at Ballochneil,
one mile east of airfield.*

**This aircraft serial number has been
incorrectly entered in the Station Record Book and on the Official Crash Card.
Hampden AE137 was lost on a bombing raid over Essen, North Rhine, Westphalia,
with 50 Squadron on 6 August 1941.*

◆

IN MEMORY OF THE PILOT OF
NORTH AMERICAN P-51 MUSTANG
AG504 241 SQUADRON

who died on Sunday 19 July 1942

Pilot Officer Peter Puleston Clarke 109113
Pilot
Royal Air Force Volunteer Reserve
of Saltburn, Yorkshire,
who died aged 21.

*While inspecting target for artillery shoot the
aircraft flew into high ground in bad visibility.
Apparently, the pilot came in from the sea and
tried to fly up a valley. Finding he was in low
cloud he endeavoured to turn and struck the
ground in hill fog.*

◆

IN MEMORY OF THE PILOT OF
HANDLEY PAGE HAMPDEN I
P2100
5(C)OTU

who died on 19 July 1942

**Sergeant William Ernest Secundus Rayner
1379510**
Royal Air Force Volunteer Reserve
of Winchester, Hampshire,
who died aged 31.

*Hampden P2100 flew into hillside at Mochrum
Wood in good visibility. The pilot appeared to
have clipped the hill with the port wing tip.
The hill rises very steeply at this point, and the
possibility of air pockets cannot be overlooked.*

◆

IN MEMORY OF THE CREW OF
BRISTOL 152 BEAUFORT I
W6480
5(C)OTU

who died on Sunday 26 July 1942

Pilot Officer Ralph Ballantyne Henry J/7615
Pilot
Royal Canadian Air Force
of Kincardine, Ontario, Canada,
who died aged 30.

and

Pilot Officer Robert Hart Pike 119357
Royal Air Force Volunteer Reserve

and

Pilot Officer Kenneth Stanley Norton 119355
WOp/AG
Royal Air Force Volunteer Reserve
of Leytonstone, Essex,
who died aged 21.

and

Pilot Officer Thomas Dean 120923
Royal Air Force Volunteer Reserve
of Liverpool,
who died aged 26.

*The aircraft hit the sea whilst on fifty-feet stick bombing
practice, was seen to climb steeply and then
crash into sea.
It was reported that the inexperience of the pilot and the
subsequent inability to judge height of the aircraft over
the smooth surface of the sea was to blame.
It was further requested that an explanation be given
as to why a pupil was allowed to do low-level bombing
exercises over a calm sea.
In such conditions no low-level bombing is to be carried
out without permission of O.C. or Chief Instructor.*

Pilot Officer Thomas Dean

Pilot Officer Thomas Dean

◆

IN MEMORY OF THE CREW OF
BRISTOL 152 BEAUFORT I
L4496
5(C)OTU

who died on Tuesday 4 August 1942

Sergeant Francis Jonathan Highmoor R/107767
Pilot
Royal Canadian Air Force
of Alberta, Canada,
who died aged 21.

and

Sergeant Douglas Burton St John Malcolm R/84388
WOp/AG
Royal Canadian Air Force
of Ontario, Canada,
who died aged 21.

*Beaufort L4496 crashed at Balkenna Bridge, one mile south of Turnberry, whilst on circuit
and landing practice at night, whilst attempting to land.
Possible cause of accident as stated on accident record card was that the air speed indicator fitted
to this aircraft was of a different type to which the pilot was accustomed.
The position of the needle on this instrument reading in advance of actual airspeed,
causing the aircraft to stall.
Although, an engine was seen to be on fire before the aircraft crashed.*

◆

IN MEMORY OF THE PILOT OF
BRISTOL 152 BEAUFORT I
L4467
5(C)OTU

Who died on Sunday 16 August 1942

Sergeant William Henry Fardoe 1234096
Royal Air Force Volunteer Reserve
of Didsbury, Manchester,
who died aged 21.

*Beaufort L4467 crashed into the sea half a mile from Girvan Pier whilst on a training flight.
The aircraft was flying at a height of 500-1,000 feet when two loud bangs were heard,
and the aircraft was seen to dive into the sea. Engine failure due to suspected bolt breakage.*

◆

IN MEMORY OF THE PILOT OF
HANDLEY PAGE HAMPDEN I AE255
5(C)OTU

who died on Tuesday 18 August 1942

Sergeant Bobbie De Savary Burns 401610
Royal Australian Air Force
of Elwood, Victoria, Australia,
who died aged 22.

Hampden AE255 crashed into ground at Milton, shortly after take-off.

◆

IN MEMORY OF THE CREW OF
BRISTOL 152 BEAUFORT I
L4480
5(C)OTU

Who died on 20 August 1942

Flying Officer Eugene Victor Roy 408525
Pilot
Royal Australian Air Force
of Croxton, Victoria, Australia,
who died aged 30.

and

Pilot Officer Denis Henry Battrick 125307
Observer
Royal Air Force Volunteer Reserve
of Wyke Regis, Weymouth, Dorset,
who died aged 21.

and

**Flight Sergeant Donald Gordon Taylor
R/73449**
WOp/AG
Royal Canadian Air Force
of Salisbury, New Brunswick, Canada,
who died aged 27.

and

**Flight Sergeant William MacDonald
Vacheresse R/76480**
WOp/AG
Royal Canadian Air Force
of Stellarton, Nova Scotia, Canada,
who died aged 26.

*Aircraft crashed into ground at Glenhead farm,
shortly after take-off on night flying exercises.
The aircraft appeared to hit the ground at great
speed and whilst turning to port.
The cause of the accident was thought to be the
aircraft assuming an abnormal flying position,
due to the pilot taking the wrong corrective
action while flying on instruments.*

♦

IN MEMORY OF THE CREW OF
BRISTOL 152 BEAUFORT I L4479
5(C)OTU

who died on 30 August 1942

Flight Sergeant Louis Orlin Glay R/101773
Pilot
Royal Canadian Air Force
of Kamloops, British Columbia, Canada,
who died aged 21.

and

Flight Sergeant John Orville MacLean R/106786
Air Obs.
Royal Canadian Air Force
of Toronto, Canada,
who died aged 21.

and

Flight Sergeant John William Leyland R/80045
WOp/AG
Royal Canadian Air Force
of Manitoba, Canada,
who died aged 28.

Beaufort L4479 crashed on Goat Fell, Arran, during night cross-country exercise. The aircraft was not flying at sufficient height to clear surrounding mountains. Probable navigational error. The number of crew given on the Flying Accident Card is four, but there are only three mentioned on the Casualty Cards.

◆

IN MEMORY OF THE CREW OF
BRISTOL 152 BEAUFORT I L9932
5(C)OTU

who died on Tuesday 1 September 1942

Flight Sergeant William Horsman Drury 1178920
Pilot
Royal Air Force Volunteer Reserve
of Kenilworth, Warwickshire,
who died aged 28.

and

Leading Aircraftman Frederick Donald Young 1298655
Radio Mechanic
Royal Air Force Volunteer Reserve
of Portsmouth,
who died aged 20.

Beaufort L9932 was on landing approach after a test flight when, at a height of 100 feet it collided with Hampden P1199, also on final approach. Both aircraft crashed in flames onto the runway. Flight Sergeant Leslie Arthur Goodson survived although seriously injured.

HANDLEY PAGE HAMPDEN I P1199
5(C)OTU

Warrant Officer James Edward Toby Jones NZ401767
Pilot
Royal New Zealand Air Force
of Wellington City, New Zealand,
who died aged 22.

and

Flight Sergeant Joseph Gerard Newman R/79294
Pilot
Royal Canadian Air Force
of Yarmouth, Nova Scotia, Canada,
who died aged 24.

Since the Beaufort and the Hampden have a different angle of approach, it appears that neither pilot saw the other until it was too late. The suggestion was made that Flying Control station a man at the end of the runway with a red Aldis lamp and red Verey cartridges in future.

◆

IN MEMORY OF THE CREW OF

BRISTOL 152 BEAUFORT I L9803
2TTU

who died on Wednesday 2 September 1942

Flight Sergeant Ronald Albert Eldon Lutes R/90709
Pilot
Royal Canadian Air Force
of Sault St Marie, Ontario, Canada,
who died aged 24.

and

Sergeant Clive Derek Hammond 1382637
Observer
Royal Air Force Volunteer Reserve
of Pinner, Middlesex,
who died aged 20.

and

Flight Sergeant Bruce Douglas Francis R/83222
WOp/AG
Royal Canadian Air Force
who died aged 26.

and

Flight Sergeant John Benedict Hargreaves R/100108
WOp/AG
Royal Canadian Air Force
of Winnipeg, Manitoba, Canada,
who died aged 19.

ORB: Beaufort 'Y' from Abbotsinch, operating from this station on night exercises, crashed on the Island of Mull, about the time indicated. All crew killed.

This aircraft flew into hills on the island of Mull, whilst off course, probably due to navigational error. During the night there was a high increase in wind velocity with an alteration in direction.

◆

IN MEMORY OF THE CREW OF
BRISTOL 152 BEAUFORT I N1180
TTU

who died on Wednesday 2 September 1942

Sergeant Albert Augustine Haydon 414287
Pilot
Royal New Zealand Air Force
of Wellington, New Zealand,
who died aged 26.

and

Pilot Officer Leonard Percy Booker 391207
Pilot
Royal New Zealand Air Force
of Taranaki, New Zealand,
who died aged 29.

and

Sergeant Tom Henry Grasswick R/60581
WOp/AG
Royal Canadian Air Force
of Calgary, Alberta, Canada,
who died aged 21.

and

Sergeant Francis John Bliss Griffin 1330126
Royal Air Force Volunteer Reserve
of Pinner, Middlesex,
who died aged 33.

ORB: Beaufort 'S' from Abbotsinch, operating from this station on night exercises crashed near Machrihanish (Mull of Kintyre), about the time indicated. All crew killed, Pilot Sgt Haydon, P/O Booker, Sgt Griffin & Sgt Grasswick

Beaufort N1180, normally stationed at Abbotsinch, had been detached to RAF Turnberry for night navigation exercises. The crew had been briefed for a general deterioration of weather conditions along their planned route. However, during the flight the weather worsened more rapidly than forecast, developing a strong westerly wind. The aircraft had drifted of course when it struck Tor Mhor, half a mile south of the Mull of Kintyre Lighthouse where it broke up and caught fire.

◆

IN MEMORY OF THE PILOT OF
BRISTOL 152 BEAUFORT II AW309
5(C)OTU

who died on 11 September 1942

Sergeant Norman James Barlow 413013
Pilot
Royal New Zealand Air Force
of Tuakau, Auckland, New Zealand,
who died aged 24.

*Beaufort AW309 crashed into the sea off Maidens Harbour during a practice flight. The port
engine failed and then the propeller fell off. The pilot approached the airfield correctly for a
forced landing when it was found that the undercarriage wouldn't lower, even with the emergency
hand-pump. The pilot having height turned away from the airfield, preparatory to firing his
undercarriage down, when the aircraft stalled and lost height in a right-hand turn. The aircraft
was then ditched successfully in the sea. Tragically the pilot was unable to leave the aircraft due
to a faulty safety harness, it was later found that the pilot had attempted to fix the harness by
using a key-ring as a makeshift 'knock-off-pin', which in the emergency he was unable to release.
The other members of the crew were rescued safely.*

◆

IN MEMORY OF THE CREW OF
HANDLEY PAGE HAMPDEN I AN125
5(C)OTU

who died on 7 October 1942

Flying Officer Charles Henry Lacerte J/10877
WOp/AG
Royal Canadian Air Force
of Winnipeg, Manitoba,
who died aged 19.

*Crashed into sea near Girvan.
Four crew picked up by Marine Craft Section, one dead.*

◆

IN MEMORY OF THE CREW OF
HANDLEY PAGE HAMPDEN I AN151
1TTU

who died on Monday 12 October 1942

Sergeant Gareth Howell 1282866
Pilot
Royal Air Force Volunteer Reserve
of Aberystwyth, Cardiganshire, Wales,
who died aged 22.

and

Pilot Officer William Thomas Walton 125309
Air Observer
Royal Air Force Volunteer Reserve
of Hastings, Sussex,
who died aged 28.

and

Sergeant Ernest William Evans 411751
WOp/AG
Royal New Zealand Air Force
of Christchurch, New Zealand,
who died aged 26.

and

Sergeant Stanley William Shadrack 137558
WOp/AG
Royal Air Force Volunteer Reserve
of Goodmayes, Essex,
who died aged 28.

Hampden AN151 struck sea during low level dummy torpedo attack at night on the Firth of Clyde. The aircraft dropped a flare and was returning to carry out attack on trawler, when pilot apparently misjudged height and crashed into the sea.

◆

IN MEMORY OF THE CREW OF
BRISTOL BEAUFORT II DD881
5(C)OTU

who died on Tuesday 27 October 1942

**Flight Sergeant Leslie Roy Sherwood
A.406493**
Pilot
Royal Australian Air Force
of West Leederville, Western Australia,
who died aged 23.

and

Sergeant George Wilkie Newman 1378510
WOp/AG
Royal Air Force Volunteer Reserve
of Harlesden, Middlesex,
who died aged 22.

and

Sergeant Gordon Hancock 1013793
WOp/AG
Royal Air Force Volunteer Reserve
of Swansea, Wales,
who died aged 27.

and

Sergeant Robert Alexander Ellis 1098276
Navigator B
Royal Air Force Volunteer Reserve
of North Shields, Northumberland,
who died aged 26.

George Wilkie 'Bill' Newman

*Beaufort DD881 went missing on a night
cross-country exercise. Accident caused
by aircraft becoming lost through faulty
navigation and crashing into sea after all
fuel was used. The wireless set was useless
prior to the flight and had not been repaired
due to a breakdown in unit organization.*

◆

IN MEMORY OF THE CREW OF
BRISTOL BEAUFORT I L9865
5(C)OTU

who died on Saturday 7 November 1942

Sergeant Roy John Sutton 1269173
Pilot
Royal Air Force Volunteer Reserve
of Dover,
who died aged 20.

and

**Sergeant Robert Ashleigh Noel St. John
Farnon 1285719**
(Bob Ashley)
WOp/AG
Royal Air Force Volunteer Reserve
of Wimbledon, Surrey,
who died aged 28.

*Beaufort L9865 took-off at 23:33 hours on a
night training flight. After take-off the aircraft
was seen to turn at a height of 200-300 feet and
dive into the sea,
one mile west south west of Turnberry.*

Bob Ashley

◆

IN MEMORY OF THE CREW OF
HANDLEY PAGE HAMPDEN I AE379
5(C)OTU

who died on Tuesday 10 November 1942

Sergeant Alan Charles Smart 412752
Pilot
Royal New Zealand Air Force
of Otago, New Zealand,
who died aged 22.

and

Pilot Officer Ernest Greaves 122118
Observer
Royal Air Force Volunteer Reserve

and

**Flight Sergeant George Kingsley Shaw
R/90100**
WOp/AG
Royal Canadian Air Force
of London, Ontario, Canada,
who died aged 21.

and

Sergeant Donald George Barnie R/117689
WOp/AG
Royal Canadian Air Force
of Collingwood, Ontario, Canada,
who died aged 26.

*Hampden AE379 dived into ground whilst night
flying at Milton Burn, 500 yards from
Turnberry Hospital.*

◆

IN MEMORY OF THE CREW OF
BRISTOL BEAUFORT II AW384
5(C)OTU

who died on 14 November 1942

Flying Officer Maurice Carson Jolly 413858
Pilot
Royal New Zealand Air Force
of Cromwell, Otago, New Zealand,
who died aged 25.

and

Sergeant Frederick Stanley Fisher 1393458
Navigator
Royal Air Force Volunteer Reserve
of Paddington, London,
who died aged 22.

and

Sergeant George Lunn 1125565
WOp/AG
Royal Air Force Volunteer Reserve
of West Hartlepool,
who died aged 22.

and

Sergeant George Maguire 1266020
WOp/AG
Royal Air Force Volunteer Reserve
of Longsight, Royton,
who died aged 33.

Maurice Carson
Jolly

Frederick Stanley
Fisher

George Lunn

George McGuire

*Beaufort AW384 crashed three miles east of
Girvan in a meadow by Killochan Castle,
Dailly, whilst on a night flying exercise.
The aircraft crashed nineteen minutes after
take-off and the accident was attributed to
the pilot losing control of the aircraft while
flying on instruments.*

◆

IN MEMORY OF
AIRCRAFTSMAN 1ST CLASS

GEORGE KIRKWOOD GARDNER 1013652

who died on Friday 27 November 1942

Royal Air Force Volunteer Reserve
of Osterley, Isleworth, Middlesex,
who died aged 22.

*Fatal accident occurred when a works flight
lorry collided with an airman.
The airman was partially deaf and did not hear
the approach of lorry when turning corner.*

◆

IN MEMORY OF THE CREW OF
BRISTOL BEAUFORT II DD871
5(C)OTU

who died on Saturday 12 December 1942

Flying Officer Joseph Ash J/10980
Pilot
Royal Canadian Air Force
of Ottawa, Ontario, Canada,
who died aged 22.

and

Flying Officer John James Earls J/11958
Air Observer
Royal Canadian Air Force
of Ottawa, Ontario, Canada.

and

Sergeant Harry Newell 1270897
WOp/AG
Royal Air Force Volunteer Reserve
of Battersea, London,
who died aged 20.

and

Sergeant Herbert William Goodman 1270737
WOp/AG
Royal Air Force Volunteer Reserve
of South Harrow, Middlesex,
who died aged 20.

Beaufort DD871 failed to return from a night navigation exercise. Missing at sea.

◆

IN MEMORY OF THE CREW OF
BRISTOL BEAUFORT II AW286*
5(C)OTU

Who died on Thursday 3 December 1942

Sergeant Bernard Moore 414727
Pilot
Royal New Zealand Air Force
of Taihape, North Island, New Zealand,
who died aged 20.

and

Sergeant Samuel Weaver 645141
Navigator
Royal Air Force
of Salford, Lancashire,
who died aged 22.

and

Sergeant Felix Henry Aldous 1266933
WOp/AG
Royal Air Force Volunteer Reserve
of South Kensington, London,
who died aged 22.

and

Sergeant Eric John Harrild 1290442
WOp/AG
Royal Air Force Volunteer Reserve

Beaufort AW386 crashed into the sea off Dowhill farm, during night flying training.
Cause of accident obscure.
*Aircraft number given on Official Crash Card as AW386

IN MEMORY OF THE CREW OF
HANDLEY PAGE HAMPDEN I P2091
5(C)OTU

who died on Tuesday 22 December 1942

**Flight Sergeant Leonard Albert Hoskins
R/106581**
Pilot
Royal Canadian Air Force
of Westbank, British Columbia, Canada,
who died aged 21.

and

**Flight Sergeant
Donald Leif Erickson R/87301**
Navigator B
Royal Canadian Air Force
of Saskatoon, Saskatchewan, Canada,
who died aged 23.

and

Sergeant Joseph George Dukick R/71382
WOp/AG
Royal Canadian Air Force

*Donald Leif
Erickson*

*Hampden P2091 was engaged on low-level
bombing practice, and having made two runs
across the target, dived into the sea, half a
mile off Culzean Castle.
The investigation reports that 'the
inexperienced pilot got into a flat turn at low
altitude and failed to recover the aircraft'.
The accident report cards states 'Another
case of the Hampden Stabilised Yaw'.
Sergeant Charles Maule WOp/AG, Royal
Canadian Air Force
was rescued with slight injuries.*

◆

IN MEMORY OF THE CREW OF
BRISTOL BEAUFORT I L9825
1TTU

who died on 19 January 1943

Flight Sergeant Tom Skinner Tamblyn 411203
Pilot
Royal Australian Air Force
of Marrickville, New South Wales, Australia,
who died aged 32.

and

Sergeant Joseph Ernest McGrath 412321
Navigator/Bomb Aimer
Royal Australian Air Force
of Grafton, New South Wales, Australia,
who died aged 22.

and

**Sergeant Robert Alfred William Quinn
1292963**
WOp/AG
Royal Air Force Volunteer Reserve
of Ilford, Essex.

*Beaufort L9825 was detailed to carry out an
authorised low flying
exercise, During the exercise the aircraft flew
into the sea 1½ miles west of Girvan at 11:45
hours. All the crew were killed.*

Tom Skinner Tamblyn

Joseph Ernest McGrath

◆

IN MEMORY OF THE CREW OF
VICKERS WELLINGTON VIII LB223
1TTU

who died on Sunday 28 March 1943

Sergeant Alexander Norman Hunter 1022817
Pilot
Royal Air Force Volunteer Reserve
of Dunfermline, Fife,
who died aged 30.

and

Sergeant Dan Norman Smallie 778818
2nd Pilot
Royal Air Force Volunteer Reserve
of Bulawayo, Southern Rhodesia,
who died aged 31.

and

Sergeant Frank Bernard Smith 1383332
Pilot
Royal Air Force Volunteer Reserve
of Plumstead, London,
who died aged 20.

and

Sergeant Humphrey William Walpole 1425750
Navigator
Royal Air Force Volunteer Reserve
of Stechford, Birmingham
who died aged 29.

and

Sergeant Wilfred Monkhouse 1381996
WOp/AG
Royal Air Force Volunteer Reserve
of Edgeware, London.

Grave of Humphrey William Walpole

and

Sergeant Harry Pike 1036504
WOp/AG
Royal Air Force Volunteer Reserve
of New Moston, Manchester,
who died aged 23.

and

Sergeant Phillip Reginald Comley 1233733
WOp/AG
Royal Air Force Volunteer Reserve
of Erdington, Birmingham,
who died aged 21.

*Wellington aircraft LB223 crashed at Drumbeg farm, between Turnberry and Girvan, at night.
Coming in to land with flaps and undercarriage down, the aircraft had partly completed circuit
when the pilot received a Red Signal.
With undercarriage and flaps still down, the aircraft stalled when it overshot the airfield,
crashed and caught fire.*

◆

IN MEMORY OF THE CREW OF
BRISTOL BEAUFORT I DX114
2TTU

who died on 7 April 1943

Sergeant Anthony Dearmer 1251401
WOp/AG
Royal Air Force Volunteer Reserve
of Chelsea, London,
who died aged 22.

Ditched off Ailsa Craig
Body found in the sea two miles North East
of Ailsa Craig

◆

IN MEMORY OF THE CREW OF
VICKERS WELLINGTON VIII LB239
1TTU

who died on Thursday 29 April 1943

Sergeant Derek Arthur Kay Treble 1335183
WOp/AG
Royal Air Force Volunteer Reserve
of Morden, Surrey,
who died aged 20.

Wellington LB239 crashed into the sea whilst
engaged in night flying exercises. The crew
were rescued safely, with the exception of
Sgt. Derek Treble, whose body was not found.

◆

IN MEMORY OF THE CREW OF
HANDLEY PAGE HAMPDEN I AT125
1TTU

who died on Sunday 18 April 1943

Sergeant Ronald Stott Cordingley 1147967
Pilot
Royal Air Force Volunteer Reserve
of Alwoodley, Leeds,
who died aged 22.
and
Sergeant Eric Vevers 1196764 Navigator
Royal Air Force Volunteer Reserve
of Timperley, Cheshire,
who died aged 23.
and
Flight Sergeant John Bertram Reid R/114641
WOp/AG
Royal Canadian Air Force

Hampden AT125 collided with Wellington
LB237 in flight and crashed into the sea.
The Wellington, pilot, Pilot Officer Ockalski,
made a safe landing although fin and rudder
were badly damaged.

◆

IN MEMORY OF THE CREW OF
HANDLEY PAGE HAMPDEN I AT117
5(C)OTU/1TTU

who died on
Monday 3 May 1943

Sergeant Leslie Frederick Parratt 1289414
Pilot
Royal Air Force Volunteer Reserve
of Twickenham, Middlesex,
who died aged 25.
and
**Flying Officer John Henry Benjamin Eyles
129407**
Navigator
Royal Air Force Volunteer Reserve
of Cleethorpes, Lincolnshire,
who died aged 20.

Hampden AT117 dived into the sea prior to
landing at Turnberry. Aircraft wobbled in
lateral plane.
No real indication as to possible cause
of accident.

◆

IN MEMORY OF THE CREW OF
VICKERS WELLINGTON III LB187
1TTU

who died on Friday 21 May 1943

Sergeant Edwin Esau 1393290
Pilot
Royal Air Force Volunteer Reserve
of Hither Green, London,
who died aged 20.

and

**Flying Officer
Kenneth Alfred Embery 133971**
Observer
Royal Air Force Volunteer Reserve
of Harpenden, Hertfordshire,
who died aged 21.

and

Sergeant Alfred Brown 1039746
WOp/AG
Royal Air Force Volunteer Reserve

*Wellington LB187 crashed into the sea on
practice torpedo drop at night off Ailsa Craig.*

◆

IN MEMORY OF THE CREW OF
VICKERS WELLINGTON VIII
LB237
1TTU

who died on Friday 21 May 1943

**Flight Sergeant John
Frederick Forsyth-Johnson
1383232**
Pilot
Royal Air Force Volunteer
Reserve
of Edmonton, Middlesex,
who died aged 20.

*John Frederick
Forsyth Johnson*

and

Sergeant Bertram Harrington Lowe 658389
Pilot
Royal Air Force
of Ewell, Surrey,
who died aged 27.

and

Sergeant William Victor Chambers 1384655
Observer
Royal Air Force Volunteer Reserve
who died aged 22.

and

**Sergeant Harry Hubert Charles Harris
1377270**
WOp/AG
Royal Air Force Volunteer Reserve
of South Kensington, London,
who died aged 31.

and

Sergeant Thomas Brian Clarke 1236744
WOp/AG
Royal Air Force Volunteer Reserve

and

Sergeant Cyril Jones 980210
WOp/AG
of Wrexham, Denbighshire,
who died aged 23.
Son of Jonathan and Mary Jane Jones
husband of Bettine Eluned Jones
(a son was born to Cyril and Bettine
the same day his father died)

*Aircraft missing on practice torpedo drop.
Presumed to have collided with Wellington LB193 and crashed into sea.*

◆

IN MEMORY OF THE CREW OF
VICKERS WELLINGTON VIII
LB193
1TTU

who died on Friday 21 May 1943

Sergeant Leslie Trevethan Gibson 1091023
WOp/AG
Royal Air Force Volunteer Reserve
of Kenton, Newcastle-on-Tyne,
who died aged 21.

and

Sergeant Ronald Dean 1087504
WOp/AG
Royal Air Force Volunteer Reserve
of Halifax,
who died aged 21.

*Presumed to have collided with Wellington
LB237 whilst on practice torpedo drop at night
off Ailsa Craig.*

◆

IN MEMORY OF THE CREW OF
BRISTOL BEAUFORT I DW923

who died on Monday 14 June 1943

Sergeant Hugh Fraser Hunter 1060264
Pilot
Royal Air Force Volunteer Reserve
of Edinburgh,
who died aged 22.

and

**Sergeant
Donald Sylvester
McDonald 421491**
Navigator
Royal Australian Air
Force
of Glen Davis,
New South Wales,
Australia,
who died aged 25.

Donald Sylvester McDonald

and

**Flight Sergeant Kenneth Hardie McDonald
411033**
WOp/AG
Royal New Zealand Air Force
who died aged 22.

and

**Sergeant Thomas Ernest Victor Doherty
1081823**
WOp/AG
Royal Air Force Volunteer Reserve
of Newtonards, County Down, Ireland,
who died aged 20.

*The aircraft was just 20 minutes into a
training exercise when it crashed into the sea
whilst making a dummy attack
on a ship.*

◆

IN MEMORY OF THE CREW OF
VICKERS WELLINGTON HX774
1TTU

who died on 7 August 1943

Flight Sergeant Derek Ralph Sorensen 1313849
Pilot
Royal Air Force Volunteer Reserve
of Leura, New South Wales, Australia,
who died aged 22.

and

Sergeant James Leonard Webber 1435307
Pilot
Royal Air Force Volunteer Reserve
of Balham,
who died aged 21.

and

Sergeant John Leonard Hiscock 1319841
Observer
Royal Air Force Volunteer Reserve
of Southampton,
who died aged 20.

and

Sergeant Eric Williams Hirst 1134186
WOp/AG
Royal Air Force Volunteer Reserve
of Birkenhead,
who died aged 21.

and

Sergeant Ernest Leslie Padfield 1384929
Royal Air Force Volunteer Reserve
who died aged 22.

and

Flight Sergeant Frederick John Wright 30881
Royal Australian Air Force
of Burnie, Tasmania, Australia,
who died aged 22.

Wellington HX774 crashed in the sea south of Brest Rocks, Turnberry.
Crew of six lost their lives

♦

IN MEMORY OF THE CREW OF
HANDLEY PAGE HAMPDEN TB.1 X3026
1TTU

Who died on Monday 07 August 1943

Flying Officer John Henry Willox Johnstone
414043
Pilot
Royal Australian Air Force
of Barcaldine, Queensland, Australia,
who died aged 24.

and

Flight Sergeant Eric Douglas (Tup) Lee
414413
Observer
Royal Australian Air Force
of Valley, Queensland, Australia,
who died aged 22.

John Henry Willox Johnstone

and

Flight Sergeant Sydney Raymond Greer
417028
WOp/AG
Royal New Zealand Air Force
of Palmerston North, Wellington, New Zealand,
who died aged 33.

and

Flying Officer Leslie Norman Buttimore
417194
WOp/AG
Royal New Zealand Air Force
of Cambridge, Auckland, New Zealand,
who died aged 23.

*Flew into Mochrum Hill, between Kirkoswald
and Maybole in low cloud and rain.*

Eric Douglas Lee

◆

IN MEMORY OF THE CREW OF
HANDLEY PAGE HAMPDEN P5431
1TTU

who died on 16 August 1943

Flight Sergeant
Alan Douglas Riordan 420489
Royal Australian Air Force
of Steve King's Plain, New South Wales, Australia,
who died aged 20.

Ditched off Dunure, following engine failure on training flight.
Three other crew injured

◆

IN MEMORY OF THE CREW OF
BRISTOL BEAUFORT I JM557
2TTU

who died on 23 August 1943

Flying Officer James Reginald Keevil 49884
Pilot
Royal Air Force
of Selly Oak, Birmingham,
who died aged 24.

and

Flight Sergeant
John Allen Kingsmill 417204
Navigator/Bomb Aimer
Royal Australian Air Force
of Loxton, South Australia,
who died aged 20.

and

Sergeant George Duncan 1367964
WOp/AG
Royal Air Force Volunteer Reserve
of St Monans, Fife,
who died aged 22.

and

Sergeant Robert Davie-Spiers 1382723
WOp/AG
Royal Air Force
of Glasgow,
who died aged 23.

Beaufort JM557 was detailed to carry out a
low flying exercise over the sea. During the
exercise the aircraft experienced starboard
engine failure and the aircraft crashed into
the Firth of Clyde about one mile off shore
near Culzean Castle.

◆

IN MEMORY OF THE CREW OF
BRISTOL BEAUFIGHTER VI
JL610
1TTU

who died on Wednesday 29 September 1943

**Sergeant John Gerald Preece Pritchard
658811**
Pilot
Royal Air Force Volunteer Reserve
of Welsh Newton Common, Herefordshire,
who died aged 25.

and

Sergeant Raymond Lucas 1477333
Nav./WOp
Royal Air Force Volunteer Reserve
of Bishop's Wearmouth, Sunderland,
who died aged 19.

and

**Aircraftsman 2nd Class John Andrews
1481652**
Royal Air Force Volunteer Reserve
of Sherwood, Nottingham,
who died aged 20.

*Beaufighter JL610 was being ferried to
RAF Turnberry when the weather closed in
suddenly. The pilot made one approach and
might have landed if AFC had not fired a
red flare. While making a second approach,
the pilot made a steep turn away from high
ground but lost height and crashed north
of Morriston farm, Maidens. A subsequent
investigation found that 'incorrect use of
signals by flying control', 'pilots' temporary
loss of bearings whilst flying part-visual,
part-instruments', 'Insufficiently clear
understanding of Met. Information' and that
'advice of experienced Airfield Controller
was not passed to Control Officer' were listed
as the causes of this 'deplorable accident'.*

◆

IN MEMORY OF THE CREW OF
BRISTOL BEAUFIGHTER VI JL715
1TTU

who died on Tuesday 8 February 1944

**Flight Lieutenant Keith Wells Peter Noble
118390**
Royal Air Force Volunteer Reserve
of Felpham, Bognor Regis, Sussex,
who died aged 22.

and

**Flying Officer Geoffrey Clay Whitehurst
141701**
Royal Air Force Volunteer Reserve

*Beaufighter JL715 engine cut, stalled on
approach to Turnberry*

◆

IN MEMORY OF LEADING AIRCRAFTMAN

Andrew Canning
son of Andrew and Janet Caning of
Dennistoun, Glasgow
who died on 15 February 1944, aged 22.

*LAC Andrew Canning died after suffering
multiple injuries in an accident on
Turnberry Airfield.*

◆

IN MEMORY OF THE PILOT OF
REPUBLIC P47-D THUNDERBOLT 42-76389
310 FRS

who died on 1 March 1944

**Second Lieutenant Hillyer Cooper Godfrey
0667692**
Pilot USAAF of Georgia,
United States of America.

Crashed at Chirmorie farm, Barrhill, Ayrshire

◆

IN MEMORY OF THE PILOT OF
HAWKER HURRICANE IV LD594
439 Squadron

who died on 18 March 1944

**Flying Officer Roswell Murray MacTavish
J/22385**
Royal Canadian Air Force
of Vancouver, British Columbia, Canada,
who died aged 24.

*Stationed at RAF Ayr (Heathfield), F/O R.
M. MacTavish died during flight formation
training when his Hawker Hurricane LD564
crashed into the forest near Loch Doon,
Dalmellington, Ayrshire, Scotland.
F/O R. M. MacTavish is buried at Ayr
Cemetery.*

◆

IN MEMORY OF THE CREW OF
AVRO ANSON I N4875

who died on Monday 3 April 1944

Captain Douglas Keith Fairweather
Pilot
Air Transport Auxiliary
son of Sir Wallace Fairweather and Dame
Margaret Eureka Fairweather,
of Glasgow; husband of Flight Captain the Hon.
Margaret Fairweather,
who died aged 53.

and

Second Officer Kathleen Mary Kershaw
Nursing Sister
of Wellington, Somerset,
who died aged 31.

*Avro Anson N4875 went missing while on
an ambulance flight from White Waltham to
Prestwick.
The body of Captain Douglas Fairweather was
washed ashore at Maidens.*

Douglas Fairweather

◆

IN MEMORY OF THE CREW OF
VENTURA V FP659

Who died on Sunday 21 May 1944

Flying Officer George John Wilcox 151803
Pilot
Royal Air Force Volunteer Reserve
of Salford, Lancashire,
who died aged 26.

and

Flight Lieutenant Stanley Park 45846
Pilot
Royal Air Force
of Manchester

and

Flying Officer William Ferguson Gray 152377
Navigator
Royal Air Force Volunteer Reserve
of Coatbridge, Lanarkshire,
who died aged 22.

and

Warrant Officer Class 1 Walter Henry Harris R/99699
WOp/AG
Royal Canadian Air Force
of Kitchener, Ontario, Canada,
who died aged 23.

and

Flight Sergeant Joseph Patrick O'Neill R/164611
WOp/AG
Royal Canadian Air Force
of Charlottetown, Canada,
who died aged 25.

*Ventura FP659 crashed on take-off at Turnberry Airfield during Flight Commander's Test prior
to first solo of pupil. After take-off the aircraft swung to port following failure of that engine, pilot
momentarily regained control, but stalled at a height of eighty feet.
Aircraft spiralled into ground nose-first, turned onto its back and caught fire.
Cause of engine failure was found to be that the plugs had oiled up due to excessive
ground-running at low revs.*

IN MEMORY OF THE PILOT OF
BRISTOL BEAUFIGHTER VI JL835
5(C)OTU

who died on 30 June 1944

Squadron Leader John Allen Bateson 37744
Pilot
Royal Air Force
of Toat Hill, Sussex,
who died aged 33.

Beaufighter JL835 returning from torpedo training demonstration, with engine failure, stalled at 100 feet and crashed into the sea 800 yards off Turnberry. After recovery of the aircraft from the sea, metal was found in the oil filter indicating the breaking up of connecting rod.
The passenger Lt. Cmdr. Carl Johan August le Maire Stansberg survived, although not wearing a Mae West life preserver and the safety boat being ten miles away.

IN MEMORY OF THE CREW OF
AVRO LANCASTER I ME729
Of 630 Squadron

who died on 18 July 1944 at Dalmellington

Sergeant Frank Helliwell 2211367
Flight Engineer
Royal Air Force Volunteer Reserve
of Lancaster

and

Flight Sergeant Frederick George Brezina R/84002
Air Bomber
Royal Canadian Air Force
of Kitchener, Ontario, Canada,
who died aged 26.

and

Sergeant James Alexander Calder R/220155
Air Gunner
Royal Canadian Air Force
of Sault Ste. Marie, Ontario, Canada,
who died aged 21.

Aircraft took off at 22:04 hrs from East Kirby on navigational training exercise. The top turret (which was driven by a belt from the

Grave of Frederick George Brezina

engine) started to spin and the pilot shut down the wrong engines in an attempt to stop it turning. Two of the crew bailed out, Sgt Helliwell bailed out too low and was killed. The aircraft crashed at 00:59 hrs at Mossdale farm by Dalmellington.
This Lancaster was not based at Turnberry; the bodies were recovered by the Station Medical Officer.

◆

IN MEMORY OF PILOT
FLIGHT CAPTAIN
THE HON MARGARET FAIRWEATHER
AIR TRANSPORT AUXILIARY

who died on Friday 4 August 1944

Daughter of the Rt. Hon. Walter Runciman,
PC, DCL, LLD, JP, 1st Viscount Runciman
of Doxford, and of the Viscountess Runciman
of Doxford, JP (nee Stevenson), of Doxford,
Northumberland; wife of Capt. Douglas Keith
Fairweather, ATA.

*Margaret Fairweather was one of the first women
to fly a Spitfire. Prior to the war she already
had 1,000 hours of civilian flying and was an
instructor with the Civil Air Guard.
Margaret was killed at Malpas in Cheshire whilst
attempting to land Percival Proctor LZ801, after
suffering engine failure, her sister Kitty was
injured in the crash.
Margaret Fairweather shares a joint grave with
her husband at Dunure.
This is unique, being the only husband and wife
Commonwealth War Grave.*

Margaret
Fairweather

Grave of Margaret and husband
Douglas Fairweather

◆

IN MEMORY OF THE CREW OF
De Havilland Mosquito KB219
Of 1655 Mosquito Training Unit

who died on 13 September 1944

Pilot Officer Thomas Alan Armstrong 177963
Navigator
Royal Air Force Volunteer Reserve
of Kinniside, Carlisle,
who died aged 24.

*Mosquito KB219 from Wyton crashed at Low Milton farm, by Maybole.
The aircraft went into uncontrollable spin at 25,000ft and disintegrated.
The pilot bailed out and was slightly injured.*

◆

IN MEMORY OF THE CREW OF
Miles M25 Martinet TTI HP335
5(C)OTU

who died on 19 November 1944

Warrant Officer Raymond Williams 1204192
Pilot
Royal Air Force Volunteer Reserve
of Cardiff, Wales,
who died aged 33.

and

LAC Thomas Douglas Laurie 1371688
Drogue Operator
Royal Air Force Volunteer Reserve
of Dumfries,
who died aged 22.

*Martinet HP335 went missing on
affiliation exercise.
Wreckage discovered.*

◆

IN MEMORY OF THE CREW OF
VICKERS WARWICK I BV340
5(C)OTU

who died on 29 December 1944

Squadron Leader Thomas Sutherland Brotherstone 90673
Pilot Instructor
Royal Air Force
of Edinburgh,
who died aged 31.

and

Flight Sergeant Harry Holmes 1458454
Pupil Pilot
Royal Air Force Volunteer Reserve
of Salford, Lancashire,
who died aged 22.

Whilst engaged on air-to-air firing exercise, Warwick BV340 overshot while making an emergency landing after suffering engine failure and crashed into the Pow Burn at the end of the runway at Prestwick.

It was found that the pupil pilot did not lower flaps, which was a contributory cause of the accident, and that the port engine had failed due to a breakdown in the oil cooler causing a loss of oil. The port oil tank was found to be empty.

♦

IN MEMORY OF THE CREW OF
LOCKHEED HUDSON V AM562
5(C)OTU

who died on
Sunday 21 January 1945

Warrant Officer John Robertson 1372529
Pilot
Royal Air Force Volunteer Reserve

and

**Flight Sergeant Norman Walter Huggins
430172**
Royal Australian Air Force
of Victoria, Australia,
who died aged 20.

and

Flight Sergeant James Vincent Jenkins 432194
Royal Australian Air Force
of New South Wales, Australia,
who died aged 21.

and

Sergeant Eric Raynor Kelsall 1676661
Royal Air Force Volunteer Reserve

and

Sergeant Peter Bernard Joseph 3204233
Met. Observer
Royal Air Force Volunteer Reserve
of Brixton, London,,
who died aged 18.

*Hudson AM562 had only just taken-off and
was in low flight when it was seen to spiral to
starboard and hit the sea, in an almost
vertical position.
Two miles off Maidens Harbour.*

♦

IN MEMORY OF THE CREW OF
LOCKHEED HUDSON VI FK502
5(C)OTU

who died on Sunday 11 February 1945

**Flying Officer Derek Mordaunt Beagley
154122**
Pilot
Royal Air Force Volunteer Reserve
of Weymouth, Dorset,
who died aged 21.

and

Pilot Officer Alan Reginald Pewsey 181348
Royal Air Force Volunteer Reserve
of Chiswick, Middlesex,
who died aged 36.

and

Sergeant Henry Mariner 1804393
Royal Air Force Volunteer Reserve
of Shadwell, London,
who died aged 20.

and

Sergeant Stanley Foster 1685236
Royal Air Force Volunteer Reserve

*Hudson FK502 took-off in poor weather
conditions on night flying practice.,
completed right-hand circuit as instructed
by flying control, and was seen three
minutes later flying low over the sea prior
to crashing into the water off Turnberry
Point. No weather test was undertaken by
the Instructing Officer I/C, and as such the
aircraft should not have been allowed
to take-off.*

◆

IN MEMORY OF THE CREW OF
BRISTOL 156 BEAUFIGHTER TF.X. LX852

who died on 19 April 1945

Flight Lieutenant Arthur Harold (Hank) Nagley J/23075
Royal Canadian Air Force
of Montreal, Canada,
who died aged 22.

and

Warrant Officer William Arthur Smith 1579873
Royal Air Force Volunteer Reserve
of Coventry,
who died aged 22.

Collided with Beaufighter RD486 (Crew Survived) over Heads of Ayr. Lost at Sea.

◆

IN MEMORY OF THE CREW OF
BRISTOL 156 BEAUFIGHTER TF.X. RD422
5(C)OTU

who died on 30 July 1945

**Flight Sergeant Harold Leonard Seabrook
1801076**
Royal Air Force Volunteer Reserve
of Lewisham, London,
who died aged 21.

and

Flight Sergeant Frederick Robert Ley 1624130
Royal Air Force Volunteer Reserve
of Tinsley, Yorkshire,
who died aged 21.

Grave of an unknown airman

The aircraft had been airborne for about fifteen minutes for a practice formation attack on shipping. The starboard engine failed and was feathered but the pilot reported that he was unable to maintain height on one engine. The aircraft was then ditched in the sea six miles south of Turnberry, but neither crew member survived.

15
Post War Period

With most of the aircraft gone and only a small administrative staff on the airfield, plans were already being formulated for its future use, even before it had been officially derequisitioned. A missive was received by the factor from a solicitor in Ayr, on behalf of his clients who 'have considerable experience of running holiday camps in England, would consider either purchasing or leasing from you the camp at Maidens when it comes to be derequisitioned. The camp would be run in an efficient manner and provision would be made for the gradual replacement of War Department buildings by proper structures as and when material and labour are more easily obtainable. We shall be glad if you will let us know if your Directors would consider such a proposal, taking into account that a high class efficiently run camp is aimed at.'

London Midland and Scottish Railway Company were also wondering what prospect there was of Turnberry Hotel being handed back. The whole of the personnel who were occupying the hotel had now gone south, and only two aeroplanes were stationed at the aerodrome. The demand for hotel accommodation was such, that the Chairman of the Scottish Local Hotels Sub-Committee was of the opinion that the hotel would be a good economic proposition under present circumstances, even without the golf course. The matter was still under consideration in December when a report was issued by the Ministry of Works on the position with regard to the release of the requisitioned LMS Hotels. (Turnberry, as well as other railway company hotels such as Gleneagles, Glasgow (St Enoch), and Strathpeffer had been requisitioned). It had been proposed that Turnberry Hotel would be used as a miners' rehabilitation centre, but this was rejected by the Scottish authorities and the War Office had abandoned a proposal to occupy

the hotel, meaning it was no longer required by the authorities and could be derequisitioned. The railway company, however, were now unwilling to take back the hotel until the Air Ministry had surrendered the golf course.

In the Ailsa Estates office, notices were now being received of the derequisition of land. In February 1946, the two large defence posts, and nine small gun sites at Jameston farm, and three small gun posts at Shanter, were released and the enormous task of clearing up began. Arrangements were made for the purchase by the estate of the small huts left behind. At Jameston farm a considerable quantity of barbed wire had been erected on and behind fences and this wire was now in a dilapidated condition and dangerous to stock grazing in the adjoining fields. The farm tenants were complaining bitterly about the barbed wire as it was preventing them grazing livestock in the fields adjacent. Near the quarry at the lighthouse an enormous dump of rubbish had been formed which was overflowing onto Jameston land, it was pointed out by the factor that it was a breeding place for vermin and most obnoxious on the shore, with the rubbish being washed in and out along the bents with each tide.

The land derequisitioned behind the houses at Maidens harbour was to be used as a site for desperately needed council houses, but a year after being released the whole of the north boundary was still a tangled mass of barbed wire and broken fences. Letters were sent to the commanding officer at Turnberry with regard to the removal of defence wire, but with the small number of men he had left, it was not likely he would be able to complete the clearing of the land.

Finally, in September 1946, LMS Scottish Hotels Committee reported that both the hotel and the golf courses were about to be derequisitioned and consideration was given to the best means of restoring them to their former use.

So far as the hotel was concerned, the structural condition was fairly good but it needed complete renovation inside. The restoration of the golf courses presented greater difficulty, however, and while it would no doubt be possible to attract visitors to the hotel, it is unlikely that people would go there if there were no facilities for golf. The Executive Committee were of the opinion that every effort should be made to re-open the hotel and replace the golf courses as soon as possible. Arrangements were made for Messrs Barr & Son, Surveyors, Glasgow, to prepare a schedule of dilapidations in regard to the hotel and grounds and any other buildings included in the requisition, excluding the golf courses, to expedite the company's claim for compensation under the Compensation (Defence) Act, 1939. It was recommended that the services of a first-class golf architect, Mr Mackenzie Ross, be obtained to prepare plans and an estimate for the rehabilitation of the golf courses. The cost of reinstating the golf courses was estimated at roughly £100,000, but as the hotel and golf courses were requisitioned at different times by different Ministries, the question of compensation became problematic. If the Government enforced that the hotel and golf courses were two separate requisitions of two separate properties, the compensation for the damage to the golf courses would be limited to an amount not exceeding the value of the land. But, if the Government accepted the argument that the hotel and golf courses are completely interdependent, the LMS Estates Manager considered the cost of restoring the courses would not be as great as the value of the hotel and courses together.

The Air Ministry were now considering purchasing approximately 27.5 acres of land from Cassillis and Culzean Estates, being part of the properties known as Jameston, Turnberry and Shanter farms, Maidens. It is not known what plans the Ministry had for this. It was stated that if a purchase was not affected, the property would remain on requisition. The property to be purchased had been valued at £22,000. This included the value of the ground, the value of two cottages at Little Turnberry which had been demolished, one cottage at Clachanton which had been demolished, and the value of one feu site at Clachanton. The estate desired to reserve the right of access over the property to be sold, to allow the tenants to collect sand and seaweed from the foreshore and also the right of wayleaves for drain, water supplies and electric lines as necessary for supplies, etc. to estate property. The factor also suggested alterations to the boundaries proposed by the Air Ministry, that the small field, next to Port Murray, by the most northerly point be excluded. This field was not a part of the larger area, but was separated by a narrow strip of land and the factor thought it better left with the estate, as if the Air Ministry purchased this small field it would be separate from the other property and would be of little value to the them, while it would form a little extra part for grazing on the bent land between the airfield boundary on Bain's Hill and the foreshore. The boundaries agreed, the Air Ministry subsequently purchased part of the lands of Little Turnberry and part of the lands of Jameston, on 11 November 1946, at the price of £18,671.

On 13 November the solicitor in Ayr again wrote enquiring if there was any chance of the RAF camp at Maidens being offered for sale. Further letters followed throughout December.

Formal notice was received from the Ministry of Works, Edinburgh, of their intention to de-requisition the Turnberry Hotel as from 28 November, and that the Air Ministry was being requested to take similar action regarding the golf courses. Messrs. Strutt, Ltd, were preparing a report and plan for the restoration of these, and arrangements had been made for the preparation of a schedule of dilapidations in regard to the hotel and grounds etc.

The derequisitioning of the airfield gave rise to the question of restoring of the highway between Maidens and Turnberry which had been closed for some years. This had resulted in considerable inconvenience to the travelling public in that they had, in travelling north and

Turnberry Airfield : 1947

south, to do so by way of a detour involving the Blawearie Road and Kirkoswald. The Air Ministry agreed that the old road should be restored to public use without any restriction. This was also thought to be an opportune time to consider the improvement of the road, by making use of one or other of the runways or perimeter tracks, and thus provide an improved road between the two communities.

On 13 October 1947, LMS executives submitted to the Ministry of Works a claim for the complete restoration of the golf courses estimated at £119,381. The claim in respect of the dilapidation to the hotel had not yet been formulated. The negotiations for compensation were rather complicated and protracted, Cassillis and Culzean Estate, who owned the land forming the golf courses, suggested that the railway should give them a guarantee to continue the payment of £187 2s 1d per annum for the use

of the land for not less than 25 years, with the condition that the company would undertake to restore the two golf courses during this period. This payment was to apply so long as the land was used for the purpose of playing golf, and in the event of its ceasing to be so used the land would automatically revert to the Cassillis and Culzean Estates. And they would consider formulating their own claim against the government in respect of damage. Alternatively, they were prepared to sell the land to the company, including all rights to compensation. The price at first mentioned was £12,000, the purchase including three cottages which were rented from the estates at £60 per annum.

The LMS Company was of the opinion that the claim in respect of the hotel and golf courses should be dealt with as one and considered that much the better way of dealing with the matter would be to acquire the land which would leave the company or their successors in the position of being able to carry out the restoration work in whatever form they desired and when they pleased.

The Ministry of Works subsequently made an offer of £85,000 compensation to cover dilapidations to both the hotel and the golf courses. The total of the offer was based upon £18,000 for the hotel and £67,000 for the golf courses. It was stated that this figure was thought to be adequate for 'sufficient restoration' of the golf courses.

On 9 December 1947 the LMS Company purchased the land forming the golf courses from Cassillis and Culzean Estates for the sum of £8,000 including all rights to the compensation money and also including the three cottages at Turnberry Warren.

left the concrete foundations and fearing that weeds would overgrow the sites and cause great havoc on the land, the factor felt these had to be removed and the ground bull-dozed, so he approached the local business of Messrs J. Faulds & Sons (Maybole) Ltd, for estimates to do the work. Arrangements were made for the purchase of some of the buildings on the airfield technical site and it was proposed that the officers' quarters be sold to the Maidens Village Committee, pending a decision from the YMCA, who had expressed an interest in using the building for a permanent camp. Neither the Village Committee nor the YMCA, decided to take over the building, so the Air Ministry were instructed to remove the buildings and restore the ground to its original condition. The town planning officer had been very diffident in allowing the buildings to remain

Turnberry Airfield : 1978

The Air Ministry had started to remove the huts on some of the sites at Turnberry. The factor was desirous of reinstating this land to its previous condition, which had formerly been excellent arable ground. But the Air Ministry

as they were, as they did not fit in with the County Planning Scheme for Maidens, so it was considered to be all to the good that neither the villagers or the YMCA wished to take over the officers' mess.

It was discovered that the five derelict blister hangars, on the south end of the de-requisitioned land at Turnberry, now belonged to the estates. As there was no stipulation regarding their use, and as proprietors of the ground, the estate could do what they liked with them. It was decided to get a contractor to remove the hangars, free of cost, but £100 was paid to the estates for the value of the material.

The protracted negotiations between the former (in 1948, the railways in Britain were nationalised and the LMS Company was absorbed into British Transport) LMS Railway Company and the Ministry of Works on the subject of compensation, had almost reached a conclusion, as a result of which a tentative settlement in the region of £101,500 was reached and plans were afoot to expend this money on the work of restoration.

In May 1948, however, the military intervened in any thoughts of post-war reconstruction, when it was announced that the Admiralty planned to use Turnberry Airfield on a permanent basis as an alternative to Heathfield. This development was viewed with great concern, because if the Admiralty were permitted to take over the site of the golf course, there would be no compensation payable, and no possibility would remain of Turnberry Hotel being reopened to the public, seeing that the hotel had little value alone. It was proposed that preliminary representations be made by the British Transport Committee through the Ministry of Transport to the Cabinet Airfields Committee, setting out the arguments against the Airfield proposal. In a report from Lord Inman against the Admiralty scheme, he writes:

10 June 1948
TURNBERRY HOTEL – AYRSHIRE
Turnberry Hotel was opened in 1906 by the Glasgow and South Western Railway Company and was taken over in 1925 by the LMS, the capital outlay standing in the books at that time being £118,000. Between 1925 and 1939 additions and improvements to the hotel were made, involving additional capital outlay of approximately £100,000.

The hotel has approximately 170 letting bedrooms.

Adjoining the hotel were two famous golf courses which, together with the hotel, made Turnberry one of the most popular resorts in Scotland.

During the three years 1935, 1936 and 1937, which may be taken as indicating normal pre-war experience, the average business done per annum was over £64,000, yielding a trading profit of £15,000 or approximately 23 per cent of the gross receipts.

On the outbreak of the war in September 1939, the hotel was immediately requisitioned for use as a hospital. Subsequently the Air Ministry requisitioned the golf courses, a large area of which was demolished in the construction of an airfield.

The hotel and golf course were de-requisitioned in November 1946, and, as a result of protracted negotiations with the Ministry of Works, in which the LMS case rested upon the inter-dependence of the hotel and golf courses, agreement was reached unofficially for a total compensation of approximately £101,500, which if confirmed, would very largely meet the total estimated cost of the complete rehabilitation of the hotel and reconstruction of the two golf courses.

It is relevant to point out that the Ministry of Works in dealing with this matter have applied Section 2(1) (b) proviso (ii) of the Compensation (Defence) Act, 1939, which provides that the sum payable in respect of damage to land shall not be greater than the value of the land at the time possession was taken. It has been understood between the LMS and the Ministry of Works throughout the negotiations that, providing a satisfactory settlement was reached on the subject of compensation, the LMS would, as soon as practicable, proceed to restore the hotel and golf courses and apply the compensation settlement to this purpose.

In view of a legal difficulty which arose with regard to the entitlement to compensation in respect of the golf courses, which before the war were held under grants of servitudes from the Marquess of Ailsa, providing for the payment by the LMS of an annual sum of £187 2s 1d for the perpetual right to use the land for the purpose of playing golf, the LMS in 1947 authorised the purchase of the land for the sum of £8,000 so as to become owners of the whole estate and entitled to receive the full amount of compensation. The completion of the purchase is now in hand between the respective solicitors.

There is strong support from Scottish interests for the earliest possible rehabilitation of the hotel and golf courses, particularly in the interests of the tourist trade of Scotland, and there is no doubt that the reconstruction of this outstanding resort would be widely welcomed.

Recently, there has arisen a wholly unexpected development in that the Admiralty have intimated that they might consider the use of Turnberry airfield, and it is understood that the decision will rest with the Cabinet Airfields Committee.

The Hotel's Executive hope that the British Transport Commission will oppose the allocation of Turnberry to the Admiralty on the following grounds:
1. It might tend to prejudice the claim to compensation already tentatively agreed.
2. It would remove the possibility of Turnberry Hotel being re-opened to the public as the hotel would have no value without the golf courses.
3. The rental payable by the Admiralty would presumably be a nominal one and not comparable with the profit that would be earned by the hotel.

4. It would deprive Scotland of an outstanding national asset in relation to the tourist industry.
5. It is understood that an alternative airfield, namely, at Heathfield, has already been allocated to the Admiralty.

In conclusion:
It should be stated that plans for the reconstruction of the golf courses have been prepared by a well-known firm of golf course architects in Scotland, and that, subject to confirmation of the claim and the granting of the necessary licences, there need be no delay in putting the rehabilitation in hand. It will be appreciated that, whilst the complete restoration of the golf courses may well take two years, there is no practical difficulty in connection with the work nor with the renovation of the hotel.

Although the RAF had given up all claim to Turnberry Airfield, the Admiralty stated that they were considering the use of Turnberry, whilst they pointed out that Heathfield had already been allocated to them by the Cabinet Airfields Committee, who alone could alter that allocation. The Regional Planning Committee considered the area of Heathfield Airfield of great importance to future developments in the Ayr and Prestwick neighbourhood, in view of plans for a new industrial zone and the overspill of population from Glasgow and the Clyde valley, and in their opinion the site of the airfield at Heathfield offered the best possibility for this purpose. It

Turnberry Hotel : 1963

seemed that the Admiralty had not formulated an exact plan for either airfield, no actual use was specified, and the airfield finally allocated would have to be kept on a care and maintenance basis (which would necessitate the concrete runways remaining) or whether the Admiralty's requirements would be met if they retained control over building and similar development in the area around the airfield.

It was decided that the whole question should be referred to the Cabinet Airfields Committee, and it was suggested that if the British Transport Commission desired to make representations regarding Turnberry hotel and golf courses, this should be done through the Minister of Transport.

The matter was brought before the Scottish Physical Planning Committee at a meeting on 26 July, the main points of consideration being a question of whether a possible industrial development of a relatively long-term nature in the Ayr district should take precedence (by the withdrawal of Heathfield from the Admiralty) over a shorter-term rehabilitation of the important tourist centre represented by Turnberry.

Prior to this hearing, Cyril Hurcomb of the British Transport Commission, wrote to Sir John G. Lang KCB, at the Admiralty in Whitehall, asking 'Is there anything you can tell me as to your position?' hoping that the Sea Lord might be on their side.

The Admiralty replied:

Dear Hurcomb,
Thank you for your letter of 17 July, about the use of the airfield at Turnberry by the Admiralty.
The facilities we already have at Heathfield meet Naval aviation requirements in that area admirably and it is only because of strong pressure from the Scottish Authorities, who would like to use Heathfield for housing and industrial development, that we have recently been considering whether we could accept Turnberry in lieu. Turnberry, as you may know, was used by the RAF during the war and was released from requisition

in 1946. Since then, no maintenance has been carried out and the runways are now in poor condition. They are also crossed by a class A road which is open for general use by the public. From a pilot's point of view, the airfield is definitely inferior to Heathfield and it also lacks many of the buildings which are available at the latter. To make Turnberry anything like the equal of Heathfield would, we are satisfied, entail substantial additional expense.
The conclusion we have reached, therefore, is that Turnberry is not acceptable to us and we intend to resist strongly all attempts to dislodge us from Heathfield unless, of course, we can be offered equivalent facilities in the area which will not involve a greater charge on Naval funds than would be involved in respect of our occupation of Heathfield.
Yours sincerely (Signed) J.S. Lang

It was not until the end of September 1948 that word filtered through that Turnberry Airfield was inferior to Heathfield and the proposal to re-open Turnberry Airfield had been dropped.

While the hotel and golf courses fought the threat posed by the Admiralty, the factor was being approached by another type of sailor. The Sea Cadet Corps wrote to say they were interested in the old Turnberry Aerodrome Officers' Mess at Maidens, thinking it might make an excellent training camp for Sea Cadets from all over the kingdom, being well situated with a small harbour where boats could be moored, and excellent accommodation and camp facilities. This did not, however, go ahead.

With all possible military claims to the land allayed, the restoration of the hotel and golf courses were progressed. In a memorandum from the Hotels Executive of the British Transport Commission, dated 5 July 1949, it was noted that firm estimates in respect of the restorations of Turnberry Hotel and Golf Courses had now been prepared and were as follows:

(1)	Rehabilitation			
Hotel		£	£	£
Builder's Work		21,750		
Engineering Work		14,050		
Redecoration		20,000	55,800	
Tennis Courts and				
Gardens		1,800		
Golf Courses				
Reconstruction of Ailsa Course				
including removal of runways		45,000		
Golf Course machinery		2,000	47,000	104,600
(2)	Modernisation and Improvements			
Hotel				
Builders' Work		9,500		
Engineering Work		2,850		
Furnishings & soft furnishings for essential improvements in staff accommodation		1,300		13,650
			TOTAL £118,250	

The examination of the furnishings and furniture stored at Turnberry during the war period had now been completed and revealed that considerable renewals were necessary due to wear and tear prior to the requisitioning of the hotel and deterioration during storage. Many of the items were not suitable for use in Turnberry Hotel when was it was to be reopened, although some of it was found could be transferred for use elsewhere and for the essential improvement of staff conditions. In addition, during the war, when supplies, particularly of carpet, were unobtainable and when the future of Turnberry Hotel was uncertain, large quantities of carpet as well as furniture were transferred from Turnberry to other establishments on the LMS system.

The total cost of making good these deficiencies was estimated at £47,650, which is additional to the sum of £118,250 quoted above. The incidence of this expenditure is shown on the next page::

(1) Replacement of articles which have deteriorated, and which are no further use:

Furnishings	£1,950	
Carpets and other floor coverings	£5,500	
Curtains, etc.,	£1,100	£8,550

(2) Replacement of items unsuitable for use at Turnberry Hotel but which will be used as replacements elsewhere, or for essential improvement of staff amenities:

Furnishings	£4,000	
Carpets and other floor coverings	£6,900	
Curtains, etc.,	£3,700	£14,600

(3) Making good of articles transferred to other hotels when supplies were unobtainable and when Turnberry Hotel was closed:

Furnishings	£9,900	
Carpets and other floor coverings	£6,200	
Curtains, etc.,	£3,700	£19,800

(4) General repair work, re-polishing, re-covering furniture, etc., on existing furniture and furnishings:

Furnishings	£ 4,700
	Total £47,650

The Executive recommended to the British Transport Commission that the rehabilitation, modernisation, etc., at a revised estimated cost of £118,250 and the making good of deficiencies in furnishings, etc., at an estimated cost of £47,650 to be approved.

Work commenced on 27 July, and in August the breaking up and removal of portions of the concrete runways began. The foundations were up to four feet deep, topped with concrete and six inches of tarmacadam, thousands of tons of excavated material had to be disposed of and fleets of lorries carried some of it to Maidens where it was used for the building of a new sea wall. These excavation works on the golf course left huge areas lacking topsoil, which had to be replaced prior to re-seeding. Work had progressed at such a pace that it was anticipated that the hotel could be open by Easter and nine holes

on the Ailsa Course would be ready for play by the summer of 1950. Eighteen holes of the Ailsa Course were publicly opened on 21 June 1951. Attention turned to the Arran Course, and in May 1951 it was felt that the reinstatement of the two golf courses was most desirable to make the hotel attractive to visitors. It had been possible to affect a considerable reduction in the estimated cost of the demolition of the runways to £25,000. Concern was raised at this expenditure, although it had been agreed that to make Turnberry a profitable proposition a second golf course was essential. Questions were asked of Sir Harry Methven, Secretary of the Executive, regarding the estimated total cost of providing the second golf course and asking for some supporting information – in particular, the business done since the Hotel was re-opened, showing the receipts from the first nine holes of the Arran

Turnberry Hotel Main Entrance: 1966

Course (which had already been re-opened). Sir Harry pointed out, 'I have had a further discussion with Lord Hurcomb on the subject of the Turnberry Golf Courses and mentioned to him that it is impracticable to give actual facts as to the necessity or otherwise for the second course. It will be recalled that this was pointed out with considerable force at the last meeting between the Commission and this Executive. There may well be some risk in this, but that is our view. Therefore, the points you raise have little bearing on the major issue – to be or not to be a second golf course at Turnberry.' He went on, 'With regard to the actual cost of restoring the second course (Arran) you will recall that we proposed to deal with this in two stages. One, was to lift the concrete and to circumscribe the amount to be lifted so that there was just room for an 18-hole course. The original estimate of £40,000 for lifting all the concrete was reduced to £25,000 and as this estimate has recently been confirmed, that would be the price for removing this concrete. Then as to the cost of constructing the golf course, we were considering how far this could be done by direct labour under the supervision of our own personnel.'

'What we are really asking for is an acceptance of our view as a matter of policy. The receipts at Turnberry in the early days of its reinstatement are not relevant to the situation.'

The Commission approved the expenditure for the demolition of the runways as a preliminary to the restoration of the second golf course at the Turnberry Hotel.

In August, Sir Harry writes rather forcefully to The Rt. Hon. Lord Hurcomb, GCB, KBE, Chairman, British Transport Commission, regarding the delay to Arran Course:

My dear Hurcomb,
As you know, we are many times frustrated in our efforts when it comes to dealing with Railway Hotel matters. Here is an example.

You will recall that I put before you a strong case for a secondary golf course at Turnberry on the basis that golf was the main artery connected with keeping the Turnberry Hotel busy. A licence had to be applied for to lift the concrete. This went through, I understand, to the Ministry of Transport, and it fell into the hands of a Miss Churchards who apparently, and as far as I can see, decides that a secondary golf course at Turnberry is not necessary, and instead of the matter being progressed towards obtaining a licence, it becomes static.

I was in Turnberry this last weekend, and it is more and more evident that to keep Turnberry going a secondary course is necessary – that is to say, that if we are to tempt conferences, clubs, and others, as indeed we are doing at Gleneagles, there is a good case for making the second course.

Turnberry, as you may remember, is used a great deal by Glasgow people at weekends, and consequently for club competitions on quite a major scale. The Women's Golf Union looked at the place only this last week or so with a view to holding their International Competition at Turnberry, next year, but one course is not sufficient for such purposes.

The trouble is that the word 'golf' savours of pleasure and probably tends to make people refuse an application for such a licence without taking all the relevant facts into consideration, but I am quite certain that in the circumstances at

Turnberry a second course is essential towards the end we have in view, and above all, to make Turnberry pay, and it is through that medium that we shall largely do so.

I am sorry to trouble you about such a matter, but I would have thought that with the support of the Commission and the recommendation from the Hotels Executive, that this would suffice. Miss Churchards gives some reason about labour – we would require about 60 people to lift the concrete and I have made enquiries and find there is labour available in Girvan and near at hand for that purpose. Is it a matter about which I can invoke your further aid? I should be most grateful.

Yours sincerely
(Signed) Harry Methven

P.S. We want to get the concrete lifted in the first place so that the ground can settle down before we tackle laying the course and which we have in mind we might do under our own auspices.

Lord Hurcomb replied:

Dear Methven,
TURNBERRY HOTEL AND GOLF COURSES
I received your letter of 1 August and have had some enquiries made about this matter. I gather that the Ministry of Transport went into the subject very carefully and do appreciate that if Turnberry is to succeed as a tournament hotel it needs two 18-hole golf courses. Apparently, however, when Hole wrote pressing for the building licence to be given for the demolition of the concrete, he said that when this work had been completed plans would be made for a start on the reconstruction of the second golf course, possibly in two years' time.

The trouble about the licence sprang therefore from two sources:

(A) That the work was not immediately necessary for the reopening of the second course.

(B) An ascertained shortage of building labour in Ayrshire is reported by the Regional Building Committee.

The Ministry were informed that there were unfilled vacancies in the district, that McAlpine's, who I gather are to do the work, would have to recruit labour locally, and that certain defence works are starting in the district shortly.

In the circumstances, it is clear that we shall have to put up a stronger case than has previously been presented to the Ministry, and perhaps you would write me again on this point. Incidentally, we have had no reply whatever to our official letter of 24 May, in which Miles Beevor informed you that the Commission had asked your Executive 'to submit as quickly as possible figures showing the actual expenditure incurred to date on the rehabilitation of the Hotel and the golf courses under the heads set out in the estimates included in Lord Inman's memorandum on this subject, dated 5 July 1949.'

Perhaps, therefore, you would now let us have this information, with any further points which may justify another approach to the Ministry of Transport.

Enquiries were made from McAlpine's, to whom it was proposed to let the contract, as to the labour they would require and the general labour situation in Ayrshire.

They stated that the work would be largely carried out by their mechanical plant and consequently most of the men would be brought to the site by them and were generally known as key men. They said they would only require 15-20 men from outside sources and, from enquiries made from the local employment exchanges, there would be no difficulty in supplying this small number. They stated further that if there should be any difficulty locally it is quite a simple matter to obtain labour of the required description from Northern Ireland, without interfering with the supply to any housing, rearmament or other urgent work.

Satisfied that there was no labour problem so far as McAlpine's were concerned, and that if a licence was forthcoming it was probable, that in view of the seasonal decline in employment in the district the local labour exchanges would welcome the scheme being put in hand.

A material point in the matter was that good weather is desirable for such a job and the sooner it could be put in hand the better.

But another blow was dealt to the Arran Course with the Chancellor's statement of policy regarding new building work which quite ruled out any prospect of authorising work on the second golf course. Sir Harry Methven having taken a great deal of personal interest and trouble over reopening the hotel, thought that he and the contractors it was proposed to employ, that this small job could be got through without impinging on national policy and hoped the matter would be reconsidered, especially in view of the constant pressure to do everything possible encourage tourist business in Scotland. But it was felt that it was a project for which the Chancellor was unlikely to give exceptional treatment.

A Mr Ryan was also pessimistic about the re-opening of the resort, writing to Sir Harry Methven after spending a fortnight at Turnberry, 'my mind was naturally stimulated by the problem of an hotel of this sort which I saw so much more clearly on the spot and one which I know is engaging your attention at so many points. I tried to allow myself not to be too depressed by the thought that results here, with the existing conditions and the problem which you have inherited, cannot hope to be satisfactory, certainly not on a short view and possibly not even on a long one either.'

Sir Harry responded in the positive:

Thank you very much for your letter. I am glad you enjoyed Turnberry. It is indeed a good hotel. I am not too worried about its future for several reasons.

I have, as you know, I believe, saved Gleneagles from further difficulties by the introduction of a large number of conferences which take place at weekends during May, June and July, and then again in September. Conferences bring money to the hotel, and Turnberry, in due course, should likewise benefit so long as we keep August free for the ordinary visitor.

Then again, Turnberry would have a much more substantial future if we had two golf courses going. This would enable us to have some of the large competitions which we have reluctantly to turn down. When there was an enquiry this season about the Women's Championship – one course is insufficient for this purpose. However, we shall probably lift the concrete from the old fairway course, preparing the ground ultimately for a second course.

Turnberry, as you know, receives an immense amount of support from Glasgow; it should not only enjoy a good weekend during the season, but a large share of important golf tournaments which are very beneficial to the hotel. I am therefore not too pessimistic about Turnberry in the long run.

It is kind of you to let me have your observations and I shall certainly keep a very close eye on the Turnberry situation in every way possible.

Not satisfied with this reply, Ryan then wrote to the Chief Secretary, British Transport Commission, enclosing correspondence from his friends regarding the situation, thinking perhaps that some Members of the Commission and the Controller may like to see them unofficially. He wrote: 'I am afraid I do not share Sir Harry Methven's optimism that its future is so assured any more than I think that Gleneagles is "saved". Financial results up to date seem to indicate to me that the present gap will not be bridged without some more fundamental developments.'

But Sir Harry was not one to give in easily. In 1953 the Ministry of Transport were still not prepared to grant a licence for the work, and he considered that the business of the hotel and its future success was being prejudiced by the absence of a second course and a scheme was prepared whereby nine holes could be laid down by the Executive's staff without taking away any of the concrete. The cost was estimated to be in the region of £8,000 plus approximately £750 for the provision of watering points. This was strongly supported by the Executive who recommend to the Commission that it be authorised. The Chief Secretary, S. B. Taylor, was not impressed and in a memorandum to the Commission wrote, 'I do not feel that a 9-hole course will add much to the attractiveness of the Turnberry Hotel, and I am surprised that existing staff will have the spare time in which to construct this extra course. Having constructed it, however, its upkeep must prove expensive as is the case with all golf courses today.'

The Commission were also unsatisfied with the proposal and felt that an expenditure of £8,750 on the provision of a nine-hole golf course would not so increase the receipts of the hotel as to yield a satisfactory return on the expenditure, and they suggested that if that amount of capital was to be spent on the hotel it might be more profitable to use it in the conversion of some single rooms to double rooms.

Mr Ryan again voices his opinion:

My views are as follows:
1. A second Golf Course (particularly a 9-hole one) will not bring any substantial addition to present patronage.
2. 'Maximum Patronage' if by this is meant making the hotel profitable, will require other actions.
3. The present course is not up to first-class championship standards. It is too easy for the first-class man and too punishing for the general golfer.
4. The 'History of Turnberry' is the wrong basis upon which to base the future – the

background which led to its erection has no place in the present economy.

Undeterred, Sir Harry defended his position:

Mr Ryan's contentions are not based on facts.
First, his comments on the present Main Course being difficult, and so forth. All the leading Amateurs and Professionals who have visited Turnberry have complimented us time and time again in respect to this particular new Course. I fear Mr Ryan is emphasising a view with which I do not agree.

A Secondary Course, even if only of nine holes, is necessary at Turnberry for several reasons. For instance, if we are to play prospective Championships at Turnberry, it is essential that we should have a smaller practice Course, as, for example, we lost the Amateur Women's Championship through these circumstances, as we were unable to provide a small ancillary Course, and have also lost other important Matches that would have greatly added to the benefit of the Turnberry Hotel. Therefore I feel that a Secondary Course is essential, if we are to gain some of these important Tournaments, but quite apart from that, many people who stay at Turnberry want a Secondary Course for their own pleasure, as it may well be that the Main Course is too strenuous for some Golfers. Therefore, after great consideration, I put this proposal before the Commission.

We cannot by any means accept Mr Ryan's contention that such a Secondary Course is not necessary. I have always thought to the main chance, and such a Course, as indicated, will bring further trade to the hotel, besides enabling us to accept some major Competitions on the Main Course.

As you know, I am a Scot, and know Turnberry extremely well, and I therefore could not agree with the suggestions that Mr Ryan has put forward. My views are always directed towards increasing the number of customers at Turnberry, and I have every reason to believe this Secondary Course will help us toward that end.

Many of the Ayrshire Courses are in bad condition – Prestwick, for example, all the turf there is worn out, and prominent Golfers have commented on this, and if we at Turnberry provide the facilities (1) the operation of the Main Course as at present, and (2) a subsidiary Course, we shall, in my opinion, gain money for the Hotel.

This has all been carefully worked out, and great consideration given to the project. I am not aware of how good a golfer Mr Ryan is, but I hope the Commission will accept my view that this extra Course will meet, or overcome, some of the difficulties which have caused us to lose not only professional but certain Amateur Tournaments, for we provide, in Hotel accommodation, the most exceptional circumstances by virtue of the Turnberry Hotel, not only in regard to people staying in the Hotel, but in catering facilities that one can offer, but the Golf Courses must not only be adequate, but of a first class character.

The proposal for the works, at an estimated cost of £8,750, were approved on 26 March 1953.

In tandem with the reconstruction of the hotel and golf courses, the factor of Cassillis and Culzean had also been busy, in 1948 taking over buildings 101 (Tracer Training Building) and 230 (Picket Post) on the Instructional site which had been on Jameston farm. Other Air Ministry buildings left on derequisitioned land were also pressed into use by the estate. Displaced persons and POWs were housed on the accommodation sites for a period during 1945/6. A longer-term problem for the factor were the squatters who had moved into the WAAF site (No. 9) on Blawearie

Road. There was a chronic housing shortage after the war, and like many other empty military installations these were quickly commandeered by local families desperate for somewhere to live. Writing to the Ministry of Works with regard to the derequisitioning of this site he was advised that it was quite clear that there was no suitable alternative accommodation in the district to which the families now living in the huts on this land, could be removed.

In these circumstances it was essential to retain the buildings for temporary housing purposes, until such time as the housing situation in the county improved and these families could then be rehoused. The Ministry regretted that it would not be possible to derequisition the land meantime. The site was transferred to the Department of Health for Scotland in November 1949. Attending to the working of the estate land, the factor arranged the planting of woodlands adjoining the site, but before planting could commence it was necessary to erect a fence round the site to replace the former fence which was entirely derelict. The squatters had broken down the fence round the site and had made a coup of the adjoining ground. A proper stob and wire fence of approximately 200 yards had to be erected and the cost to the estates for materials and erection was £40. The fence between the aforementioned site and the lands of Kirklands was entirely derelict and the farm tenant complained he could not keep his stock from straying. The paling which was erected between the former wireless station and the lands of Kirklands was also in a derelict state. It was the belief of all that the damage to the fencing had been caused by the squatters in the WAAF site, taking the paling for firewood. The farm tenant had continually repaired the wood paling, but the only cure would be to remove it and erect a proper stob and wire fence.

The factor wished to repair these fences using estate labour and wanted to know if the Ministry of Works or the Department of Health would meet this expenditure. The squatters were still there in May 1950, when the factor was informed that there was just a possibility that the proposed new accommodation for the squatters in the plasterboard huts may not be quite ready by the 28th instant, and presuming he would have no objection to their remaining on for a short period. The site was now the sole responsibility of the Department of Health for Scotland, passing to Ayr County Council, who then charged the squatters rent.

Another type of inhabitant moved in temporarily in July 1950 with the arrival of international students to study Scottish methods of agriculture. Christian Mølle from Denmark recalls his time there:

I arrived at Turnberry in the summer of 1950, when I, a young Dane, had the luck to be admitted to the camp for four weeks with a group of twelve fellow students. All of us were boys, 20-30 years old, and we had been studying agriculture at the Royal Veterinary and Agricultural University in Copenhagen. We easily found the camp which was situated some hundreds of metres north west of Maidens, rather near the seafront. The buildings stood, a little scattered, on the former aerodrome area. In a big, rather high-roofed building-complex, was the leader's office, the kitchen, the refectory, the roomy assembly hall and the bathroom and dormitory for the women. A more modest building was fitted out for the men's bathroom and dormitory. Both buildings were situated about 500 metres from the seafront with a distance between them of 200-300 metres. The camp leader was a friendly man, about 30-35 years old. I think his name was Mr Duncan. He and the other inhabitants of the camp made us Danes very welcome. There were about fifty students altogether, girls as well as boys. A good deal had come from England and Scotland, others from Poland, the Netherlands, Norway and Denmark. The camp had a welfare committee, it held meetings where the students could state their complaints, wishes etc. I remember

only a single complaint, concerning the lunch packets. Otherwise the conditions seemed to be acceptable. The work, as I already mentioned was unpaid, Saturday forenoons were included in the working week. The students had free conveyance to the workplaces. Bed and breakfast and all other meals were also gratis.

In the early morning, a very substantial breakfast was served, and all students got a packed lunch. After breakfast the camp-leader divided us into groups of 6-8 persons. Then the groups were transported by lorries to the farmers who had asked for manpower in order to have any odd jobs carried out, for instance:

Pulling/digging docks etc., in grass fields
Hoeing weeds in fields with marrow-stem kale, turnip or swede
Lifting and sorting early potatoes
Sheaving reaped grass grown for seed
Gathering stones in the fields

The farmers provided us with tea at lunchtime. At Mr Marshall's farm in Kirkhill, Ballantrae, we got tea with bread in an extra break in the forenoon and – to our great, positive surprise, supper at noon!

Some of the students had an inquiring mind, and wished to learn more about Scottish farming, and the employers and foremen that we contacted were always willing to help when we asked questions concerning their farm and farming methods.

During the working day some funny events could occur. One day a foreman had allocated a certain number of potato rows that they should lift. These rows were marked with sticks, forming a border with another plot, to be lifted by another group consisting of Irish boys. The Irish lads kept moving the stick and so moving the extra rows to the Danish boys, our protests seemed to be ineffective. One of the Danes rose to his feet, lifted his arms, with a deadpan expression and recited in a very loud voice an epic poem about an illustrious Danish navy officer, who was called Tordenskjold. He became ennobled for his brilliant strategy while fighting against the enemy. This peculiar performance proved to be 100 per cent effective, the Irish boys' inclination for moving sticks disappeared straight away. My group was sent to a farm belonging to Mr Duncan. The farm had 500 acres of land, three tractors and an Aberdeen Angus herd, he grew amongst other crops, early potatoes.

Some other days the job at Mr Duncan's farm was to pull/dig up docks in some new grass fields planned for sheep grazing. I really was astonished when I was told not to pull up the thistles, in Denmark this greedy weed is hated as poison! I wonder if the thistle, the Scottish national flower, was protected by law at the time? Or maybe it was left as a special 'pick your own' crop for the sheep.

About 16:30 the lorries brought us back to the camp, at 18:00 we had trimmed ourselves and met in the refectory for dinner, which at Turnberry was called 'supper'. Thereafter we could spend some amusing hours in the assembly hall, conversing, taking a game of cards, playing ping-pong etc.

Some nights there were arranged funny evening parties and celebrations, then we danced to good music – not à la heavy metal! In a pause we had ice cream and juice, but no alcohol – a fine custom, I think! On other evenings films were shown, often informative, for instance dealing with the challenges of the FAO organization (The Food and Agriculture Organization of the United Nations). Saturday afternoons and Sundays were often taken up with outings in the surrounding country and the nearby small towns. The old imposing Culzean Castle

offered an unforgettable experience for me. The Danish group stayed in the camp from 11 July to 7 August, and in these four weeks I worked on five farms, Mr Paton's, Mr Duncan's, Mr Neil's, Mr Robertson's, Mr Marshall's and a farm in Lendalfoot, the farmer's name I have forgotten.

One day I hunted weeds in the front buildings around a big building complex. It was called Mauchline Camp *(Author's note: possibly Kingencleugh POW Camp)*, its buildings definitely didn't look like ordinary farm buildings.

When I returned to the Turnberry camp in 2006, I found the camp area with its small rumpled sand dunes and wind-blown shrubberies had undergone a metamorphosis into a modern well-kept 'Holiday Paradise', with green lawns and a lot of people enjoying the sun.

I spent an enjoyable couple of days exploring the region and saw the comprehensive technical development that had taken place in Scottish agriculture. Very large fields, advanced machinery and promising crops. Presumably the Scottish farmers who employed young, unskilled students from abroad have passed away many years ago now. I wondered what the typical local farmer dressed like nowadays, I did not meet one person armed with a walking stick and wearing a jacket, dress shirt and tie with lace-up shoes. Did the job of farm foreman still exist, with his flat cap, waistcoat and always smoking a pipe which was charged with a fat black-brownish tobacco, the smell of which was sickly sweet. I saw no one matching this description on my visit. I have many happy memories of my stay at the International Student Camp at Turnberry.

I think the camp was established in the early post war period as part of several extraordinary efforts to increase the agricultural production of the UK. If that was the reason, it seems likely that the student activities in the camp had been initiated by Scottish agricultural associations, farm schools etc. I guess that the club of Danish agricultural students had made some 'unofficial' arrangement for exchange students. Another thought is that the establishment of the camp had a double purpose, it should 1) particularly support the Scottish farmers, providing them with unpaid student labour. 2) give students from abroad the chance to meet English and Scottish students and so promote international understanding and exchange of knowledge.

Even though some of the students had little knowledge of agriculture, it is not unlikely that their work had a modest effect on the economic output for some Scottish farmers – those who dared ask for free shifts from the more or less skilled and hard-working students from the camp!

Redgates Caravan Park.
618 Squadron Hangar in the background.

The 'Holiday Paradise' Christian Mølle mentions is Redgates Caravan Park, built on the Communal Site between Jameston and Maidens. The site was bought from the estate after derequisition in or around 1950, and with the existing roadways, power supply and ablution blocks, was an ideal place to establish a site for touring caravans. Site No. 1 was also transformed into a holiday park, and later the disused Maidens Railway Station site.

Redgates Caravan Park. Maidens Railway Goods Station was at top right of the photograph.

Redgates Caravan Park showing Barley Bree Public House with WWII Picket Post and Link Trainer buildings bottom left.

The estate land of Turnberry Aerodrome which had been sold to the Air Ministry in November 1946 had lain vacant and derelict, and in May 1949, and in light of this the factor was interested in re-purchasing the ground.

It was first proposed to purchase it seven or eight months before, and in the meantime the factor had been watching the ground go from bad to worse, the damage to the drainage had caused the soil to become sour and it was overgrown with many noxious weeds which would be very difficult to eradicate. It would require a great deal of labour before it was to be brought back to proper grazing or arable ground. During the preceding year the ground had deteriorated very quickly, and it would be a long while before it could be brought back into heart. At least £2,000 would be required to fence and drain it properly without counting the cost of manuring and cultivating. There were some buildings on the site which the Air Ministry were required to remove,

and it was suggested that a time limit of one year be stated in the purchase offer, giving ample time to remove the hangars etc. The estate were of the opinion that £3,500 was really the highest sum they could offer, making it quite clear to the Air Ministry that they did not consider the ground worth more than £3,500 and that they would not increase their bid beyond that figure.

The Air Ministry suggested increasing the offer to £4,000 but were later prepared to split the difference of £400 and sell at £3,800, or thereabouts, implying that the land was more valuable because it was 'located in the heart of the Culzean Estates.' The factor thought this added nothing to the value from the estate point of view, and that it would not be advisable to increase the sum of £3,500 originally proposed as any greater expenditure on the land could be better employed in improving some of the existing farms rather than re-purchasing what was really a derelict piece of land. In July the factor wrote to the Air Ministry stating that the estates were no longer interested in the land and asking them to treat the offer of £3,500 as withdrawn, it was not to the estates benefit to purchase the area, which was in a hopeless state, and commented 'Probably by the time the Air Ministry get around to disposing of the ground in question it will be even more valueless from an agricultural point of view.'

The Air Ministry had already been making other arrangements regarding the disposal of the land in question, when it was sold to wealthy entrepreneur Niall Dingwall Hodge. Born in Glasgow in 1914, Niall Hodge, with John Blackwood, founded the company of Blackwood Hodge in 1938, dealing principally in agricultural machinery and contract ploughing, although other contracts included the painting of white lines on roads in Scotland for which they charged £11 per mile. During and after the war the business flourished, and it was John Blackwood who brought the first heavy earth moving plant and construction machinery to Britain from America. Niall Hodge left the company in the late 1940s, and founded other businesses such as Runways Ltd., and Scottish Land Development Corporation in 1947 (later becoming Scottish Land Development Ltd.) which grew to become the largest privately-owned plant hire firms in Scotland. Niall Hodge had become very wealthy by realising the need for heavy earth-moving equipment during the period of post-war reconstruction and filled that market. He was a founder member of the Scottish Plantowners' Association, of which he was the first president.

He had advertised for derelict land and, one day, got a call from an Air Ministry surveyor about the airfield at Turnberry. He fell in love

Niall Hodge

with the place, bought the airfield and made his home there. He transformed the disused control tower into his private residence, keeping the runway for his private use and that of his many friends and business associates who were keen fliers. With his expertise in land reclamation and

machinery, he set about removing the buildings and installations left by the Air Ministry. Ever the astute businessman, it was said that he dug up and sold enough copper wire from the underground communication system to cover the cost of the purchase of the land.

Niall Hodge had many and varied interests – his love of show jumping and horse racing, led him to attempt a breeding programme at Lands of Turnberry. One of his racehorses, called Malin Court had some success at Ayr Races and his other horses won many trophies. Another passion was sailing, and his vast fortune allowed him to indulge in commissioning a 200-tonne steam yacht launched in 1961, the *Maureen Mhor*, in which he enjoyed entertaining guests as they sailed on the Clyde. In 1968 he bought a T Type, Mulliner Park Ward convertible Bentley, one of only 41 to be manufactured.

Niall Hodge was not the sort of man to selfishly enjoy his enormous wealth, he was always interested in people, was a true philanthropist, and was a generous donor for the benefit of others. He took an active interest in the Quarrier Orphan Homes, building a complex of four holiday cottages at Lands of Turnberry, including a swimming pool, the use of which he also granted to Ayr County Council for holidays for the disabled people of the burgh. He was a devout Christian, and much admired the charitable work of all denominations including the Salvation Army, and the Roman Catholic, Daughters of Charity of the Order of St Vincent, providing them with accommodation in a house at Turnberry for their care of mentally handicapped children. He set up a foundation, the Turnberry Trust, to concentrate his charitable efforts. His altruistic mind turned to the problem of helping the elderly. With an ever-increasing ageing population, there was no provision for their care south of Ayr. With his astute business acumen, he set about investigating the possibility of establishing a nursing home in the area and took advice from experts in all aspects of geriatric care. It was soon apparent that the nursing home idea would be better served by the building of sheltered housing, enabling the residents to retain as much of their independence and dignity as possible. It had been planned to build the complex at Lands of Turnberry, but after deliberation decided on the present site, in the place of the 618 Squadron hangar.

Plans were drawn up, with Niall Hodge insisting that all aspects of the project were carried out to a very high specification, aiming for the ambience of a first-class hotel. He stated at the time that Malin Court was 'to provide a way and style of life commensurate with dignity and independence of spirit.' The scheme was to provide accommodation for about thirty residents, within a first-class hotel environment, with a full range of services including a dining room, a six-bed nursing ward, and care and supervision in place for when the residents became frailer. It was anticipated that the profits from the hotel would go towards containing the costs for the permanent residents. Niall Hodge was convinced that his plan for a sheltered housing complex within a hotel and restaurant would not only be of financial benefit but would be of great social and psychological benefit too, bringing in the general public which helped to maintain a lively atmosphere for the 'permanent guests'.

To expedite the process of planning and design, a full-sized model of a proposed flat was built in one of the disused airfield buildings, complete with furnishings. The intention was to provide self-contained rooms with en-suite facilities and a small kitchen area, ease of access and comfort was of primary importance.

The Malin Housing Association met for the first time in June 1971, with Niall Hodge as chairman. At this meeting the objectives of the association were established. The members of the management committee came from a variety of backgrounds but bringing to the project a great deal of experience and expertise. Malin Court Hotel opened on 22 August 1973. In 1974 an article which appeared in the *Carrick Gazette* stated that the combination of a sheltered housing complex and the functions of a first-class hotel was in effect 'a revolutionary concept in the care of the elderly.'

In 1981 Niall Hodge was granted an Honorary Fellowship of the Royal College of Physicians and Surgeons in Glasgow and the Order of the British Empire. Sadly, he died on 6 July, the day before he was due to be presented with the award. Niall Dingwall Hodge is buried in the Byne Hill Cemetery in Girvan, leaving behind as his legacy of one of the best hotels and private care homes in Ayrshire.

In the immediate post war period disused airfields were ideal sites for motor racing tracks, the most famous example of these is of course Silverstone. On 1 September 1951, the opening meeting at Turnberry, organised by the Scottish Sporting Car Club and sponsored by the Scottish Daily Express attracted sixteen entrants in the 500 cc Formula 3 race over the 1.7-mile circuit. Other meetings followed in May and August 1952. There were ten races at the May event, including saloon-cars of any capacity, sports cars up to 1500 cc, sports cars over 1500 cc, sports cars over 2500 cc, vintage sports cars over 3000 cc, sports cars of any capacity and an eight lap Formula Libre event. The cars raced were Bentleys, Healey and Jaguar, reaching speeds of almost 85 mph.

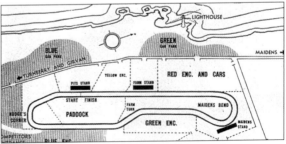

Racetrack

In September, the National Trophy meeting was held, attracting a crowd of over 50,000 people to Turnberry. Top drivers of the day included Stirling Moss, Reg Parnell and Mike Hawthorn, Ken Wharton, Ninian Sanderson and Ian Stewart. Competing in such illustrious cars like the Ferrari Thinwall Specials, ERA (English Racing Automobiles), Connaught GP, Jaguar and V16 BRM (British Racing Motors).

Ninian Sanderson, driver of the Ecurie Ecosse Bristol F2 on the Turnberry grid 1952

The incredible sound of the supercharged V16 BRMs, it was reported, could be heard as far away as Maybole and Girvan and even ten miles away over the hill in Dailly. The airfield circuit however, had its problems, the back leg was extremely bumpy, this was solved by introducing a chicane between Farm Bend and Hodge's Corner, to slow down the faster cars. There were some undulating sections of the track which caused some of the machines to become airborne, much to the delight of the spectators. An exciting day was had with cars spinning off the track on the corners, and Bill Dobson driving a Jaguar, put his foot to the floor on Farm Corner, when the throttle stuck open and he went flying off the track. The racing was not without personal cost, Freddy Strang was taken to Ayr County Hospital with suspected spinal injuries after his car overturned when a leaf spring snapped. Turnberry's short life as a racecourse was ended by local rating impositions which made the venue financially unviable. Turnberry is one of only two venues in Scotland where a formula one team has entered a race. Unofficial motorcycle racing also took place on the aerodrome and beach in 1950 and 1951.

The RAF made a return surprise visit in February 1955 when two Meteor jet fighters made an emergency landing on the deserted runway, startling a bus load of pupils from Wellington School, Ayr, who were on their way to a swimming class at Turnberry Hotel baths.

The aircraft had taken off from an RAF base near Newcastle, heading for Northern Ireland, but on arrival found they were unable to land due to bad weather. Running short of fuel, the planes tried to reach Prestwick Airport but were advised to land at Turnberry.

Just as the girls in their double-decker bus were about to cross the main runway on their way to the hotel, the first aircraft screamed past only a few yards from them and touched down safely, the shocked bus driver braked and stopped as the second aircraft flew across their path and landed seconds later.

The history of the struggle for the survival of Turnberry Hotel continued long after the war. Ownership by the nationalised British Transport

Hotels was fraught with the constant challenge of the maintenance and upkeep of such a prestigious hotel, coupled with poor receipts over the years, it was put up for sale in 1983.

Mike Hawthorn driving the V12 Ferrari Thinwall Special

Japanese company Nitto Kogyo, bought the hotel and golf courses in 1987, investing $21m in the resort before falling victim to the financial crisis in the Far East, selling Turnberry for $32m in 1997 to Starwood Hotels & Resorts Worldwide, Inc. just days before filing for bankruptcy. Starwood operated the resort under the Westin brand until 2008, when it was sold to Leisurecorp.

The Trump Organization purchased the hotel and golf courses from Leisurecorp in 2014, spending over $200m in renovation and restoration, including the addition of an opulent ballroom to the main hotel building and the luxurious conversion of the iconic Turnberry Lighthouse into the half-way house and renaming the resort "Trump Turnberry', shortly before owner Donald Trump became President of the USA. The clubhouse was renovated, a new golf course, King Robert the Bruce, was built, and the Ailsa course has undergone major alterations and improvements, being named 'Redevelopment of the Year' by Golf Inc. Magazine and '2019 Best Golf Resort in Europe by LeadingCourses.com.'

Over a century after Turnberry Hotel and Golf Course was opened, it remains an iconic image of historical importance. Guests have included heads of state and Hollywood's elite, Prince Edward, Prince Andrew, Bing Crosby, Bob Hope, Rod Stewart, Bruce Forsyth, Jack Nicholson,

Turnberry Racing at Farm Corner.

Luciano Pavarotti, the pop group Abba and many other well-known visitors to Ayrshire.

In 1998, amidst stringent security turning the hotel into a five-star fortress, Turnberry was the venue for the 46th Bilderberg Meeting. This was a private conference with over 120 participants, compromising a broad cross-section of leading citizens, in and out of government, from North America and Europe. Some of those attending to take part in the discussions on subjects such as the 'Atlantic Relationship in a Time of Change, NATO, Asian Crisis, EMU, Growing Military Disparity, Japan, Multilateral Organizations, Europe's social model, Turkey and the EU/US Market Place' were William Hague, Leader of the Opposition, Kenneth Clarke, ex-Chancellor the Exchequer, Henry Kissenger, UK Defence Secretary George Robertson, Tony Blair, Lord Carrington and American banker David Rockefeller. For four days the whole of the surrounding area was on lock-down with security personnel, marksmen, bomb-disposal experts and sniffer dogs on patrol. Even the most lowly of hotel staff had to obtain security clearance from Special Branch. One waitress reported that when at the end of her shift, going to unlock her car in the car park, she distinctly heard the nearby shrub say goodnight to her!

So, what of this beautiful and historic corner of Carrick today, little changed in spite of the major upheavals during both world wars. The agriculture remains, but the fishing fleet of Maidens is all but disappeared. Individual craftsmen and women are established in the area, some ensconced in the old airfield buildings. The tourism envisaged by the 3rd Marquis of Ailsa flourishes, hosting visitors from all over the world.

Grass now covers the land where aircraft once roared into the sky, and all is still. The screeching gulls can be seen swooping by the lapping water's edge, on the beach the soughing sound of pebbles can be heard sliding noisily with the ebbing tide, and there in the eeriness of the night one can sense the presence of those who came in planes to Turnberry and never went away, either in body or spirit.

Post Script

When I began my quest of researching the story of Turnberry Airfield, I was ill prepared for the emotional journey I was to undertake. What was a seemingly straightforward task of writing a book, grew to such proportions as to almost engulf me. When asked what prompted me to write about the airfield, I replied, 'Did I really have a choice?' Listening to the veterans telling me their stories, and sharing with me, what they had never shared before, I was compelled to capture them for posterity. To bear witness to their service and in so doing was given the honour of great friendships, evolving over many letters, phone calls and visits, only ending when they hung up their flying boots forever.

As one airman wrote to me: 'We, the war veterans are grateful for you sharing with us, your heartfelt thanks and understanding of that traumatic period when our lives were balanced on a razor's edge ... The love you have shown us is more valuable than medals.' As I said, 'Did I really have a choice?'

Bibliography

Best, Robert Dudley, *A Short Life and a Gay One, Frank Best, 1893-1917*, unpublished manuscript, Birmingham Public Library.

Boyd, Jack, *The Bonnie Links of Turnberry*, Turnberry Golf Club, Turnberry, 2004.

Cole, Christopher, *McCudden VC*, Kimber, London, 1967.

Connan, Peter, *History of Dumfries and Galloway Vol: II*, St Patrick's Press, Penrith, 1984.

Crundall, Wing Commander E. D., DFC, AFC, *Fighter Pilot on the Western Front*, Harper Collins, London, 1975.

Fife, Malcolm, *British Airship Bases of the Twentieth Century*, Fonthill Media, Stroud, 2015.

Hayward, Roger, *The Beaufort File*, Air Britain Historians, Slough, 1990.

Hearn, Peter, *Flying Rebel*, TSO, London, 1994.

Heide, Cecil LeRoy, *Whispering Death*, Trafford, Victoria, British Columbia, Canada, 2006.

Janes Fighting Aircraft of WW1, Random House, London, 2001.

Janes Fighting Aircraft of WW2, Random House, London, 1989.

Johnstone, AVM Sandy, *Spitfire into War*, Kimber, London, 1986.

Jones, Hugh Gwyn, *A Welsh Journey*, Amazon, 2001.

Loch Doon Scandal, Doon Valley Family History Society, Dalmellington, 2018.

Love, Dane, *Ayrshire: Discovering a County*, Fort Publishing, Ayr, 2003.

Love, Dane, *Scottish Ghosts*, Robert Hale, London, 1995.

MacArthur, Wilson, *The River Doon*, Cassell, London, 1952.

Macmillan, Capt. Norman, *Sir Sefton Brancker*, William Heinemann, London, 1935.

McCudden, James, VC, *Flying Fury*, Aviation Book Club, London, 1939.

Moss, Michael, *The Magnificent Castle of Culzean*, Edinburgh University Press, Edinburgh, 2002.

Moyle, Harry, *The Hampden File*, Air Britain Historians, Slough, 1998.

Oughton, Frederick, *The Aces*, Neville Spearman, London, 1961.

Parsons, Les and Battams, Samantha, *The Red Devil: The Story of South Australian Aviation Pioneer, Captain Harry Butler AFC*, Wakefield Press, Mile End, South Australia, 2019.

Robertson, Bruce, *British Military Aircraft Serials 1878-1987*, Midland Counties Publications, Leicester, 1998.

Rossano, Geoffrey L., *Hero of the Angry Sky*, Ohio University Press, Ohio, United States, 2013.

Shaw, James E., *Ayrshire 1745-1950*, Oliver & Boyd, Edinburgh, 1953.

Sloan, James J., *Wings of Honor: American Airmen in WWI*, Schiffer Publishing, Atglen, Pennsylvania, United States, 2004.

Sperberg-McQueen, Marian, *Somewhere in France*, http://parr-hooper.cmsmcq.com, 2016.

Springs, Elliot White, *War Birds: Diary of an Unknown Aviator*, Temple, Brighton, 1966.

Strange, L. A., *Recollections of an Airman*, John Hamilton, Oxford, 1935.

Taylor, Carole McEntee, *From Colonial Warrior to Western Front Flier*, Pen & Sword, Barnsley, 2015.

Todd, Robert Miles, *Sopwith Camel Fighter Ace*, Ajay Enterprises, Falls Church, Virginia, 1978.

Wright, Sam, *Up, Up and Away*, unpublished manuscript.

Other Sources

Introduction

Ayr Advertiser, 20 November 1896
Cassillis Estate Archive
The National Archives (UK)
The Bonnie Links of Turnberry, Jack Boyd
The Magnificent Castle of Culzean, Michael Moss

WWI Airfield

Cassillis Estate Archive
Recollections of an Airman, L. A. Strange
The National Archives (UK) AIR 2/72/A7794
The National Archives (UK) AIR 1/30/15/1/152
The National Archives (UK) AIR 1/32/15/1/188
The National Archives (UK) MT 6/2588/5
Sir Sefton Brancker, Capt. Norman Macmillan
Ayrshire 1745-1950, James E. Shaw

The River Doon, Wilson MacArthur

Loch Doon Scandal, Doon Valley Family History
    Society

History of Dumfries and Galloway Vol: II,
    Peter Connan

Tramps Across Watersheds, A. S. Alexander

Scottish Aerodromes of the First World War,
    Malcom Fife

Lieut. Gwilym Hugh Lewis letters

Daily Record

Drogheda Argus and Leinster Journal, 12 October
    1918

Dumfries and Galloway Standard, 23 February
    1918

Dundee Evening Telegraph

Northern Whig, 3 May 1918

Pall Mall Gazette, 22 March 1918

Pall Mall Gazette, 20 May 1918

Sheffield Daily Telegraph, 20 May 1918

Pall Mall Gazette, 5 June 1918

Sunday Post, 18 February 1923

The Irish Independent, 20 May 1918

The Scotsman, 2 July 1918

The Scotsman, 19 September 1918

The Scotsman, Wednesday 30 October 1918

Daily Record and Mail, 20 December 1918

The Scotsman, Tuesday 25 March 1919

Dundee Evening Telegraph, 23 March 1923

A Short Life and a Gay One, Frank Best, 1893-1917,
    Robert Dudley Best

'Line', Henry Frederick Vulliamy Battle

Sopwith Camel Fighter Ace, Robert Miles Todd

*War Birds: Diary of an Unknown Aviator/John
    McGavock Grider*, Elliot Springs

*From Colonial Warrior to Western Front Flier
    (Captain Sydney Harris)*, Carole McEntee Taylor

Jack Henry Weingarth, letter January 1918

L.A.C. Arthur Charles 'Pop' Lewcock Diary

National Records of Scotland

WWI Airmen

The Aerodrome Forum

London Gazette

National Records of Scotland

The National Archives (UK) - Reports on aeroplane
    and personnel casualties, 11 June 1918 – 20 June
    1918

The National Archives (UK) - PP/MCR/62 George
    Buxton

RAF Museum Archive London

Australian War Memorial Collection

British Royal Air Force, Officers' Service Records,
    1912-1920

Wikipedia

Ancestry.com

Lives of The First World War 1914-1918

Recollections of an Airman, L. A. Strange

*The Red Devil: The Story of South Australian Aviation
    (Harry Butler)*, Les Parsons, Samantha Battams

*From Colonial Warrior to Western Front Flier (Sydney
    Herbert Bywater Harris)*, Carole McEntee Taylor

A Short Life and a Gay One, Frank Best, 1893-1917,
    Robert Dudley Best

Blairgowrie Advertiser, 31 August 1918

Imperial War Museum Archive

Ancestry.com - British Army WWI Service Records,
    1914-1920

Ancestry.com - UK Soldiers Died in the Great War,
    1914-1919 (online database)

Ancestry.com - British Army WWI Medal Rolls Index
    Cards, 1914-1920

Ancestry.com - British Army WWI Service Records,
    1914-1920 (online database)

Ancestry.com - UK De Ruvigny's Roll of Honour,
    1914-1919 (online database)

British Commonwealth War Graves Registers,
    1914-1918

Ancestry.com - Canada WWI CEF Attestation Papers,
    1914-1918 (online database)

Ancestry.com - U.S. World War I Draft Registration
    Cards, 1917-1918 (online database)

Ancestry.com - UK Royal Air Force Airmen Records,
    1918-1940 (online database)

Great Britain, Royal Aero Club Aviators' Certificates,
    1910-1950

British Army WWI Service Records, 1914-1920

The Men of the Second Oxford Detachment (Website)
    - Marian Sperberg-McQueen

Harrow School War Memorials, Volume V (Wilfred
    Allan Fleming)

R.F.C. Communiques

Birmingham Daily Gazette, 20 March 1920 (Charles
    Ernest Ottiwell 'Otto' Cowell)

Portsmouth Evening News, 23 March 1920 (Charles
    Ernest Ottiwell 'Otto' Cowell)

Edinburgh Evening News, 31 March 1920 (Charles
    Ernest Ottiwell 'Otto' Cowell)

Yorkshire Post (Motor Cycling News), April 1922
    (Geoffrey Clapham)

East Aberdeenshire Advertiser, 13 June 1916 (Oswald
    Douglas Hay)

The New Zealand Herald, 28 August 1917 (Allan
    Adolphus Veale)

The New Zealand Herald, 2 February 1918 (Allan
    Adolphus Veale)

'Table Talk', Melbourne, Victoria, Australia 9 May

1918 (Cyril Whelan)
Jewish Herald, Victoria, Australia 17 May 1918
    (Cyril Whelan)
https://www.wdl.org/ (Diary of Gustav Hermann
    Kissel)
Wings of Honor: American Airmen in WWI, (Kenneth
    McLeish), James J. Sloan Jr.
The Journal Tribune, 11 November 1981 (Robert
    Miles Todd)
Osceola Times, 5 October 1917 (John McGavock
    Grider)
Osceola Times, 12 Jul 1918 (John McGavock
    Grider)
Evening Missourian, 15 July 1918 (Edward Russell
    Moore)
Melbourne Grammar School Magazine, June 1918
    (Howard Richmond Henry Butler)
The Age, Melbourne, Victoria 5 June 1918 (Howard
    Richmond Henry Butler)
Flight Global, 20 June 1918 (Howard Richmond
    Henry Butler)
London Times, 25 September 1912 (Cyril Edgar
    Foggin)
Scottish Ghosts, Dane Love (Cyril Edgar Foggin)
McCudden VC, Christopher Cole
Flying Fury, James McCudden VC
The Aces, Frederick Oughton
'Line', Henry Frederick Vulliamy Battle
Essex Newsman, 21 December 1918 (Thomas Allister
    King)
Ulster Press, January 1919 (James Millikin
    (Milligan))
RFC Communiques
RFC 1915 -16   Christopher Cole
RFC 1917 -18   Chaz Bowyer
http://www.theaerodrome.com/aces

WWI Aircraft
Janes Fighting Aircraft of WW1
http://www.theaerodrome.com/The Aircraft of WWI
British Military Aircraft Serials 1878-1987

WWI Training
The National Archives (UK)
Notes on Guns, Gears and Sights, Turnberry 1918,
    Bell-Irving, R.

WWI Accidents
Carrick Gazette, 4 May 1923
National Records of Scotland (Statutory Registers of
    Deaths)
British Commonwealth War Graves Registers,
    1914-1918

RAF Museum Archive London Casualty Cards

WWI Memorial
Carrick Gazette, 4 May 1923
National Records of Scotland
British Commonwealth War Graves Registers,
    1914-1918
RAF Museum Archive London, Casualty Cards
Airmen Died in The Great War, 1914-1919
British Army, De Ruvigny's Roll of Honour
    1914-1918

Interwar
The National Archives (UK)
Evening Telegraph, 12 October 1926
Dundee Courier, 16 August 1930
The Scotsman, April 1938
Cassillis Estate Archive

WWII Airfield
Spitfire into War, AVM Sandy Johnstone
The National Archives (UK) AIR 28/866
London Metropolitan Archives: City of London
    LCC/PH/HOSP/6/01B
London Metropolitan Archives: City of London
    LCC/PH/HOSP/6/02
London Metropolitan Archives: City of London
    LCC/PH/HOSP/6/01A
The National Archives (UK) AN 109/724
The National Archives (UK) AIR 29/705/2
The National Archives (UK) AIR 29/702/3
The National Archives (UK) AIR 27/2130
The National Archives (UK) AIR 29/631/3
Cassillis Estate Archive
A Most Secret Squadron, Des Curtis DFC

WWII Airmen
Personal Letters and Interviews
www.britishpathe.com/video/robert-ashley
National Records of Scotland
Up, Up and Away, Sam Wright
Whispering Death, Cecil LeRoy Heide
A Welsh Journey, Hugh Gwyn Jones

WWII Aircraft
British Military Aircraft Serials 1878-1987, Midland
    Counties Publications.
Janes Fighting Aircraft of WW2
Wikipedia
The Beaufort File, Air Britain Historians, Slough,
    1990.
The Hampden File, Air Britain Historians, Slough,
    1998.

WWII Training
Personal Letters
History of the RAF, Chaz Bowyer

WWII Accidents
National Records of Scotland (Statutory Registers of
    Deaths)
The National Archives (UK) AIR 28/866
British Commonwealth War Graves Registers,
    1914-1918
Ministry of Defence Air Historical Branch

WWII Memorial
National Records of Scotland (Statutory Registers of
    Deaths)
The National Archives (UK) AIR 28/866
British Commonwealth War Graves Registers, 1914-
    1918
Ministry of Defence Air Historical Branch
Ayrshire Post

Post War
The National Archives (UK)AN 13/1443
Cassillis Estate Archive
The National Archives (UK) AN 109/726
The National Archives (UK) AN 109/725
The National Archives (UK) AN 109/724
Christian Mølle
Redgates Caravan Park
https://blackwood-hodge.typepad.com
Malin Court Hotel: The First Twenty-Five Years,
    Iain Hall
Malin Court Hotel
https://www.oldracingcars.com/f1/1952/
Autosport Magazine, 9 May 1952
Autosport Magazine, 29 August 1952
Ayrshire Post, 18 February 1955
https://www.turnberry.co.uk/

Abbreviations

A/Gnr	Air Gunner		CO	Commanding Officer
AC	Aircraftman		Col.	Colonel
AC1	Aircraftman 1st Class		Commdre	Commodore
AC2	Aircraftman 2nd Class		Cpl	Corporal
ACW	Aircraftwoman		CPO	Chief Petty Officer
ADC	Aide De Camp		CWGC	Commonwealth War Graves
ADU	Aircraft Delivery Unit			Commission
AEF	American Expeditionary Force		DFC	Distinguished Flying Cross
AFC	Australian Flying Corps		DFM	Distinguished Flying Medal
AIB	Accident Investigation Branch		DFW	Deutsche FlugzeugWerke
AIF	Australian Imperial Force		DH	de Havilland
ALT	Attack Light Torpedo		DP	Displaced Person
AM	Air Mechanic		DRO	Daily Routine Orders
AMDW	Air Ministry, Directorate of Works,		DSC	Distinguished Service Cross
ANZAC	Australian and New Zealand Army		DSO	Distinguished Service Order
	Corps		EA	Enemy Aircraft
AOC	Air Officer Commanding		EFTS	Elementary Flying Training School
ARP	Air Raid Precautions		EMS	Emergency Medical Services
ART	Attack Running Torpedo		ENSA	The Entertainments National Service
ASD	Aircraft Storage Depot			Association
ASR	Air Sea Rescue		F/Lt	Flight Lieutenant
ASV	Air-Surface Vessel		F/O	Flying Officer
ATA	Air Transport Auxiliary		FAA	Fleet Air Arm
ATC	Air Training Corps		FALT	Formation Attack Light Torpedo
ATS	Auxiliary Territorial Service		FART	Formation Attack Running Torpedo
Aux	Auxiliary		FIS	Fighter Instructors School
AVM	Air Vice Marshall		FTU	Ferry Training Unit
BA	Bachelor of Arts		G&SW	The Glasgow and South Western
BCATP	British Commonwealth Air Training			Railway
	Plan		GCVO	Knight Grand Cross
BD	Base Depot		Gp/Capt	Group Captain
BE	Blériot Experimental		GPO	The General Post Office
BEF	British Expeditionary Force		GR	General Reconnaissance
BOAC	British Overseas Airways Corporation		GS	General Service
Brig Gen	Brigadier General		HFDF	High Frequency Direction Finder
BSc	Bachelor of Science		HM	His Majesty
CAG	Civil Air Guard		HMS	His Majesty's Ship
Capt	Captain		HMT	His Majesty's Transport
CC	A synchronization gear		HQCC	Head Quarters Coastal Command
CCFIS	Coastal Command Flying Instructors		HS	Home Service
	School		HSL	High Speed Launch
CEF	Canadian Expeditionary Force		KBE	Knight (Commander of the Order) of
CFI	Chief Flying Instructor			the British Empire
CFS	Central Flying School		KCMG	Knight Commander (of the Order) of St
CI	Chief Instructor			Michael and St George
CIE	Companion of the Indian Empire		KCVO	Knight Commander of the Royal

	Victorian Order	RCPS	Royal College of Physicians and Surgeons of Glasgow
L&NW	The London and North Western Railway	RDFS	Radio Direction Finding School
LAC	Leading Aircraftman	RE	Royal Engineers
LACW	Leading Aircraftwoman	REME	Royal Electrical and Mechanical Engineers
LMF	Lack of moral fibre	RFC	Royal Flying Corps
LMS	London Midland and Scottish Railway Company	RN	Royal Navy
Lt Col	Lieutenant-Colonel	RNACD	Armoured Car Division
Lt Gen	Lieutenant-General	RNAS	Royal Naval Air Service
Lt	Lieutenant	RNoAF	Royal Norwegian Air Force
Ltn	Leutnant (German)	RNVR	Royal Naval Volunteer Reserve
LVG	Luftverkehrsgesellschaft	RNZAF	Royal New Zealand Air Force
Maj	Major	RSF	Royal Scots Fusiliers
Maj Gen	Major-General	S/Ldr	Squadron Leader
MBChB	Bachelor of Medicine and Bachelor of Surgery	SAAF	South African Air Force
		SASO	Senior Air Staff Officer
MC	Military Cross	SE	Scout Experimental
MCS	Marine Craft Section	SE	South East
MD	Medicinae Doctor (Doctor of Medicine)	SFTS	Service Flying Training School
METS	Multi-engine Training Squadron	Sgt Maj	Sergeant Major
MG	Machine Gun	Sgt	Sergeant
MM	Military Medal	SHQ	Station Head Quarters
MO	Medical Officer	SMO	Senior Medical Officer
MOD	Ministry of Defence	SNCO	Senior Non Commissioned Officer
MT	Motor Transport	SOE	Special Operations Executive
MU	Maintenance Unit	SofAF	School of Aerial Fighting
N/Ord	Nursing Orderly	SofAG	School of Aerial Gunnery
NAAFI	Navy, Army, Air Force Institute	SP	Service Policemen
NCO	Non Commissioned Officer	Spec Res	Special Reserve
NE	North East	SSQ	Station Sick Quarters
Notts	Nottinghamshire	SW	South West
OAPU	Overseas Aircraft Preparation Unit	TA	Territorial Army
OBE	Order of the British Empire	TAT	Torpedo Attack Trainer
OFE	Operational Flying Exercise	TB	Torpedo Bomber
OIC	Officer in Charge	TDS	Training Depot Station
OR	Other Ranks	TF	Territorial Force
ORB	Operations Record Book (Form 540)	TNT	Trinitrotoluene (explosive compound used in dynamite)
OTC	Officer Training Corps		
OTU	Operational Training Unit	TS	Training Squadron
P/O	Pilot Officer	TT	Tourist Trophy
PC	Police Constable	TTU	Torpedo Training Unit
POW	Prisoner of War	U/S	Unserviceable
PPI	Planned Position Indicator	U/T	Under Training
PU	Permanently Unfit	USAF	United States Air Force
QDM	Magnetic bearing to a DF station	VAD	Voluntary Aid Detachment
QGH	Controlled descent through cloud	VC	Victoria Cross
RAAF	Royal Australian Air Force	VE	Victory in Europe Day
RAF	Royal Air Force	VJ	Victory over Japan
RAFVR	Royal Air Force Volunteer Reserve	W Op	Wireless Operator
RAMC	Royal Army Medical Corps	W Op/AG	Wireless Operator Air Gunner
RASC	Royal Army Service Corps	W Ops/DF	Wireless Operator direction finding
RAuxAF	Royal Auxiliary Air Force	W/Cdr	Wing Commander
RCAF	Royal Canadian Air Force	W/T	Wireless Telegraphy

WAAC	Women's Army Auxiliary Corps	WO/M	Wireless Operator Mechanic
WCTU	Woman's Christian Temperance Union	WRAF	Women's Auxiliary Air Force
WLA	Women's Land Army	WVS	Women's Voluntary Service
WO	Warrant Officer	YMCA	Young Men's Christian Association

Acknowledgements

Special thanks are due to the many people have been with me on this journey, there are far more people than I can ever list by name, and they have all believed in me and supported me in my 'quest', which is extremely humbling. History books require a lot of help, especially one of this size, and I am eternally grateful to the many archivists across Britain and beyond who gave me a great deal of assistance. I am indebted to The National Archive, Royal Air Force Museum Archive at Hendon, the Imperial War Museum, Cassillis and Culzean Estate Archive, Reference Department, Carnegie Library, and many, many more.

I would like to extend particular gratitude to Chris Savage, Factor of Cassillis and Culzean Estates, Les and Kathleen Henshew for the gift of the WWI photographs, also Les Parsons, Greg Ward, and Theo Steel.

Thanks are also due to Carole McEntee Taylor and Marian Sperberg-McQueen
for allowing me to quote passages from their work.

My completion of this project could not have been accomplished without the support of my family, first and foremost my husband Anthony and children Edward, Douglas, and Jessica. Thank you for allowing me to hide away from you all to research and write. Finally, thanks to Dane Love of Carn Publishing for proof reading and editing, and for having faith that there was a book in there somewhere!

I am also indebted to copyright holders who have allowed me to quote passages. In some cases it has not been possible to contact the holder; if anyone feels their copyright has been inadvertently infringed, a correction will appear in subsequent editions of this book.

I dedicate this book about Turnberry Airfield to the airmen who told me their stories and gave me their precious gift of friendship, sadly most of them have now passed on…. I will remember them.

Index

284–88, 296–98, 345, 349, 358, 359-60, 375, 381, 387, 391, 392, 395, 397, 399, 403, 428, 438-9, 447, 450, 459, 460, 466, 474, 475, 478, 482, 489, 491, 517

Anti-submarine, 64, 287, 295, 391, 440, 475

Anzac, 159, 415, 553

Anzani, 162

Ardlochan, 53, 311

Ardrossan, 10, 279, 297, 402, 465

Argatarla, 421

Argyll, 86, 155, 157, 241, 249, 260

Armentières, 73, 112, 119, 126-7, 178

Armistice, 55, 65, 85, 123, 124, 145, 171, 186, 229, 249, 354, 379

Armstrong Siddeley, 387

Armstrong Whitworth, 22, 71, 86, 162, 195-6, 472

Armstrong, F/O, 110, 290

Armstrong, Thomas, 520

Arnhem, 378, 421

Arras, 67, 71, 72, 83, 88, 91, 92, 96, 98, 104, 105, 124, 128, 180, 199, 203

Atkinson, Joseph, 164, 179, 188

Atlantic, 287, 311, 325, 326, 330, 385, 427, 435, 437, 441, 445, 447, 448, 454, 461, 547

Atteridge, Brian, 76

Aubigny-en-Artois, 84

Auchenfail, 176

Auchey, 67

Auchincruive, 319

Auchinleck, 4

Auckland, 100, 320, 432, 500, 514

Australia, 62, 65, 66, 76, 77, 87, 95–99, 102, 105, 119, 156, 158, 159, 170, 203, 236, 240, 242, 300, 302, 304, 317, 318, 324, 453, 482, 488, 495, 502, 508, 512–15, 522, 549–51

Australian, 43, 62, 65, 66, 76, 87, 95–98, 102, 105, 108, 154, 156, 158, 159, 170, 171, 185, 230, 231, 240, 242, 286, 318,

322, 364-5, 381, 384, 394, 414, 416, 421, 436, 439, 442, 458, 488, 495, 502, 508, 512–15, 522, 549, 550, 553, 554

Auxi-le-Château, 169

Avonweir, 353–55

Avro aircraft, 46, 65, 66, 68, 90, 93, 102, 106, 108, 115, 118, 133-4, 136-40, 142, 163, 164, 168, 169, 175, 179, 187, 190-2, 201, 221–26, 299, 344, 349, 355, 358, 387, 395, 438, 447, 450, 466, 474, 475, 478, 489, 491, 517, 519

Ayr, 10, 13, 17, 22, 27, 36, 45, 51, 60, 74, 82, 86, 91, 103, 106, 108, 110, 113, 115, 117–19, 122, 124, 126, 130–37, 139, 140, 142–45, 149, 151, 156, 157, 162, 163, 165, 167–70, 176, 177, 179, 184, 186, 208, 213, 226, 234, 248, 250–52, 257, 259, 260, 264, 274, 276, 278, 279, 282, 284–86, 289, 290, 293, 294, 296, 297, 299, 311, 317, 329, 333, 342, 343, 347, 357, 371, 373, 376, 380, 387–91, 394, 397, 398, 403, 405, 421, 426, 429, 430, 435, 437, 441–43, 448, 457, 458, 461, 465, 480, 517, 523, 525, 526, 530, 531, 539, 543, 545, 549

B

Babbington, P., 276

Baghdad, 70

Baikie, L. S., 147

Bailey, 200, 345

Bailkie, Mr, 168

Bailleul, 88, 101, 105, 162, 204

Baird, J. H., 273

Baird, Major, 25, 31,

Balaam, Augustus, 84

Balado, 313

Balfour, Lord, 26, 28

Balkenna, 270, 319, 345, 494

Ballantrae, 12, 287, 330, 331, 380, 539

Ballochmyle, 388, 442

Buttner, Karl, 199
Buxton, George, 88, 297
Byne Hill, 331, 350, 544

C

Cairo, 361, 367, 377, 404, 421, 468
Calcutta, 97, 414
Calder, James, 487, 519
Calgary, 74, 160, 499
California, 77, 112, 123, 152
Callaghan, Casey, 80
Callaghan, Joseph, 103-4
Callahan, Lawrence, 119, 125
Calthrop, R. E., 222
Calveley, 453
Cambrai, 92, 150, 180
Cambridge, 79, 100, 101, 177, 259, 386, 452, 514
Camel, 46, 70, 76-7, 83, 84, 86, 99, 100, 105, 108-12, 115–23, 133, 138, 145-7, 149-50, 153–55, 159–61, 163-65, 170, 171, 180, 182, 184, 190, 196–98, 200, 201, 205–7, 213, 227, 228, 235, 238–40, 242–44, 549, 550
Cameron, Anne, 348
Cameron, Eric, 404, 407
Cameron, Gordon, 439-40
Camlarg, 18, 19
Campbell, James, 487, 491
Campbell, Jesse, 111
Campbell, Mrs (Kirkton Inn), 340-1
Campbell, Sally, 449
Campbell, Sanny, 330
Canada, 25, 60, 73–75, 77, 101, 104–6, 108, 119, 145, 149, 152, 153, 158, 160, 181, 185, 242, 243, 245, 254, 317, 318, 325, 326, 344, 367, 368, 370, 371, 373, 374, 399, 400, 402, 403, 406, 407, 409–11, 414, 417, 426, 432, 439, 441, 442, 447, 461, 478, 480, 491, 493–99, 503, 505, 507, 517–19, 523, 549, 550

Canadian, 60, 73, 74, 77, 101, 106, 129, 153, 160, 161, 172, 181, 187, 286, 290, 304, 313, 316, 367, 369, 372, 400–402, 407, 410, 419, 421, 436, 439, 446, 461, 478, 487, 491, 493–500, 503, 505, 507, 510, 517–19, 523, 553, 554
Canning, Andrew, 516
Canteen, 32, 274, 285, 314, 317, 330, 348, 350, 402
Canterbury, 100, 187
Canterbury, HMS, 363, 397
Canton, HMS, 397
Canton-Unne, 24
Capetown Castle, 379
Cardiff, 96, 237, 321, 377, 418, 480, 521
Caribbean, 410
Carleton House, Ayr, 134, 136, 137, 140, 142
Carlisle, 19, 346, 520
Carmania, 109, 110, 112, 124, 125, 127, 129, 143, 152
Carrick, 9, 12, 13, 249, 256, 258, 279, 317, 544, 547
Carrickfergus, 184, 185, 245
Carrington, Lord, 547
Cassillis, 11, 24, 44, 49, 225, 226, 256, 258–60, 264, 274, 278, 526–28, 538, 549, 551, 552, 555
Castle Kennedy, 287-8, 290, 294-5, 391, 481
Catalina, 287, 352, 426, 440, 469, 482
Catfoss, 415
Cathcart, 177
Cathcart, Major F., 18
Cavers, James, 106-7
Cazaux, 16, 18, 24
Cessna aircraft, 399
Chadwick, R., 85
Challis, P/O, 285
Chapelton, 404, 407
Chaplain, 146, 148, 156, 248, 293, 299, 353
Charlottetown, 414, 518
Charterhall airfield, 306, 307, 483

Chaulnes, 128, 129, 150
Cheley, Sgt, 286
Chelsea, 243, 350, 510
Chirmorie, 297, 517
Chiswick, 367, 522
Churchards, Miss, 534, 535
Cincinnati, 120, 123
Cinema, 19, 267, 290, 312, 316, 344, 348,
 349, 380, 397, 398, 433, 447, 457
Clachanton, 285, 317, 526
Claggenfort, 314
Clapham, Geoffrey, 93-4
Clark, Archie, 365, 434
Clark, Thomas, 487, 511
Clarke, Pauline, 381-3
Clarke, Peter, 279, 492
Clarke, S/Lr, 273, 275
Clausnitzer, Ernst, 203
Clayson, Percy, 180-1
Clayton, John, 197
Cleghorn, Lt, 248
Clifford, G, 274, 275, 278
Clydebank Blitz, 316
Cockburn, Maj, 225
Colclough, Sgt, 290
Coldstream Guards, 11, 256
Coleman, W/Cdr, 285, 288
Collett, C., 203
Collis, E. W., 178
Colney, 99, 125, 129, 130, 135, 143
Cologne, 351, 472
Colquhoun, Billie, 366
Connaught Rangers, 154, 240
Connaught, Duke of, 24
Constantinesco, 90, 173
Cooke, Dennis, 404–8
Cooke, Douglas, 73
Cornwall, 351, 392, 403, 414, 422, 448, 490
Cowan, R. J., 23
Cowell, Charles 'Otto', 92, 93, 550
Cowie, Sgt, 287

Cracroft, G/Capt, 297
Craigencolon, 17
Craigengillan, 16, 18, 22, 23, 26, 27, 33, 35,
 36
Craigton Cemetery, 181, 463
Crail, 297, 299, 313, 480
Cranwell, 329, 352–55, 377, 415, 450, 452,
 458, 469
Crashes, 8, 21, 66, 70, 78, 83-4, 86, 88, 90,
 91, 93, 94, 97-100, 103-4, 106, 111, 116,
 123, 129, 143-4, 147-50, 151, 153-4,
 155-6, 158, 161-4, 180-2, 188, 202-5,
 223, 234, 236, 239-40, 242-3, 273,
 275-81, 283-90, 292-5, 297, 297-301,
 303, 315-7, 319, 331, 340, 344-5, 349-50,
 356, 358-60, 362, 363, 366, 376, 381,
 388, 398, 402, 406, 416, 423, 425, 427,
 432-4, 447-8, 451, 455, 456, 458, 460,
 467, 471, 483–85, 488-501, 506, 509-13,
 515-21
Craven, R. E., 285, 290–92
Crawford, Sgt-Major, 65
Cronin, F/Lt, 454
Crossraguel Abbey, 342
Cullendoch, 17, 27
Culzean Castle and Estate, 11–13, 44, 47, 49,
 225, 252, 256–58, 260, 264, 274, 277,
 278, 280, 338, 340, 359, 360, 364, 398,
 419, 438, 441, 442, 445, 447, 449–51,
 457, 507, 515, 526–28, 538, 540, 542
Cumbrae Islands, 458
Cunningham, C., 299
Cunningham, Isa, 449
Cunningham, Lyman, 120
Curacoa, HMS, 347
Curragh, The, 79
Curzon, Earl, 29, 32
Custerson, S/Ldr, 295

D
Dailly, 48, 93, 263, 504, 545

Dakota, 297, 312, 329, 367, 378, 421, 460, 481

Dalfarson, 22, 23, 25, 27

Dallachy Airfield, 433

Dalmellington, 16, 19–24, 27, 28, 32–35, 37, 38, 517, 519, 549

Dalquhat, 88, 236, 243

Dalrymple-Hamilton, Lady Marjorie, 225

Dambuster, 292, 412

Dankeith House, 467

Dardanelles, 100, 353

Davie-Spiers, Robert, 515

Davis, Randall, 77, 83

Davis, Willie, 356

De Havilland aircraft, 160, 190, 192, 195, 200, 202, 205, 292, 293, 422, 477, 520

De Havilland, John, 354

De Lacey, Matthew, 96

De Lambert propeller, 24

De Rochie, 77-8

Denham, 103, 182

Denmark, 430, 539, 540

Dennison, W/Cdr, 286

Dennistoun, J. A., 73

Derby, 71, 87, 236, 324, 329, 441

Derclach, Loch, 133

Devey, S/Ldr, 275

Devitt, W/Cdr, 296, 297

Devon, 178, 179, 344, 345, 351, 353, 379, 423, 446, 447, 465, 480

Devonport Hospital, 154

Dewey, S/Ldr, 273

Dixon-Spain, Gerald, 84-5

Dobson, Bill, 545

Docherty, T. E. V., 487

Doggett, B., 423

Donaldson, John, 124

Doncaster, 91, 378

Donkersley, 'Hank', 420

Dorset, 58, 59, 163

Douglas, Sir Sholto, 297

Douglas-Hamilton, Lord Malcolm, 94

Douglaston, 10, 11

Doune Cemetery, 83, 88, 89, 112, 148, 149, 153, 156–58, 160, 182, 186, 221, 248, 249

Dover, 58, 68, 79, 503

Dowhill, 241, 506

Down Ampney, 378

Drogue, 194, 212, 301, 365, 385, 447, 454, 521

Drumbeg, 247, 509

Dumbarton, 384, 471

Dumfries, 63, 406, 450, 521

Dunanhill, 288

Dunaskin Ironworks, 27

Dunbar, John, 158, 230, 242

Dundee, 37, 386

Dunfermline, 160, 242, 312, 509

Dungavel, 382

Dunure, 8, 11–13, 40, 226, 257, 278, 281, 286, 294, 315, 317, 357, 367, 432, 466, 487, 488, 515, 520

Durham, 183, 244

Durrad, F. A., 85

Dusseldorf, 79

Dwyer, G. J., 149

Dyas, Major, 107

Dyce Airfield, 297, 450

Dymchurch Redoubt, 44

E

Eagleton, S/Ldr, 277

Earne, Lough, 345

Eastbourne, 87, 161, 162, 181, 490

Eastchurch, 70, 184, 446

Ebblinghem Cemetery, 105

Edgeware, 102, 106, 374, 509

Edinburgh, 11, 49, 60, 67, 68, 70, 93, 187, 259, 260, 277, 286, 294, 302, 350, 357, 383, 386, 414, 438, 440, 453, 512, 521, 526, 549, 550

Edwards, W/Cdr, 286
Egger, Jack, 399-401
Eglinton, Earl of, 258
Ehmann, Vizefeldwebel Friedriech, 119
Eindecker, 190, 193
Elgin, 464
Ellis, D. C., 23
Ellis, Robert, 284, 502
Elrick, George, 376, 380
Elton, G., 288, 290, 293, 294, 416, 485
Enniskillen, 304
Entwhistle, A. M., 159
Entwistle, John, 461
Erickson, Donald, 507
Esau, Edwin, 511
Escanaffles Cemetery, 172
Essex, 66, 67, 161, 243, 430, 493, 501, 508
Essex, Sir R., 31
Estevelles Cemetery, 98
Etaples, 97, 159
Eton, 11, 68, 72, 73, 92, 256, 259
Ewart, Sir Spencer, 32
Exmouth, Viscount, 23
Eyre, Fred, 415-6

F
Faber, Reg, 374-9
Fahrenholtz, F/Off, 303
Failford, 139, 140, 168, 169, 176
Fairey aircraft, 68, 344, 479
Fairfield Shipyard, 181
Fairweather, Douglas and Margaret, 466, 467, 487, 517, 520
Fairweather, Jimmy, 172
Fardoe, William, 280, 494
Farman aircraft, 58, 67, 68, 71, 75, 83, 85, 88, 90, 93, 101, 103, 144, 163, 165, 176, 191, 195
Farnborough, 88, 190, 193, 200, 438
Farnon, Robert, 284, 375–77, 379, 503
Farquhar, W. A., 249

Farrer, Hon. Kitty, 467
Fasson, J. R., 71
Felixstowe, 67, 332, 333
Ferguson, Sandy, 380
Fergusson, Sir Charles, 226, 250, 258
Fernie, Tom, 224, 226, 232
Fernie, Will, 14
Ffrench, Evelyn, 98–100
Fisherton Cemetery, 317, 432, 466, 487
Fleming, Ian, 73
Fleming, Wilfred, 78, 79
Fletcher, Charles, 182, 183, 230, 244
Fokker aircraft, 72, 73, 98, 109, 112, 124, 128, 129, 150, 164, 178, 190, 193, 198, 203
Folley, W/O, 306
Forsyth-Johnson, John, 293, 469, 483, 485, 487, 511
Fox, George, 453-7
France, 13, 16, 18, 24, 58, 64, 67–69, 71, 73, 75, 77–79, 83, 84, 87–89, 91–93, 95–101, 103, 105, 106, 108, 110, 119, 121, 123, 124, 126–28, 131–33, 137, 142–45, 147, 151, 152, 159–63, 165, 169–71, 174–80, 182–85, 188, 189, 191, 210, 211, 218, 231, 258, 314, 330, 354, 373, 421, 430, 432, 476, 549
Fraserburgh, 280, 463
Frazer, A. A., 290
Fresnes-sur-Escaut POW camp, 124
Fullard, Capt., 214
Furgerson, Mr., 289

G
Gallipoli, 76, 95, 96, 101, 153, 163, 182, 353
Galloway Engineering Company, 205
Gamesloup, 380
Gaskell, R. G., 290
Germany, 75, 86, 123, 124, 144, 145, 189, 206, 314, 315, 330, 345, 358, 369, 429, 438, 442

Gibraltar, 294, 361, 392, 402, 423, 425, 445, 448, 452, 468

Gilchrist, Euan, 128

Gilmour, P/O, 286

Gilson, W/Cdr, 275

Girvan, 8, 13, 44, 48, 51, 82, 83, 88, 89, 112, 135, 138, 148, 149, 153, 156–58, 160, 168, 182, 186, 221, 224, 226, 241, 243, 248–51, 257, 260, 263, 276, 278–86, 288, 293–95, 301, 312–14, 316, 319, 330, 331, 342, 345, 348–50, 352, 357, 359, 363, 364, 366, 372, 373, 381, 387, 390, 391, 394, 397, 398, 405, 419, 429, 430, 439, 442, 443, 455–57, 483, 489, 491, 494, 500, 504, 508, 509, 535, 544, 545

Glamis, Lady of, 282, 477

Glasgow, 4, 12, 26, 27, 35, 38, 49, 63, 69, 82, 86, 89, 119, 136, 139, 142, 161, 170, 174, 177, 181, 183, 225, 243, 249, 250, 257, 258, 264, 271, 279, 300, 307, 313, 319, 321–24, 330, 331, 333, 347, 348, 380, 385, 387, 421, 429, 430, 434, 442, 457, 462, 463, 465, 466, 490, 515–17, 525, 526, 529, 530, 534, 536, 543, 544, 553, 554

Glenapp Hospital, 287

Gleneagles Hotel, 254, 525, 534, 536

Glenhead, 158, 241, 495

Glenside, 53

Gloucester aircraft, 422

Gloucestershire, 92, 108, 160, 378, 489

Goddard, Sgt, 365, 367

Godfrey, Charles, 183-4, 230, 244

Godfrey, Hillyer, 517

Goldspink, Charles, 435-6

Golf, 8, 13, 14, 40, 42–44, 48, 49, 66, 67, 80, 82, 114–18, 121, 122, 126, 130, 151, 170, 173, 174, 186, 187, 208, 224, 247, 250, 253, 254, 256, 258, 260, 261, 263–65, 270–72, 274, 305, 307–9, 311, 324, 331, 349, 351, 384, 387, 399, 406, 426, 429, 458–60, 525–38, 546, 549

Goodman, Herbert, 286, 505

Goodson, Leslie, 280, 388, 497

Gordon, Evelyn, 22, 24-5

Gordon, James, 89-90, 113, 155, 165, 186

Gordon, Lindsay, 74

Gorton, Rev William, 489

Gosport, 58, 68, 74, 119, 165, 208, 300, 479

Grangemouth, 273

Grantham, 70, 125, 129, 131, 143

Grantown-on-Spey, 463, 464

Grasswick, Tom, 281, 499

Greece, 70, 106, 107

Greenan POW Camp, 316

Greenburgh, F/Lt, 287

Greenfield, F/O, 296

Greenhalgh, J., 285, 290

Greenock, 290, 338, 347, 378, 385, 461

Gregory, Maj, 99

Griffin, Francis, 281, 293, 483, 485, 499

Guernsey, Isle of, HMS 363, 394, 397, 414, 467

Guynemeyer, Capt, 214

H

Haig, Douglas, 80, 95, 101

Halberstadt, 149

Halifax, 127, 299, 312, 329, 370, 399, 448, 463, 512

Hamar, Don, 437

Hamburg, 312, 388

Hamelincourt, 100

Hamilton, Duke of, 382

Hamilton, Georgia, 449

Hamilton, Lloyd, 110–12, 115

Hammond, Clive, 281, 498

Hampden, 275, 279, 280, 283–96, 317, 318, 327, 333-4, 344, 355, 357–59, 362-4, 368, 370–72, 374-5, 377, 383, 388, 395, 397, 398, 404-7, 414-5, 420, 428, 439, 442, 451, 472, 480, 481, 492, 495, 497, 500, 501, 503, 507, 510, 514, 515

Hampshire, 58, 77, 79, 93, 165, 492

Houthulst, 94, 177

Hoylake, 425

Hudson aircraft, 273-4, 277, 285, 296–304, 306, 318, 324-6, 359, 386, 388, 392, 411, 425, 430, 434-5, 438, 440, 442, 448, 454–56, 458, 461, 477-8, 481-2, 522

Hudson, Sgt, 276

Hudswell Clarke, 21, 23

Huggins, Norman, 488, 522

Hughes, John, 183, 230, 244

Hughes, Peter, 353, 355–6

Hughes, S/Ldr, 291, 357, 400-1

Hull, Edwin, 153, 230, 238

Hullavington airfield, 306, 354

Hurricane, 273, 279, 297, 477, 517

Hutchinson, W/Cdr, 300

Hutchison, W/Cdr, 293, 299

Hylands, Tom, 345

Hyslop, Annie, 349

Hyslop, Murray, 362

Hythe, 15, 18, 44, 45, 62–64, 75, 85, 176, 184, 209, 211, 216

I

Ibsley, 378

Iceland, 278, 359, 386, 388, 389, 392, 432, 434, 441, 445

Illinois, 119, 150, 152

India, 68, 76, 89, 90, 160, 343, 367, 402, 403, 410, 414, 440, 453, 458

Inglis, F/Lt, 385

Inglis, Thomas, 186, 230, 245

Ingram, Campbell, 314

Inman, Lord, 529, 535

Invergordon, 460

Iraq, 70, 174, 181, 367

Ireland, 16, 22, 29, 74, 79, 119, 130, 184, 185, 193, 240, 245, 279, 284, 286–88, 304, 308, 326, 345, 356, 364, 374, 377, 383, 411, 414, 421, 426, 427, 435, 439, 440, 456, 492, 512, 536, 545

Ireland, Douglas, 288

Iroquois engine, 344

Ismailia, 88, 95

Italians, 316, 319, 404, 406, 418, 457, 463, 466

Italy, 70, 76, 77, 100, 109, 124, 182, 303, 314, 345, 481

Ithaca, 112, 157

J

Jackson, Ben, 440

Jackson, H. C., 23

Jameston, 10, 270, 525, 526, 538, 541

Japan, 88, 547

Japanese aircraft, 315

Japanese Fleet, 295

Japanese, 76, 343, 367, 411, 416, 432, 433, 453, 458, 546

Jarrett, Ken, 436-7

Jarvis, F/Lt, 296

Jasta (Jagdstaffel squadrons), 71, 72, 79, 82, 84, 86, 91, 92, 96, 103, 105, 107, 119, 121, 129, 144, 176, 199, 203

Jersey, 78, 179, 180, 392, 394

Johnson, D. S., 85

Johnson, F/Lt, 306, 308

Johnson, O. P., 172

Johnstone, 263, 302, 375, 488, 514, 549, 551

Jones, Cyril, 487, 511

Jones, Fred, 384

Jones, Harry, 441

Jones, Hugh, 426, 427

Jones, James, 488, 497, 511

Jones, W/O, 280, 281

Joynson-Hicks, 28, 31

K

Kaldaoarnes, 386, 389

Kallang, 379

Karachi, 361, 362, 367, 414

Karcher, Bruno, 202

Paton, Marie, 449
Patton, General, 145
Pawsey, Bert, 360
Payne, Frederick, 277, 460, 461, 490
Pemberton-Billing, Noel, 190, 207
Pembrey, 365
Pembroke, 177, 180, 460
Pendlebury, S/Ldr, 286, 292
Pengally, Major, 297
Penman, Helen, 449
Pennsylvania, 112, 143, 145, 239
Pershore airfield, 459
Perth, 77, 491
Pewsey, Alan, 302, 522
Pfalz aircraft, 73, 105, 128, 129, 178, 182
Phillips, David, 329
Phillips, Rear/Adml, 285
Pittsfield, 112
Plattsburg, 143
Pollock, Sgt, 401
Pont-Du-Hem Cemetery, 105
Poona, 414
Port Murray, 11, 12, 526
Portreath airfield, 351, 403, 414, 468
Portsmouth, 497
Potez aircraft, 425
Potocki, 'Spud', 344
Pozières, Battle of, 76, 178
Prestwick, 116, 117, 258, 264, 273–75, 277,
    278, 281, 283–85, 291, 294, 296, 297,
    301, 303, 307, 309, 311, 313, 315, 324,
    358, 365, 375, 437, 445, 447, 450, 453,
    466, 467, 488, 490, 517, 521, 530, 538,
    545
Prettyman, Major, 24
Prince-Smith, Donald, 84
Princess Victoria, MV, 428
Prisoners, 22, 24, 33, 35, 38, 71, 75, 82, 85,
    86, 103, 109, 118, 121, 123, 127, 143–45,
    161, 169, 173, 199, 203, 319, 331, 356,
    457, 463, 464, 554

Pritchard, John, 295, 516
Pullan, F/O, 296

Q
Queensland, 440, 514
Quinn, Robert, 508

R
Racetrack, 544
Ransley, Frank, 96
Rastatt POW camp, 86
Rawes, Eric, 388
Rayner, William, 279, 296, 492
Redgates Caravan Park, 316, 541, 542
Redler, Harold, 84, 157-8, 173, 230, 241
Redmund, Sgt, 313
Redruth airfield, 403
Reed, Richard, 156-7, 230, 241
Rees, Lionel, 51–53, 86-7, 116, 133, 148-9,
    155, 159, 168, 175, 192, 226, 232
Reid, 2nd/Lt, 84
Reid, David, 357
Reid, John, 510
Reneau, P/O, 284
Rhodesia, 86, 386, 417, 480, 491, 509
Richardson, Lancelot, 85
Richmond Hall, 316, 318
Richmond Hospital, 93
Riordan, Allan, 488, 515
Roberts, Alexander, 144, 146, 148
Roberts, Sir S., 28, 35
Robertson, George, 547
Robertson, John, 522
Robertson, Robert, 359
Robinson, Eric, 430-1
Robinson, H. N. C., 182
Robinson, Harry, 105-6
Robroyston Hospital, 463
Rodger, Kenneth, 86
Rommel, Gen Erwin, 423, 452, 472
Ross, Dr Jacob, 111

Ross, Mackenzie, 526
Ross, S/Ldr, 273–75, 287
Ross, Sgt, 280
Rosyth, 67, 307
Rothermere, Harold Harmsworth, 1st Lord, 29-30
Roylace, F/Lt, 288
Rozelle House, 36
Ruddick, P/O, 273
Runciman, Hon. Margaret, 466, 520
Ruscoe, 'Rusty', 375
Russell, J. C., 143
Russians, 206, 249, 316, 415
Russon, E., 297
Ryan, Mr, 536–38

S
Salisbury (Canada), 495
Salisbury, 108, 118, 151, 154
Salmond, Sgt, 287
Salonika, 89, 106, 107
Samoa, 100
Sandeman, W/Cdr, 302, 434
Sanderson, Ninian, 544
Sandgren, P/O, 286
Sandilands, Nan, 449
Saskatchewan, 101, 158, 242, 316, 451, 507
Scaramanga, John, 72-3
Schlachtstaffel, 202
Schleuder, Paul, 202
Scholte, Owen, 162
Scott-Moncrieff, Sir George, 18, 30
Seabrook, Harold, 307, 523
Seafield Hospital, 373
Seaforth Highlanders, 60, 95
Sedgeford, 90
Sedgwick, F/Sgt, 454
Shacklady, H. J., 298
Shackleton aircraft, 460
Shalloch Park, 243
Shalloch-on-Minnoch, 279, 491

Shanter farm, 49, 194, 200, 205, 216, 272, 315–17, 319, 387, 525-6
Shanter Inn, 336, 337, 342, 364, 398, 406, 412
Shawbury, 154, 295
Shelley, Brian, 420-1
Sheppard, William, 488
Shropshire, 108
Shropshire, 163, 182, 183, 244
Sicily, 424, 452, 459, 481
Siddeley-Deasy, 198, 205
Simmonds aircraft, 59
Simms, Mr, 291
Simpson, Capt, 201
Simpson, F/Sgt, 296
Simrock, Sgt, 280
Sinclair, Chrissie, 449
Sinclair, Sir Archibald, 282
Singapore, 76, 379, 463
Siskin aircraft, 86
Sissmore, F/Lt, 274
Skellingthorpe, 299
Skidmore, John, 414
Slingsby glider, 303
Sloan, Jan, 317, 441, 442
Sloane, Granny, 338, 341
Sloane, Wullie, 338, 340
Sloss, William, 249
Smith, Alex, 457
Smith, C. Hodgkinson, 107
Smith, David, 19, 21
Smith, Donald, 490
Smith, F., 487
Smith, F/Lt, 306
Smith, Frank, 509
Smith, Howard, 146, 148, 152, 230, 240
Smith, J., 17
Smith, Tom, 11, 39, 47, 52, 53, 270
Smith, W/O, 303
Smith, William, 523
Smith-Barry Gosport Training Scheme, 208

Thomson, Sammy, 337
Thorburn, Merv, 320, 323
Thornaby, 299, 482
Tidworth, 78, 97
Tillinghast, Theose, 124
Tirpitz, 292, 295, 481
Torbeau aircraft, 474, 481
Toronto, 106, 149, 152, 158, 181, 185, 243,
    245, 421, 496
Towlson, Harry, 89, 230, 236
Townsley, Annie, 349
Trenchard, Lord, 25, 231, 280
Troax, 380
Trudeau, Pierre, 75
Trump Organization, 546
Trump, Donald, 546
Turnhouse airfield, 67, 68, 273, 275, 277,
    280, 285, 297
Twickenham, 93, 106, 510
Tyler, 'Ty', 414

U

U-boats, 294-5, 317, 347, 402, 417, 427,
    442, 448, 458
Upavon, 15, 87, 105, 108, 163, 355
Upper Hayford, 280
Uxbridge, 92, 178, 181, 375

V

Vacheresse, William, 280, 487, 495
Valheureux, 128
Vancouver, 60, 104, 105, 399, 414, 439, 517
Varssenaere, 112
Vaughn, George, 119
Veale, Allan, 100-1
Ventura, 278, 295–99, 317, 430, 481, 482,
    518
Verona, 100
Vevers, Eric, 510
Vickers Warwicks, 432, 436, 475, 476, 521
Vickers Wildebeest, 471

Vickers, 44-5, 65, 69–74, 87, 92, 95, 108,
    122, 129, 143, 154, 167, 169, 173, 175,
    189, 192-3, 195–97, 200–202, 205, 206,
    211, 212, 215, 292, 406, 407, 472, 480,
    509–13, 521
Vickers, O. H. D., 92
Vickers-Armstrong, 436, 474–77
Victoria Cross, 87, 166-7, 197
Victoria, MV, 98, 423
Villiers, Lt-Col, 278
Voigt, Bruno von, 176
Vollendam, SS, 360
Von Tutschek, Adolf, 158
Voormezeele, 176
Voss, Werner, 72, 96

W

Waddington airfield, 381, 448
Wales, 87, 183, 237, 244, 312, 358, 360,
    364, 367, 377, 383, 391, 403, 411, 414,
    423, 436, 440, 448, 501, 502, 515, 521
Wallace, H. R., 225
Wallace, J., 261
Wallace, William, 10
Wallis, Barnes, 292, 475, 481
Wallis, Irwin, 84
Walney Island, 460
Walpole, Humphrey, 487, 509
Walters, Sgt, 284
Warburton, Adrian, 452
Waring, H., 290, 294, 297, 299, 301
Warwick aircraft, 296, 298–304, 306–9, 432,
    436-7, 440-1, 443-5, 447, 464, 475-6,
    481-2, 521
Warwickshire, 68, 92, 178, 497
Washington, 34, 145
Waterlow, Eric, 101-2
Watt, Sgt, 289
Wavans Cemetery, 169
Waxman, Cyril, 102
Weatherley, Sgt, 401